Successful Writing at Work

Successful Writing at Work

TENTH EDITION

Philip C. Kolin
University of Southern Mississippi

WADSWORTH
CENGAGE Learning

Australia • Brazil • Japan • Korea • Mexico • Singapore • Spain • United Kingdom • United States

WADSWORTH
CENGAGE Learning·

Successful Writing at Work, Tenth Edition
Philip C. Kolin

To Kristin, Erica, and Theresa
Evan Philip and Megan Elise
Julie and Loretta
Diane
and
MARY

Senior Publisher: Lyn Uhl

Publisher: Michael Rosenberg

Associate Development Editor: Megan
Garvey

Assistant Editor: Erin Bosco

Editorial Assistant: Rebecca Donahue

Media Editor: Janine Tangney

Marketing Project Manager: Gurpreet Saran

Senior Marketing Communications Manager:
Linda Yip

Senior Content Project Manager: Michael
Lepera

Art Director: Marissa Falco

Senior Print Buyer: Betsy Donaghey

Senior Rights Acquisition Specialist: Jennifer
Meyer Dare

Production Service: Books By Design, Inc.

Text Designer: Books By Design, Inc.

Cover Designer: Riezebos Holzbaur
Design Group

Cover Image: LWA/Getty Images

Compositor: S4Carlisle Publishing Services

Design images
Part opening image: Rawcaptured/
Shutterstock.com
Chapter opening image: dimitris_k/
Shutterstock.com

For product information and technology assistance, contact us at
Cengage Learning Customer & Sales Support, 1-800-354-9706
For permission to use material from this text or product,
submit all requests online at **www.cengage.com/permissions.**
Further permissions questions can be emailed to
permissionrequest@cengage.com.

Library of Congress Control Number: 2011942964

ISBN-13: 978-1-111-83479-1

ISBN-10: 1-111-83479-2

Wadsworth
20 Channel Center Street
Boston, MA 02210
USA

Cengage Learning is a leading provider of customized learning solutions with office locations around the globe, including Singapore, the United Kingdom, Australia, Mexico, Brazil and Japan. Locate your local office at **international.cengage.com/region**

Cengage Learning products are represented in Canada by Nelson Education, Ltd.

For your course and learning solutions, visit **www.cengage.com.**

Purchase any of our products at your local college store or at our preferred online store **www.cengagebrain.com.**

Instructors: Please visit **login.cengage.com** and log in to access instructor-specific resources.

Printed in the United States of America
1 2 3 4 5 6 7 15 14 13 12 11

Contents

v

Preface

Successful Writing at Work, Tenth Edition, is a comprehensive introductory text for business, technical, professional, and occupational writing courses. Writing is a vital part of virtually every job, and as readers of earlier editions have learned, *Successful Writing at Work* can help students develop key communication skills essential for their career. This practical textbook will guide students to become better writers while they also learn to develop and design effective workplace documents for multicultural, global audiences. *Successful Writing at Work*, Tenth Edition, is organized to take students step-by-step from the basic concepts of audience analysis, purpose, message, style, and tone to the processes of researching, drafting, revising, formatting/designing, and editing. Students will learn to write a variety of job-related documents, from emails to more complex proposals, reports, and presentations.

Building on Past Editions

Benefiting from the feedback of instructors, students, and employers over many editions, this revised Tenth Edition continues to give students detailed, clear guidelines for preparing well-organized and readable emails, blogs, correspondence, instructions, procedures, surveys, proposals, reports, websites, and presentations intended for a variety of readers. Moreover, because effective models are critical to learning new skills, students will find a wide range of realistic, up-to-date, and rhetorically diverse examples (all of them annotated and visually varied) demonstrating the function, scope, format, and organization of numerous documents for audiences with differing needs. Each of these model documents focuses directly on practical issues in the world of work and portrays employees as successful writers, either individually or as part of a collaborative writing group. Furthermore, this new edition fully covers a broad spectrum of current workplace technologies and considerations, such as instant messaging, Google Docs, Track Changes, webinars, and storyboarding.

As in past editions, this Tenth Edition is as versatile as it is comprehensive. Full enough for a sixteen-week semester, it can also be easily adapted to a shorter six-, eight-, or ten-week course. *Successful Writing at Work*, Tenth Edition, is designed to go beyond classroom applications: It is a ready reference that students can easily carry with them as they begin or advance in the workplace. As students will discover, this edition maintains the reputation of former editions by including numerous practical applications in each chapter. It can be as useful to readers with little or no job experience as to those with years of experience in one or several fields. This edition also addresses the needs of students re-entering the job market or changing careers.

Distinctive Features of *Successful Writing at Work*

The distinctive features that in the past have made *Successful Writing at Work* a student-friendly text in the contemporary workplace continue to be emphasized and expanded in this Tenth Edition. These features, stressing up-to-date approaches to teaching business, technical, and professional writing, can be found throughout this new edition.

- **Analyzing audiences.** The Tenth Edition once again focuses on the importance of audience analysis and the writer's obligation to achieve the "you attitude" in every workplace document. In addition, the concept of audience extends to readers worldwide, as well as to non-native speakers of English, whether as co-workers, employers, clients, or representatives of various agencies and organizations. Memos, emails, letters, résumés, presentations, and other documents are written, designed, organized, and introduced with the intended audience(s) in mind.
- **Approaching writing as a problem-solving activity.** The Tenth Edition continues to emphasize workplace writing not merely as a set of rules and formats but as a problem-solving activity in which employees meet the needs of their employers, co-workers, customers, clients, community groups, and vendors worldwide. This approach to writing, introduced in Chapter 1 and woven through subsequent chapters, helps students understand the varied situations and problems they will have to address and highlights the rhetorical and design options available to them to solve these problems. As in earlier editions, this new Tenth Edition teaches students how to develop the critical skills necessary for planning, researching, drafting, revising, designing, and editing business documents. In-depth case studies throughout, as well as boxed inserts offering professional advice from experts in the world of work, guide writers in finding resources and developing rhetorical strategies to solve problems in the business world.
- **Being an ethical employee.** Stressing that companies expect their employees to behave and write ethically, the Tenth Edition reinforces and expands discussions of ethical writing practices in almost every chapter. Beginning with a discussion of ethical writing at work, Chapter 1 introduces a study of ethics in the workplace with enhanced sections on ethical behavior and solving ethical dilemmas in the workplace. Subsequent chapters offer practical guidelines on and numerous examples of documents that illustrate the types of ethical choices workers must make in the business world. Special attention to ethics can be found in discussions on editing to avoid sexism and biased language in Chapter 2; working cooperatively with a collaborative writing team in Chapter 3; writing accurate memos, emails, instant messages, and blogs in Chapter 4; drafting diplomatic letters in Chapters 5 and 6; preparing honest, realistic résumés and portfolios/webfolios in Chapter 7; conducting truthful, objective, and carefully documented research in Chapter 8; using and constructing unbiased visuals and ethical websites in Chapters 10 and 11; preparing safe, legal instructions and procedures in Chapter 12; writing honest proposals and reports in Chapters 13 to 15; and making clear and accurate presentations in Chapter 16.

- **Writing for the global marketplace.** In today's international workplace, effective employees must know how to write for a variety of readers, both in the United States and across the globe. Consequently, almost every chapter in this new edition has increased coverage of writing for international readers and non-native speakers of English. The needs of these audiences receive special attention starting with a discussion of International English in Chapter 1 and continuing through the chapters on correspondence, visuals, instructions, websites, proposals, short and long reports, and presentations. Especially important, too, is the revised, updated long report in Chapter 15 on the role international workers play as members of a diverse workforce in the global corporate world.

New and Updated Material in the Tenth Edition

To meet the growing and changing needs of employees in today's global workplace, this new edition is one of the most extensively revised editions of *Successful Writing at Work* yet. It has been carefully streamlined and updated to meet the needs of today's students. Throughout this Tenth Edition you will find expanded coverage of key topics, writing guidelines, and a wealth of new annotated examples of workplace documents, plus many new Tech Notes, case studies, and extended box commentaries featuring advice from professionals in the business world. Numerous exercises after each chapter make teaching and learning workplace writing relevant and current.

With its full-color palette, this new edition also exemplifies a wide range of professional design and layout choices that writers make in the world of work, thus giving students models to help them prepare their own documents. Not only has the layout of the text been streamlined and redesigned, but many of the examples and figures have also benefited from the full-color adaptation and have been revised to show the variety of layouts, logos, and visual designs used in the world of work.

The next section outlines, chapter by chapter, the specific additions found in the Tenth Edition.

Chapter-by-Chapter Updates

Chapter 1 Getting Started: Writing and Your Career

- New section on how writing relates to other skills in the world of work
- New case study on adapting technical information to meet the needs of diverse audiences
- New section, "Know Your Company's Culture and Codes"
- New figure, "A Letter from the CEO of IBM"
- Expanded discussion, "Ethical Requirements on the Job"
- New section, "'Thinking Green': Making Ethical Choices About the Environment"
- Further attention to solving ethical dilemmas in the workplace
- Additional exercises on diversity in the workplace and on audience analysis

Chapter 2 The Writing Process at Work

- Enhanced coverage of drafting, revising, and editing on the job
- Expanded case study, "A 'Before' and 'After' Revision of a Short Report" on writing for a general reader
- Additional advice on avoiding stereotypical language, including eliminating sexism
- New section, "The Writing Process: Some Final Thoughts"

Chapter 3 Collaborative Writing and Meetings at Work

- Increased emphasis on being a team player at work
- Greater attention to collaborative communication technologies
- New sections and figures illustrating using Track Changes and Google Docs for collaborative team writing
- Updated discussion of conferencing tools and presentation software

Chapter 4 Writing Routine Business Correspondence: Memos, Faxes, Emails, IMs, and Blogs

- Substantially revised sections on email and instant messaging in the workplace
- Expanded discussion of business blogs, blog sponsors, and blog formats with two new examples, including a blog post from Whole Foods emphasizing eco-friendly aquaculture and building consumer confidence

Chapter 5 Writing Letters: Some Basics for Communicating with Audiences Worldwide

- Further emphasis on the importance of letters in the age of the Internet
- New sections on envelopes and formatting letters
- Greater attention to the needs of international audiences
- New exercises

Chapter 6 Types of Business Letters

- Strengthened discussion on the contexts for correspondence
- New section on cover letters
- Numerous redesigned letters and logos
- Use of Web-based media in connection with correspondence
- Enhanced case study on adapting letters to international readers
- New conclusion, "Writing Business Letters That Matter: A Summary"

Chapter 7 How to Get a Job: Searches, Networking, Dossiers, Portfolios/ Webfolios, Résumés, Letters, and Interviews

- Expanded section on identifying marketable job skills
- Updated coverage on where to look for a job, with further examples of and advice on using job-posting sites
- New section, "Using Online Social and Professional Networking Sites in Your Job Search"

- New section, "LinkedIn Network" with a new annotated figure of a LinkedIn profile with commentary
- Updated and redesigned résumés
- Eight different résumés for print and digital formats, presented as models for students
- Greater attention to online résumés, with two new model résumés
- New case study, "Developing an Electronic Résumé and Building a Network"
- New Tech Note, "Skype Interviews"
- Expanded coverage of interview do's and don'ts
- Updated, practical advice on finding salary ranges and inquiring about salary
- Additional exercises on preparing online résumés and application letters

Chapter 8 Doing Research and Documentation on the Job

- Updated coverage and examples of databases used for research
- New section on using social networking sites to recruit participants in research projects
- Greater attention to evaluating websites, with a new figure illustrating FDA advice on assessing online veterinary pharmacies
- Coverage of latest MLA and APA documentation styles, with additional guidelines and updated examples
- An extensively revised and updated long business report with new information on m-commerce and customizing a website used in a marketing campaign
- New exercises

Chapter 9 Summarizing Information at Work

- New section, "Summaries in the Information Age"
- New sample abstracts

Chapter 10 Designing Clear Visuals

- Expanded section, "Visual Thinking in the Global Workplace"
- Greater attention to documenting and citing visuals in written work
- Further guidelines on using pictographs
- New models throughout chapter (e.g., maps, clip art, tables, drawings)
- Enhanced discussion of using visuals for international readers
- Additional discussion and examples of icons
- New exercise on using visuals and creating corporate logos

Chapter 11 Designing Successful Documents and Websites

- New figure, "Designing a Company Newsletter: Advice from a Publications Manager"
- New section on creating and incorporating reader-friendly headings and subheadings in a document

- Additional emphasis on writing for the Web
- Further attention to understanding differences between writing a print document and writing for the Web
- New annotated sample CDC (Centers for Disease Control) and ICC (International Chamber of Commerce) websites
- Revised section on storyboarding for Web design

Chapter 12 Writing Instructions and Procedures

- Increased coverage of preparing legally and ethically proper instructions and procedures
- Greater attention to writing, formatting, and illustrating online instructions
- Two new figures on repairing a leaky faucet and improving a house's energy efficiency
- New section and figure on troubleshooting guides
- New, fully annotated model of a full set of instructions, illustrating use of both print and online formats—instructions for assembling and using Epson's Workforce 610 Printer

Chapter 13 Writing Winning Proposals

- Updated sales and internal proposals
- New section on legal and ethical issues to consider when writing a proposal
- New sample proposal on writing a research report on the ethical issues involved in m-commerce

Chapter 14 Writing Effective Short Reports

- Expanded discussion of how different audiences read a report, and why
- Enhanced coverage of designing reader-friendly reports
- New section, "Employee Activity/Performance Reports," with new figure and advice on preparing performance reports
- New progress report on researching ethical issues involved in m-commerce
- New section on sales reports
- New exercises

Chapter 15 Writing Careful Long Reports

- Further attention to evaluating research sources
- Revised discussion of transmittal letters, with new model letter
- Expanded discussion of abstracts and audience needs
- New section on paginating a long report
- Additional coverage on writing conclusions and recommendations
- Completely revised and updated model long report on meeting the needs of multinational workers

Chapter 16 Making Successful Presentations at Work

- New section with a new figure on informal briefings
- New guidelines/table listing characteristics of an ineffective speaker
- New Tech Note on webinars
- Revised slides for a PowerPoint presentation
- New section on specific audience needs for different types of presentations
- Enhanced discussion of using different types of presentation software

Additional Resources

Resources for Students

English CourseMate. Cengage Learning's English CourseMate brings course concepts to life with interactive learning, study, and exam-preparation tools that support the printed textbook. CourseMate includes the following:

- An interactive eBook
- Interactive teaching and learning tools, including
  - Interactive quizzes for each chapter in the text
  - Online exercises that help students enhance their understanding of chapter topics and improve their technology skills
  - Simulations that provide practice handling typical workplace situations
  - Documents for analysis
  - Web links that expand on topics in the text

Learn more at **www.cengage.com/coursemate**.

Resources for Instructors

Instructor's Edition (IE). Examination and desk copies of the Instructor's Edition of *Successful Writing at Work*, Tenth Edition, are available upon request.

  Online Instructor's Resource Manual. The Instructor's Manual contains resources designed to streamline and maximize the effectiveness of your course preparation. This helpful manual provides a sample course syllabus; suggestions for teaching job-related writing, with ideas for simulating real-world experience in the classroom; suggested approaches to exercises; and test items for each chapter.

  Instructor's Website. This password-protected website includes chapter-level PowerPoint lecture slides, as well as the Online Instructor's Resource Manual, available for download.

  English CourseMate. Cengage Learning's English CourseMate brings course concepts to life with interactive learning, study, and exam-preparation tools that support the printed textbook. CourseMate includes the following:

- An interactive eBook
- Interactive teaching and learning tools, including
  - Interactive quizzes for each chapter in the text

- Online exercises that help students enhance their understanding of chapter topics and improve their technology skills
- Simulations that provide practice handling typical workplace situations
- Documents for analysis
- Web links that expand on topics in the text
- Engagement Tracker, a first-of-its-kind tool that monitors student engagement in the course

Learn more at **www.cengage.com/coursemate**.

Write Experience. Write Experience is a new technology product that allows you to assess written communication skills without adding to your workload. Write Experience utilizes artificial intelligence to not only score student writing instantly and accurately but also provide students with detailed revision goals and feedback on their writing to help them improve. Two key features of Write Experience, MYTutor and MYEditor, provide students with real-time, simultaneous feedback in their native language *while* they write! Learn more at **www.cengage.com/writeexperience**.

Please contact your local Cengage sales representative for more information, to evaluate examination copies of any of these instructor or student resources, or for product demonstrations. You may also contact the Cengage Learning Academic Resource Center at 800-423-0563, or visit us at **www.cengagebrain.com**.

Acknowledgments

In a very real sense, the Tenth Edition of *Successful Writing at Work* has profited from my collaboration with various reviewers. I am, therefore, honored to thank the following reviewers who have helped me improve this edition significantly: Etta Barksdale, North Carolina State University; Jonathan Lee Campbell, Valdosta State University; Don Cunningham, Radford University; Linda Eicken, Cape Fear Community College; Wolfgang Lepschy, Tallahassee Community College; Sabrina Peters-Whitehead, University of Toledo; Mary E. Shannon, California State University–Northridge; and Pinfan Zhu, Texas State University.

I would also like to thank the peers who have graciously worked on the supplement package for the Tenth Edition: Leslie St. Martin, College of the Canyons, for her work on the Instructor's Manual and Test Bank; Miriam Mattsey, Indiana State University, for the creation of the PowerPoint presentation; and David Leight, Reading Area Community College, for creating the student tutorial quizzes and Web links.

I am also deeply grateful to the following individuals at the University of Southern Mississippi for their help: Danielle Sypher-Haley and Jeremy DeFatta (Department of English), David Tisdale (University Communications), Sandra Leal (Department of Biological Sciences), Mary Lux (Department of Medical Technology), and Cliff Burgess (Department of Computer Science). My thanks also go to Andreas Skalko (Office of Professional, Developmental, and Educational Outreach) and Naofumi Tatsumi (Department of Foreign Languages). I am also grateful to Dean of the Library Carole Kiehl for her assistance. To the following librarians at Cook Library goes my gratitude for their help as well: Edward McCormack, Steven Turner, Tracy Englert, and Robert Fowler. My special thanks go to Ann Branton, head of Bibliographic Services. I am also grateful to Sherry Laughlin at William Carey University Library as well. I am especially grateful to Steven Moser, Interim Dean of the College of Arts and Letters, for his continued appreciation of my work.

I also want to thank Terri Smith Ruckel at Pearl River Community College for assistance with Chapter 15, and Jianqing Zheng at Mississippi Valley State University and Fatih Uzuner for their help with the Chinese and Turkish translations used in Chapter 12. I also appreciate the assistance of Billy Middleton at Stevens Institute, Erin Smith at the University of Tennessee, Knoxville, and Daniel Vollaro at Georgia Gwinnett College.

Several individuals from the business world also gave me wise counsel, for which I am thankful. They include Sally Eddy at Georgia Pacific; Kirk Woodward at Visiting Nurses Services of New York; Jimmy Stockstill at Petro Automotive Group; Nancy Steen from Adelman and Steen; Theresa Rogers and Rachel Sullivan at Regions Bank, Inc.; Staff Sergeant Scott Jamison of the U.S. Army; and Gloria Fontana of the Fontana Group.

I am especially grateful as well to Father Michael Tracey for his counsel and his contributions to Chapter 11 on document design, particularly the design of websites.

My gratitude goes to the team at Cengage Learning for their assistance, encouragement, and friendship: Michael Rosenberg, Michael Lepera, Megan Garvey, Erin Bosco, Samantha Ross Miller, and Karen Judd. I am especially grateful to Nancy Benjamin at Books By Design for her assistance and advice as this new edition went through the painstaking production cycle. I also express my thanks to Sue Brekka, Jennifer Meyer Dare, and Tim McDonough, who handled the permissions for this edition of *Successful Writing at Work*.

My prayers and love go to my extended family—Margie and Al Parish; Sister Carmelita Stinn, SFCC; Mary and Deacon Ralph Torrelli; and Lois and Norman Joseph Dobson—for sustaining me.

Finally, I remain deeply grateful to my son, Eric, and my daughter-in-law, Theresa, for their enthusiastic and invaluable advice as I revised Chapter 8; to my grandson, Evan Philip, and granddaughter, Megan Elise, for their love and encouragement. My daughter, Kristin, also merits loving praise for her help throughout this new edition by conducting various searches and revisions and by offering keen practical advice on a variety of workplace documents. And to Diane Dobson, my wife, I say thank you for bringing your prayers and love of music into my life.

P.C.K.
January 2012

Successful Writing at Work

PART I

Backgrounds

CHAPTER 1

Getting Started

Writing and Your Career

Access chapter-specific interactive learning tools, including quizzes and more in your English CourseMate, accessed through www .cengagebrain.com.

Writing—An Essential Job Skill

Writing is a part of every job, from your initial letter of application conveying first impressions to memos, emails, blogs, letters, websites, proposals, instructions, and reports. Writing keeps businesses moving. It allows employees to communicate with one another, with management, and with the customers, clients, and agencies a company must serve to stay in business.

How Writing Relates to Other Skills

Almost everything you do at work is related to your writing ability. Deborah Price, a human resource director with thirty years of experience, stresses that "without the ability to write clearly an employee cannot perform the other duties of the job, regardless of the company he or she works for." Here is a list of the common tasks you will be expected to perform in the workplace and that require clear and concise writing to get them done well.

- Do computer programming and be familiar with recent software.
- Assess a situation, a condition, a job site, etc.
- Research and record the results accurately.
- Summarize information concisely and identify main points quickly.
- Work as part of a team to collect, to share, and to evaluate information.
- Exhibit cultural sensitivity in the workplace.
- Network with individuals in diverse fields outside your company and across the globe.
- Set priorities and stick to them.
- Tackle and solve problems and explain how and why you did.
- Meet customer or client needs and expectations for information.
- Prepare and test instructions and procedures.
- Justify financial, personnel, or other actions and decisions.
- Make persuasive presentations to co-workers, employers, and clients.
- Learn and apply new technologies in your field.

To perform each of these essential workplace tasks, you have to be an effective writer—clear, concise, accurate, ethical, and persuasive.

The High Cost of Effective Writing

Clearly, then, writing is an essential skill for employees and employers alike. According to Don Bagin, a communications consultant, most people need an hour or more to write a typical business letter. If an employer is paying someone $30,000 a year, one letter costs $14 of that employee's time; for someone who earns $50,000 a year, the cost for the average letter jumps to $24. Mistakes in letters are costly for workers as well as for employers. As David Noble cautions in his book *Gallery of Best Cover Letters*, "The cost of a cover letter (in applying for a job, for instance) might be as much as a third of a million dollars—even more if you figure the amount of income and benefits you don't receive, say, in a 10-year period for a job you don't get because of an error that got you screened out."

Unfortunately, as the Associated Press (AP) reported in a recent survey, "Most American businesses say workers need to improve their writing . . . skills." Yet that same report cited a survey of more than 400 companies that identified writing as "the most valuable skill employees can have." In fact, the employers polled in that AP survey indicated that 80 percent of their workforce needed to improve their writing. Illustrating a company's keen interest in employee writing, Figure 1.1 shows an email from a human resources director, Rowe Pinkerton, offering an incentive to employees to improve their writing abilities by taking a college writing course. Beyond a doubt, your success as an employee will depend on your success as a writer. The higher you advance in an organization, the more and better writing you will be expected to do. Promotions, and other types of job recognition, are often based on an employee's writing skills.

How This Book Will Help You

This book will show you, step by step, how to write clearly and efficiently the job-related communications you need for success in the world of work. Chapter 1 gives you some basic information about writing in the global marketplace and raises major questions you need to ask yourself to make the writing process easier and the results more effective. It also describes the basic functions of on-the-job writing and introduces you to one of the most important requirements in the business world— writing ethically.

Writing for the Global Marketplace

The Internet, email, express delivery, teleconferencing, and e-commerce have shrunk the world into a global village. Accordingly, it is no longer feasible to think of business in exclusively regional or even national terms. Many companies are multinational corporations with offices throughout the world. In fact, many U.S. businesses are branches of international firms. A large, multinational corporation

FIGURE 1.1 An Employer's View of the Importance of Writing

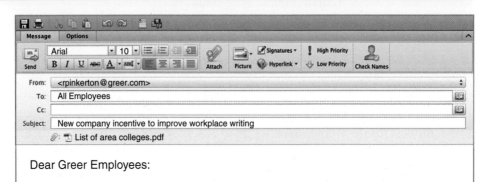

Dear Greer Employees:

Stresses the importance of writing

I am pleased to announce a new company incentive approved by the administration last week. In its continuing effort to improve writing in the workplace, Greer, Inc., will offer tuition reimbursement to any employee who takes a college-level course in business, technical, or occupational writing, starting this fall.

Three Requirements to Receive Tuition Reimbursement:
To qualify for this benefit, employees must do the following:

(1) Submit a two-page proposal on how such a course will improve the employee's job performance here at Greer.

Clearly explains how to take advantage of company offer

(2) Take the class at one of the approved colleges or universities in the Cleveland area listed on the attachment to this email.

(3) Provide proof (through a transcript or final grade report) that he or she has successfully completed the course.

To apply, please submit your proposal to Dawn Wagner-Lawlor in Human Resources (<u>dwlawlor@greer.com</u>) at least one month before you intend to enroll in the course.

Closes on an upbeat note

Here's to productive writing!

Sincerely,

Rowe Pinkerton
Human Resources Director
781-555-3692
<rpinkerton@greer.com>

may have its equipment designed in Japan; built in Bangladesh; and sold in Detroit, Atlanta, and Los Angeles. Its stockholders may be in Mexico City as well as Saudi Arabia—in fact, anywhere. In this global economy, every country is affected by every other one, and all of them are connected by the Internet.

Know Your Computer at Work

A major part of any job is knowing your workplace technology, which can include PCs, smartphones, and tablet computers. You need to know not just how to use the applications installed on your computer or other device but also what to do if there is a computer emergency.

Given the kinds of security risks businesses face today, employees have to be especially careful. As Kim Becker cautions in *Nevada Business*, "With malware, spyware, adware, viruses, Trojans, worms, phishing, and server problems, it's time for every business to review its IT strategy and security before a loss occurs."*

Here are some guidelines on how to use your computer effectively on the job:

■ **Understand how to use the software programs required for your job.** Your office will most likely require employees to use Microsoft Word or Corel WordPerfect. But make sure you know how to use the entire software package—not just the word-processing application, but also the filing, formatting, spreadsheet, presentation, and tables/graphics programs.

■ **Get training on how to use company-specific applications.** You will be expected to know how to use company-created databases, templates, and other customized applications on the job. If your company offers classes on how to use these programs, take them. Otherwise, ask for the advice of a co-worker or someone in your company's information technology (IT) department who knows the programs.

■ **Learn how to back up your files.** You will save yourself, your boss, your co-workers, and your clients time and stress by backing up your essential files regularly to prevent losing them in the event that your computer crashes.

■ **Set up an alternate email account.** If you cannot access your email on the job because a server is down, sign up for a free email account with a provider such as Hotmail, Yahoo! Mail, or Gmail. You can use this alternate account until the server is up again.

*Kim Becker, "Security in the Workplace: Technology Issues Threaten Business Prosperity," *Nevada Business*, July 2008.

Competing for International Business

Companies must compete for international sales to stay in business. Every business, whether large or small, has to appeal to diverse international markets to be competitive. Each year a larger share of the U.S. gross national product (GNP) depends on global markets. Some U.S. firms estimate that 40 to 50 percent of their business is conducted outside of the United States. Walmart, for example, has opened hundreds of stores in mainland China, and General Electric has plants in over fifty countries. In fact, estimates suggest that 75 percent of the global Internet population lives outside the United States. If your company, however small, has a website, then it is an international business.

Communicating with Global Audiences

To be a successful employee in this highly competitive global market, you have to communicate clearly and diplomatically with a host of readers from different cultural backgrounds. Adopting a global perspective on business will help you communicate and build goodwill with the customers you write to, no matter where they live—across town, in another state, or on other continents, miles and time zones away. As a result, don't presume that you will be writing only to native speakers of American English. As a part of your job, you may communicate with readers in Singapore, Jamaica, and South Africa, for example, who speak varieties of English quite different from American English, as illustrated in the first pie chart in Figure 1.2. You will also very likely be writing to readers for whom English is not their first (or native) language, as shown in the second pie chart in Figure 1.2. These international readers will have varying degrees of proficiency in English, from a fairly good command (as with many readers in India and the Philippines, where English is widely spoken), to little comprehension without the use of a foreign language dictionary and a grammar book to decode your message (as in countries where English is widely taught in schools and recommended for success in the business world but not spoken on a regular basis). Non-native speakers, who may reside either in the United States or in a foreign country, will constitute a large and important audience for your work.

FIGURE 1.2 Native and Non-Native Speakers of English Across the World

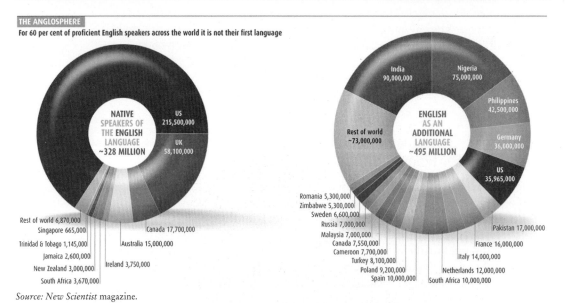

Source: New Scientist magazine.

Seeing the World Through the Eyes of Another Culture

Writing to international readers with proper business etiquette means first learning about their cultural values and assumptions—what they value and also what they regard as communication taboos. They may not conduct business exactly the way it is done in the United States, and to think they should is wrong. Your international audience is likely to have different expectations of how they want a letter addressed or written to them, how they prefer a proposal to be submitted, how they wish a business meeting to be conducted, or how they think questions should be asked and agreements reached. Their concepts of time, family, money, the world, the environment, managers, and communication itself may be nothing like those in the United States. Visuals, including icons, that are easily understood in the United States may be baffling elsewhere in the world. If you misunderstand your audience's culture and inadvertently write, create, or say something inappropriate, it can cost your company a contract and you your job.

Cultural Diversity at Home

Cultural diversity exists inside as well as outside the company you work for. Don't conclude that your boss or co-workers are all native speakers of English, either, or that they come from the same cultural background that you do. In the next decade, as much as 40 to 50 percent of the U.S. skilled workforce may be composed of recent immigrants who bring their own traditions and languages with them. These are highly educated, multicultural, and multinational individuals who have acquired English as a second or even a third language.

For the common good of your company, you need to be respectful of your international colleagues. In fact, multinational employees can be tremendously important for your company in making contacts in their native country and in helping your firm understand and appreciate ethical and cultural differences among customers. The model long report in Chapter 15 (pages 702–718) describes some ways in which a company can both acknowledge and respect the different cultural traditions of its international employees. Businesses want to emphasize their international commitments. A large corporation such as Citibank, for instance, is eager to promote its image of helping customers worldwide, as Figure 1.3 on page 10 shows.

Using International English

Whether your international readers are customers or colleagues, you need to adapt your writing to respect their language needs and cultural protocols. To communicate with non-native speakers, use "international English," a way of writing that is easily understood, culturally appropriate, and diplomatic. International English is user friendly in terms of the words, sentences, formats, and visuals you choose.

To write international English means you re-examine your own writing. The words, idioms, phrases, and sentences you select instinctively for U.S. readers may not be appropriate for an audience for whom English is a second, or even a third,

FIGURE 1.3 A Company's Dedication to Globalization

How Citigroup Meets Banking Needs Around the World

WITH A BANKING EMPIRE that spans more than 100 countries, Citigroup is experienced at meeting the diverse financial services needs of businesses, individuals, customers, and governments. The bank is headquartered in New York City but has offices in Africa, Asia, Central and South America, Europe, the Middle East, as well as throughout North America. Live or work in Japan? You can open a checking account at Citigroup's Citibank branch in downtown Tokyo. How about Mexico? Visit a Grupo Financiero Banamex-Accival branch, owned by Citigroup. Citigroup owns European American Bank and has even bought a stake in a Shanghai-based bank with an eye toward attracting more of China's $1 trillion in bank deposits. Between acquisitions and long-established branches, Citigroup covers the globe from the Atlantic to the Pacific and the Indian Oceans.

Individuals can use Citigroup for all the usual banking services. . . . Personalized service is the hallmark of . . . the bank, which can help prepare customized financial plans for . . . customers, manage their securities trading activities, provide trust services, and much more. What's more, Citigroup is active in communities around the world through . . . financial literacy seminars, volunteerism, and supplier diversity programs. This financial services giant strives for the best of both worlds, wielding its global presence and resources to meet banking needs locally, one customer at a time.

Source: From William M. Pride, Robert J. Hughes, and Jack R. Kapoor, *Business,* 8th ed. (Boston: Houghton Mifflin, 2005), 587. Copyright © 2005 by Houghton Mifflin Company. Used by permission.

language. If you find the set of instructions accompanying your software package confusing, imagine how much more intimidating such a document would be for non-native speakers of English. You can eliminate such confusion by making your message clear, straightforward, and appropriately polite for readers who are not native speakers.

Here are some basic guidelines to help you write international English:

- Use clear, easy-to-understand sentences, not rambling, complex ones. That does not mean you write insultingly short and simple sentences but that you take into account that readers will find your message easier to translate if your sentences do not exceed fifteen to twenty words.
- Do not try to pack too much information into a single sentence; consider using two or more sentences instead. (See pages 67–72.)
- Avoid jargon, idioms (e.g., "to line one's pockets"), and abbreviations (e.g., "FEMA") that international readers may not know.
- Choose clear, commonly used words that unambiguously translate into the non-native speaker's language. Avoid flowery or pretentious language ("amend" for "change").
- Select visuals and icons that are free from cultural bias and that are not taboo in the non-native speaker's country. (For more on this, see pages 503–507.)
- When in doubt, consult someone from the native speaker's country—a co-worker or an instructor, for example.

Because it is so important, international English is discussed in greater detail on pages 183–189. Later chapters of this book will also give you additional practical guidelines on writing correspondence, instructions, proposals, reports, websites, PowerPoint presentations, and other work-related documents suitable for a global audience.

Four Keys to Effective Writing

Effective writing on the job is carefully planned, thoroughly researched, and clearly presented. Its purpose is always to accomplish a specific goal and to be as persuasive as possible. Whether you send a routine email to a co-worker in Cincinnati or Shanghai or a commissioned report to the president of the company, your writing will be more effective if you ask yourself these four questions:

1. Who will read what I write? (Identify your audience.)
2. Why should they read what I write? (Establish your purpose.)
3. What do I have to say to them? (Formulate your message.)
4. How can I best communicate? (Select an appropriate style and tone.)

The questions *who, why, what,* and *how* do not function independently; they are all related. You write (1) for a specific audience (2) with a clearly defined purpose in mind (3) about a topic your readers need to understand (4) in language appropriate for the occasion. Once you answer the first question, you are off to a good start toward answering the other three. Now let's examine each of the four questions in detail.

Identifying Your Audience

Knowing *who* makes up your audience is one of your most important responsibilities as a writer. Keep in mind that you are not writing for yourself but for a specific reader or group of readers. Expect to analyze your audience throughout the composing process.

Look at the advertisements in Figures 1.4, 1.5, and 1.6. The main purpose of all three documents is the same—to discourage people from smoking. The underlying message in each ad—smoking is dangerous to your health—is also the same. But note how the different details—words, photographs, situations—have been selected to appeal to three different audiences.

The advertisement in Figure 1.4 is aimed at fathers who smoke. As you can see, it shows an image of a father smoking next to his son, who is reaching for his pack of cigarettes. Note how the headline "Will your child follow in your footsteps?" plays on the fact that the father and son are both literally sitting on steps, but at the same time it implies that the son will imitate his father's behavior as a smoker. The statistic at the bottom of the advertisement reinforces both the headline and the

FIGURE 1.4 No-Smoking Advertisement Aimed at Fathers Who Smoke

Peter Poulides/Getty Images.

FIGURE 1.5 No-Smoking Advertisement Directed at Pregnant Women

**Smoking
Puts
Both
Mother**

**and Child
at Risk**

Photo by Bill Crump/Brand X Pictures/Fotosearch/Royalty-Free Image

image, hitting home the point that parental behavior strongly influences children's behavior. The child in the photograph already is following his father by showing a clear interest in smoking, picking up his father's pack of cigarettes.

The advertisement in Figure 1.5, however, is aimed at an audience of pregnant women and shows a member of this audience with a lit cigarette. The words on the advertisement appeal to a mother's sense of responsibility, encouraging pregnant women to stop smoking to avoid harm to their unborn children.

Figure 1.6 is directed toward still another audience: young athletes. The word *smoke* in this advertisement is aimed directly at their game and their goal. The headline includes a pun. The writer aptly made the goal the same for the game as well as for the players' lives. Note, too, how this image with its four photos is suitable for an international audience.

The copywriters who created these advertisements have chosen appropriate details—words, pictures, captions, and so on—to persuade each audience not to smoke. With their careful choices, they successfully answered the question "How can we best communicate with each audience?" Note that details relevant for

FIGURE 1.6 No-Smoking Advertisement Appealing to Young Athletes

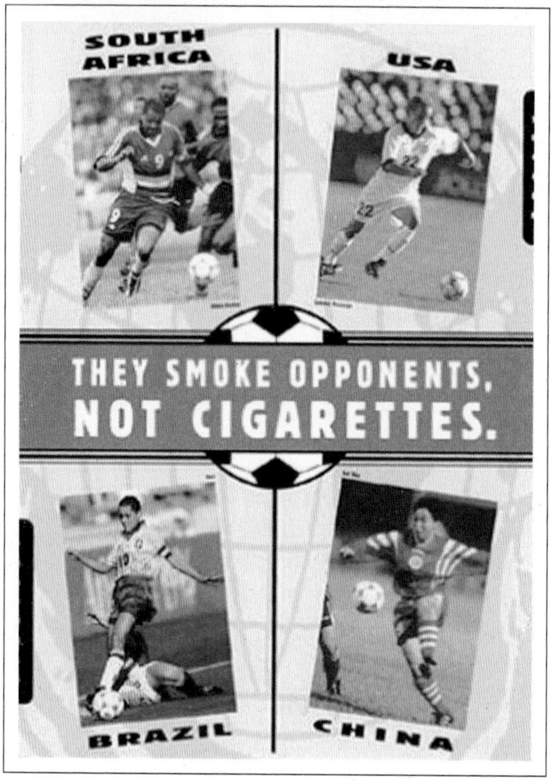

CDC, Tobacco Free Sports Initiative

one audience (athletes, for example) could not be used as effectively for another audience (such as fathers).

The three advertisements in Figures 1.4, 1.5, and 1.6 illustrate some fundamental points you need to keep in mind when identifying your audience:

- Members of each audience differ in their backgrounds, experiences, and needs.
- How you picture your audience will determine what you say to them.
- Viewing something from the audience's perspective will help you to select the most relevant details for that audience.

Some Questions to Ask About Your Audience

You can form a fairly accurate picture of your audience by asking yourself key questions before you write. For each audience you need to reach, consider the following questions:

1. **Who is my audience?** What individual(s) will most likely be reading my work?

If you are writing for colleagues or managers at work:

- What is my reader's job title? Is he or she a co-worker? Immediate supervisor? Vice president?
- What kinds of job experience, education, and interests does my reader have?

If you are writing for clients or consumers (a very large, often diverse audience):

- How can I find out about their interest in my product or service?
- How much will this audience know about my company? About me?

2. **How many people will make up my audience?**

- Will just one individual read what I write (the nurse on the next shift, the production manager), or will many people read it (all the consumers of my company's product or service)?
- Will my boss want to see my work (say, a letter to a consumer in response to a complaint) to approve it?
- Will I be sending my message to a large group of people sharing a similar interest in my topic?

3. **How well does my audience understand English?**

- Are all my readers native speakers of English?
- Will I be communicating with people around the globe?
- Will some of my readers speak English as a second or even a third language and thereby require extra sensitivity on my part to their needs as non-native speakers of English?
- Will some of my readers speak no English and instead use an English grammar book, a foreign language dictionary, or perhaps an online translator, such as Google Translate, for email or webpages where you just copy and paste the text into the translate window?

4. **How much does my audience already know about my topic?**

- Will my readers know as much as I do about the particular problem or issue, or will they need to be briefed, be given background information, or be updated?
- Are my readers familiar with, and do they expect me to use, technical terms and descriptions, or will I have to provide definitions and easy-to-understand, nontechnical wording and visuals?

5. **What is my audience's reason for reading my work?**

- Is my communication part of their routine duties, or are they looking for information to solve a problem or make a decision?
- Am I writing to describe benefits that another writer or company cannot offer?
- Will my readers expect complete details, or will a short summary be enough?
- Are they looking at my work to make an important decision affecting a co-worker, a client, a community, or the environment?
- Are they reading something I write because they must (a legal notification or an incident report, for instance)?

Writing to Different Audiences in a Large Corporation

Jan Melius works in the Communication Department of GrandCo, a firm that manufactures large heavy-duty equipment. As a regular part of her job, Melius has to prepare documents for several different audiences, including the management and staff at GrandCo, current and potential customers, and the larger community of Fairfield where the company is located. Each group of her readers will have different requirements and expectations, and she has to understand those differences if she wants to meet their needs. She first has to do research, gathering the right type and amount of information from many sources, including engineers, accountants, and management at GrandCo, Web searches, government documents, etc., as well as consulting with clients and community officials. Often the documents that she prepares are a result of her collaborations with these various individuals. She also has to select the right type of document to send to her readers.

Below is a list of the audiences that Melius writes for or to, along with the kinds of documents they need with examples of appropriate information found in these documents. Note how Melius has successfully analyzed her audiences according to the guidelines on pages 12–17.

Audience	Types of Information/Documents to Supply
Customer	Proposals urging customers to buy a GrandCo model, stressing its state-of-the-art advantages over the competition's and emphasizing the specific benefits it offers (cost, service, quality, efficiency)
Principal Executive	Short reports on sales, cash flow, productivity, market trends; research about potential competition
Production Engineer	Detailed reports on design and manufacturing models, including spec sheets, diagrams, etc., on transmissions, strength of materials; status reports following Environmental Protection Agency (EPA) or Occupational Safety and Health Administration (OSHA) guidelines
Production Supervisor	Service reports about schedules, staffing needs, and employee activity reports; availability of parts from vendors
Operator	Instructions in manuals, including visuals on how to operate equipment safely and responsibly; warnings about any type of precautions; information on any special training necessary
Maintenance Worker	Reports and guidelines about maintenance procedures; schedules; checklists of items to be inspected; troubleshooting procedures on dealing with any problems
Community Residents	News releases about GrandCo contributions to Fairfield— sponsoring events, offering tours or demonstrations; blogs on how GrandCo is greening the workplace; articles on GrandCo's dedication to community environment and safety

As these examples show, to succeed in the world of work, give each reader the details he or she needs to accomplish a given job.

6. **What are my audience's expectations about my written work?**

   - Do they want an email, or will they expect a formal letter?
   - Will they expect me to follow a company format and style?
   - Are they looking for a one-page memo or for a comprehensive report?
   - Should I use a formal tone or a more relaxed and conversational style?

7. **What is my audience's attitude toward me and my work?**

   - Will I be writing to a group of disgruntled and angry customers or vendors about a sensitive issue (a product recall, the discontinuation of a service, a refusal of credit, or a shipment delay)?
   - Will I have to be sympathetic while at the same time give firm, convincing reasons for my company's (or my) decision?
   - Will my readers be skeptical, indifferent, or accepting about what I write?
   - Will my readers feel guilty that they have not answered an earlier message of mine, not paid a bill now overdue, or not kept a promise or commitment?

8. **What do I want my audience to do after reading my work?**

   - Do I want my readers to purchase something from me, approve my plan, or send me additional documentation?
   - Do I expect my readers to acknowledge my message, save it for future reference, or review and email it to another individual or office?
   - Do my readers have to take immediate action, or do they have several days or weeks to respond?
   - Do I simply want my readers to get my message and not respond at all?

As your answers to these questions will show, you may have to communicate with many different audiences on your job. Each group of readers will have different expectations and requirements; you need to understand those audience differences if you want to supply relevant information.

Establishing Your Purpose

By knowing *why* you are writing, you will communicate better and find the writing process itself to be easier. The reader's needs and your goal in communicating will help you to formulate your purpose. It will guide you in determining exactly what you can and must say.

Make sure you follow the most important rule in occupational writing: *Get to the point right away.* At the beginning of your message, state your goal clearly. Don't feel as if you have to entertain or impress your reader.

> I want new employees to know how to log on to the computer.

Think over what you have written. Rewrite your purpose statement until it states precisely why you are writing and what you want your readers to do or to know.

> I want to teach new employees the security code for logging on to the company computer.

Since your purpose controls the amount and order of information you include, state it clearly at the beginning of every email, memo, letter, and report.

> This memo will acquaint new employees with the security measures they must take when logging on to the company computer.

In the opening purpose statement that follows, note how the author clearly informs the reader what the report will and will not cover.

> As you requested at last month's organizational meeting, I have conducted a survey of how well our websites advertise our products. This survey describes users' responses but does not prioritize them.

The following preface to a publication on architectural casework details contains a model statement of purpose suited to a particular audience.

> This publication has been prepared by the Architectural Woodwork Institute to provide a source book of conventional details and uniform detail terminology. For this purpose a series of casework detail drawings, . . . representative of the best industry-wide practice, has been prepared and is presented here. By supplying both architect and woodwork manufacturer with a common authoritative reference, this work will enable architects and woodworkers to communicate in a common technical language. . . . Besides serving as a basic reference for architects and architectural drafters, this guide will be an effective educational tool for the beginning drafter-architect-in-training. It should also be a valuable aid to the project manager in coordinating the work of many drafters on large projects.[1]

After studying that preface, readers have a clear sense of why they should read the publication and what to do with the information they find in it.

Formulating Your Message

Your message is the sum of the facts, responses, and recommendations you put into writing. A message includes the scope and details of your communication.

- *Scope* refers to how much information you give readers about key details.
- *Details* are the key points you think readers need to know.

Some messages will consist of one or two phrases or sentences: "Do not touch; wet paint." "Order #756 was sent this afternoon by FedEx. It should arrive at your office on March 22." At the other extreme, messages may extend over many pages. Messages may carry good news or bad news. They may deal with routine matters, or they may handle changes in policy, special situations, or problems.

Keep in mind that you will need to adapt your message to fit your audience. For some audiences, such as engineers or technicians, you may have to supply a complete report with every detail noted or contained in an appendix. For other readers—busy executives, for example—include only a summary of financial or managerial significance. (See page 704 for an example of an abstract.)

---

[1]Reprinted by permission of Architectural Woodwork Institute.

Selecting Your Style and Tone

Style

Style refers to *how* something is written rather than what is written. Style helps to determine how well you communicate with an audience and how well your readers understand and receive your message. It involves the choices you make about

- the construction of your paragraphs
- the length and patterns of your sentences
- your choice of words

You will have to adapt your style to take into account different messages, different purposes, and different audiences. Your words, for example, will certainly vary with your audience. If all your readers are specialists in your field, you may safely use the technical language and symbols of your profession. Nonspecialists, however, will be confused and annoyed if you write to them in the same way. The average consumer, for example, will not know what a potentiometer is; but if you write "volume control on a radio" instead, you will be using words that the general public can understand. And as we saw, when you write for an international audience you have to take into account their proficiency in English and choose your words and sentences with their needs in mind (see pages 8–11).

Tone

Tone in writing, like tone of voice, expresses your attitude toward a topic and toward your audience. Your tone can range from formal and impersonal (a scientific

Case Study

Adapting the Technical Details to Meet Your Audience's Needs

The excerpt in Figure 1.7 on page 20 comes from a section called "Technology in the Grocery Store" in a consumer handbook. The message provides factual information and a brief explanation of how a grocery store clerk scans an item, informing consumers about how and why they may have to wait longer in line. It also tells readers that the process is not as simple as it looks. This message is appropriate for consumers who do not need or desire more information.

Individuals responsible for entering data into the computer or doing inventory control, however, would need more detailed instructions on how to program the supermarket's computer so that it automatically tells the point-of-sale (POS) terminal what price and product match each bar code.

The graphic designers responsible for product packaging, including affixing the bar codes at the manufacturing plant, would require much more detailed information than would consumers or store cashiers. These designers must be familiar with the Universal Product Code (UPC), which specifies bar codes worldwide. They would also have to know about the UPC binary code formulas and how they work—that is, the number of lines, the width of spacing, and the framework to indicate to the scanner when to start reading the code and when to stop. Such formulas, technical details, and functions of photoelectric scanners are necessary and appropriate for this audience.

FIGURE 1.7 An Appropriately Formulated Message for Consumers

Bar Code Readers

Every time you check out at a grocery or retail store, your purchases are scanned to record the price. An optical scanner uses a laser beam to read the bar codes, those zebra-striped lines imprinted on packages or canned goods. These codes are fed into the store's computer, which provides the price that matches the product code. The product and its price are then recorded on your receipt. Simultaneously, the information is fed into the store's computer to keep track of the inventory on hand.

Direction in which clerk moves
product over scanner

Whether it is you or the store clerk that does the scanning, the process starts with lining up the bar code with the red light of the scanner. An electric beep sounds when the scan is successful. The scanner may be handheld for items too heavy to lift and line up.

Scanning an item requires more skill than you might think. To make sure that the scanner accurately reads bar codes, you or the clerk must pay attention to the following:

Readable bar codes: Make sure that the codes are clearly printed and are not faded, torn, or crumpled.

Scanner glass: Be sure that the glass plate is clean and clear; it should not be coated with anything sticky that would leave a film.

Speed: Pass the item across the scanner at a steady speed. If you move the item too slowly, the bars will look too long and too wide and the computer will reject them. If you move the item too quickly, the scanner will not be able to identify the code.

Angle: Hold the bar code at a right angle so that the item reflects as much of the laser as possible. If the angle is wrong, there will be an insufficient reflection of the laser beam back to the scanner.

Distance: It is best to hold the item 3 to 4 inches away from the glass. If you hold the object out much farther, say, 8 to 9 inches away from the glass, the code will be out of focus for the scanner to identify.

Rotation: Be sure the code faces the scanner so that the lines can be read correctly.

You can also scan bar codes with a camera-equipped mobile phone. Thanks to the software available through one of your apps, you can read the UPC bar code and access a manufacturer's database that will give you information about the item you may wish to purchase. For instance, if you are shopping for a coffeemaker, you would open the bar code reader app, scan the bar code using the camera on your mobile phone, and, after a few seconds, find out the price, the location of stores near you that sell the coffeemaker, and links to product reviews. A scan might also give you information about a nearby store that has the same model you want to purchase on sale at a lower price.

report) to informal and personal (an email to a friend or a how-to article for consumers). Your tone can be unprofessionally sarcastic or diplomatically agreeable.

Tone, like style, is indicated in part by the words you choose. For example, saying that someone is "interested in details" conveys a more positive tone than saying the person is a "nitpicker." The word *economical* is more positive than *stingy* or *cheap*.

The tone of your writing is especially important in occupational writing because it reflects the image you project to your readers and thus determines how they will respond to you, your work, and your company. Depending on your tone, you can appear sincere and intelligent or angry and uninformed. Of course, in all your written work, you need to sound professional and knowledgeable. The wrong tone in a letter or a proposal might cost you a customer. Sarcastic or hostile language will alienate you from your readers, as the letters in Figures 5.5 and 5.7 demonstrate (see pages 180 and 184).

Case Study

Adapting a Description of Heparin for Two Different Audiences

In the workplace you will often be faced with the problem of presenting the same information to two completely different audiences. To better understand the impact that style and tone can have when you have to solve this problem, read the following two descriptions of heparin, a drug used to prevent blood clots. In both descriptions, the message is basically the same. Yet because the audiences differ, so do the style and the tone.

The first description of heparin appears in a reference work for physicians and other health care providers and is written in a highly technical style with an impersonal tone appropriate for the contexts in which this medicine is discussed.

The writer has made the appropriate stylistic choices for the audience, the purpose, and the message. Health care providers understand and expect the jargon and the scientific explanations, which enable them to prescribe or administer heparin correctly. The writer's authoritative, impersonal tone is coldly clinical, which, of course, is also appropriate because the purpose is to convey the accurate, complete scientific facts about this drug, not the writer's or reader's personal opinions or beliefs. The writer sounds both knowledgeable and objective.

Technical Description

Heparin Sodium Injection, USP Sterile Solution

Description: Heparin Sodium Injection, USP is a sterile solution of heparin sodium derived from bovine lung tissue, standardized for anticoagulant activity.

Each ml of the 1,000 and 5,000 USP units per ml preparations contains heparin sodium 1,000 or 5,000 USP units; 9 mg sodium chloride; 9.45 mg benzyl alcohol added as preservative. Each ml of the 10,000 USP units per ml preparations contains heparin sodium 10,000 units; 9.45 mg benzyl alcohol added as a preservative.

(Continued)

When necessary, the pH of Heparin Sodium Injection, USP was adjusted with hydrochloric acid and/or sodium hydroxide. The pH range is 5.0–7.5.

Clinical pharmacology: Heparin inhibits reactions that lead to the clotting of blood and the formation of fibrin clots both *in vitro* and *in vivo*. Heparin acts at multiple sites in the normal coagulation system. Small amounts of heparin in combination with antithrombin III (heparin cofactor) can inhibit thrombosis by inactivating activated Factor X and inhibiting the conversion of prothrombin to thrombin.

Dosage and administration: Heparin sodium is not effective by oral administration and should be given by intermittent intravenous injection, intravenous infusion, or deep subcutaneous (intrafrat, i.e., above the iliac crest or abdominal fat layer) injection. **The intramuscular route of administration should be avoided because of the frequent occurrence of hematoma at the injection site.**[2]

The second description of heparin below, however, is written in a nontechnical style and with an informal, caring tone. This description is similar to those found on information sheets given to patients about the medications they are receiving in a hospital.

The writer of this patient-centered description has also made appropriate choices for nonspecialists, such as patients or their families, who do not need elaborate descriptions of the origin and composition of the drug. Using familiar words and adopting a personal, friendly tone help to win the patients' confidence and enable them to understand why and how they should take the drug.

Nontechnical Description

Your doctor has prescribed a drug called *heparin* for you. This drug will prevent any new blood clots from forming in your body. Since heparin cannot be absorbed from your stomach or intestines, you will not receive it in a capsule or tablet. Instead, it will be given into a vein or the fatty tissue of your abdomen. After several days, when the danger of clotting is past, your dosage of heparin will be gradually reduced. Then another medication you can take by mouth will be started.

Characteristics of Job-Related Writing

Job-related writing characteristically serves six basic functions: (1) to provide practical information, (2) to give facts rather than impressions, (3) to supply visuals to clarify and condense information, (4) to give accurate measurements, (5) to state responsibilities precisely, and (6) to persuade and offer recommendations. These six functions tell you what kind of writing you will produce after you successfully answer the *who, why, what,* and *how.*

1. Providing Practical Information

On-the-job writing requires a practical "here's what you need to do or to know" approach. One such practical approach is *action oriented*. You instruct the reader to do something—assemble a ceiling fan, test for bacteria, perform an audit, or create a website. Another practical approach of job-related writing is *knowledge oriented*. You explain what you want the reader to understand—why a procedure was changed, what caused a problem or solved it, how much progress was made on a job site, or why a new piece of equipment should be purchased.

The following description of the Energy Efficiency Ratio combines both the action-oriented and knowledge-oriented approaches of practical writing.

> Whether you are buying window air-conditioning units or a central air-conditioning system, consider the performance factors and efficiency of the various units on the market. Before you buy, determine the Energy Efficiency Ratio (EER) of the units under consideration. The EER is found by dividing the BTUs (units of heat) that the unit removes from the area to be cooled by the watts (amount of electricity) the unit consumes. The result is usually a number between 5 and 12. The higher the number, the more efficiently the unit will use electricity.[3]

2. Giving Facts, Not Impressions

Occupational writing is concerned with what can be seen, heard, felt, tasted, or smelled. The writer uses *concrete language* and specific details. The emphasis is on facts rather than on the writer's feelings or guesses.

The discussion below, addressed to a group of scientists about the sources of oil spills and their impact on the environment, is an example of writing with objectivity. It describes events and causes without anger or tears. Imagine how much emotion could have been packed into a paragraph by the residents of the coastal states who watched massive spills come ashore in 2010.

> The most critical impact results from the escapement of oil into the ecosystem, both crude oil and refined fuel oils, the latter coming from sources such as marine traffic. Major oil spills occur as a result of accidents such as blowout, pipeline breakage, etc. Technological advances coupled with stringent regulations [can] reduce the chances of such major spills; however, there is [still] a chronic low-level discharge of oil associated with normal drilling and production operations. Waste oils discharged through the river systems and practices associated with tanker transports dump more significant quantities

---

[3]Reprinted by permission of New Orleans Public Services, Inc.

FIGURE 1.8 Use of a Visual to Convey Information

Using Your Computer Safely

By following the bulleted guidelines below, and illustrated in the photo to the right, you can avoid workplace injuries when using your computer.

- **To reduce the possibility of eye damage,** maintain a distance of 18 to 24 inches between your eyes and the computer screen and always make sure to keep your work area well lit.

- **To minimize neck strain,** position your computer screen so that the top of the screen is at or just below your eye level.

- **To avoid back and shoulder strain,** sit up straight at a right angle in your chair with your shoulders relaxed and your lower back firmly supported (with a cushion, if necessary).

- **To lessen leg and back strain,** adjust your chair height so that your upper body and your legs form a 90-degree angle and that your feet are either flat on the floor or on a footrest.

---

of oils into the ocean, compared to what is introduced by the offshore oil industry. All of this contributes to the chronic low-level discharge of oil into world oceans. The long-range cumulative effect of these discharges is possibly the most significant threat to the ecosystem.[4]

3. Supplying Visuals to Clarify and Condense Information

Visuals are indispensable partners of words in conveying information to your readers. On-the-job writing makes frequent use of visuals—such as tables, charts, photographs, flow charts, diagrams, and drawings—to clarify and condense information. Thanks to various software packages, you can easily create and insert visuals into your writing. The use of visuals is discussed in detail in Chapters 10 and 11, and PowerPoint presentations are covered in Chapter 16.

Visuals play an important role in the workplace. Note how the photograph in Figure 1.8 can help computer users to better understand and follow the accompanying written ergonomics guidelines. A visual like this, reproduced in an employee handbook or displayed on a website, can significantly reduce physical stress and increase a worker's productivity.

---

[4]*Source:* The Offshore Ecology Investigation. Reprinted by permission of Gulf Universities Research Consortium.

TABLE 1.1   Ten Most Populous Countries, 2010 and in 2050 (Projected)

2010		2050	
Country	**Population (Millions)**	**Country**	**Population (Millions)**
China	1,330	India	1,808
India	1,173	China	1,424
United States	310	United States	420
Indonesia	243	Nigeria	356
Brazil	201	Indonesia	313
Pakistan	177	Bangladesh	280
Bangladesh	158	Pakistan	278
Nigeria	152	Brazil	228
Russia	139	Kinshasa (Congo)	203
Japan	136	Mexico	148

Source: U.S. Census Bureau, Jan. 2010.

Visuals are extremely useful in making detailed relationships clear to readers. The information in Table 1.1 on the world's ten most populous countries in 2010 and those projected for 2050 would be very difficult to discuss and follow if it were not in twin tables. When that information is presented in two tables, the writer makes it easy for the reader to see and understand relationships. If such information were just written in prose, it would be much harder to compare, contrast, and summarize.

In addition to the visuals already mentioned, the following graphic devices in your letters, reports, and websites can make your writing easier to read and follow:

- headings, such as "Four Keys to Effective Writing" or "Characteristics of Job-Related Writing"
- subheadings to divide major sections into parts, such as "Providing Practical Information" or "Giving Facts, Not Impressions"
- numbers within a paragraph, or even a line, such as (1) this, (2) this, and (3) also this
- different types of s p a c i n g
- CAPITALIZATION
- *italics* (easily made by a word processing command or indicated in typed copy by <u>underscoring</u>)
- **boldface** (darker print for emphasis)
- symbols (visual markers such as →)
- <u>hypertext</u> (Internet links, often presented underscored, in boldface, or in a different color)
- asterisks (*) to separate items or to note key information
- lists with bullets (like those before each entry in this list)

Keep in mind that graphic devices should be used carefully and in moderation, not to decorate a letter or report. When used properly, they can help you to

- organize, arrange, and emphasize your ideas
- make your work easier to read and to recall
- preview and summarize your ideas, for example, through boldface headings
- list related items to help readers distinguish, follow, compare, and recall them—as this bulleted list does

4. Giving Accurate Measurements

Much of your work will depend on measurements—acres, bytes, calories, kilometers, centimeters, degrees, dollars and cents, grams, percentages, pounds, square feet, and so on. Numbers are clear and convincing. However, you must be sensitive to which units of measurement you use when writing to international readers. Not every culture computes in dollars or records temperatures in degrees Fahrenheit. See pages 190–191.

The following discussion of mixing colored cement for a basement floor would be useless to readers if it did not supply accurate quantities:

> Including permanent color in a basement floor is a good selling point. One way of doing this is by incorporating commercially pure mineral pigments in a topping mixture placed to a 1-inch depth over a normal base slab. The topping mix should range in volume between 1 part portland cement, 1¼ parts sand, and 1¼ parts gravel or crushed stone and 1 part portland cement, 2 parts sand, and 2 parts gravel or crushed stone. Maximum size gravel or crushed stone should be 3/8 inch.
>
>   Mix cement and pigment before aggregate and water are added and be very thorough to secure uniform dispersion and the full color value of the pigment. The proportion varies from 5 to 10 percent of pigment by weight of cement, depending on the shade desired. If carbon black is used as a pigment to obtain grays or black, a proportion of from ½ to 1 percent will be adequate. Manufacturers' instructions should be followed closely; care in cleanliness, placing, and finishing are also essential. Colored topping mixes are available from some suppliers of ready mixed concrete.[5]

5. Stating Responsibilities Precisely

Because it is directed to a specific audience, your job-related writing should make absolutely clear what it expects of, or can do for, that audience. Misunderstandings waste time, cost money, and can result in injuries. Directions on order forms, for example, should indicate how and where information is to be listed and how it is to be routed and acted on. The following directions show readers how to perform different tasks:

- Enter agency code numbers in the message box.
- Items 1 through 16 of this form should be completed by the injured employee or by someone acting on his or her behalf, whenever an injury is sustained on the job. The term *injury* includes occupational disease caused by the employment. The form should be given to the employee's official superior within 12–24 hours following the injury. The official superior is that individual having responsible supervision over the employee.

[5]Reprinted by permission of *Concrete Construction Magazine*, World of Concrete Center, 426 S. Westgate, Addison, IL 60101.

Other kinds of job-related writing deal with the writer's responsibilities rather than the reader's, for example, "Tomorrow I will meet with the district sales manager to discuss (1) July's sales, (2) the opportunities of expanding our market, and (3) next fall's production schedule. I will send a PDF of our discussion by August 3."

6. Persuading and Offering Recommendations

Persuasion is a crucial part of writing on the job. In fact, it is one of the most crucial skills you can learn in the business world. Persuasion means trying to convince your reader(s) to accept your ideas, approve your recommendations, or order your products. Convincing your reader to accept your interpretation or ideas is at the heart of the world of work, whether you are writing to someone outside or inside your company.

Writing Persuasively to Clients and Customers

Much of your writing in the business world will promote your company's image by persuading customers and clients (a) to buy a product or service, (b) to adopt a plan of action endorsed by your employer, or (c) to support a particular cause or campaign that affects a community. You will have to convince readers that you (and your company)—your products and services—can save them time and money, increase efficiency, reduce risks, or improve their image and that you can do this better than your competitors can.

Expect also to be called on to write convincingly about your company's image, as in the case of product recalls or discontinuances (see Figure 4.11, page 161), customer complaints, or damage control after a corporate mistake affecting the environment. You may also have to convince customers around the globe that your company respects cultural diversity and upholds specific ethnic values.

A large part of being a persuasive writer is supporting your claims with evidence. You will have to conduct research; provide logical arguments; supply appropriate facts, examples, and statistics; and identify the most relevant information for your particular audience(s). Notice how the advertisement in Figure 1.9 offers a bulleted list of persuasive reasons—based on cost, time, efficiency, safety, and convenience—to convince correctional officials that they should use General Medical's services rather than those of a hospital or clinic.

Writing Persuasively to In-House Personnel

As much as 70 percent of your writing may be directed to individuals you work with and for. In fact, your very first job-related writing will likely be a persuasive letter of application to obtain a job interview with a potential employer.

On the job, you may have to persuade a manager to buy a new technology or lobby for a change in your office or department. To be successful, you will have to evaluate various products or options by studying, analyzing, and deciding on the most relevant one(s) for your boss. Your reader will expect you to offer clear-cut, logical, and convincing reasons for your choice, backed up with persuasive facts.

FIGURE 1.9 An Advertisement Using Persuasive Arguments to Convince Potential Customers to Use a Service

---

Visual stresses the need for a more efficient way to transport prisoners for medical attention

Bulleted list conveniently and persuasively uses factual data to convince

GENERAL MEDICAL WILL STOP THE UNNECESSARY TRANSPORTING OF YOUR INMATES.

- We'll bring our X-ray services to your facility, 7 days a week, 24 hours a day.
- We can reduce your X-ray costs by a minimum of 28%. X-ray cost includes radiologist's interpretation and written report.
- Same-day service with immediate results telephoned to your facility.
- Save correctional officers' time, thereby saving your facility money.
- Avoid chance of prisoner's escape and possible danger to the public.
- Avoid long waits in overcrowded hospitals.
- Reduce your insurance liabilities.
- Other Services Available: Ultrasound, Two-Dimensional Echocardiogram, C.T. Scan, EKG, Blood Lab and Holter Monitor.

General Medical Services Corp.

A subsidiary of

Federal Medical Industries, Inc. O.T.C.
950 S.W. 12th Avenue, 2nd Floor Suite, Pompano, Florida 33069
(305) 942-1111   FL WATS: 1-800-654-8282

General Medical Is Your On-Site Medical Problem Solver

© Cengage Learning 2013

As part of your job, too, you will be asked to write convincing memos, emails, letters, blogs, and websites to boost the morale of employees, encourage them to be more productive, and compliment them on a job well done.

The following summary concludes that it is better for a company to lease a truck rather than to purchase one. Note how the writer uses a persuasive tone and presents information logically. You can also expect to write persuasively to explain and solve budget, safety, or marketing problems your company faces, as in Figure 4.2 (see page 135).

After studying the pros and cons of buying or leasing a company truck, I recommend that we lease it for the following five reasons.

1. We will not have to expend any of our funds for a down payment, which is being waived.
2. Our monthly payments for leasing the vehicle will be at least $150 less than the payments we would have to make if we purchased the truck on a three-year contract.
3. All major and minor maintenance (up to 36,000 miles) is included as part of our monthly leasing payment.
4. Insurance (theft and damage) is also part of our monthly leasing payment.
5. We have the option of trading in the truck every sixteen months for a newer model or trading up for a more expensive model in the line every twelve months.

FIGURE 1.10 A Persuasive Email from an Employee to a Business Manager

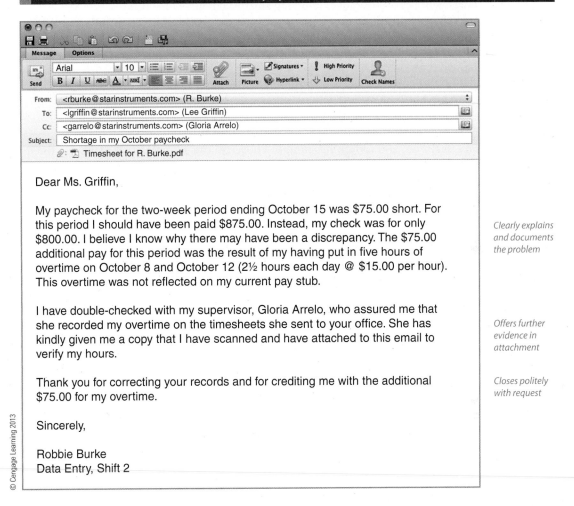

From: <rburke@starinstruments.com> (R. Burke)
To: <lgriffin@starinstruments.com> (Lee Griffin)
Cc: <garrelo@starinstruments.com> (Gloria Arrelo)
Subject: Shortage in my October paycheck
📎: 📄 Timesheet for R. Burke.pdf

Dear Ms. Griffin,

My paycheck for the two-week period ending October 15 was $75.00 short. For this period I should have been paid $875.00. Instead, my check was for only $800.00. I believe I know why there may have been a discrepancy. The $75.00 additional pay for this period was the result of my having put in five hours of overtime on October 8 and October 12 (2½ hours each day @ $15.00 per hour). This overtime was not reflected on my current pay stub.

Clearly explains and documents the problem

I have double-checked with my supervisor, Gloria Arrelo, who assured me that she recorded my overtime on the timesheets she sent to your office. She has kindly given me a copy that I have scanned and have attached to this email to verify my hours.

Offers further evidence in attachment

Thank you for correcting your records and for crediting me with the additional $75.00 for my overtime.

Closes politely with request

Sincerely,

Robbie Burke
Data Entry, Shift 2

© Cengage Learning 2013

Figure 1.10 is a persuasive email from an employee to a manager reporting a payroll mistake and persuading the reader to correct it. The email contains many of the other characteristics of job-related writing we have discussed. Note how the writer provides factual, not subjective, information; attaches a time sheet (a type of visual); gives accurate details; and identifies her own and her immediate supervisor's responsibilities. The writer's tone is suitably polite yet direct.

Ethical Writing in the Workplace

One of your most important job responsibilities is to ensure that your writing and behavior are ethical. Writing ethically means choosing language that is right and fair, honest, and complete in all documents prepared for your employer, co-workers, and customers. Your reputation and character plus your employer's corporate image will depend on your following an ethical course of action.

Many of the most significant phrases in the world of business reflect an ethical commitment to honesty and fairness: *accountability, public trust, equal opportunity employer, core values, global citizenship, good-faith effort, truth in lending, fair play, honest advertising, full disclosure, high professional standards, fair trade, community involvement*, and *corporate responsibility*.

Unethical business dealings, conversely, are stigmatized in *cover-ups, dodges, stonewalling, shady deals, spin-doctoring, foul play, bid rigging, employee raiding, misrepresentations, kickbacks, hostile takeovers, planned obsolescence, price gouging, bias*, and *unfair advantage*. Those are the activities that make customers angry and that the Better Business Bureau and local, state, and federal agencies may investigate.

Employers Insist on Ethical Behavior

Ethical behavior is crucial to your success in the workplace. Your employer will insist that you are honest, follow professional standards, show integrity, and exhibit loyalty in your professional relationships with clients, co-workers, supervisors, and vendors. These ethical values are stressed in orientation and training sessions and through every level of management in the business world. You will be expected to know and comply with your company policies and procedures, as outlined in the employee or agency handbook (see pages 577–596), and you will also have to follow the professional codes, regulations, and methods that affect your job. Operating with ethical behavior and sound judgment is necessary so that companies can

- adhere to all industry-specific rules and regulations
- follow all state, local, and federal requirements
- provide a safe, healthy, and productive workplace
- create strategic advantages for healthy competition
- champion change and innovation
- create customer loyalty
- promote global perspective of their "brand"

Employers Monitor Ethical Behavior

On the job, employers can legally monitor their employees' work—electronically, through cameras, or by personal visits. Some of these visits are not announced (such as the "secret shoppers" who report on the customer service they receive). How many times have you made a call to an organization and heard, "This call may be monitored for quality assurance"? According to a 2010 survey conducted by the American Management Association, the monitoring of employees has risen 45 percent in the past few years and extends to voice mail, email, instant messages (IMs), and Web use.

Employers monitor the behavior of their employees for several reasons:

- to determine if a worker is doing his or her job correctly
- to find out how well a worker is performing
- to identify employee wrongdoing
- to improve service, production, communication, or transportation
- to ensure compliance with federal, state, and municipal codes
- to limit liability
- to adhere to and even heighten security measures

Monitoring gives management solid facts about employee training, performance reviews, and promotions. But working with integrity means doing the right thing—even when no one is watching.

Know Your Company's Culture and Codes: A Letter from IBM's CEO

You need to know your company's history, mission, image, accomplishments, and ethical code of conduct—that is, what it expects of you in the workplace and in the marketplace.

Figure 1.11 on page 32 contains a letter from IBM's CEO introducing the company's Business Conduct Guidelines that employees are required to follow. The letter, and the list of guidelines, gives IBM employees a clear sense of how IBM defines its corporate identity through the high ethical standards it maintains. Knowing the core values that IBM fosters and understanding the actions it will not tolerate, employees can better perform their duties. In such a corporate culture, both IBM and its employees can successfully fulfill IBM's mission, uphold the company's image, and serve customers ethically around the globe.

Ethical Requirements on the Job

In the workplace, you will be expected to meet the highest ethical standards by fulfilling the following eight requirements:

1. Be professionally competent. Know your job. Your company will expect you to be well prepared through your education, internships, experience, in-house training, continuing education units (CEUs), professional conferences, discussions with co-workers, and reading. You need to use equipment safely and efficiently, produce high-quality products, deliver up-to-date and accurate service, and represent your company as a knowledgeable professional.

2. Be honest. Never misrepresent yourself on a résumé, at an interview, or on a networking site such as LinkedIn (see Figure 7.2, pages 266–267), by lying about your background, inflating a job title, or exaggerating your responsibilities at a previous job. The résumé (see pages 272–289) is just one place you must make decisions with honesty and candor. At work, honesty is equally crucial. You need to

FIGURE 1.11 A Letter from the CEO of IBM

Dear IBMer:

In 2003, we undertook a global, company-wide discussion about the values that define IBM. In addition to finding a common set of qualities that characterize "an IBMer," we also learned something equally important: Almost every one of us thinks our work and choices should be determined by what we value.

This is particularly relevant to what we agree explicitly to do and not do as individuals when conducting IBM's business. Each one of us makes decisions that could affect our company and its reputation—whether with one person or with millions of people.

At one level, the IBM Business Conduct Guidelines are a document of conduct we establish for ourselves to help us comply with laws and good ethical practices. We regularly review and update it as business and the world at large become more complex, and as the need for such guidelines becomes greater.

But this is not just about compliance with the law and general standards of ethics. By establishing these guidelines and giving them the weight of a governing document, we are acknowledging that our choices and actions help define IBM for others. We are ensuring that our relationships—with clients, investors, colleagues and the communities in which we live and work—are built on trust.

In other words, the Business Conduct Guidelines are a tangible example of our values and an expression of each IBMer's personal responsibility to uphold them.

I hardly find it necessary to remind IBMers to "act ethically." I know you feel as strongly as I that anyone doing otherwise does not belong at IBM. But I do ask you to read these Business Conduct Guidelines and commit yourself to them. In addition to establishing a baseline for behavior throughout IBM, they provide some excellent examples of how we live out our values as a company. They are an important part of what it means to be an IBMer.

Sam Palmisano
Chief Executive Officer

January 2009

Source: From IBM Business Conduct Guidelines, January 12, 2009. Reprint courtesy of International Business Machines Corporation, © 2009 International Business Machines Corporation.

acknowledge and correct all mistakes and make sure you submit complete, accurate, and truthful reports. Never falsify a document by padding an expense account, covering up a problem, or wrongly accusing a co-worker. Inventing or falsifying information is fraud. In your dealings with customers, honor all guaranties and warranties and respond to customer requests promptly and fairly. It would be neglectful and dishonest to allow an unsafe product to stay on the market just to spare your company the expense and embarrassment of a product recall.

3. Maintain confidentiality. Never share sensitive or confidential information with individuals who are not entitled to see or hear it. You violate your employer's trust by telling others about your company's research, financial business, marketing strategies, sales records, personnel decisions, or customer interactions. In fact, your employer may rightfully insist that you sign a binding confidentiality agreement when you are hired. You also have to respect an individual's right to privacy. For example, according to Health Insurance Portability and Accountability Act (HIPAA) guidelines, health care professionals are not allowed to share a patient's records with unauthorized individuals. It is also unethical to divulge personal information that a co-worker or supervisor has asked you to keep confidential.

4. Be loyal. Observe all your employer's instructions to conform with company policies. Your employer has every right to expect you to be a team player striving for the good of the company; its products, service, and image; your department; and your co-workers. Cooperate fully with your collaborative team; do not neglect your responsibilities to contribute to the group effort (see pages 84–86, 92). Working secretly for a competitor is a clear conflict of interest. Also, criticizing a boss, product, service, or event, or engaging in malicious gossip at work are examples of disloyalty that companies will not tolerate.

5. Follow the chain of command. You need to know and follow your company's or department's chain of command—for example, whom you report to, who gets copies of your written work and who does not, how work is to be submitted and routed, and whom you need to go to with problems. Always direct your correspondence to the appropriate person(s) in the company. Nothing infuriates a manager more than having an employee go over his or her head without authorization. All businesses, large and small, operate according to protocols. To help you identify the proper chain of command at work, find or construct an organizational chart (see pages 482–485).

6. Respect your employer, co-workers, customers, and vendors. Avoid intimidation, bullying, spreading rumors, discrimination, defamation, or any other unfair, unprofessional action that would harm someone or tarnish his or her reputation. It is unethical and illegal to use language that excludes others on the basis of gender, national origin, religion, age, physical ability, or sexual orientation (see pages 73–77). Never use racial slurs or obscene language.

7. Research and document your work carefully. Your boss will expect you to do your homework to provide the hard evidence he or she needs. You do your homework by studying codes, specifications, agency handbooks, and websites;

by keeping up to date with professional literature found in journals and books in your field; by conferring with experts in your company; by interviewing clients; by making a site visit; by performing a test, and so on. (See pages 331–332 for a description of the different types of research you will be expected to do on the job.) You are also ethically obligated to admit when you did not do the work by yourself. Always give credit to your sources—whether print or Web sources or individuals whose discussions contributed to your work. Not documenting your sources makes you guilty of plagiarism (see pages 383–385).

8. Maintain accurate and up-to-date records. Remember, "If it isn't written, it didn't happen." A vital part of your job may be writing about it. You have a responsibility to your employer to prepare and store documents, keep backup files, and submit your work by the deadline. Disregarding a deadline could jeopardize a group's effort or prevent your company from receiving a key permit or license or a government contract. Moreover, not maintaining timely, accurate records may constitute a safety hazard according to the Food and Drug Administration (FDA) or the Occupational Safety and Health Administration (OSHA).

Computer Ethics

Computer ethics are essential in the world of e-commerce. A good rule to follow is never to do anything online that you wouldn't do offline. For instance, never use a company computer for any activity not directly related to your job. Moreover, it would be grossly unethical to erase a computer program intentionally, violate a software licensing agreement, or misrepresent (by fabrication or exaggeration) the scope of a database. Also, posting anything that attacks a competitor, a colleague, your boss, or your company is considered unethical. Follow the Ten Commandments of Computer Ethics prepared by the Computer Ethics Institute listed in Figure 1.12.

FIGURE 1.12 The Ten Commandments of Computer Ethics

1. Thou shalt not use a computer to harm other people.
2. Thou shalt not interfere with other people's computer work.
3. Thou shalt not snoop around in other people's computer files.
4. Thou shalt not use a computer to steal.
5. Thou shalt not use a computer to bear false witness.
6. Thou shalt not copy or use proprietary software for which you have not paid.
7. Thou shalt not use other people's computer resources without authorization or proper compensation.
8. Thou shalt not appropriate other people's intellectual output.
9. Thou shalt think about the social consequences of the program you are writing or the system you are designing.
10. Thou shalt always use a computer in ways that ensure consideration and respect for your fellow humans beings.

Source: Computer Ethics Institute, London.

You are also ethically bound to protect your computer at work from security risks and possible system malfunctions. Never be afraid to ask for advice from a co-worker or someone in your firm's IT department who knows what to do if there is a computer emergency.

Here are some other specific guidelines to follow when using your computer at work:

- Protect passwords that allow access to your company's documents as well as its proprietary databases, templates, and other customized applications. Do not share your password, and never use a password belonging to someone else.
- Always save sensitive emails, IMs, blogs, memos, letters, and so on, that you or your employer may need to document actions or decisions.
- Protect your computer from viruses, spyware, and malware by checking with your IT department to make sure the most recent updates to your antivirus programs are installed on your computer. They will protect against spyware and malware but not against phishing schemes conducted by phone or email.
- Be especially careful in opening attachments or anything you suspect may be infected, such as spam. Never forward a document you think may have a virus.
- Do not use your work email account for personal emails (see pages 142–150). Instead, use an alternate email address; you can sign up for a free email account through Hotmail, Yahoo! Mail, Gmail, and similar services. If you cannot access your email on the job because of a computer emergency, you can use this alternate email address until the problem is solved.

"Thinking Green": Making Ethical Choices About the Environment

Be respectful of the environment—whether at the office, at a work site, in the community, or in the global marketplace. Many companies are proud of their ethical commitments to the environment. Starbucks, for example, tells customers that its "10 percent post-consumer recycled . . . paper cups helped conserve enough energy to supply your homes for a year and save approximately 110,000 trees."

Like Starbucks, companies around the globe have adopted a green philosophy that they encourage their employees to support and put into practice. Note how in Figure 1.13 the Southern Company and its employees are proud of their ethical commitment to the environment, the community, and the country. The Southern Company thereby projects an image of itself as being concerned about pollution and dedicated to the highest ethical goal of preserving and protecting the environment.

You can "think green" in several ways. At your office, conserve energy by turning off all computers, copiers, and other machines when you leave work; replace incandescent lightbulbs with energy-efficient ones; recycle paper; copy and print your documents on both sides of paper; view documents on your computer screen instead of printing them; adjust thermostats when you are gone for the day

FIGURE 1.13 A Company's Commitment to Ethical Responsibility

Our Environmental Responsibility

Emphasizes corporate commitment to ethical conduct

Southern Company is not only a leader in the energy market, but also a leader in protecting the environment. We believe our environmental initiatives and our strong compliance record will give us a competitive advantage.

The Southern Company's environmental policy spells out each company's commitment to protecting the environment. The first and foremost goal is to meet or exceed all regulatory requirements for domestic and international operations. To do that, we're using a combination of the best technologies and voluntary pollution-prevention programs. We also set aggressive environmental goals and make sure employees are aware of their individual environmental responsibilities. We are good citizens wherever we serve.

Links good business practices with good ethical behavior

As an affiliate of Southern Company, Mississippi Power's environmental issues are business issues. In addition to regulatory obligations, our employees carry out a most active grassroots environmental program. It's this employee involvement and strong environmental commitment that gives our commitment life and promises future generations a healthy environment.

Praises employees for their contributions to both the community and the company

For example, one employee's concern that motor oil is properly discarded led to the founding of a countrywide annual household hazardous waste collection program. Thousands of tons of waste have been collected, including jars of DDT, mercury, paint, batteries, pesticides, and other poisons.

Scores of employees participate in island, beach, and river cleanups throughout Mississippi Power's 23-county service area. More than 30 employees compiled "The Wolf River Environmental Monitoring Program."

This report is the first-ever historical, biological assessment completed on the Wolf River by scientists and engineers. Employees volunteered countless hours to compile the statistical data. Today, Mississippi Power employees continue to support the Wolf River Project by producing photographs and slides as an educational and community awareness project.

Assures readers that corporate ethical behavior extends to the entire community

Our commitment to the environment goes beyond our business. By sponsoring a variety of programs, we're helping to teach the public, students, and teachers about environmental responsibility.

Source: Reprinted by permission of Mississippi Power Company.

or weekend, and car pool to and from work. You can also reduce toxic chemicals in the atmosphere by using soy-based ink, by inspecting vehicles regularly, and by maintaining them properly to reduce or eliminate pollution.

International Readers and Ethics

Communicating in the world of multinational corporations places additional ethical demands on you as a writer. You have to make sure that you respect the ethics of all of the countries where your firm does business. Some behaviors regarded as normal or routine in the United States might be seen as highly unethical elsewhere, and vice versa. In many countries, accepting a gift to initiate or conclude a business agreement is considered not only proper but also honorable. This is not the case in the United States, where a "bribe" is seen as bad business or may be illegal. Moreover, you should be on your ethical guard not to take advantage of a host country, such as allowing or encouraging poor environmental control because regulatory and inspection procedures are not as strict as those of the United States, or by using pesticides or conducting experiments outlawed in the United States. It would also be unethical to conceal something risky about a product from international clients that you would disclose to U.S. customers.

Some Guidelines to Help You Act Ethically

The workplace presents conflicts over who is right and who is wrong, what is best for the company and what is not, and whether a service or product should be changed and why. You will be asked to make decisions and justify them. Here are some guidelines to help you comply with the ethical requirements of your job:

1. **Follow your conscience, and "to thine own self be true."** Do not authorize something that you believe is wrong, dangerous, unfair, contradictory, or incomplete. But don't be hasty. Leave plenty of room for diplomacy and for careful questioning. Recall the story of Chicken Little, who always cried that the sky was falling. Don't blow a small matter out of proportion.

2. **Be suspicious of convenient (and false) appeals that go against your beliefs.** Watch out for these red flags: "No one will ever know." "It's OK to cut corners every once in a while." "We got away with it last time." "Don't rock the boat." "No one's looking." "As long as the company makes money, who cares?" These excuses are traps you must avoid.

3. **Take responsibility for your actions.** Saying "I don't know" when you do know can constitute a serious ethical violation. Keep your records up-to-date and accurate, sign and date your work, and never backdate a document to delete information or to cover an error that you made. Always do what is expected in terms of documentation and notification. Failing to test a set of instructions thoroughly, for example, might endanger readers around the globe.

4. **Keep others in the loop.** Confer regularly with your collaborative writing team (see Chapter 3) and any other co-workers affected by your job. Report to your boss as required to give progress reports, to alert him or her about problems, and to help you coordinate your duties with co-workers. If you experience a problem at work, don't wait until it gets worse to tell your supervisor or co-workers. Prompt and honest notifications are essential to the safety, security, progress, and ultimate success of a company. Also, never keep a customer or vendor waiting; call in advance if you are going to be delayed.

5. **Weigh all sides before you come to a conclusion.** You may think a particular course of action is right at the time, but don't overlook the possibility that your decision may create a bigger problem in the future. For example, you hear that a co-worker is involved in wrongdoing; you report it to your boss, and a reprimand

is placed in that worker's file. Later you learn that what was reported to you was malicious gossip or only a small part of a much larger but very ethical picture. Give people the benefit of the doubt until you have sufficient facts to the contrary. Giving incomplete information on an incident report may temporarily protect you but may falsely incriminate someone else or unfairly increase your company's liability insurance rates.

6. Treat company property respectfully. Use company supplies, computer and other equipment, networks, and vehicles responsibly and only for work-related business. Taking supplies home, making personal long-distance calls on a company cell phone, charging non-work-related expenses (meals, clothes, travel) on a company credit card, surfing the Internet when you are at work—these are just a few examples of unethical behavior. Being wasteful (of paper, supplies, and ink) also disrespects your company's resources.

And never take company property (equipment or supplies) with you when you leave a job.

Ethical Dilemmas

Sometimes in the workplace you will face situations where there is no clear-cut right or wrong choice. Here are a few scenarios that cover gray areas, ethically speaking, along with some possible solutions:

- You work with an office bully who often intimidates co-workers, including you, by talking down to them, interrupting them, or insulting their suggestions. At times, this bully has even sent sarcastic emails and IMs. You are upset that this behavior has not been reported to management, but you are concerned that if the bully finds out that you have reported the situation, the entire office may suffer. How should you handle the problem?

  You cannot allow such rude, insulting behavior to go unreported. But first you need to provide documentation about where, when, and how often the bullying has occurred. You may want to speak directly to the bully, but if you feel uncomfortable doing this, go directly to your boss, report how the bully's actions have negatively affected the workplace, and ask for assistance. You may also get help from your company's employee assistance program or from someone in human resources. In accordance with state and federal laws, companies must provide a safe work environment, free from intimidation, harassment, or threats of dismissal for reporting bullying.

- You work very closely with an individual who takes frequent extended lunch breaks, often comes in late and leaves early, and even misses deadlines. As a result, you are put in an awkward position. Sometimes you cover for him when he is not at the office to answer questions, and often you take on additional work he should be doing to keep your department running smoothly and efficiently. Your department is under minimal supervision from an off-site manager, so there is no boss looking over your colleague's shoulder. You like your co-worker and do not want him to be reprimanded or, worse yet, fired, but he is taking advantage of your friendship and unfairly expecting you to cover for him. What should you do?

The best route is to take your co-worker aside and speak with him before informing management. Let him know you value working with him, but firmly explain that you no longer will cover for him or take on his workload. If he does not agree with you, let him know that you will be forced to discuss the problem with your manager. If the problem persists, and you go to your boss, bring documentation—dates, duties not performed, and so on—with you.

- You see an opening for a job in your area, but the employer wants someone with a minimum of two years of field experience. You have just completed an internship and had one summer's experience, which together total almost seven months. Should you apply for the job, describing yourself as "experienced"?

Yes, but honestly state the type and the extent of your field experience and the conditions under which you obtained it.

- You work for a company that usually assigns commissions to the salesperson for whom the customer asks. One afternoon a customer asks for a salesperson who happens to have the day off. You assist the customer all afternoon and even arrange to have an item shipped overnight so that she can have it in the morning. When you ring up the sale, should you list your employee number for the commission or the off-duty employee's?

You probably should defer crediting the sale to either of you until you speak to the absent employee and suggest a compromise—splitting the commission, for instance.

- You are passed over for a promotion to assistant manager in favor of a co-worker whom you often see flirting with the manager. You have seniority over this co-worker and have received more commendations from the central office for your sales record and contributions to team efforts. Should you be a whistle-blower and write to the head of your human resources department to report the co-worker for unprofessional behavior in the workplace?

This is a tricky and complex workplace issue. Promotions can intensify personal rivalries, leading to low morale and even legal problems. Before you write an accusing letter about your co-worker, be sure of the facts. What you see as flirting may not be characterized that way by other co-workers or the manager. Unless you have clear evidence of favoritism, corroborated by other workers with specific instances, times, places, and violations of company policy, it would be more prudent to inquire about why you were passed over, citing reasons why your qualifications (education, experience, track record with the company) entitled you to the position.

- A piece of computer equipment, scheduled for delivery to your customer the next day, arrives with a damaged part. You decide to replace it at your store before the customer receives it. Should you inform the customer?

Yes, but assure the customer that the equipment is still under the same warranty and that the replacement part is new and also under the same warranty. If the customer protests, agree to let him or her use the computer until a new unit arrives.

As these brief scenarios suggest, sometimes you have to make concessions and compromises to be ethical in the world of work.

Writing Ethically on the Job

Your writing as well as your behavior must be ethical. Words, like actions, have implications and consequences. If you slant your words to conceal the truth or to gain an unfair advantage, you are not being ethical. False reporting and advertising are unethical. Bias and omission of facts are wrong. Strive to be fair, reliable, and accurate in reporting products, services, events, environmental issues, statistics, and trends.

Unethical writers are usually guilty of one or more of the following faults, which can conveniently be listed as the three *M*'s: misquotation, misrepresentation, and manipulation. Here are nine examples:

1. Plagiarism is stealing someone else's words and claiming them as your own without documenting the source. Do not think that by changing a few words of someone else's writing here and there you are not plagiarizing. Give proper credit to your source, whether in print, in person (through an interview), or online. The penalties for plagiarism are severe—a reprimand or even the loss of your job. See pages 383–385 for further advice on how to avoid plagiarism.

2. Selective misquoting deliberately omits damaging or unflattering comments to paint a better (but untruthful) picture of you or your company. By picking and choosing only a few words from a quotation, you unethically misrepresent what the speaker or writer originally intended.

> **Selective Misquotation:** I've enjoyed . . . our firm's association with Technology, Inc. The quality of their service was . . . excellent.
>
> **Full Quotation:** I've enjoyed at times our firm's association with Technology, Inc., although I was troubled by the uneven quality of their service. At times, it was excellent while at others it was far less so.

The dots, called *ellipses*, unethically suggest that only extraneous or unimportant details were omitted.

3. The arbitrary embellishment of numbers unethically misrepresents, by increasing or decreasing percentages or other numbers, statistical or other information. It is unethical to stretch the differences between competing plans or proposals to gain an unfair advantage or to express accurate figures in an inaccurate way.

> **Embellishment:** An overwhelming majority of residents voted for the new plan.
>
> **Ethical:** The new plan was passed by a vote of 53 to 49.

Embellishment: Our competitor's sales volume increased by only 10 percent in the preceding year, while ours doubled.

Ethical: Our competitor controls 90 percent of the market, yet we increased our share of that market from 5 percent to 10 percent last year.

4. Omitting key information about a product, service, or location intentionally deprives readers of the facts they need to reach a decision.

Omitting Information: You will save thousands of dollars when buying the Model 2400T, the least expensive four-wheeler on the market.

Key Information Supplied: Although the model 2400T is the least expensive four-wheeler you can purchase, it is the most expensive four-wheeler to operate and to repair, making it the most costly model to choose.

5. The manipulation of information or context, which is closely related to the embellishment of numbers, is the misrepresentation of events, usually to put a good face on a bad situation. The writer here unethically uses slanted language and intentionally misleading euphemisms to misinterpret events for readers.

Manipulation: Looking ahead to 2014, the United Funds Group is exceptionally optimistic about its long-term prospects in an expanding global market. We are happy to report steady to moderate activity in an expanding sales environment last year. The United Funds Group seeks to build on sustaining investment opportunities beneficial to all subscribers.

Ethical: Looking ahead to 2014, the United Funds Group is optimistic about its long-term prospects in an expanding global market. Though the market suffered from inflation this year, the United Funds Group hopes to recoup its losses in the year ahead.

The writer who manipulates information minimizes the negative effects of inflation by calling it "an expanding sales environment."

6. Using fictitious benefits to promote a product or service seemingly promises customers advantages but delivers none.

False Benefit: Our bottled water is naturally hydrogenated from clear underground springs.

Truth: All water is hydrogenated because it contains hydrogen.

False Benefit: All our homes come with construction-grade fixtures.

Truth: Construction-grade fixtures are the least expensive and least durable a builder can use.

7. Unfairly exaggerating or minimizing hiring or firing conditions is unethical.

Unethical: One of the benefits of working for Spelco is the double pay you earn for overtime.

Truth: Overtime is assigned on the basis of seniority.

Unethical: Our corporate restructuring will create a more efficient and streamlined company, benefiting management and workers alike.

Truth: Downsizing has led to 150 layoffs this quarter.

Companies faced with laying off employees want to protect their corporate image and maintain their stockholders' good faith, so they often put the best face on such an action.

8. Manipulating international readers by adopting a condescending view of their culture and economy is unethical.

> Unethical: Since our product has appealed to U.S. customers for the last sixteen months, there's no doubt that it will be popular in your country as well.
>
> Fair: Please let us know if any changes in product design or construction may be necessary for customers in your country.

9. Misrepresenting through distorted or slanted visuals is one of the most common types of unethical writing. Making a visual appear bigger, smaller, or more or less favorable is all too easy with graphics software. Making warning or caution statements the same size and type font as ingredients or directions or enlarging advertising hype ("Double Your Money Back") is unethical if major points are then reduced to small print. (See pages 496–503 in Chapter 10 for guidelines on how to prepare ethical visuals.)

Ethical writing is clear, accurate, fair, and honest. These are among the most important goals of any workplace communication. Because ethics is such an important topic in writing for the business world, it will be emphasized throughout this book.

Successful Employees Are Successful Writers

As this chapter has stressed, being a successful employee means being a successful writer at work. The following guidelines, which summarize the key points of this chapter, will help you to be both:

1. Know your job—assignments, roles, responsibilities, goals, what you need to write, and what you *shouldn't*.
2. Be prepared to give and to receive feedback from co-workers, managers, vendors, and customers.
3. Work toward and meet all deadlines.
4. Analyze your audience's needs and what they will expect to find in your written work.
5. Make sure your written work is accurate, relevant, and practical, and include culturally appropriate visuals to help readers understand your message.
6. Document, document, document. Submit everything you write with clear-cut evidence based on factual details and persuasive, logical interpretations.
7. Use your computer only for company business. Never share your password, and keep your computer safe from viruses.
8. Follow your company's policy, and promote your company's image, culture, and traditions.
9. Be ethical in what you say, write, illustrate, and do.

✓ Revision Checklist

At the end of each chapter is a checklist you should review before you submit the final copy of your work, either to your instructor or to your boss. The checklists specify the types of research, planning, drafting, editing, and revising you should do to ensure the success of your work. Regard each checklist as a summary of the main ideas in the chapter as well as a handy guide to quality control. You may find it helpful to check each box as you verify that you have performed the necessary revision and review. Effective writers are also careful editors.

- [ ] Showed respect for and appropriately shaped my message for a global audience.
- [ ] Identified my audience—background, knowledge of English, reason for reading my work, and likely response to my work and me.
- [ ] Tailored my message to my audience's needs and background, giving them neither too little nor too much information.
- [ ] Pushed to the main point right away; did not waste my readers' time.
- [ ] Selected the most appropriate language, technical level, tone, and level of formality.
- [ ] Did not waste my readers' time with unsupported generalizations or opinions; instead gave them accurate measurements, facts, and carefully researched material.
- [ ] Selected appropriate visuals to make my work easier for my audience to follow.
- [ ] Used persuasive reasons and data to convince my readers to accept my plan or work.
- [ ] Ensured that my writing and visuals are ethical—accurate, fair, honest, a true reflection of the situation or condition I am explaining or describing, for U.S. as well as global audiences.
- [ ] Followed the Ten Commandments of Computer Ethics.
- [ ] Adhered to the ethical codes of my profession as well as the policies and regulations set down by my employer.
- [ ] Gave full and complete credit to any sources I used, including resource people.
- [ ] Avoided plagiarism and unfair or dishonest use of copyrighted materials, both written and visual, including all electronic media.

Exercises

1. Write a memo (see pages 133–141 for format) addressed to a prospective supervisor to introduce yourself. Your memo should have four headings: **Education**—including goals and accomplishments; **Job Information**—where you have worked and your responsibilities; **Community Service**—volunteer work, church

work, youth groups; and **Writing Experience**—your strengths and weaknesses as a writer, the types of writing you have done, and the audiences for whom you have written.

2. Write a memo responding to Rowe Pinkerton's email in Figure 1.1. Explain how you will use the skills you learn in the tuition-reimbursed writing course on the job.

3. Bring to class a set of printed instructions, a memo, a sales letter, a brochure, or the printout of a home page. Comment on how well the example answers the following questions:
   a. Who is the audience?
   b. Why was the material written?
   c. What is the message?
   d. Are the style and tone appropriate for the audience, the purpose, and the message? Explain.
   e. Discuss the use of any visuals and color in the document. For instance, how does color (or the lack of it) affect an audience's response to the message?

4. Find an advertisement in a print source or online that contains a drawing or photograph. Bring the ad to class along with a paragraph of your own (75–100 words) describing how the message of the ad is directed to a particular audience and commenting on how the drawing or photo is appropriate for that audience.

5. Select one of the following topics, and write two descriptions of it. In the first description, use technical details and vocabulary. In the second, use language and details suitable for the general public.
   a. iPad
   b. blood pressure cuff
   c. flash drive
   d. computer chip
   e. Bluetooth headset
   f. legal contract
   g. electric sander
   h. cyberspace
   i. muscle
   j. protein
   k. smartphone
   l. social networking
   m. bread
   n. money
   o. iPod
   p. soap
   q. blogging
   r. computer virus
   s. swine flu
   t. thermostat
   u. trees

     v. food processor

     w. earthquake

     x. recycling

6. Select another topic from Exercise 5, and write two more descriptions as a collaborative writing project.

7. Select one article from a newspaper and one article from a professional journal in your major field or from one of the following journals: *Advertising Age, American Journal of Nursing, Business Marketing, Businessweek, Computer, Computer Design, Construction Equipment, Criminal Justice Review, E-Commerce, Food Service Marketing, Journal of Forestry, Journal of Soil and Water Conservation, National Safety News, Nutrition Action, Office Machines, Park Maintenance, Scientific American*. State how the two articles you selected differ in terms of audience, purpose, message, style, and tone.

8. Assume that you work for Appliance Rentals, Inc., a company that rents TVs, microwave ovens, stereo components, and the like. Write a persuasive letter to the members of a campus organization or civic club urging them to rent an appropriate appliance or appliances. Include details in your letter that might have special relevance to members of this specific organization.

9. Read the article, "Microwaves," on pages 47–49, and identify its audience (technical or general), purpose, message, style, and tone.

10. How do the visuals and the text of the Digital World Technologies advertisement on page 46 stress to current (and potential) employees, customers, and stockholders that the company is committed to diversity in the workplace? Also explain how the ad illustrates the functions of on-the-job writing as defined on pages 23–29.

11. Write a letter to a phone company that has mistakenly billed you for caller ID equipment that you never ordered, received, or needed.

12. The following statements contain embellishments, selected misquotations, false benefits, omitted key information, and other types of unethical tactics. Revise each statement to eliminate the unethical aspects. Make up details as needed.
   a. Storm damage done to water filtration plant #3 was minimal. While we had to shut down temporarily, service resumed to meet residents' needs.
   b. All customers qualify for the maximum discount available.
   c. "The service contract . . . on the whole . . . applied to upgrades."
   d. We followed the protocols precisely with test results yielding further opportunities for experimentation.
   e. All our costs were within fair-use guidelines.
   f. Customers' complaints have been held to a minimum.
   g. All the lots we are selling offer relatively easy access to the lake.
   h. Factory-trained technicians respond to all our calls.

13. You work for a large international company, and a co-worker tells you that he has no plans to return to his job after he takes his annual two-week vacation. You

Photos, top to bottom: © iStockphoto.com/Pali Rao; © iStockphoto.com/Wilson Valentin; © iStockphoto.com/Pali Rao; © iStockphoto.com/Jacob Wackerhausen

know that your department cannot meet its deadlines shorthanded and that your company will need at least two or three weeks to recruit and hire a qualified replacement. You also know that it is your company's policy not to give paid vacations to employees who do not agree to work for at least three months following their return. What should you do? What points would you make in a confidential email to your boss? What points would you raise to your co-worker?

14. Your company is regulated and inspected by the Environmental Protection Agency (EPA). In 90 days, the EPA will relax a regulation about dumping occupational waste. Your company's management is considering cutting costs by relaxing the standard now, before the new, less demanding regulation is in place. You know that the EPA inspector probably will not return before the 90-day period elapses. What do you recommend to management?

15. You and your co-workers have been intimidated by an office bully, a twelve-year employee who has seniority. As a collaborative writing project (see pages 88–91), draft a letter to the head of your human resources department documenting instances of the bully's actions and asking for advice on how to proceed.

16. Write a memo to your boss about being passed over for promotion. Diplomatically and ethically compare your work with that of the individual who did receive the promotion.

17. Write a memo to your boss about one of the following unethical activities you have witnessed in your workplace. Your memo must be carefully documented, fair, and persuasive—in short, ethical.

   a. bullying
   b. surfing pornography websites
   c. using workplace technology for personal matters (shopping, dating, buying stocks)
   d. falsifying compensatory or travel time
   e. telling sexist, off-color jokes
   f. concealing the use of company funds for personal gifts for fellow employees
   g. misdating or backdating company records
   h. sharing privileged information with individuals outside your department or company
   i. fudging the number of hours worked
   j. lying about family illnesses
   k. exaggerating a workplace-related injury
   l. not reporting a second job to avoid scheduled weekend work
   m. misrepresenting, by minimizing, a client's complaint

Microwaves

Much of the world around us is in motion. A wave-like motion. Some waves are big like tidal waves and some are small like the almost unseen footprints of a waterspider on a quiet pond. Other waves can't be seen at all, such as an idling truck sending out vibrations our bodies can feel. Among these are electromagnetic waves. They range from very low frequency sound waves to very high frequency X-rays, gamma rays, and even cosmic rays.

Energy behaves differently as its frequency changes. The start of audible sound—somewhere around 20 cycles per second—covers a segment at the low end of the electromagnetic spectrum. Household electricity operates at 60 hertz (cycles per second). At a somewhat higher frequency we have radio, ranging from shortwave and marine beacons, through the familiar AM broadcast band that lies between 500 and 1600 kilohertz, then to citizen's band, FM, television, and up to the higher frequency police and aviation bands.

Even higher up the scale lies visible light with its array of colors best seen when light is scattered by raindrops to create a rainbow.

Lying between radio waves and visible light is the microwave region—from roughly one gigahertz (a billion cycles per second) up to 3000 gigahertz. In this region the electromagnetic energy behaves in special ways.

Microwaves travel in straight lines, so they can be aimed in a given direction. They can be *reflected* by dense objects so that they send back echoes—this is the basis for radar. They can be *absorbed*, with their energy being converted into heat—the principle behind microwave ovens. Or they can pass *through* some substances that are transparent to the energy—this enables food to be cooked on a paper plate in a microwave oven.

Microwaves for Radar

World War II provided the impetus to harness microwave energy as a means of detecting enemy planes. Early radars were mounted on the Cliffs of Dover to bounce their microwave signals off Nazi bombers that threatened England. The word *radar* itself is an acronym for *RAdio Detection And Ranging.*

Radars grew more sophisticated. Special-purpose systems were developed to detect airplanes, to scan the horizon for enemy ships, to paint finely detailed electronic pictures of harbors to guide ships, and to measure the speeds of targets. These were installed on land and aboard warships. Radar—especially shipboard radar—was surely one of the most significant technological achievements to tip the scales toward an Allied victory in World War II.

Today, few mariners can recall what it was like before radar. It is such an important aid that it was embraced universally as soon as hostilities ended. Now, virtually every commercial vessel in the world has one, and most larger vessels have two radars: one for use on the open sea and one, operating at a higher frequency, to "paint" a more finely detailed picture, for use near shore.

Microwaves are also beamed across the skies to fix the positions of aircraft in flight, obviously an essential aid to controlling the movement of aircraft from city to city across the nation. These radars have also been linked to computers to tell air traffic controllers the altitude of planes in the area and to label them on their screens.

A new kind of radar, phased array, is now being used to search the skies thousands of miles out over the Atlantic and Pacific oceans. Although these advanced radars use microwave energy just as ordinary radars do, they do not depend upon a rotating antenna. Instead, a fixed antenna array, comprising thousands of elements like those of a fly's eye, looks everywhere. It has been said that these radars roll their eyes instead of turning their heads.

High-Speed Cooking

During World War II Raytheon had been selected to work with M.I.T. and British scientists to accelerate the production of magnetrons, the electron tubes that generate microwave energy, in order to speed up the production of radars. While testing some new, higher-powered tubes in a laboratory at Raytheon's Waltham, Massachusetts, plant, Percy L. Spencer and several of his staff engineers observed an interesting phenomenon. If you placed your hand in a beam of microwave energy, your hand would grow pleasantly warm. It was not like putting your hand in a heated oven that might sear the skin. The warmth was deep-heating and uniform.

Spencer and his engineers sent out for some popcorn and some food, then piped the energy into a metal wastebasket. The microwave oven was born.

From these discoveries, some 35 years ago, a new industry was born. In millions of homes around the world, meals are prepared in minutes using microwave ovens. In many processing industries, microwaves are being used to perform difficult heating or drying jobs. Even printing presses use microwaves to speed the drying of ink on paper.

In hospitals, doctors' offices, and athletic training rooms, that deep heat that Percy Spencer noticed is now used in diathermy equipment to ease the discomfort of muscle aches and pains.

Telephones Without Cable

The third characteristic of microwaves—that they pass undistorted through the air—makes them good messengers to carry telephone conversations as well as live television signals—without telephone poles or cables—across town or across the country. The microwave signals are beamed via satellite or by dish reflectors mounted atop buildings and mountain-top towers.

Microwaves take their name from the Greek *mikro* meaning very small. While the waves themselves may be very small, they play an important role in our world today: in defense; in communications; in air, sea, and highway safety; in industrial processing; and in cooking. At Raytheon the applications expand every day.

Reprinted by permission of *Raytheon Magazine*.

CHAPTER 2

The Writing Process at Work

Access chapter-specific interactive learning tools, including quizzes and more in your English CourseMate, accessed through www .cengagebrain.com.

In Chapter 1 you learned about the different functions of writing for the world of work and also explored some basic concepts all writers must master. To be a successful writer, you need to

- identify your audience's needs
- determine your purpose in writing to that audience
- make sure your message meets your audience's needs
- use the most appropriate style and tone for your message
- format your work so that it clearly reflects your message to your audience

Just as significant to your success is knowing how effective writers actually create their work for their audiences. This chapter gives you some practical information about the strategies and techniques careful writers use when they work. These procedures are a vital part of what is known as the *writing process*. This process involves such matters as how writers gather information, how they transform their ideas into written form, and how they organize and revise what they have written to make it relevant for their audiences.

What Writing Is and Is Not

As you begin your study of writing for the world of work, it might be helpful to identify some notions about what writing is and what it is not.

What Writing Is

- **Writing is a dynamic process; it is not static.** It enables you to discover and evaluate your thoughts as you draft and revise.
- **A piece of writing changes as your thoughts and information change and as your view of the material changes.**
- **Writing takes time.** Some people think that revising and polishing are too time consuming. But poor writing actually takes more time and costs more

money in the end. It can lead to misunderstandings, lost sales, product recalls, and even damage to your reputation and that of your company.

- **Writing means making a number of judgment calls.**
- **Writing grows sometimes in bits and pieces and sometimes in great spurts.** It needs many revisions; an early draft is never a final copy.

What Writing Is Not

- **Writing is not a mysterious process, known only to a few.** Even if you have not done much writing before, you can learn to do it effectively.
- **Writing is not simply following a magical formula.** Successful writing requires hard work and thoughtful effort, not simply following a formula, as if you were painting by numbers. Writing does not proceed in some predictable way, in which introductions are always written first and conclusions last.
- **Writing is not completed in a first attempt.** Just because you put something down on paper or on a computer screen does not mean it is permanent and unchangeable. Writing means *re*writing, *re*vising, and *re*thinking. The better a piece of writing is, the more the writer has reworked it.

The Writing Process

The writing process we have just discussed is something fluid, not static. Think of it as a back and forth process rather than following a formula—do this, then do that. To move from a blank sheet of paper or computer screen to a successful piece of writing, you need to follow a process. The parts of that process include researching, planning, drafting, revising, and editing. See how each part of the process is illustrated in Figures 2.2 through 2.5. As these figures show, office manager Melissa Hill asked employee Marcus Weekley to write a short report recommending the easiest and most economical way to improve office efficiency. In doing so, Weekley followed the steps in the writing process from researching and planning to drafting, revising, and editing.

To go through these steps successfully, Weekley first had to have a clear idea of what he had to write and why. Through a meeting with Melissa Hill and several follow-up emails and phone calls, he learned (1) what she expected to find in his report, (2) how she intended to use that information, (3) where he needed to look to find and evaluate the most relevant information, (4) how she expected the report to be prepared, and (5) when he needed to submit it.

Researching

Before you start to compose any email, memo, letter, report, proposal, or website, you'll need to do research. Research is crucial because it enables you to obtain the right information for your audience. Information must be factually correct and relevant. The world of work is based on conveying information—the logical presentation and sensible interpretation of facts. Chapter 8 will introduce you to the variety of research strategies and tools you can expect to use in the world of work.

Don't ever think you are wasting time by doing some research before starting to write any document. Actually, you will waste time and risk doing a poor job if you do not find out as much as possible about your topic (and your audience's interest in it). Find out about your readers' needs and how to meet them.

Then you can determine the kind of research you must do to gather and interpret the information your audience needs. Depending on the length of your written work and on your audience's needs, your research may include

- interviewing people inside and outside your company
- reviewing similar or related company documents
- consulting notes from conferences or meetings
- collaborating in person, by email, or by instant messaging (IM)
- doing Internet searches
- locating and evaluating websites
- searching abstracts, indexes, and other references on the Internet or in print
- reading current periodicals, trade journals, reports, and other documents
- evaluating reports, products, and services
- getting briefings from sales or technical staff
- conferring with co-workers, customers, or vendors
- participating in a focus group
- surveying customers' views
- visiting a work site
- performing a test
- observing a condition, event, or process

Keep in mind that research is not confined to the beginning of the writing process; it is an ongoing process.

As Figures 2.2 through 2.4 demonstrate, Marcus Weekley knew he would have to do research to make his recommendations to manager Melissa Hill about how the office could run more efficiently. To locate information about different and more recent information technologies that might improve office efficiency, he searched for and read relevant materials in print and online sources, spoke with co-workers as well as with individuals in other departments, and contacted office supply representatives. As he researched various options, he discovered that switching to a multifunction printer would save the office time and money. He determined that purchasing this piece of equipment would be the most sensible and economically feasible way to improve efficiency, and so he based his report on the research he found about this new technology. The results of his research are reflected throughout the drafts of his work.

Planning

At the planning stage in the writing process, your goal is to get something—anything—down on paper or on your screen. For most writers, getting started is the hardest part of the job. But you will feel more comfortable and confident once you begin to see your ideas before your eyes. It is always easier to clarify and criticize something you can see.

Getting started is also easier if you have researched your topic, because then you have something to say and to build on. Each part of the process relates to and supports the next. Careful research prepares you to begin writing.

Still, getting started is not easy. Take advantage of a number of widely used strategies to develop, organize, and tailor the right information for your audience. Use the following techniques, alone or in combination.

1. **Clustering.** In the middle of a sheet of paper, write the word or phrase that best describes your topic, then start writing other words or phrases that come to mind. As you write, circle each word or phrase and connect it to the word from which it sprang. Note the cluster diagram in Figure 2.1 for a report encouraging a manager to switch to flextime, a system in which employees work on a flexible time schedule. The diagram gives the writer a rough sense of some of the major divisions of the topic and where they may belong in the report. (Figure 12.13 on pages 595–596 shows the final workplace document that evolved from this early stage of planning on flextime schedules.)

2. **Brainstorming.** At the top of a sheet of paper or on your computer screen, describe your topic in a word or phrase, and then list any information you know

---

FIGURE 2.1 Clustering on the Topic of Flextime

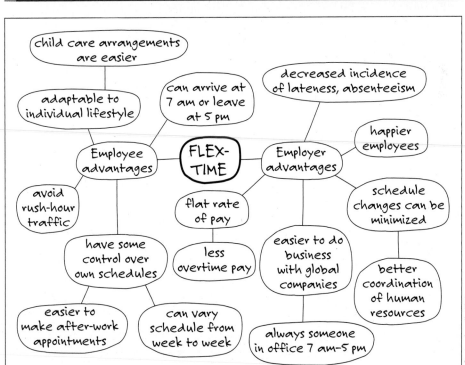

© Cengage Learning 2013

or have found out about that topic—in any order and as quickly as you can. Brainstorming is like thinking aloud except that you are recording your thoughts.

- Don't stop to delete, rearrange, or rewrite anything, and don't dwell on any one item.
- Don't worry about spelling, punctuation, grammar, or whether you are using words and phrases instead of complete sentences.
- Keep the ideas flowing. The result may well be an odd assortment of details, comments, and opinions.
- Step away from the list for a while. When you return with fresh eyes, expect to add and delete ideas or combine and rearrange others as you start to develop them in more detail.

Figure 2.2 shows Marcus Weekley's initial brainstormed list of how and why purchasing a new laser all-in-one printer would help his office. After he began to revise his list, he realized that some items were not relevant for his audience (6, 8, and 13) and that another was pertinent but needed to be adapted for his reader (5). He also recognized that some items were repetitious (1, 2, and 11). Further investigation revealed that his company could purchase the new equipment for far less than his initial high guess (17).

3. Outlining. For most writers, outlining may be the easiest and most comfortable way to begin or to continue planning a report or letter. Outlines can go through stages, so don't worry if your first attempt is brief and messy. It does not have to be formal (with lots of Roman and Arabic numerals), complete, or pretty. It is intended for no one's eyes but yours. Use your preliminary outline as a quick way to sketch in some ideas, a convenient container into which you can put information. You might simply jot down a few major points and identify a few subpoints. Note how Weekley organized his revised brainstormed list into an outline (Figure 2.3).

Drafting

If you have planned carefully, you will find it easier to start your first draft. When you draft, you convert the words and phrases from your outline, brainstormed list, or clustered grouping into paragraphs. During drafting, as elsewhere in the writing process, you will see some overlap as you look back over your list or outline to shape your text.

Don't expect to wind up with a polished, complete version of your letter or report after working on only one draft. In most cases, you will have to work through many drafts, but each draft should be less rough and more acceptable than the preceding one.

Key Questions to Ask as You Draft

As you work on your drafts, ask yourself the following questions about your content and organization:

- Am I giving my readers too much or too little information?
- Do I need to do more research—where and why?
- Should I confer further with my boss or co-workers?

FIGURE 2.2 Marcus Weekley's Initial, Unrevised Brainstormed List

1. combines four separate pieces of equip—printer, copier, fax, and scanner

2. more comprehensive than our current configuration of four pieces of equip

3. would coordinate with office furniture

4. energy efficiency increased due to fewer machines being used

5. more scalable fonts

6. one machine interfaces with all others in same-case housing

7. scanner makes photographic-quality pictures

8. print capabilities are a real contribution to technology

9. increased communication abilities through fax machine

10. new scanner picture quality better than current scanner

11. only have to buy one machine as opposed to four

12. speed of fax allows quick response time

13. stock is doing better on Wall Street compared to other equities

14. reducing our advertising costs through use of color printer

15. increased work area available

16. would help us do our work better

17. top-of-line models can be bought for $4,500

List is not organized but simply reflects writer's initial ideas about possible topics

1, 2, and 11 are repetitious

6, 8, and 13 are not relevant for audience or purpose

4, 9, 12, and 14 are of special interest to decision makers concerned about costs

Research will show cost estimate is too high

© Cengage Learning 2013

- Does this point belong where I have it, or would it more logically follow or precede something else?
- Is this point necessary and relevant?
- Am I repeating or contradicting myself?
- Have I ended appropriately for my audience?

FIGURE 2.3 Marcus Weekley's Early Outline After Revising His Brainstormed List

Outline form helps writer group ideas/ topics and go to next step in the process

Headings correspond to major sections of report

Outline reflects scope and details of writer's research

Last section of outline also functions as a conclusion

I. Convenience/capabilities of all-in-one laser printer

    A. Would reduce number of machines having to be serviced

    B. Can be configured easily for our network system

    C. Easy to install and to operate

    D. 35.6 Kbps fax machine would increase our communication

    E. 2,400 × 1,200 dpi copier means better copy quality

    F. 4,800 × 4,800 dpi scanner means higher quality pictures than current scanner provides

II. Time/efficiency

    A. 40 ppm color printer is nearly twice as fast as current printer

    B. 35.6 Kbps fax allows quick response time

    C. Greater graphics capability—130 scalable fonts

    D. Scanner compatible with our current PhotoEdit imaging/graphics software

    E. 100,000-page monthly duty cycle means less maintenance

III. Money

    A. Costs less overall for multitasking printer than combined four machines

    B. Reduced monthly power bill by using one machine rather than four

    C. Included ScanText and WordPort software means not buying new software to network office computers

    D. Save on service costs

    E. Reduced advertising costs through printer's 50–400% enlargement/ reduction options, which allow for more in-house advertising

Planning on Your Computer

You can brainstorm, outline, and create clusters on paper, or you can use your computer to help you plan your document. Here are some tips for using these text-generating strategies on your computer:

- To use your word processor to brainstorm, simply key in your ideas as quickly as you can. Your word processor makes it easy for you to go back and add, delete, combine, and rearrange your ideas when you revisit the document later.
- You can make easy-to-read and easy-to-change outlines on your computer as well. Use a word-processing program to create an outline with main points and subpoints that help you to organize your thoughts when you start to write a draft. This is a good way to get your ideas down fast and efficiently. These programs provide outline views allowing you to select the number of levels to display in an outline and to change levels in the hierarchy (for example, from *I* to *A* or from *a* to *i*). You can also add notes to remind yourself where you may have to fill in gaps or where you may need to transition from one idea to another. Keep in mind that there is no one right way to outline. You might simply write down large concepts, as in the list in Figure 2.2, or use complete sentences for your major headings.
- A "mind-mapping" software program such as FreeMind, MindMapper, or NovaMind will help you to brainstorm ideas. The software automatically produces a cluster diagram as you type. (Figure 2.1 shows a brainstorming cluster created in FreeMind.) You can then convert your cluster diagram into an outline by exporting it into your word processor.

To answer the questions successfully, you may have to continue researching your topic and re-examining your audience's needs. But in the process new and even better ideas will come to you, and ideas that you once thought were essential may in time appear unworkable or unnecessary.

Guidelines for Successful Drafting

Following are some suggestions to help your drafting go more smoothly and efficiently:

- Find a comfortable place where you can draft and allow yourself enough time—at least an hour or more—to work without interruptions on each draft.
- Get as much down on paper or on the screen as you can. The more text you generate, the more ideas you can explore.
- In an early draft, write the easiest part first. Some writers feel more comfortable drafting the body (or middle) of their work first.
- As you work on a later draft, write straight through. Do not worry about spelling, punctuation, or the way a word or sentence sounds. Save those concerns for later stages. Flag any sections that may require more research. Don't stop now; just get it done.

Drafting on Your Computer

Here are a few tips to make the process go more smoothly when you draft your document using your word processor:

- Get your thoughts down on the screen as quickly as possible, without stopping to worry about spelling, punctuation, or spacing. If you stop to correct these small errors, you may forget to write down larger points.
- Save your document regularly (every few minutes) so that you do not lose valuable work and time in the event of a power outage or computer malfunction. Back up your document at the end of each day.
- If you run into trouble completing a point or writing a transition, or if you know you will need to add documentation later, do not lose momentum by stopping. Instead, add notes to yourself (in parentheses, in italics, or in a different color) reminding you to "add transition," "back up this point," "clarify," "check with boss," or "add documentation."
- Use the Save As option to save each draft of your document. That way, you will be able to salvage versions of sentences or paragraphs from one draft and incorporate them into another. Give each draft a clear title, such as "Office Equipment, 3rd draft," so you can tell at a glance which version is which.
- If you intend to share your document with others for collaborative editing, use cloud computing tools like Google Docs (see page 113) and Adobe Buzzword to create your draft. These applications will allow you to save a complete revision history of your document, add comments and notes to your work, share your work with others for their review, and create presentation-ready documents.

- Allow enough time between drafts so that you can evaluate your work with fresh eyes and a clear mind.
- Get frequent outside opinions. Show, email, or fax a draft to a co-worker or a supervisor for comment. A new pair of eyes will see things you missed. Collaboration is essential in the workplace, as you will see in Chapter 3.
- Consider whether visuals would enhance the quality of your work, and if so, decide on what types of visuals to use and where.

Figure 2.4 shows one of the several drafts that Marcus Weekley prepared. Because he wisely recognized that his outline was not final, he continued to work on it during the drafting stage. Note that he added an introduction and a conclusion, which were not part of his original outline (Figure 2.3), to convince Melissa Hill to purchase a new laser printer. Even so, Weekley recognized that this draft was still not ready for his boss to see, so he showed it to co-workers for suggestions.

From discussions with co-workers and vendors, and after further work on his draft, Weekley realized that he had placed one of the most important considerations for his audience (savings) last. In his final version, shown in Figure 2.5, he moved that section to the beginning of his report. Weekley thus paid attention to his audience's priorities and needs.

FIGURE 2.4 Intermediate Draft of Marcus Weekley's Report

To: Melissa Hill, Office Manager
From: Marcus Weekley
Date: May 25, 2012
Subject: Improving Efficiency

As you requested, I have been researching what to do about improving our office efficiency. One of the most beneficial and immediate solutions I have found is to replace our current laser printer with a new all-in-one laser printer. More and more businesses today are incorporating multifunction printers into their information technology systems because of ease and efficiency. With the advances in printer technology in the past three years, it makes good business sense to replace our Van Eisen 4200 laser printer with a Lightech 520 multifunction printer. This printer would reduce business costs and increase business efficiency at a total cost significantly lower than the combined price of a laser printer, fax machine, copier, and scanner.

The printer, fax machine, copier, and scanner in the new all-in-one model are compatible with our operating systems. The addition of a 35.6 Kbps fax machine to our office, as opposed to our current 20.5 Kbps, would increase operating efficiency by allowing quicker response times. With the new multifunction color printer, our office can print up to 40 high-resolution color pages per minute, compared with our current printer's 20 color ppm. And the new printer's 3,000 MB of memory helps to manage multiple jobs easily. The Lightech 520 printer also holds 3,100 sheets of paper, as opposed to our current printer's 500-page capacity.

The multifunction printer would provide higher quality printing through a 4,800 dpi printer, whereas our current Image 4200 laser printer only has 900 dpi. The new printer also offers over 90 scalable type fonts, whereas our current printer only offers 60. Also, our current scanners scan images in at $1,200 \times 1,200$ dpi, whereas the new scanner operates at $4,800 \times 4,800$ dpi, so scanned images will be of an even better quality.

The greatest benefit a multifunction printer would provide our company is monetary. The price of a new multifunction business printer ranges from $3,000 to $9,000 depending on the model. Lightech's 520 multifunction color laser printer (including $2,400 \times 1,200$ dpi copier, $4,800 \times 4,800$ dpi scanner, and 35.6 Kbps fax) costs only $3,175 not including shipping and handling purchased from Computerbuyers.com. This cost nearly equals the price of our own Van Eisen 4200 printer and West 400 scanners, but combines the equipment into one more efficient machine. Purchasing the multifunction printer would not only save our business money on the initial purchase, but use of a multitasking unit that combines four machines into one would also save on subsequent servicing and maintenance, as well as decreasing our monthly electric bill by $50–$150 per month. Use of the multifunction printer's 50–400% enlargement/reduction options should also save us an additional $300–$500 each month by using less outside advertising. Computerbuyers.com also offers a two-year warranty on all products sold through its website. What better way to begin improving business efficiency than through the purchase of a new multifunction color printer?

Vague subject line

Wordy opening paragraph takes too long to get to the point

No headings or bullets make information hard to find

Does not supply source of research

Includes most important point for reader— costs—too late

Paragraph is long and hard to follow. Break into 2 or 3 paragraphs

Ends with question rather than plan for how to make change

FIGURE 2.5 Final Version of Marcus Weekley's Report

To: Melissa Hill, Office Manager
From: Marcus Weekley
Date: June 12, 2012
Subject: Advantages of Purchasing a New All-in-One Printer

*More precise
subject line*

As you requested, I have investigated some ways to improve office efficiency.
The best solution I have found is to replace our current Van Eisen 4200 laser
printer, two West 400 scanners, and XL290 fax machine with a new-generation
laser all-in-one printer.

*Introduction
gets to the
point quickly*

With advances in printer technology over the last two years (enclosed is a copy of
a review article "New Technology Means Office Efficiency" from *Computer
World* [Mar. 2012]: 96–98), it makes good business sense to replace our
less-efficient laser printer with a Lightech 520 all-in-one printer. Our laser printer
does only one task, while the Lightech will give us higher-resolution color
printing, a high-speed fax machine, a copier, and a scanner all in one. This new
unit is economical and more efficient and will significantly improve the
transmission and design of our documents.

*Documentation
shows research
on subject*

Cost
The greatest benefit of the all-in-one Lightech 520 is cost. We can purchase this
printer for only $3,175, plus shipping and handling, totaling $3,298 when ordered
through **Computerbuyers.com**. Purchasing a Lightech 520 would allow us to
recoup that cost easily in just a few months because we would

*Excellent use of
headings and
bulleted lists*

- realize a savings in the purchase price—one Lightech costs less than the four
  current machines combined
- receive ScanText and WordPort software free with the printer
- decrease our monthly electricity bill by $50–$150 by reducing four pieces of
  office equipment to one
- have less maintenance, saving at least $150 per month in service calls
- receive a two-year warranty with guaranteed overnight service, which should
  decrease downtime
- be able to trade in our current printer for $375 and our two scanners for $200
  each
- save an additional $300–500 each month by not having to use outside
  advertising thanks to Lightech's 50–400% enlargement/reduction capabilities

*Cites specific
model and
costs*

*Justification for
new purchase
clearly and
persuasively
laid out*

FIGURE 2.5 (Continued)

Page 2

Efficiency

This all-in-one printer accomplishes all four tasks—printing, faxing, scanning, copying—simultaneously, and each task is compatible with our current networking system. Other key features include

- allows quicker faxing (3 seconds per page) with 35.6 Kbps and greater storage (600 pages)
- prints twice as fast—40 high-resolution color pages per minute as opposed to our current printer's 20
- enhanced memory of 3,000 MB manages multiple jobs easily
- expanded paper capacity (3,100 sheets) to handle our heavy quarterly mailings easily
- 100,000-page monthly duty cycle means less maintenance
- all four tasks compatible with our PhotoEdit software

Quality

The Lightech 520 prints and processes higher quality and quantity of work because it

- has a 4,800 dpi printer, whereas our current printer offers only 900 dpi
- exhibits same color density on the 1,000th copy as on the first; solid ink sticks ensure no toner spills
- offers a 2,400 × 2,400 dpi high-resolution copier
- would increase the quality of manipulated scanned images with a 4,800 × 4,800 dpi, whereas our current scanner is only 1,200 × 1,200
- offers over 90 scalable fonts, as opposed to the 60 we now have, which will give our publications a more varied and professional appearance

I recommend that we purchase a Lightech 520 laser printer from **Computerbuyers.com.** It will unquestionably save us money, improve office efficiency, and help us to integrate our office systems to better project our corporate image.

Gives reader appropriate amount and type of detail

Contrasts current equipment with new model

Selects only most relevant points for reader

Recommendation ends report persuasively, highlighting benefits to employer

He also added headings and bulleted lists to help his reader find information. The format of his earlier draft (Figure 2.4) did not assist his readers in finding information quickly, nor did it reflect a convincing organizational plan.

Revising

Revision is an essential stage in the writing process. It requires more than giving your work one more quick glance. Do not be tempted to skip the revision stage just because you have written the required number of words or sections or because you think you have put in too much time already. Revision is done *after* you produce a draft that you think conveys the appropriate message for your audience. The quality of your letter or report depends on the revisions you make. Revision gives you a second (or third or fourth) chance to get things right for your audience and to clarify your purpose in writing to them, as Marcus Weekley did.

Like planning and drafting, revision is not done well in one big push. It evolves over time. Allow yourself enough time to do it carefully.

- Avoid drafting and revising in one sitting. If possible, wait at least a day before you start to revise. (In the busy work world, waiting a couple of hours may have to suffice.)
- Ask a co-worker or supervisor familiar with your topic to comment on your work, as Marcus Weekley did to make sure his report was convincing.
- Plan to read your revised work more than once.

Key Questions to Ask as You Revise

By asking and successfully answering the following questions as you revise, you can discover gaps or omissions, points to change, and errors to correct in your draft.

Content

- Is it accurate? Are my facts (figures, names, addresses, dates, costs, references, warranty terms, statistics) correct? Have I documented them?
- Is it relevant for my audience and purpose? Have I included information that is unnecessary, too technical, or not appropriate?
- Have I given enough evidence to explain things adequately and to persuade my readers?
- Have I left anything out? (Too little information will make readers skeptical about what you are describing or proposing.)

Organization

- Have I clearly identified my main points and shown my readers why those points are important?
- Is everything in the correct, most effective order? Should anything be switched or moved—closer to the beginning or toward the end of my document?

Revising on Your Computer

Here are some suggestions on how you can effectively revise a document in your word processor:

- During this part of the writing process, respond to the notes you left yourself when you were drafting the document. Move sentences and paragraphs to more appropriate places, add clarifications and transitions, insert parenthetical documentation, create headers, or add visuals. At this stage, do not worry about spelling and punctuation, which are fine-tuning matters you can resolve during the editing stage.
- As you did when you drafted, save your document regularly (every few minutes) so that you do not lose your work if there's a power outage or your computer malfunctions. Back up your document at the end of each day.
- Use the Save As option for each revision of your document, and give each revision a clear title, such as "Office Equipment, Revision #1."
- Take advantage of change-tracking options (see pages 109–111) that allow you to see your edits (additions, cuts, and moves) in a different color, rather than deleting existing text. Use the Comment option to insert notes to yourself that will appear in the margin or at the bottom of your document.
- Choose the Print Preview option to see the final version of your document. Check to see if your document is too long or too short or if some sections of your document appear to be too dense with text. If your text is too dense, you may need to make deletions, add headings or headers, or include visuals to make your document attractive and easy to read and follow. (See pages 528–535 on document design.)
- If you are using collaborative editing tools like Google Docs and Adobe Buzzword, share the document with the colleagues, managers, vendors, and so on, who can best help you to revise it. Then inform your editing partners about how you want them to use this technology.

- Do I provide transitions from one section to another?
- Have I spent too much (or too little) effort on one section? Do I repeat myself? What can be cut?
- Have I grouped related items in the same part of my report or letter, or have I scattered details that really need to appear together in one paragraph or section?

Tone

- How do I sound to my readers—professional and sincere, or arrogant and unreliable? What attitude do my words or expressions convey?
- How will my readers think I perceive them—honest and intelligent or unprofessional and uncooperative?

A "Before" and "After" Revision of a Short Report

Mary Fonseca, a staff member at Seacoast Labs, was asked by her supervisor to prepare a short report for the general public on the lab's most recent experiments. Conferring with her supervisor, Fonseca learned that the report was intended to attract favorable publicity for the lab's commitment to conserving energy and lowering marine fuel costs. She also realized that a large part of her job would involve making technical information about the lab's research easily understandable to an audience of nonspecialists. Figures 2.6 and 2.7 show her "before" and "after" revisions of one section of her report.

When she began to revise her first draft (Figure 2.6), Fonseca realized that it lacked focus. It jumped back and forth between drag on ships and drag on airplanes. Because Seacoast Labs did not work on planes, she wisely decided to drop that idea. She also realized that the information on the effects of drag, something Seacoast was working on, was so important it deserved a separate paragraph. In light of this key idea, she realized that her explanation of molecules, eddies, and drag needed to be made more reader-friendly, and so she added the graphic analogy about spoons/ships and honey/drag. Researching further, she decided to add a new paragraph on the causes and effects of drag, which became paragraph 2 in her second draft (Figure 2.7).

FIGURE 2.6 Unorganized Opening Paragraphs of Mary Fonseca's "Before" Draft

Information hard to follow and not relevant for audience

Does not explain process very well

> Drag is an important concept in the world of science and technology. It has many implications. Drag occurs when a ship moves through the water and eddies build up. Ships on the high seas have to fight the eddies, which results in drag. In the same way, an airplane has to fight the winds at various altitudes at which it flies; these winds are very forceful, moving at many knots per hour. All these forces of nature are around us. Sometimes we can feel them, too. We get tired walking against a strong wind. The eddies around a ship are the same thing. These eddies form various barriers around the ship's hull. They come from a combination of different molecules around the ship's hull and exert quite a force. Both types of molecules pull against the ship. This is where the eddies come in.

Important point not developed

> Scientists at Seacoast Labs are concerned about drag. Dr. Karen Runnels, who joined Seacoast about three years ago, is the chief investigator. She and her team of highly qualified experts have constructed some fascinating multilevel water tunnels. These tunnels should be useful to ship owners. Drag wastes a ship's fuel.

What Is Drag?

We cannot see or hear many of the forces around us, but we can certainly detect their presence. Walking or running into a strong wind, for example, requires a great deal of effort and often quickly leaves us feeling tired. When a ship sails through the water, it also experiences these opposing forces known as **drag**. Overcoming drag causes a ship to reduce its energy efficiency, which leads to higher fuel costs.

Effective use of headings and definition

How Drag Works

It is not easy for a ship to fight drag. As the ship moves through the water, it drags the water molecules around its hull at the same rate the ship is moving. Because of the cohesive force of those molecules, other water molecules immediately outside the ship's path get pulled into its way. All the molecules become tangled rather than simply sliding past each other. The result is an eddy, or small circling burst of water around the ship's hull, which intensifies the drag. Dr. Jorge Fröes, a highly respected structural engineer, explains the process using this analogy: "When you put a spoon in honey and pull it out, half the honey comes out with the spoon. That's what is happening to ships. The ship is moving and at the same time dragging the ocean with it."

Describes cause and effect of drag

Reducing Drag

At Seacoast Labs, scientists are working to find ways to reduce drag on ships. Dr. Karen Runnels, the principal investigator, and a team of researchers have constructed water tunnels to simulate the movement of ships at sea. The drag a ship encounters is measured from the tiny air bubbles emitted in the water tunnel. Dr. Runnels's team has also developed the use of polymers, or long carbon chain molecules, to reduce drag. The polymers act like a slimy coating for the ship's hull to help it glide through the water more easily. When asbestos fibers were added to the polymer solutions, the investigators measured a 90 percent reduction in drag. The team has also experimented with an external pump attached to the hull of a ship, which pushes the water away from a ship's path, saving fuel and time.

Uses an easy-to-follow analogy for her audience

Clearly explains the lab's research and its importance for readers concerned about the economy and environment

Yet by pulling ideas about drag and its effects from the longish first paragraph in Figure 2.6, Fonseca was left with the job of finding an opening for this section of her report. Buried in her original opening paragraph was the idea that we cannot always see the forces of nature, but we can feel them. She thought this comparison of walking against the wind and fighting drag would work better for her audience than the original wooden remarks she had started with.

Although her organization and ideas were far better in her second draft than in her "before" draft in Figure 2.6, she realized she had said very little about her employer, Seacoast Labs, and the corporate image it wanted to project through its experiments. Doing more research, she found additional information about Seacoast's experiments and why they were so important in conserving fuel and saving money. This information was far more significant and relevant than saying that Dr. Runnels had been at Seacoast for three years.

Through revision and further research, then, Mary Fonseca transformed two poorly organized and incomplete paragraphs into three separate yet logically connected ones that highlighted her employer's work. In her revision (Figure 2.7), she came up with three very helpful headings—"What Is Drag?" "How Drag Works," and "Reducing Drag"—for her nonspecialist readers.

Editing

Editing is quality control for your reader. This last stage in the writing process might be compared to detailing an automobile—the preparation a dealer goes through to ready a new car for prospective buyers. Editing is done only after you are completely satisfied that you have made all of the big decisions about content, organization, and format—that you have said what you wanted to, where and how you intended, for your audience.

When you edit, you will check your work to make sure it is readable and correct. At this stage, pay close attention to

- sentences
- word choices
- punctuation
- spelling
- grammar and usage
- tone

As you edit, check to be sure your message is clear and concise so that readers will be able to understand it quickly and find it persuasive. The sentences you write and the words and tone you choose play a major role in how your message is received.

As with revising, don't skip or rush through the editing process, thinking that once your ideas are down, your work is done. If your work is hard to read or contains mistakes in spelling or punctuation, readers will think that your ideas and your research are also faulty.

The following sections will give you basic guidelines about what to look for when you edit your sentences and words. The appendix, "A Writer's Brief Guide to Paragraphs, Sentences, and Words" (pages A-1–A-19), also contains helpful suggestions on using correct spelling and punctuation.

Editing on Your Computer

Word-processing programs make editing easy and efficient. These programs flag errors in spelling, punctuation, word choice, subject-verb agreement, wordiness, and sentence construction. But keep the following guidelines in mind when you use these programs. Be aware of what your computer can and cannot do for you.

- Customize your spell-checker. You can set your spell-checker to flag words that you frequently misspell and to ignore words that aren't in its dictionary (such as proper names, brand names, technical terms or concepts, etc.).
- Don't accept everything your spell-checker or grammar-checker tells you. While these tools are helpful, do not rely on them exclusively. For instance, a grammar-checker may fail to recognize homonyms (their/there or its/it's), and a spell-checker may highlight a proper name or industry jargon that is spelled correctly but is not included in your spell-checker's dictionary. When in doubt, use a dictionary or consult a grammar handbook to verify the accuracy of your spelling and grammar.
- Use global find and replace for unique character strings only. For example, trying to change the word "on" to "at" will also change "only" to "atly," "once" to "atce," and "content" to "cattent." If your spell-checker spots a word that is in fact misspelled or used incorrectly in one place, don't assume that all occurrences of the word are misspelled. For instance, if your spell-checker tells you to capitalize the word "South" (as in "South Carolina") in one instance, the word may not need to be capitalized elsewhere (e.g., "south of the highway"). Check each correction individually.
- To avoid having to make time-consuming repetitive changes to style for such things as paragraph indents or heading style, set up your document using autoformatting for accuracy and consistency throughout your document.
- Pay attention to the verbs and adjectives you use. Do not keep repeating the same ones throughout your document. This makes for very boring reading for your audience.
- When you edit online, be careful not to focus only on the lines you can see on the screen and neglect the larger organization of your document. Scroll down or print out your document to see how a change in one place may affect (by contradicting, duplicating, or weakening) something earlier or later.

Editing Guidelines for Writing Lean and Clear Sentences

Here are four of the most frequent complaints readers voice about poorly edited writing in the world of work:

- **The sentences are too long.** I could not follow the writer's ideas easily.
- **The sentences are too complex,** making it hard to understand what the writer meant the first time I read the work; I had to reread it several times.
- **The sentences are unclear.** Even after I reread them, I was not sure I understood the writer's message.
- **The sentences are too short and simplistic.** The writing felt "dumbed down."

Writing clear, readable sentences is not always easy. It takes effort, but the time you spend editing will pay off in rich dividends for you and your readers. The seven guidelines that follow should help with the editing phase of your work.

1. Avoid needlessly complex or lengthy sentences. Do not pile words on top of words. Instead, edit one overly long sentence into two or even three more manageable ones.

Too long:	The planning committee decided that the awards banquet should be held on March 15 at 6:30, since the other two dates (March 7 and March 22) suggested by the hospitality committee conflict with local sports events, even though one of those events could be changed to fit our needs.
Edited for easier reading:	The planning committee has decided to hold the awards banquet on March 15 at 6:30. The other dates suggested by the hospitality committee—March 7 and March 22— conflict with two local sports events. Although the date of one of those sports events could be changed, the planning committee still believes that March 15 is our best choice.

2. Combine short, choppy sentences. Don't shorten long, complex sentences, only to turn them into choppy, simplistic ones. A memo, an email, or a letter written exclusively in short, staccato sentences sounds immature.

When you find yourself looking at a series of short, blunt sentences, as in the following example, combine them where possible and use connective words similar to those italicized in the edited version.

Choppy:	Medical transcriptionists have many responsibilities. Their responsibilities are important. They must be familiar with medical terminology. They must listen to dictation. Sometimes physicians talk very fast. Then the transcriptionist must be quick to transcribe what is heard. Words could be missed. Transcriptionists must forward reports. These reports have to be approved. This will take a great deal of time and concentration. These final reports are copied and stored properly for reference.
Edited:	Medical transcriptionists have many important responsibilities. *These* include transcribing physicians' orders using correct medical terminology. *When* physicians talk rapidly, transcriptionists have to keyboard accurately *so that* no words are omitted. *Among their most demanding* duties are keyboarding and forwarding transcriptions *and then*, after approval, storing copies properly for future reference.

3. Edit sentences to tell who does what to whom or what. The clearest sentence pattern in English is the subject-verb-object (s-v-o) pattern.

s   v   o
Sue booted the computer.

s   v   o
Our website contains a link to key training software programs.

Readers find this pattern easiest to understand because it provides direct and specific information about the action. Hard-to-read sentences obscure or scramble information about the subject, the verb, or the object. In the following unedited sentence, the subject is hidden in the middle rather than being placed in the most crucial subject position.

> **Unclear:** The control of the ceiling limits of glycidyl ethers on the part of the employers for the optimal safety of workers in the workplace is necessary. (Who is responsible for taking action? What action must they take? For whom is such action taken?)
>
> **Edited:** Employers must control the ceiling limits of glycidyl ethers for their workers' safety.

4. Use strong, active verbs rather than verb phrases. In trying to sound important, many bureaucratic writers avoid using simple, graphic verbs. Instead, these writers use a weak verb phrase (for example, *provide maintenance of* instead of *maintain, work in cooperation with* instead of *cooperate*). Such verb phrases imprison the active verb inside a noun format and slow readers down. Note how the edited version here rewrites the weak verb phrases.

> **Weak:** The city provided the employment of two work crews to assist the strengthening of the dam.
>
> **Strong:** The city employed two work crews to strengthen the dam.

5. Avoid piling modifiers in front of nouns. Putting too many modifiers (words used as adjectives) in the readers' path to the noun is confusing for readers, who will have trouble deciphering how one modifier relates to another modifier or to the noun. To avoid that problem, edit the sentence to place some of the modifiers after the nouns they modify.

> **Crowded:** The vibration noise control heat pump condenser quieter can make your customer happier.
>
> **Readable:** The quieter on the condenser for the heat pump will make your customer happier by controlling noise and vibrations.

6. Replace wordy phrases or clauses with one- or two-word synonyms.

> **Wordy:** The college has parking zones for different areas for people living on campus as well as for those who do not live on campus and who commute to school.
>
> **Edited:** The college has different parking zones for resident and commuter students. (Twenty words of the original sentence—everything after "areas for"—have been reduced to four words: "resident and commuter students.")

7. Combine sentences beginning with the same subject or ending with an object that becomes the subject of the next sentence.

> **Wordy:** I asked the inspector if she planned to visit the plant this afternoon. I also asked her if she would come alone.
>
> **Edited:** I asked the inspector if she planned to visit the plant alone this afternoon.
>
> **Wordy:** Homeowners want to buy low-maintenance bushes. These low-maintenance bushes include the ever-popular holly and boxwood varieties. These bushes are also inexpensive.

Edited: Homeowners want to buy low-maintenance and inexpensive bushes such as holly and boxwood. (This revision combines three sentences into one, condenses twenty-four words into fourteen, and joins three related thoughts.)

Editing Guidelines for Cutting Out Unnecessary Words

Too many people in business think the more words, the better. Nothing could be more self-defeating. Your readers are busy; unnecessary words slow them down. Make every word work. Cut out any words you can from your sentences. If the sentence still makes sense and reads correctly, you have eliminated wordiness.

1. **Replace wordy phrases with precise ones.** See how in Table 2.1 wordy phrases on the left are replaced with their much more concise equivalents on the right. Many of these wordy phrases have slowed business writing down for decades.

TABLE 2.1   Wordy Phrases and Their Concise Equivalents

Wordy	Concise
at a slow rate	slowly
at an early date	soon
at this point in time	now
based on the fact	because
be in agreement with	agree
bring to a conclusion	conclude, end
come to terms with	agree, accept
despite the fact that	although
due to the fact that	because
during the course of	during
express an opinion that	affirm
for the period of	for
has the capability to	can
in communication with	communicate
in connection with	about
in such a manner	so that
in the area/case/field of	in
in the event that	if
in the month of May	in May
in the neighborhood of	approximately, about
it is often the case that	often
it is our understanding that	we understand that
look something like	resemble
of the opinion that	think that

Wordy	Concise
on the grounds that	because
open the conversation with	open, begin with
serve in the capacity of	serve as
until such time as	until
with reference to	regarding, about
with the result that	so

2. Use concise, not redundant, phrases. Another kind of wordiness comes from using redundant expressions—saying the same thing a second time, only in different words. "Fellow colleague," "component parts," "corrosive acid," and "free gift" are phrases that contain this kind of double speech; a fellow *is* a colleague, a component *is* a part, acid *is* corrosive, and a gift *is* free. In the examples below, the suggested changes on the right are preferable to the redundant phrases on the left.

Redundant	Concise
absolutely essential	essential
advance reservations	reservations
basic necessities	necessities, needs
close proximity	proximity, nearness
end result	result
final conclusions/final outcome	conclusions/outcome
first and foremost	first
full and complete	full, complete
personal opinion	opinion
tried-and-true	tried, proven

3. Watch for repetitious words, phrases, or clauses within a sentence. Sometimes one sentence or one part of a sentence needlessly duplicates another.

> Redundant: To provide more room for employees' cars, the security department is studying ways to expand the employees' parking lot.
>
> Edited: The security department is studying ways to expand the employees' parking lot. (Since the first phrase says nothing that the reader does not know from the independent clause, it can be cut.)

4. Avoid unnecessary prepositional phrases. Adding a prepositional phrase can sometimes contribute to redundancy. The italicized words below are unnecessary. Be on the lookout for these phrases and delete them.

audible *to the ear*	light *in weight*	short *in duration*
bitter *in taste*	loud *in volume*	soft *in texture*
fly *through the air*	orange *in color*	tall *in height*
hard *to the touch*	rectangular *in shape*	twenty *in number*
honest *in character*	second *in sequence*	visible *to the eye*

Figure 2.8 shows an email that Trudy Wallace wants to send to her boss, Lee Chadwick, about issuing tablet PCs to the sales force. Her unedited work is bloated with unnecessary words, expendable phrases, and repetitious ideas.

After careful editing, Trudy Wallace streamlined her email to her boss, Lee Chadwick; see Figure 2.9. She pruned wordy expressions and combined sentences to cut out duplication. The revised version shown in Figure 2.9 is only 182 words,

FIGURE 2.8 A Wordy, Unedited Email

From: <twallace@transtech.org>
To: <lchadwick@transtech.org>
Cc:
Subject: Smartphones and Today's Technology

Wordy and unfocused subject

Dear Lee,

One long, unbroken paragraph is hard to follow

Due to the inescapable reliance on technology, specifically on email and Internet communications, within our company, I believe it would be beneficial to look into the possibility of issuing tablet PCs to our employees. Issuing these devices would have a variety of positive implications for the efficiency of our company. Unlike smartphones, tablet PCs have many more capabilities that will help our employees in their daily work since these tablets expand the technological features of a smartphone and a laptop's ability to transfer documents. On the road their portability would assist the sales staff to participate in online meetings both locally and at distant sites. It should be pointed out, too, that the tablet's screens are two and one-half to three times bigger than a smartphone's, so our employees could actually make better presentations to clients on these devices. With a tablet PC, employees would also benefit by having access to more extensive business documents such as manuals, contracts, and invoices, even when they are out on the road traveling or at our local office. By means of tablet PCs, I feel quite certain that our company's correspondence would be dealt with much more speedily, since not only will these devices allow our employees to access their email and full documents at all times, it will enable them to actually create documents in a more efficient manner because of the size of the virtual keyboard. I think it would be absolutely essential for the satisfaction of our customers and to the ongoing operation of our company's business today to respond fully and completely to the possibility such a proposal affords us. It would, therefore, appear safe to conclude that with reference to issuing tablet PCs every means at our disposal would be brought to bear on aiding our sales staff.

Repeats same idea in two or three sentences

Uses awkward and wordy sentences

Does not specifically indicate what writer will do about problem

Thanks,

Trudy

FIGURE 2.9 A Concise Version of the Wordy Email in Figure 2.8

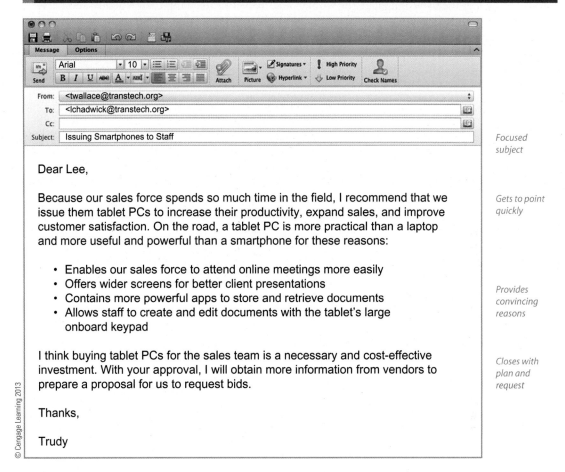

From: <twallace@transtech.org>
To: <lchadwick@transtech.org>
Cc:
Subject: Issuing Smartphones to Staff

Dear Lee,

Because our sales force spends so much time in the field, I recommend that we issue them tablet PCs to increase their productivity, expand sales, and improve customer satisfaction. On the road, a tablet PC is more practical than a laptop and more useful and powerful than a smartphone for these reasons:

- Enables our sales force to attend online meetings more easily
- Offers wider screens for better client presentations
- Contains more powerful apps to store and retrieve documents
- Allows staff to create and edit documents with the tablet's large onboard keypad

I think buying tablet PCs for the sales team is a necessary and cost-effective investment. With your approval, I will obtain more information from vendors to prepare a proposal for us to request bids.

Thanks,

Trudy

Focused subject

Gets to point quickly

Provides convincing reasons

Closes with plan and request

© Cengage Learning 2013

as opposed to the 383 words in the draft. Not only has Wallace shortened her message, but she has also made it easier to read.

Editing Guidelines to Eliminate Sexist Language

Editing involves far more than just making sure that your sentences are readable. It also reflects your professional style—how you see and characterize the world of work and the individuals in it, not to mention how you want your readers to see you. Your words should reflect a high degree of ethics and honesty, free from bias, offense, and stereotype. They need to be sensitive to the needs of your international audience as well (see pages 187–189).

Sexist language in particular offers a distorted view of a job force and discriminates in favor of one sex at the expense of another, usually women. Sexist language portrays men as having more powerful, higher-paying jobs than women do. Using sexist language offends and demeans female readers by depriving them of their equal rights. It may cost your company business as well. You can avoid gender bias by using inclusive language for women and men alike, treating them equally and fairly.

Sexist language is often based on stereotypes that depict men as superior to women. For example, calling politicians *city fathers* or *favorite sons* follows the stereotypical picture of seeing politicians as male. Such phrases discriminate against women who do or could hold public office at all levels of government. Never assume or imply a person's gender based on his or her profession.

As these examples show, such language prejudiciously labels some professions as masculine and others as feminine. Keep in mind that sexist phrases assume engineers, physicians, and pilots are male (*he*, *his*, and *him* are often linked with these professions in descriptions), while social workers, nurses, administrative assistants, and secretaries are often portrayed as female (*she*, *her*), although members of both sexes work in all these professions. Sexist language also wrongly points out gender identities when such roles do not seem to follow biased expectations—*lady lawyer*, *male secretary*, *female surgeon*, or *male nurse*. Such offensive distinctions reflect prejudiced attitudes that you should eliminate from your writing.

Always prune the following sexist phrases: *every man for himself, gal Friday, little woman, lady of the house, old maid, women's intuition, the best man for the job, to man a desk* (or *post*), *the weaker sex, woman's work, working wives, a manly thing to do*, and *young man on the way up*.

Finally, don't assume all employees are male. Instead of writing, "All staff members and their wives are invited to attend," simply say, "All staff members and their guests are invited to attend."

Ways to Avoid Sexist Language

Here are five ways you can eliminate sexist writing from your work.

1. **Replace sexist words with neutral ones.** Neutral words do *not* refer to a specific sex; they are genderless. The sexist words on the left in the following list can be replaced by the neutral nonsexist substitutes on the right.

Sexist	Neutral
alderman, assemblyman	representative
businessman	businessperson
cameraman	photographer
chairman	chair, chairperson
congressman	representative
craftsman	skilled worker
divorcée	divorced person
fireman	firefighter
foreman	supervisor
housewife	homemaker
janitress	custodian, cleaning person
landlord, landlady	owner
maiden name	family name
mailman, postman	mail carrier
man-hours	work-hours
mankind	humanity, human beings

manmade	synthetic, artificial
manpower	strength, power
man to man	candidly
men	human beings, people
policeman	police officer
repairman	repair person
salesman	salesperson, clerk
spokesman	spokesperson
weatherman	meteorologist
women's intuition	intuition
workman	worker

2. Watch masculine pronouns. Avoid using the masculine pronouns (*he, his, him*) when referring to a group that includes both men and women.

> *Every worker must submit his travel expenses by Monday.*

Workers may include women as well as men, and to assume that all workers are men is misleading and unfair to women. You can edit such sexist language in several ways.

a. Make the subject of your sentence plural and thus neutral.

> *Workers must submit their travel expenses by Monday.*

b. Replace the pronoun *his* with *the* or *a* or drop it altogether.

> *Every employee is to submit a travel expense report by Monday.*
> *Every worker must submit travel expenses by Monday.*

c. Use *his or her* instead of *his*.

> *Every worker must submit his or her travel expenses by Monday.*

d. Reword the sentence using the passive voice.

> *All travel expenses must be submitted by Monday.*

Moreover, in some contexts exclusive use of the masculine pronoun might invite a lawsuit. For example, you would be violating federal employment laws prohibiting discrimination on the basis of sex if you wrote the following in a help-wanted notice for your company.

> *Each applicant must submit his transcript with his application. He must also supply three letters of recommendation from individuals familiar with his work.*

The language of such a notice implies that only men can apply for the position.

Keep in mind that international readers may find these guidelines on avoiding masculine pronouns confusing since many languages (e.g., French, Spanish) follow grammatical instead of natural gender. In French, the word for *doctor* is masculine, for example.

3. Avoid using sexist words that end in -ess or -ette. Use gender-neutral alternatives for words like *stewardess (flight attendant), poetess (poet), waitress (server), usherette (usher),* and *drum majorette (drummer).*

4. Eliminate sexist salutations. Never use the following salutations when you are unsure of who your readers are:

- Dear Sir
- Gentlemen
- Dear Madam

Any woman in the audience will surely be offended by the first two greetings and may also be unhappy with the pompous and obsolete *madam.* It is usually best to write to a specific individual, but if you cannot do that, direct your letter to a particular department or group: *Dear Warranty Department* **or** *Dear Selection Committee.*

Be careful, too, about using the titles *Miss, Mr.,* and *Mrs.* Sexist distinctions are unjust and insulting. It is preferable to write *Dear Ms. McCarty* rather than *Dear Miss or Mrs. McCarty.* A woman's marital status should not be an issue. Try to find out if the person prefers *Ms.* to another courtesy title (e.g., *Editor Hawkins, Supervisor Jones*). If you are in doubt, write *Dear Indira Kumar.* Chapter 5 shows acceptable salutations to use in your letters (see pages 173 and 189).

5. Never single out a person's physical appearance. Sexist physical references negatively draw attention to a woman's gender. Sexist writers would not describe a male manager using physical terms as in the sentence below:

The manager is a tall blonde who received her training at Mason Technical Institute.

Avoiding Other Types of Stereotypical Language

In addition to sexist language, avoid any references that stereotype an individual because of race, national origin, age, disability, or sexual orientation. Not only are such references almost always irrelevant in the workplace (except for Equal Employment Opportunity Commission reports or health care), they are discriminatory, culturally insensitive, and ethically wrong.

To eliminate biased language in your workplace writing, follow the guidelines below.

1. Do not single out an individual because of race or national origin or stereotype him or her because of it. Be especially sensitive when referring to someone's ethnic identity.

Wrong: Bill, who is African American, is one of the company's top sales reps.
Right: Bill is one of the company's top sales reps.

Wrong: The Chinese computer whiz was able to find the problem.
Right: The programmer was able to find the problem.

It is just as prejudiced to associate someone's ethnicity with a profession or skill (as in the second example above) as it is to point out an individual's ethnicity or race as if it were unexpected to see someone from that background hired for such a position (as in the first example above).

2. Identify members of an international community accurately. Not every native Spanish speaker is Latin American or Hispanic. Be sensitive to significant cultural differences among groups (Cuban Americans and Mexican Americans, for example).

3. Avoid words or phrases that discriminate against an individual because of age. For example, do not use *elderly, up in years, geezer, old-timer, over the hill, senior moment,* or the adjectives *spry* or *frail* when they are applied to someone's age: "a spry sixty-seven." Similarly, don't refer to employees as *kids, youngsters, juveniles, wet behind the ears,* or *middle-aged supervisor.* Making someone's age an issue is unfair, whatever it may be.

> Wrong: Jerry Fox, who will be fifty-seven next month, comes up with obsolete plans from time to time.
> Right: Some of Jerry Fox's plans have not been adopted.

> Wrong: Our company keeps hiring youngsters who lack experience.
> Right: Our firm recruits individuals with little or no experience.

4. Respect individuals who may have a disability. Do not discriminate against someone who has a disability. Keep in mind that the Americans with Disabilities Act (1990) prohibits employers from asking if a job applicant has a disability. Avoid derogatory words such as *amputee, crippled, handicapped, impaired,* or *lame* (physical disabilities) or *retarded* or *slow* (mental disabilities). Stay away from terms such as these because they identify the entire individual rather than just the aspects that the disability affects. Put the person first. Emphasize the individual instead of the physical or mental condition as if it solely determined the person's abilities.

> Wrong: Tom suffers from MS.
> Right: Tom is a person living with MS.

> Wrong: Sarah, who is crippled, still does an excellent job of keyboarding.
> Right: Sarah's disability does not prevent her from keyboarding.

Also, do not use such phrases as *wheelchair-bound* or *confined to a wheelchair,* which wrongly and unfairly imply that a person in a wheelchair cannot move around on his or her job. Avoid using discriminatory expressions in your writing, such as a *crippled economy, lame excuse, mentally challenged, mental midget,* or *crazy scheme.*

5. Don't stereotype based on sexual orientation. Avoid unnecessarily labeling a person by sexual orientation as if that is the only significant aspect of that person's life.

> Wrong: Paula Smith, a strong supporter of gays and lesbians, hosts a successful daytime talk show.
> Right: Paula Smith hosts a successful daytime talk show.

In addition, avoid derogatory innuendos, comments, or jokes about gay men, lesbians, or bisexuals (e.g., "That's so gay"), and don't assume that all of your readers are heterosexual (e.g., don't ask about marital status).

The Writing Process: Some Final Thoughts

To make sure that all your researching, planning, drafting, revising and editing—the writing process at work—are successful, follow these final guidelines:

1. Be sure that your document focuses clearly and consistently on your audience's needs—that is, meets their expectations and answers their questions.

2. Verify that all the information you have used in your document is current, accurate, ethical, and relevant.

3. Confer one more time with any co-workers, members of your collaborative writing team, your supervisor, etc., who worked on the document with you to make sure everyone agrees with the final version.

4. Proofread the final version of your document, paying close attention to spelling, punctuation, formatting, and factual content, including names, dates, models, costs, and places. Misspelled words, comma splices, inconsistent dates, or incorrect costs can seriously undermine all your hard work in developing your document.

5. Make sure that you send the final revised, edited, and proofread version—and not an earlier draft or revision.

6. Send the document in the correct format—hard copy, Word file, PDF, and so on.

7. Be sure you send the document to the right individual(s).

✓ Revision Checklist

☐ Investigated the research, planning, drafting, revising, and editing benefits of my computer software.

☐ Researched my topic carefully to obtain enough information to answer all my readers' questions—online searches, interviews, questionnaires, personal observations.

☐ Before writing, determined the amount and kinds of information needed to complete my writing task.

☐ Spent enough time planning—brainstorming, outlining, clustering, or a combination of these techniques. Produced substantial material from which to shape a draft. Documented sources.

☐ Prepared enough drafts to decide on the major points in my message to readers. Made major changes and deletions where necessary in my drafts to strengthen the document.

☐ Revised drafts carefully to successfully answer readers' questions about content, organization, and tone. Formatted the text to make it easy to follow.

☐ Made time to edit my work so that the style is clear and concise and the sentences are readable and varied. Checked punctuation, sentences, and words to make sure they are spelled correctly and are appropriate for my audience.

☐ Eliminated sexist and other biased language that unfairly stereotypes individuals because of race, ethnicity, disability, or sexual orientation.

☐ Made sure final edited copy was carefully proofread and transmitted.

Exercises

1. Following is a writer's initial brainstormed list for a report on stress in the workplace. Revise the brainstormed list, eliminating repetition and combining related items.

   leads to absenteeism
   high costs for compensation for stress-related illnesses
   proper nutrition
   numerous stress-reduction techniques
   good idea to conduct interviews to find out levels, causes, and extent of stress in the workplace
   low morale caused by stress
   higher insurance claims for employees' physical ailments
   myth to see stress leading to greater productivity
   various tapes used to teach relaxation
   environmental factors—too hot? too cold?
   teamwork intensifies stress
   counseling
   work overload
   setting priorities
   wellness campaign
   savings per employee add up to $6,150 per year
   skills to relax
   learning to get along with co-workers
   need for privacy
   interpersonal communication
   employee's need for clear policies on transfers, promotion
   stress-management workshops very successful in California
   physical activity to relieve stress
   affects management
   breathing exercises

2. Prepare a suitable outline from your revised list in Exercise 1 for a report to a decision maker on the problems of stress in the workplace and the necessity of creating a stress-management program.

3. From the revised brainstormed list in Exercise 1, write a one-page memo to a decision maker about how the problems of stress negatively affect workplace production.

4. You have been asked to write a short report (2–3 pages) to the manager of the small company you work for on a topic of your choice. Prepare a cluster diagram similar to that on flextime in Figure 2.1. Add, delete, or rearrange anything in the diagram to complete your outline. Submit your final outline along with your report to your instructor.

5. Compare the draft of Marcus Weekley's report in Figure 2.4 with the final copy of his report in Figure 2.5. What kinds of changes did he make? Were they appropriate and effective for his audience and purpose? Why or why not?

6. Assume you have been asked to write a short report (similar to Marcus Weekley's in Figure 2.5) to a decision maker (the manager of a business you work for or have worked for; the director of your campus union, library, or security force; a city official) about one of the following topics:

   a. recruitment of more specialists in your field
   b. Internet resources
   c. security lighting
   d. food service
   e. health care plans
   f. public transportation
   g. sporting events/activities
   h. team building
   i. greening the workplace or community
   j. hiring more part-time student workers

Do relevant research and planning about one of the topics and the audience for whom it is intended by answering the following questions:

- What is my precise purpose in writing to my audience?
- What do I know about the topic?
- What information will my audience expect me to know?
- Where can I obtain relevant information about my topic to meet my audience's needs?

7. Using one or more of the planning strategies discussed in this chapter (clustering, brainstorming, outlining), generate a group of ideas for the topic you chose in Exercise 6. Work on your planning activities for about 15–20 minutes or until you have 10–15 items. At this stage, do not worry about how appropriate the ideas are or even if some of them overlap. Just get some thoughts down on paper.

8. Go through the list you prepared in Exercise 7, and eliminate any entries that are inappropriate for your topic or audience or that overlap. Try to see how many entries you might expand or rearrange into categories or subcategories. Then create an outline similar to the one in Figure 2.3.

9. Using your outline from Exercise 8, prepare some drafts of your memo report. Submit at least two drafts to your instructor.

10. Revise your drafts as much as necessary to create your final report.

11. In a few paragraphs, explain to your instructor the changes you made between your early drafts and your final revision. Explain why you made them. Concentrate on major changes—adding and moving paragraphs—as well as matters of style, tone, and even format.

12. In an email addressed to your instructor, describe any problems you encountered while working on your report. Also point out any planning, drafting, and revising strategies that worked especially well for you.

13. The following paragraphs are wordy and full of awkward, hard-to-read sentences. Either as an individual or a group exercise, edit these paragraphs to make them more readable by using clear and concise words and sentences.

    a. It has been verified conclusively by this writer that our institution must of necessity install more bicycle holding racks for the convenience of students, faculty, and staff. These parking modules should be fastened securely to walls outside strategic locations on the campus. They could be positioned there by work crews or even by the security forces who vigilantly and constantly patrol the campus grounds. There are many students in particular who would value the installation of these racks. Their bicycles could be stationed there by them, and they would know that safety measures have been taken to ensure that none of their bicycles would be apprehended or confiscated illegally. Besides the precaution factor, these racks would afford users maximized convenience in utilizing their means of transportation when they have academic business to conduct, whether at the learning resource center or in the instructional facilities.

    b. On the basis of preliminary investigations, it would seem reasonable to hypothesize that among the situational factors predisposing the Smith family toward showing pronounced psychological identification with the San Francisco Giants is the fact that the Smiths make their domicile in the San Francisco area. In the absence of contrariwise considerations, the Smiths' attitudinal preferences would in this respect interface with earlier behavioral studies. These studies, within acceptable parameters, correlate the fan's domicile with athletic allegiance. Yet it would be counterproductive to establish domicility as the sole determining factor for the Smiths' preference. Certain sociometric studies of the Smiths disclose a factor of atypicality, which enters into an analysis of their determinations. One of these factors is that a younger Smith sibling is a participant in the athletic organization in question.

14. Following are very early drafts of memos that businesspeople have sent to their bosses or co-workers. Revise and edit each draft, referring to the checklist on page 78. Turn in your revision and the final, reader-ready copy. As you revise, keep in mind that you may have to delete and add information, rearrange the order of information, and make the tone suitable for the readers. As you edit, make sure your sentences are clear and concise and the tone of your words is well chosen.

    a. TO: All workers
         FROM: B. J. Blackwell
         DATE: February 3, 2012
         RE: Parking

The parking violations around here have gotten very, very bad. And the administration is provoked and wants some action taken. I don't blame them. I have been late for meetings several times in the last month because inconsiderate folks from other divisions have parked their cars in our zone. That just is not fair, and so we in our department must not be the only ones who are upset. No wonder the management finds things so bad they have asked me to prepare this memo.

A big part of the problem it seems to me is that employees just cannot read signs. They park in the wrong zones. They also park in visitors' spots. The penalties are going to be stiff. The administration, or so I was led to believe, is thinking of fining any employee who does not obey the parking policies. I know for a fact that I saw someone from the research department pull right into a visitor parking area last week just because it was 8:55 and he did not want to be late for work. That gives our business a bad name. People will not want to do business with us if they cannot even find a parking spot in the area that the company has reserved for them.

Vice President Watson has laid the law down to me about all this and told me to let each and every one of you know that things have to improve. One of the other big problems around here is that some employees have even parked their cars in loading zones, and security had to track them down to move.

As part of the administration's new policy, each employee is going to be issued a company parking policy and will have to come in and sign for it verifying that he received it. I think things really have gotten out of hand and that some drastic action has to be taken. We will all have to shape up around here.

b. TO: All Employees
FROM: George Holmes
DATE: October 19, 2012
RE: Travel

Every company has its policies regarding travel and vouchers. Ours strike me as important and fairly straightforward. Yet for the life of me I cannot fathom why they are being ignored. It is in everyone's best interest. When you travel, you are on company time, company business. Respect that, won't you. Explain your purpose, keep your receipts, document your visits, keep track of meals. Do the math.

If you see more than one client per day, it should not be too hard or too much to ask you to keep a log of each, separate, individual visit. After all, our business does depend on these people, and we will never know your true contributions on company trips unless you inform us (please!) of whom you see, where, why, and how much it costs you. That way we can keep our books straight and know that everything is going according to company policy.

Please review the appropriate pages (I think they are pages 23–25) about travel procedures. Thanks. If you have questions, give me a call, but check your employee handbook or with your office/section manager, first. That will save everyone more time. Good luck.

15. Find a piece of writing—email, memo, letter, brochure, short report, or website—that you believe was not carefully drafted or revised. In a short memo or email, point out to your instructor what is wrong with the piece of writing—for example, it is not logically organized, it has an inappropriate tone, it is incomplete, or the information is too technical. Attach a copy of the poor example to your memo or email.

16. Revise the piece of poor writing you analyzed in Exercise 15. Submit your improved version to your instructor.

17. The following sentences contain sexist and other biased language. Edit them to remove these errors.
   a. Every intern had to record his readings daily for the spokesman.
   b. Although Marcel was an amputee, he still could hunt and peck at the keyboard.
   c. She saw a woman doctor, who told her to take an aspirin every day.
   d. Our agency was founded to help mankind.
   e. John, who is a diabetic, has an excellent attendance record.
   f. Every social worker found her schedule taxing—not enough days in the week to help out man-to-man.
   g. Maria, who is Cuban, always adds spice to company events.
   h. To be a policeman, each applicant had to pass a rigorous physical and prove himself in the manly art of self-defense.
   i. It's a wise man who can rise to the top in this cutthroat, volatile stock market.
   j. Sandy Frain, a middle-aged Irish-born woman, came by this morning wanting an appointment to discuss the new policies on energy efficiency.
   k. Mrs. Johnson is in charge of safety issues.
   l. He made his PowerPoint presentation as emphatically as an Italian opera singer on stage.
   m. Sanji, a longtime member of the IT department, is naturally adept at crunching the figures.
   n. Team B tried to disable our proposal by introducing irrelevant references to the many foreigners living on the north side.
   o. The average consumer spends at least two to three hours a week on her computer looking for coupons and other bargains.
   p. How many of our customers don't speak English well?
   q. Our office installed new power doors to assist those employees who are unable to leave their wheelchairs.
   r. More mentally retarded individuals are being hired to do menial tasks on the job.

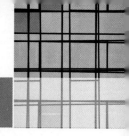

CHAPTER 3

Collaborative Writing and Meetings at Work

Access chapter-specific interactive learning tools, including quizzes and more in your English CourseMate, accessed through www.cengagebrain.com.

In the workplace, you will not always have to write alone, isolated from co-workers or managers. In fact, much of your business writing time may be spent working as part of a team. One survey estimates that in the world of work, 90 percent of all business people spend some time writing as part of a collaborative team. Thanks to teleconferencing, instant messaging, blogs, and other interactive media, collaboration is one of an employee's most valuable contributions to corporate culture. You can expect to prepare a document with other employees or managers who will collaborate or, at the very least, review your work and revise it. Collaborative writing occurs when a group of individuals—from two to seven or more—do the following:

- combine their efforts to prepare a single document
- share authorship
- work together for the common good, maybe even the survival, of a department, a company, or an organization

Collaboration is responsible for preparing many types of writing in the world of work, including brochures, manuals, websites, proposals, and short and long reports. And collaboration can be done in a variety of ways—face-to-face, over the phone, or electronically (see "Computer-Supported Collaboration" on pages 108–118 later in this chapter).

This chapter introduces you to successful ways to collaborate with co-workers in your office as well as others with whom you do business around the globe. You'll receive practical advice on how to write as part of a group and how to solve communication problems within a group setting. It takes much work and skillful negotiation to be a member of an effective collaborative writing group, but the result is worth the effort.

Teamwork Is Crucial to Business Success

Teamwork is essential for success within the world of work. Being a part of a writing team is a major responsibility in a world where each employee is connected to co-workers, managers, and customers around the globe. See, for instance,

Figure 1.11, in which the CEO of IBM characterizes employees as a group of "IBMers." Collaboration is networking, and collaborative writing is a vital part of the global network in which individuals depend on one another's expertise, experience, and viewpoints.

Successful collaboration hinges on being a team player, one of the most highly valued skills in the workplace. Being a team player means you

- interact successfully on an interpersonal level
- network to access and archive information
- participate and provide feedback
- give and take constructive criticism
- raise important and relevant questions
- get assistance from resource experts in other departments and fields
- put the good of your company above your ego
- work toward consensus
- contribute to customer service and satisfaction

Collaboration builds teamwork as it helps get writing done more easily and more efficiently. Figure 3.1 shows a collaborating team at work, planning and interacting on a company project.

FIGURE 3.1 A Collaborating Team at Work

Yuri Arcurs/Shutterstock.com

Keep in mind that teamwork must always take into consideration your employer's corporate image, goals, and politics. Your employer determines the subject that your group will discuss and the formats in which your work will appear. Moreover, everything your team does is subject to review and revision by your employer. Viewing collaboration in this broader corporate context will get you off to a good start in the world of work.

Advantages of Collaborative Writing

Collaborative writing teams benefit both employers and employees. Specific advantages of collaboration include the following. Many of these advantages are highlighted in Lee Booker's report on his team's interaction with other groups of employees at Keeton Pharmaceuticals in Figure 3.2.

1. **Collaboration builds on collective talents.** Many heads are better than one. A writing team profits from the diverse backgrounds and skills of its individual members. Collaboration joins individuals from diverse disciplines in the corporate world—scientists, lawyers, designers, and security experts, among others.

2. **It allows for productive feedback and critique.** Collaborative writing encourages all members of the team to get involved, offering both their agreement as well as their critical viewpoints in a shared effort to produce the highest-quality document possible.

3. **It increases productivity and saves time and money.** When a group has planned its strategies carefully, collaboration actually cuts down on the number of meetings and conferences, saving a company time and employee travel expenses.

4. **It ensures overall writing effectiveness.** The more people there are involved in developing a document, the greater the chances are for thoroughness and cohesion. Guided by the shared principles of company style, documentation, and mission, a collaborative team can better guarantee uniformity and consistency in a piece of writing.

5. **It accelerates decision-making time.** A group investigating problems and offering pertinent solutions can cut down considerably on the time it takes to prepare a document.

6. **It reduces corporate risk.** With a diverse group of individuals assessing options and discussing possible outcomes, a company has a better chance of success than if only one or two individuals were involved in strategic planning.

7. **It boosts employee morale and confidence while decreasing stress.** Working as part of a team relieves an individual of some job stress because he or she is not solely responsible for planning, drafting, and revising a document. A team, therefore, provides a safety net by assuring individual members that they can always talk over problems and will have help in meeting deadlines.

8. **It contributes to customer service and satisfaction.** By pooling their knowledge of an audience's needs, a collaborative writing team is much more likely—through discussions, interactions, even disagreements—to anticipate a customer's or an agency's requests and complaints.

FIGURE 3.2 The Advantages of Collaboration in the Workplace: A Manager's Notes

Keeton Pharmaceuticals

———————— Notes for Human Resources File ————————

Collaborative writing is essential to the success and morale of the Environmental Testing Lab here at Keeton. I supervise four microbiologists, each of whom is responsible for testing a specific area of our plant for environmental contaminants by sampling the air, water, and surfaces (e.g., belts, floors, vents). When we find unacceptable ranges of bacteria, we must report our findings to management and, ultimately, to the FDA.

The report that emerges is the collaborative effort of microbiologists, production personnel, and management. First, the microbiologist for that area interviews the production supervisor as well as line personnel to locate the potential causes of contamination. Based on their feedback, the microbiologist then prepares a memo report in collaboration with the other three microbiologists on how these unacceptable ranges could be eliminated or reduced. This collective brainstorming makes sure our lab does not overlook key information.

After receiving the report, I share it with my boss, the Quality Lab Manager. We then confer with the microbiologists and production staff to ensure that the report follows all company and government protocols.

The report is then forwarded to the Quality Control Manager and, finally, to the plant manager for their signatures. The report is then archived in the plant's Document Center where it is subject to periodic FDA inspections to verify its accuracy.

These collaborative efforts from production, microbiologists, and management guarantee that our report will be clear, precise, and ethical. In the last three years, management has frequently commended the lab for our team spirit.

Lee Booker
Lab Supervisor

9. It affords a greater opportunity to understand global perspectives. By working with a multinational workforce, such as that described in the long report in Chapter 15 (pages 702–718), individuals can develop greater sensitivity to and appreciation of the needs and problems of an international audience.

Collaborative Writing and the Writing Process

The writing process described in Chapter 2 also applies to collaborative writing. Groups use the same strategies and confront the same problems that individual writers do. Writing teams must brainstorm, plan, research, draft, and edit. Like the individual writer, too, a team must move through the writing process, going back and forth as necessary between the various stages. But although the essential steps in the writing process are the same, there are some differences you need to know about. Below is a rundown of how collaborative groups tackle the process of preparing a document.

1. Groups must plan before they write. As in individual writing, the planning phase includes brainstorming and outlining (see pages 53–56), as well as identifying the audience, purpose, format, and scope of a document. Unlike individual writers, though, the group must also set ground rules for the collaborative process to work, select a leader, assign individual responsibilities within the group, and create a schedule that respects individual group members' schedules.

2. Groups must do research. Using a variety of research methods to gather data, as discussed in Chapter 8, the group documents the findings of the report. Insufficient or incorrect research can jeopardize the success of a collaborative document just as easily as it can derail the efforts of an individual writer. In a group context, the responsibility of researching is usually shared by several individuals who must share and coordinate the results of their research prior to the group drafting process.

3. Groups must prepare drafts. Based on our research and company goals, the drafting process in group writing can be done in two ways. The first way is similar to individual writing in that only one complete draft of the document is prepared at a time, benefiting from the feedback and critique of several individuals. The second, more common, way is unique to group writing: Individual members of the group draft separate portions of the document, later to be combined and edited by the entire group.

Case Study

Collaborative Writing and Editing

Figure 3.3 contains an email written for a listserv by Tara Barber, the Documentation Manager of CText, Inc., a large firm in Ann Arbor, Michigan, that develops software for the publishing industry. Barber describes the writing process followed by her team of collaborative writers. Her approach sheds light on the practical, day-to-day process of group writing and editing. Group harmony is essential to completing documents successfully and on time.

Study Barber's approach. She effectively works *with* and *not against* her staff. She manages the collaborative effort without being heavy-handed or suppressing individual creativity. Her strategy offers a good model to follow.

FIGURE 3.3 Collaborative Editing: Advice from a Pro

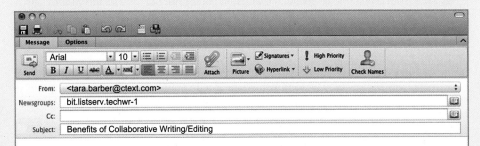

I would like to describe some constructive ways to edit a document to help new writers improve their skills.

I agree that the more projects you make collaborative, the better results you will get. Here's how we do it in my department. Although all my writers are experienced, they come from different backgrounds, which means I have to coordinate editing activities similar to the ways I would interact with new writers. It's my job to make sure all the pieces fit.

When I first took this job, several documentation styles were used for the company's manuals, but the department was small, and so the senior writer and I prepared a consistent style manual. We experienced some tension, but by compromising, we ironed out our differences. We developed a style guide and a set of manual conventions that gave writers precise guidelines to follow. The new manual helped us eliminate glaring editing problems such as inconsistencies in spelling, capitalization, use of formats, headings, and documentation.

But the department has grown since that time, and on at least two occasions all of us have reassessed the content and format of the style guide and manual. By doing this, we are able to get input from everyone and to allow for new ideas. When a problem arises now, we are comfortable addressing it as a group.

In addition to the usual reviews by subject-matter experts, all of us read each other's materials. This collaborative review helps us to become familiar with each other's projects, homogenizes our writing styles, and keeps our documentation consistent. It's also a great way to get a good "clean-eyes edit."

In my role as documentation manager, I try to edit early and late, but not in the middle. Our materials usually go through several edits before they're ready. I try to look over early material found in outlines and first drafts to make sure everything follows our house style and procedures. I edit final drafts because I am responsible for the work my department does. Peer review works best during this intermediate phase, and I don't want to step on individual creativity.

(Continued)

FIGURE 3.3 (Continued)

If I have to make comments, I make them at various levels.

a) I point out problems, or sections that seem confusing, but then let the writer suggest fixes.

b) I make suggestions and provide examples—more than one if I can.

c) I actively work with the writer to develop new ways around a problem, such as finding another way of approaching the documentation. This issue is then addressed at our next style meeting, which involves the rest of the department in a brainstorming session.

d) If nothing else works, I play the heavy manager and say, "Do it this way because I say so." I try to avoid this, however, if I possibly can.

Working as a collaborative team, we rarely have editing problems that we can't solve, with the result that everyone is happy with the final document. And, even more important, our customers find the documents usable and valuable. You can't really ask for more than that.

Tara Barber
CText, Inc.
<tara.barber@ctext.com>

© Cengage Learning 2013

Source: Reprinted with the permission of Tara Barber.

---

4. Groups must revise and edit. The revising and editing processes in group writing are also similar to what an individual would do. But group writing offers the added benefit of multiple writers being involved, each of whom may spot organizational problems, inconsistencies in format and tone, as well as grammatical and mechanical mistakes. Groupware, discussed on pages 108–116, is an invaluable tool in helping writing teams access, share, and comment on one another's files.

Seven Guidelines for Establishing a Successful Writing Group

As discussed earlier in this chapter, a group not only has to plan the document but it must also plan how to work together most efficiently and productively. This planning involves a great deal of thought, organization, and cooperation. Following the guidelines below will increase your group's chances of successfully moving forward to its goal.

1. Get to know one another. To establish group rapport, each member of the group should learn as much as he or she can about the other members. For instance, group members should learn about one another's schedules, professional backgrounds, special competencies, national or cultural perspectives, work habits, and even pet peeves. By being concerned and friendly, you create an environment that enhances productivity.

2. Set up a preliminary meeting to establish guidelines. At an introductory meeting, the group should discuss the objectives, audience, scope, format, importance of the document, and, most important, the deadline. It is helpful at the first meeting to identify and discuss any directions or directives from management and to set up priorities.

3. Agree on the group's leadership. In some firms, the leader of the group is chosen according to his or her seniority or experience. In other companies, the group may vote on which member they want to lead them. Regardless of the selection process, the leader should be chosen to keep the group on track. An effective group leader must be skillful in initiating discussions, providing encouragement, and helping the group reach consensus. As Lee Booker points out in Figure 3.2, management also plays a critical role by giving the group its feedback.

The group may also need to decide on the leader's responsibilities and level of authority. The leader may take on any number of roles, including setting schedules, presiding over meetings, coordinating the activities of the team, evaluating the work of individual members, writing progress reports, communicating with management, and transmitting the finished document.

4. Identify each member's responsibilities and roles, but allow for individual talents and skills. A key question to ask and to answer is "Who gets credit for what and where?" Members need to acknowledge one another's strengths—in writing, providing graphics, developing software, marketing, editing, and even presenting the document at a meeting (see pages 118–125). Assign each member a task appropriate for his or her background.

5. Establish the times, places, and length of group meetings. The group must decide on a calendar of scheduled meetings and ensure that each member has a copy. Various software programs can help groups schedule meetings.

6. Follow an agreed-on timetable, but leave room for flexibility. The group should estimate a realistic time to complete the various stages of the work—when drafts are due or when editing must be concluded, for example. The group's timetable, with major **milestones** (dates when key parts are to be completed), should be sent to each member. **But remember: Projects always take longer than initially planned.** Prepare for a possible delay at any one stage. The group may have to submit progress reports (see pages 654–661) to management.

7. Use a standard reference guide for matters of style, documentation, and format. Establishing guidelines about style (spelling, abbreviations, dates, capitalization, symbols), documentation, format, graphics, and so forth, makes the group's task easier. Be consistent; write *email* or *e-mail*, not both. Many large companies have policy or style manuals that their teams must use, as Tara Barber notes in Figure 3.3. If such a document is not available, select a manual or other reference work to which members should adhere, such as the American Psychological Association (APA) guide (see pages 385–397).

Ten Proven Ways to Be a Valuable Team Player

As we saw, being a team player is one of the most highly prized skills in the world of work. By following the ten guidelines below, many of which Tara Barber's team follows in Figure 3.3, you can advance the work of your writing team and earn the respect of your colleagues and supervisor:

   1. Think collectively. Adopt a "we can get this done together" attitude. Do not treat some team members as favorites and ignore others. Both actions nurture resentment.

   2. Participate. Avoid being a passive observer. Recognize that you have valuable ideas to contribute to the group.

   3. Be clear and precise. Express your ideas clearly and appropriately. Avoid veering off the point, being vague, or making unsupported generalizations. Do not be afraid to ask for clarification of anything you don't understand.

   4. Set aside your ego. Do not view the group as a battleground or contest in which someone wins and someone else loses. Put the team ahead of self.

   5. Be enthusiastic. Avoid sounding uninterested in or inconvenienced by the group process. Replace cynicism with a "can do" spirit, and your team will value you as a vital player.

   6. Listen. Give everyone a chance to be heard. Pay attention to what others say when they are speaking, and avoid monopolizing the conversations. Develop a "third ear" to discern the meanings behind group members' words.

   7. Be open-minded. Be receptive to other points of view. Give serious consideration to each member's suggestions and ideas. Weigh the pros and cons before firing off a question or an objection.

   8. Follow company and group protocols. Adhere to company and group protocols regarding the organization, style, and format of the document. Follow the group's rules about meeting times, places, and order of business.

   9. Compromise. After ample discussion and deliberation, you may need to compromise for the overall benefit of the group.

10. Go with group consensus. Don't be a lone ranger. Accept the group's decisions as final. Stubbornness only leads to delay and hard feelings.

Sources of Conflict in Collaborative Groups and How to Solve Them

The success of collaborative writing depends on how well the team interacts. Discussion and criticism are essential to discover ideas, results, and solutions. Members must build on one another's strengths and eliminate or downplay weaknesses. Inevitably, members will have different perspectives or viewpoints out of which conflicts may arise.

But *conflict*—in the sense of conflicting opinions, a healthy give-and-take— can be positive if it alerts the group to problems (inconsistencies, redundancies, incompleteness, and outright errors) and provides ways to resolve them. Conflict in the sense of creating problems for the group dynamic, however, is another matter entirely.

Following are some common problems in group dynamics, with suggestions on how to avoid or solve them. A few of these problems involve serious ethical violations in the workplace.

Common Problems, Practical Solutions

1. Resisting constructive criticism. No one likes to be criticized, yet criticism can be vital to the group effort. But individuals who insist on "their way or no way" can become hostile to any change or revision, no matter how small.

> Solution: When emotions become heated, the group leader may wisely move the discussion to another section of the document or to another issue to allow for some cooling-off time. It is always better to defuse a situation before it escalates into a major conflict. The leader should also discuss (in private) the importance of listening to constructive criticism with the individual or with the entire group if the problem is widespread. Fair play, and being open to others' opinions, is essential to ensure maximum team effort.

2. Giving only negative criticism. At times, a member may saturate a meeting with nothing but negatives. He or she faults any new idea or refuses to work toward consensus, both of which can sap the group's enthusiasm. Using condescending phrases like "Don't you realize that . . . ," "You haven't taken into account . . . ," or "What you really need to do is . . . " only embarrasses or irritates the person being criticized. Even more serious is anger and name-calling, which are never acceptable in the world of work.

> Solution: The leader should diplomatically (and again in private) remind the individual about the importance of working as a team member toward a common goal. The leader should also insist that the person act courteously even though he or she may disagree. Rather than torpedoing others' ideas, this individual might be encouraged to offer a more helpful and relevant revision or solution to the problem. If the individual does have a valid point to make, he or she should express it diplomatically. The bottom line: Individuals need to encourage one another, too, for the sake of group harmony.

3. Dominating a meeting. The group process is about sharing and responding to ideas, not about taking over. When one member dominates the discussion and becomes aggressive and territorial, the group process suffers.

> Solution: The leader should let the group know that the participation of all members is valued but then say, "We need to hear from the rest of the group." Some groups follow a three-minute rule—each member has three minutes to make comments and does not get the floor again until everyone has had a chance to speak. If one group member still continues to dominate, the leader may (in private) have to speak to him or her.

4. Refusing to participate. Some individuals are afraid to express their opinions because of a lack of self-confidence and so deprive the group of hearing another viewpoint.

> Solution: If a group member lacks confidence about an idea or a suggestion, he or she should safely test it out on an individual member before presenting it to the group as a whole. When a leader perceives that a group member lacks confidence, the leader should strive to encourage that member to participate at key junctures and assure him or her of a welcome forum to do so. The leader may also want to appoint another member of the team to serve as a listening partner for a member who lacks confidence, giving that individual a chance to voice and test his or her views before the group convenes.

5. Interrupting with incessant questions. Some people interrupt a meeting so many times with questions that all group work stops. The individual may simply be unprepared or may be trying to exercise control of the group.

> Solution: When that happens, a group leader can remark, "We appreciate your interest, but would you try an experiment, please, and attempt to answer your own questions?" or if the person claims not to know, the leader might then say, "Why don't you think about it for a while and then get back to us?"

6. Inflating small details out of proportion. Nitpickers can derail any group. Some individuals waste valuable discussion and revision time by dwelling on relatively insignificant points (e.g., an optional comma, the choice of a single word, etc.) or steer the group away from larger, more important issues (e.g., costs, schedules, etc.).

> Solution: If there is consensus about a matter, leave it alone and turn to more pressing issues. The leader should remind the group (without singling anyone out) about the bigger picture and caution them to stay on track.

7. Being overly deferential to avoid conflict. This problem is the opposite of that described in Point 1. Group members who agree to everything out of fear of conflict do not help the group. Moreover, they only appease a strong-willed member. Hiding their feelings, these "yes people" may actually be depriving the group of valuable input.

> Solution: The group leader needs to encourage all members to express their feelings and opinions freely, and to do so courteously and professionally. If tempers flare, the group leader needs to intervene. Again, the leader needs to promote and protect meetings as a safe place to express ideas and air differences for the good of the project.

8. Not respecting cultural differences. You may have individuals from different countries or cultures on your team. Disregarding or misjudging the way they interact with the group can seriously threaten group success and harmony. In addition, discrimination of any type—based on race, age, religious beliefs, or nationality, for example—is unacceptable in a collaborative group or workplace.

> Solution: As the long report example in Chapter 15 illustrates, some companies offer employees seminars on cultural sensitivity that emphasize respecting diversity in the workplace. A group leader must ensure that diversity is

honored in the group. If anyone does not respect cultural differences, the leader may write that person up, refer him or her to a company handbook, or remind that person of antidiscriminatory legislation.

9. Violating confidentiality. Leaking confidential information about a personal issue, product, procedure, or operation is a serious violation in the world of work. Sometimes meetings are closed to everyone in the company except for those who are a part of the group.

> Solution: Violating confidentiality is often grounds for dismissal. At a preliminary meeting, and periodically during the writing process, the leader should emphasize the whys, hows, and whens of confidentiality.

10. Not finishing on time or submitting an incomplete document. Meeting established deadlines is the group's most important obligation to individual members and to the company. When some members are not involved in the planning stages or when they skip meetings or ignore group communications, deadlines are invariably missed. If you miss a meeting, get briefed by an individual who was there.

> Solution: The group leader can institute networking through email to announce meetings, keep members updated, or provide for ongoing communication and questions. See page 109.

Models for Collaboration

There are as many types of collaborative writing methods as there are companies. The process can range from relatively simple phone calls or emails to a much more extensive use of groupware (see pages 108–116).

The scope, size, and complexity of your document as well as your company's organization will determine what type of (and how much) collaboration is necessary. A shorter assignment (say, a memo) will not require the same type of group participation as would a policy handbook, a proposal, or a long report. At some large companies, for example, a staff of professional editors (such as Tara Barber's team in Figure 3.3) revises the final draft prepared by a departmental team. The more important and more complex a document is, the more extensive collaboration will be.

The following sections describe four models for collaboration used in the world of work.

Cooperative Model

The cooperative model is one of the simplest and most expedient ways to write collaboratively in the business world. An individual writer is given an assignment and then goes through the writing process to complete it (see Chapter 2). Along the way, he or she may show a draft to a peer to get feedback or to a supervisor for a critique. That is what Randy Taylor did in Figure 3.4. He shared a draft of a letter to a potential client with his boss, Felicia Krumpholtz, who made changes in content, wording, and format and then sent it back to Taylor. Following his boss's suggestions, Taylor then created the revised letter in Figure 3.5.

FIGURE 3.4 A Draft of Randy Taylor's Letter, Edited by His Supervisor, Felicia Krumpholtz

servitron

4083 Randolph Street
Houston, TX 77016
(713) 555-6761

November 15, 2012

Terry Tatum
Manager
Consolidated Solutions
Houston, TX   *add zip code*

Dear ~~Sir,~~ *Terry Tatum:*

Thank you for asking
~~I am taking the opportunity of answering your request~~ for a price list of
Servitron products. Servitron has been in business in the Houston area for
more than ~~twenty-two~~ *22* years and we offer unparalleled equipment and
service ~~to any customer.~~ *Consolidated Solutions* Using a Servitron product will give you both *can*
efficiency and economy.

boldface   *boldface*
Whatever Servitron model you choose carries with it a full one-year
warranty on all parts and labor. After the expiration date of your warranty
you ~~should~~ *might* purchase our service contract for $75,000 a year.

~~Here are the models Servitron offers~~

Model Name	Number	Price
Zephyr	81072	$459.95 — *wrong price*
Colt	86085	$629.95
Meteor	88096	$769.95

Depending on your needs, one of these models should be right for you.

If I might be of further assistance to you, please call on me. I am also
enclosing a brochure giving you more information, including specifications,
on these Servitron products.

*add phone number
and email*

~~Truly,~~ *Sincerely yours,*

Servitron   *leave 4 spaces*
Randy Taylor   *sign your name*
Sales Associate

*reverse
the order
of these
two
sentences*

www.servitron.com

FIGURE 3.5 The Edited, Final Copy of Figure 3.4

servitron

4083 Randolph Street
Houston, TX 77016
(713) 555-6761

November 15, 2012

Terry Tatum
Manager
Consolidated Solutions
Houston, TX 77005-0096

Dear Terry Tatum:

Thank you for asking for a price list of Servitron products. Servitron has
been in business in the Houston area for more than 22 years and we can
offer unparalleled equipment and service to Consolidated Solutions. Using
a Servitron product will give you both **efficiency** and **economy**.

Depending on your needs, one of these models should be right for you.

Model Name	Number	Price
Zephyr	81072	$469.95
Colt	86085	$629.95
Meteor	88096	$769.95

Whatever Servitron model you choose carries with it a full **one-year
warranty** on all parts and labor. After your warranty expires, you can
purchase our service contract for only $75.00 a year.

I am also enclosing a brochure giving you more information, including
specifications, about these Servitron products. If I can help you further,
please call me at (713) 555-6761 or email me at **rtaylor@servitron.com**.

Sincerely yours,

SERVITRON

Randy Taylor

Randy Taylor
Sales Associate

www.servitron.com

If Taylor had sent his first draft, the customer would hardly have been impressed with his company's professionalism and might have reconsidered placing an order. If Terry Tatum were a woman, she might have been offended by being addressed as "Sir." But thanks to Krumpholtz's revisions, Taylor made his letter much more effective. Careful writers always benefit from constructive criticism. Even though Krumpholtz helped Taylor improve his work, strictly speaking it was not a case of group writing. Although the two interacted, they did not share the final responsibility for creating the letter. Even so, Servitron profited from their interaction.

Sequential Model

In the sequential model, each individual is assigned a specific, nonoverlapping responsibility—from brainstorming to revising—for a section of a proposal, report, or other document. There is a clear-cut, rigid division of labor. If four people are on the team, each will be responsible for his or her part of the document. For example, one employee might write the introduction; another, the body of the report; another, the conclusion; and the fourth, the group's recommendation.

Team members may discuss their individual progress and even exchange their work for group review and commentary. They may also choose a coordinator to oversee the progress of their work. When each team member finishes his or her section, the coordinator then assembles the individual parts to form the report.

Functional Model

The division of labor in the functional collaborative model is assigned not according to parts of a document but by skill or job function of the members. For example, a four-person team may be organized as follows:

- The **leader** schedules and conducts meetings, issues progress reports to management, and generally coordinates everyone's efforts to keep the project on schedule.
- The **researcher** collects data, conducts interviews, searches the literature, administers tests, classifies the information, and then prepares notes on the work.
- The **designated writer/editor**, who receives the researcher's notes, prepares outlines and drafts and circulates them for corrections and revisions.
- The **graphics expert** obtains and prepares all graphics/illustrations, specifying why, how, and where visuals should be placed, and might even suggest that visuals replace certain sections of text. The graphics expert may also be responsible for the design (layout) and production of the document (see Chapter 11).

This organizational scheme fosters much more group interaction than the sequential model does. It also allows each member to do what he or she knows best.

Figure 3.6 illustrates how a functional model works. It describes the behind-the-scenes joint effort that went into Joycelyn Woolfolk's proposal for her boss to authorize a new journal, which would incorporate a newsletter her office currently prepares. A publications coordinator for a large, regional health maintenance organization

FIGURE 3.6 How a Proposal Is Collaboratively Written

Status Report: Coordinating Regional Magazine Project

The idea to start a regional magazine was first expressed in passing by our vice president, who is interested in getting more and higher-level visibility for our regional office. Several other regional offices in our company have created fairly attractive magazines, and one office in particular has earned a lot of good publicity for its online publication.

The public affairs manager (my boss) and I quickly picked up on the vice president's hint and began to formulate ways to investigate the need for such a publication and ways to substantiate our recommendation. For several weeks the public affairs manager and I discussed such points as the kind of documentation our proposal would need, what our resources for researching the question were, what our capabilities would be for producing such a publication, and so on. We were guided by the twofold goal of getting the vice president's approval and, ideally, meeting a genuine market need.

After discussions with my boss, I met with members of my staff to ask them to do the following tasks:

1. Review existing HMO publications and report on whether there was already a regional magazine for the Northwest

2. Develop, administer, and analyze a readership survey for current subscribers to our newsletter, which would potentially be incorporated into the new magazine

3. Formulate general design concepts for the magazine in print and online that we can implement without increasing staff, while still producing the quality magazine the vice president wants

4. Prepare a detailed budget for projected costs

5. Consult with experts on our staff (actuaries, physicians, nurses) about topics of interest

6. Confer with IT about online designs and problems

I requested emails, Web sources, and other documentation from my staff members about most of these tasks, and then I used that information to draft the proposal that eventually would go to the vice president, and perhaps even to the president's office. And I communicated with my staff often, through emails, blogs, personal meetings, and group sessions, both face-to-face and online.

© Cengage Learning 2013

(Continued)

FIGURE 3.6 (Continued)

After I revised my draft several times, I asked my staff to look over each draft for their feedback. Based on their comments, I made further revisions and did careful editing. My boss also reviewed my proposal, revising and editing it several times. Ultimately, the proposal will go out as a memo from me to my boss, who will then send it under her name to the vice president.

The copy of my proposal may be further revised in response to the vice president's comments when she gets it. She may use some portion or all of it in another report from her to the president, or she may write her own proposal to the president supporting her argument with the specifics from my proposal. The vice president will then word the proposal to fit the expressed values and mission of our regional office as well as provide information that addresses budgetary and policy concerns for which her readers (our company president and board members) have final responsibility.

This is how a proposal started and where it will eventually end.

(HMO), Woolfolk supervises a small staff and reports directly to the public affairs manager, who in turn is responsible to the vice president of the regional office.

As you will see from Woolfolk's functional approach, collaboration can move up and down the chain of command, with participation at all levels. Often, individual employees will pull together information from their separate functional areas (such as finance, information systems, marketing, and sales), and someone else will put that information into a draft that others read and revise until the document is ready to send to the boss. By the time the boss reviews the document, it has been edited and revised many times and by many individuals to ensure accuracy and consistency. (Review Figure 3.3.) This is a common business practice.

Integrated Model

In the integrated model, all members of the team are engaged in planning, researching, and revising. Each shares the responsibility of producing the document. Members participate in every stage of the document's creation and design, and the group goes back to each stage as often as needed. This model offers intense group interaction. Even though individual writers on the team may be asked to draft different sections of the document, all share in drafting, revising, and editing that document. Depending on the scope of the document and company policy, the group may also go outside the team to solicit reviews and evaluations from experts—both inside and outside the company.

Evolution of a Collaboratively Written Document

Figures 3.7, 3.8, and 3.9 show the evolution of a memo that follows an integrated model of collaboration. This memo informed employees that their company was enhancing its recycling program. Alice Schuster, the vice president of Fenton Companies, which manufactures appliances, asked two employees in the human resources department—Abigail Chappel and Manuel Garcia—to prepare a memo ("a few paragraphs" is how Schuster put it) to be sent to all Fenton employees.

Schuster had an initial conference with Chappel and Garcia, at which she stressed that their memo had to convey Fenton's renewed commitment to the environment and greening the workplace and that, as part of that commitment, the employees had to intensify their recycling efforts. Chappel and Garcia thus shared the responsibility of convincing co-workers of the importance of recycling and educating them about practicing it.

Chappel and Garcia also had the difficult job of writing for several audiences simultaneously—the boss, whose name would not appear on the memo, other managers at Fenton, and the Fenton workforce itself.

First Draft

Figure 3.7 is the first draft that Chappel and Garcia collaborated on and then presented to the manager of the human resources department—Wells McCraw—for his comments and revisions. As you can see from McCraw's remarks, written in ink, he was not especially pleased with their first attempt and asked them to make a number of revisions. As a careful reader (conscious of the document's audience), McCraw found Chappel and Garcia's paragraphs to be rambling and repetitious—the writers were unable to stick to the point. Specifically, he pointed out that they included too much information in one paragraph and not enough in others.

As a good editor/manager, McCraw also directed their attention to factual mistakes, irrelevant and even contradictory comments, and essential information they had omitted. Finally, McCraw offered some advice on using visual devices (see pages 24–26) to make their information more accessible to readers.

Subsequent Drafts

Figure 3.8 shows the next stage in the collaboration. In this version of the memo, prepared through several revisions over a two-day period, the authors incorporated McCraw's suggestions as well as several changes of their own. It was this revision that they submitted to Vice President Schuster, who also made some comments on the memo. Schuster's suggestions—all valid—show how different readers can help writing teams meet their objectives.

Note that in the process of revising their memo Chappel and Garcia had to make major changes from the first draft in Figure 3.7. Their work went through several versions to arrive at the document in Figure 3.8. Those changes—shortening and expanding paragraphs, adding and deleting information, and refocusing their approach to meet the needs of their audience—are the essential revisions a collaborative writing team, like an individual writer, would expect to make.

(Continued)

FIGURE 3.7 Early Draft of the Chappel and Garcia Collaborative Memo, with Revisions in Blue Suggested by Wells McCraw, Their Manager

FENTON COMPANIES

TO:	All Employees
FROM:	Abigail Chappel; Manuel Garcia
DATE:	February 6, 2012
RE:	Our Recycling Program

Vague—"Improving"

This ¶ is too long. Too many topics—costs, protecting the environment. Keep it short—say what we are doing and why

An in-house study has shown that Fenton sends approximately 26,000 pounds of paper to the landfill. The landfill charge for this runs about $2,240, which we could save by recycling. Fenton Companies is conscious of our responsibility to save and protect the environment. Accordingly, starting March 1 we will begin a more intensive paper recycling program.

Start off with this key idea

Our program, like many others nationwide, will use the latest degradable technology to safeguard the air, trees, and water in our community. It has been estimated that of the 250 million tons of solid waste, three-quarters of goes to landfills. These landfills across the country are becoming dangerously overcrowded. Such a practice wastes our natural resources and endangers our air and drinking water. For example, it takes 10 trees to make 1 ton of paper, or roughly the amount of paper Fenton uses in four weeks. If we could recycle that amount of paper, we could save those trees. Recycling old paper into new paper involves less energy than making paper from new trees. Moreover, waste sent to landfills can, once broken down, leach, seep into our water supply, and contaminate it. The dangers are great.

Word "it" left out

Delete — not relevant to our purpose

Check your facts; I think it is closer to 16–17

No cap

By enhancing our recycling, we will not be sending so much to the Springfield Landfill and so help alleviate a dangerous condition there. We will keep it from overflowing. Fenton will also be contributing to transforming waste products into valuable reusable materials. Recycling paper in our own office shows that we are concerned about the environmental clutter. By having an improved paper recycling program, we will establish our company's reputation as an environmentally conscious industry and enhance our company's image.

Add the fact about our saving trees in this ¶

Delete — makes us look bad

Start ¶ with this point

Fenton is primarily concerned with recycling paper. The 200 old phone books that otherwise would be tossed away can get our recycling program off to a good start.

When? How? Implications for saving/costs?

Give some examples

We encourage you to start thinking about the additional kinds of paper around your office/workspace that needs to be earmarked for recycling. When you start to think about it, you will see how much paper we as a company use.

FIGURE 3.7 (Continued)

↑ not into your wastebaskets

Starting the last week of February, paper bins will be placed by each office door inside the outer wall. These bins will be green—not unsightly and blending with our decor. Separate your waste paper (white, colored, and computer) and put it into these bins. You do not need to remove paper clips and staples, but you must remove rubber bands, tape, and sticky labels. They will be emptied each day by the clean-up crew. There will also be large bins at the north end of the hallway for you to deposit larger paper products. The crucial point is that you use these specially marked bins rather than your wastebasket to deposit paper.

Delete — repetitious

¶ lacks effective "call to action" and incentives

Fenton Companies will deeply appreciate your cooperation and efforts. Thanks for your cooperation.

This memo needs more work
  — add at least one ¶ on how recycling will save us money
  — insert headings to separate sections

  — make directions clearer and easier to follow; try using numbered steps

  — end on a more upbeat note; tell employees about the benefits coming to them for recycling — i.e., a bonus/our contribution to their favorite charity

  — why don't we say something about America Recycles Day, Nov. 15

© Cengage Learning 2013

Final Copy

With McCraw's further input and their own revision, Chappel and Garcia submitted the final, revised memo found in Figure 3.9 to the vice president a few days later. This final copy received Schuster's approval and was then routed to the Fenton staff.

Thanks to an integrated model of collaboration and careful critiques by McCraw and Schuster, Chappel and Garcia successfully revised their work. Effective team effort and shared responsibility were at the heart of Chappel and Garcia's assignment.

(Continued)

FIGURE 3.8 Revision of the Chappel and Garcia Memo, with Changes Suggested by Vice President Schuster

FENTON COMPANIES

TO: All Employees
FROM: Abigail Chappel; Manuel Garcia
DATE: February 10, 2012
RE: Improving Our Recycling Program

Make headings more precise

To save money and protect our environment, Fenton Companies will begin a more intensive waste paper recycling program on March 1. It has been estimated that three-quarters of the 250 million tons of solid waste dumped annually in America could be recycled. Our program will still use the latest recycling technology to safeguard trees, air, and water.

New Program

Say strengthen or continue

¶ needs more information

This new program will ~~establish~~ Fenton's reputation as an environmentally conscious company. Fenton now uses one ton of paper every four weeks. This represents 17 trees that can be saved just by recycling our paper waste. Recycling also means we will send less waste to the landfill, alleviating problems of overfill and reducing the potential for contamination of the water supply. Waste sent to large landfills can leach and seep into water systems. By recycling paper, Fenton will also reduce the risk of long-term environmental pollution.

Add that we no longer use Styrofoam and that our suppliers use only biodegradable products

Saving Money

Indicate how much we save

Recycling saves us money. Right now Fenton pays $180 per month to dump 2,000 pounds (one ton) of paper waste at the landfill. However, scrap paper is worth $100 per ton. Recycling will generate $100 each month in new revenue while eliminating the $180 dumping expense.

This sentence more logically goes in ¶ 3

Ultimately, the success of our project depends <u>on renewed awareness of the variety of office paper suitable for recycling.</u> Paper products that can be recycled include newspapers, scrap paper, computer printouts, letters, envelopes (without windows), shipping cartons, old phone books, and uncoated paper cups.

Put in itemized bulleted list to stand out better

Do not start new ¶ here; keep as part of previous ¶

<u>Recycling 200 or so old phone books each year alone will save 22 cubic yards of landfill space (or close to $100) and will generate $25 in scrap paper income for our company.</u> The old 2011 phone books, which will be replaced by new 2012 ones on March 1, will give us an excellent opportunity to intensify our recycling.

FIGURE 3.8 (Continued)

Directions

Here are some easy-to-follow directions to make our recycling efforts even more effective:

Boldface these words 1. Starting the last week in February, paper bins will be placed inside each office door. Put all waste paper into these bins, which will be emptied by maintenance.
2. Place larger paper products—such as cartons or phone books—in the green bigger paper bins at the end of each main corridor.
3. Put white, colored, computer printout, and newspapers into separate marked bins. Remove all rubber bands, tape, and sticky notes.

Thanks for your cooperation. To show our appreciation for your help, 50 percent of all proceeds from the recycled paper will go toward employee bonuses and the other 50 percent will be given to the office's favorite charity. The benefits of our new recycling program will more than outweigh the inconvenience it may cause. We will conduct another study in time for America Recycles Day on November 15 to determine how our company has improved in this area.

End with the incentive for emphasis

Add another sentence to this ¶ on how a safer environment will benefit our company and the employees, too

(Continued)

FIGURE 3.9 Final Copy of the Memo Prepared by Chappel and Garcia Using an Integrated Model of Collaboration

FENTON COMPANIES

TO:	All Employees
FROM:	Abigail Chappel; Manuel Garcia
DATE:	February 13, 2012
RE:	Improving Our Recycling Program

Why We Need to Increase Recycling Efforts

To save additional money and to protect the environment, Fenton will begin a more intensive waste paper recycling program on March 1. For the new program to succeed, we need to increase our paper recycling efforts. It has been estimated that three-quarters of the 250 million tons of solid waste dumped annually in America is still not being recycled. Our program will continue to rely on the latest recycling technology for better greening of the workplace and to safeguard trees, air, and water.

Recycling has already saved us money. Right now Fenton pays $180 per month to dump 2,000 pounds (one ton) of paper waste at the landfill, compared to $360 last year. However, scrap paper is worth $100 per ton. Increased recycling will generate $100 each month in revenue while eliminating the $180 dumping expense.

Fenton's Image as a Green Company

This new, more intensive program will further strengthen Fenton's reputation as an environmentally conscious company. Three years ago, we stopped using Styrofoam products and asked suppliers to use biodegradable materials for all our shipping containers. Our efforts have proved successful, but Fenton still uses one ton of paper every four weeks. This paper represents 17 trees that can be saved just by recycling our paper waste. We need to send even less waste to the landfill, alleviating problems of overfill and reducing the potential for contamination of the water supply. Waste sent to large landfills can leach and seep into water systems.

Paper Products to Recycle

Ultimately, the success of this project depends on our being aware of the variety of office paper suitable for recycling. An expanded list of paper products that can be recycled includes

- newspapers
- letters
- envelopes (without cellophane windows)
- phone books
- uncoated paper cups
- scrap paper
- computer printouts
- junk mail (only black print on white paper)
- shipping cartons

FIGURE 3.9 (Continued)

The old 2011 phone books, which will be replaced by new 2012 ones on March 1, will give us an excellent opportunity to launch our intensified recycling program.

Steps to Follow

Here are some easy-to-follow directions to make our recycling efforts even more effective:

1. Starting the last week in February, put all waste paper into the **paper bins** that will be placed **inside each office door.** These bins will be emptied by maintenance.

2. Place **larger paper products**—such as bulky cartons—in the **green bigger paper bins** at the end of the main corridor.

3. Remove all rubber bands, tape, and sticky notes and put white, colored, computer printout, and newspapers into separate marked bins.

We will conduct another study in time for America Recycles Day on November 15 to determine how our company has improved in this area. The benefits of our new recycling program will more than outweigh the inconvenience it may cause. A safer environment—and a more cost-effective way to run our company—benefits us all.

Thanks for your cooperation. To show our appreciation for your increased efforts, 50 percent of all proceeds from the recycled paper will go into an annual employee bonus fund and the other 50 percent will be given to the office's favorite charity.

Computer-Supported Collaboration

To be successful writers, employees must be proficient in using multiple types of collaborative software applications, otherwise known as *groupware*. Using groupware such as Microsoft Word and Google Docs, for example, you can plan, draft, revise, and edit a document collaboratively at the same time or from different locations. By sharing files, every team member can read, comment on, alter, and add text at any stage in the writing process. Groupware allows you to track all the changes you and others on your team make. You will be expected to know—or at least to be willing to learn—how to use email for collaborative communications, navigate document tracking systems such as Microsoft Word's Track Changes feature, and manage group editing and writing projects using Web-based collaboration systems.

Advantages of Computer-Supported Collaboration

Computer-supported collaboration offers significant advantages to both employers and workers. Here are some of those benefits:

1. **Increased opportunities to "meet."** Today's meeting-heavy and travel-filled work environments often make collaborating in person difficult and costly. Various types of groupware provide a virtual shared workplace where team members can ask questions, share information, make suggestions and revisions, and troubleshoot.

2. **Reduced stress in updating new group members.** Groupware can efficiently and quickly bring new team members into the ongoing conversation, saving time and effort and reducing the stress of face-to-face meetings.

3. **Expanded options for communicating worldwide.** Groupware ensures the flexibility to contact group members anywhere—at a home office (telecommuters, freelancers), on the road, at another office, or in another country.

4. **Improved feedback and accountability.** Because it is accessible and easy to use, groupware helps team members participate actively in the collaboration process. Groupware encourages team members to get to the point and make sure other members understand their message clearly and quickly. Because each member's contributions are documented, everyone's participation becomes a matter of record.

5. **Enhanced possibility of complete and clear information.** Groupware allows everyone involved in the team to truly be "on the same page." It provides a running record, thus saving all information and preventing misinterpretation.

Groupware and Face-to-Face Meetings

Groupware does not, of course, completely eliminate the need for a group to meet in person to discuss priorities, clarify issues, or build team spirit. But in the world of work, face-to-face communications, although sometimes essential, are frequently accompanied by computer-supported collaboration via groupware. In fact, groupware can enhance meetings by allowing people to draft and discuss documents online *before* they arrive at a meeting, thus saving time and increasing efficiency. This technology also allows teams to extend their collaboration beyond the confines of the workplace, whether at an off-site location or at a telecommuting employee's home. For instance, you and your colleagues might use Track Changes to revise a progress report after an important meeting, or you might use Google Docs to help your supervisor make last-minute edits to a PowerPoint presentation before an important trade show.

Types of Groupware

There are essentially three types of groupware commonly used to produce collaboratively written documents in today's workplace: (1) email, (2) document tracking software, and (3) Web-based collaboration systems, which include wikis and online word processors. Email is widely used in the workplace to share brief comments and suggestions among team members, but for more complicated collaborative writing and editing tasks, document tracking software allows teams to more efficiently "show" rather than "tell" how a document might be improved. A team might also use a wiki to share ideas about how to draft a document, but an online word processor like Google Docs would better allow those ideas to be quickly turned into a usable document. Depending on your employer's instructions and the needs of your team members, you will have to decide which groupware tool is the most appropriate for the kinds of collaboration you must do in the workplace.

Email

Email is used for many jobs in the world of work, as you will see in Chapter 4. It has an important role to play in collaborative writing online as well. While email is not the place to create or revise a collaboratively written document (because edits and other changes are hard to incorporate and keep track of), email nevertheless makes online collaboration possible for the following reasons:

1. Email is used to send collaboratively written documents as text files or PDF files. Team members from around the globe can then access the same document and share their feedback.
2. Email helps group members to conduct group business—setting up meeting times, notifying members about the status of a report, alerting members to a problem, and so on.
3. Email can be used to make sure every member of the team is working on the same document by identifying each document by name and number in the "subject field" (*Report on Parking, Rev. 4* or *Proposal on Recycling, Draft 2*).
4. Email saves the group time by decreasing the number of face-to-face meetings it must have. Email cannot take the place of a face-to-face exchange, however.

Document Tracking Software

Document tracking software, such as Microsoft Word's Track Changes feature or Adobe Acrobat, provides another way for collaborative writing teams to share, comment on, and revise their work online. Figure 3.10 shows an example of a collaboratively written document using Microsoft Word—the first draft of a section of a report on increasing parking spaces at a hospital—and how it has been revised and edited by several team members using the Track Changes feature. (Keep in mind, though, that depending on the document tracking system your company is using, your tracked documents may look somewhat different.)

Sent as an email attachment to every team member, Figure 3.10 preserves all the original text of Draft 1 while automatically showing and identifying comments from each individual. Notice that each of the three readers' initials and his or her changes and comments are "tracked" in the markup "balloons" on the right. (Microsoft Word automatically assigns a different color to each team member so that his or her edits and comments are easily distinguishable from each other.) This tracked document is then shared by other team members, who can tag areas of the document that they would like to comment on. Team members can insert headings, clarify and verify factual data, call for visuals, and ask for and even supply new text, as Mary Ming (MM)

FIGURE 3.10 Collaborative Editing Using a Document Tracking System

Draft 1 — ~~Proposal to Expand Hospital Parking Facilities~~ How We Can Expand the Hospital's Parking

~~CGH~~ Community General Hospital needs to expand its parking facilities. Right now there is just too little room for visitors, staff, and patients. These inadequate parking facilities are a ~~detrime~~ detriment to the overall growth of pt. care. They have been an important talking point since the inception of this committee. Maybe they were ok when the hospital opened its doors in 1973, but not today. ~~CGH is a good place to work and t~~The new parking facilities would definitely benefit the present staff and visitors from walking ~~long distances~~ two blocks in the rain and ice.

The exact number of new spots is hard to estimate now but I am thinking around 500 might be just right. The problem is the traffic flow around the hospital. While the new parking facilities would alleviate it, it also raises a central question about how to get it done. Perhaps Wentworth Avenue, East to West, might be turned into a one-way street. That way we could add up to 11 new spots in the front of the ER and thus resolve the congestion that has hampered easy ~~egress~~ entrance and ~~ingress~~ exit. Another possibility worth considering is changing Taylor Street—right now it is a two-way street and we could make it one-way West to East.

At any rate, the traffic flow is a key issue the hospital needs to solve if it is to expand its parking facilities. But there are other important engineering problems that must be solved. Eleanor Yi, the hospital engineer, has studied the stress points, pre-cast concrete, and the slope of vehicular access ramps that would accommodate increased traffic flow. As you can see, she believes that the hospital does not have the space to locate all the new parking spots in one plane area. She recommends a two-story structure and believes the North side of the ER might be the best place.

LB 5/5/11 10:13 AM
Comment: Centered and rephrased the title to match house style for internal proposals.

KT 5/5/11 8:58 AM
Comment: Let's use the full name of the hospital here.

MM 5/5/11 11:28 AM
Comment: We should cite source and statistics. See hospital report CGH-GR-2010

KT 5/5/11 9:00 AM
Comment: We need to take out abbreviations in the final copy.

LB 5/5/11 10:15 AM
Comment: We must be precise. It was 417, according to the engineering proposal.

MM 5/5/11 11:38 AM
Comment: I think it might be best to start a new section here and title it "Increased Traffic Flow."

MM 5/5/11 11:39 AM
Comment: We want to include an alternate plan. I've inserted text here to address this.

LB 5/5/11 10:17 AM
Comment: Let's put her documentation in an appendix.

MM 5/5/11 11:40 AM
Comment: That's a good idea; her solution appears workable and falls within our budget.

KT 5/5/11 9:04 AM
Comment: I don't understand this terminology. Should we use a different phrase here?

LB 5/5/11 10:19 AM
Comment: We have to provide further information about the number of ramps and perhaps confirmation from engineering consulting firm to corroborate Yi's findings.

does with her newly inserted text in Figure 3.10. Team members can also edit sentences in the document, and Word will automatically track these changes.

The team leader or group member responsible for working on revising the next version of this section of the report can accept or reject the tracked changes after they are agreed on by the group. This is where the dynamics of collaborative writing must work for the good of the entire group and to produce the document on time. After the changes have been made to Draft 1, the group will then receive a clean document of Draft 2 (free of tracked changes/comments) and can continue with the writing process leading to a final draft.

Web-Based Collaboration Systems

Professionals use a wide variety of Web-based applications to write collaboratively at work. These include wikis and online word-processing applications like Google Docs.

Web-based systems are quite different from stand-alone word-processing software. When you edit the work of another person using Microsoft Word, for example, you complete the work on your computer and then share the file electronically with your editing partners, usually by sending the document as an attachment to an email. When you use a Web-based collaboration system, however, the work is automatically shared between you and your various collaborators—no emails or attachments are required. Web-based collaboration systems create a virtual room where you and your collaborators can deposit files, gather to review one another's work, offer suggestions, and communicate ideas, and store your files and revisions.

This common space makes communication between collaborators more direct; it also makes collaboration between more than two people more efficient. An email exchange among three or four team members over a round of edits could generate more than a dozen emails as well as several electronic versions of a document, making the team's job very difficult. Most Web-based collaboration systems, however, will consolidate all of this communication into a single site on the Web for every team member to see. Workplace professionals use these systems to manage complicated editing projects and receive feedback from people who might be at different branches of a company.

Wikis   The wiki was one of the first online applications that allowed for collaboration on the Web. Wikis are similar to document tracking systems, but they have a few crucially different characteristics. First, wikis are not features found within a software package such as Microsoft Office. They are websites for which team members are given access passwords, enabling them to check documents in and out of the site. Second, wikis typically do not show tracked changes right on the document. Instead, when an edited document is uploaded back to the website, each version is assigned a new version number. Versions of the document can be compared and the differences between versions can be viewed, but these differences will not show up within a single version.

The advantage of wikis over tracked documents is that subsequent versions of a document are as easy to read from start to finish as the first draft. Each wiki version is a clean document free of complicated tracked edits. Therefore, when team members are revising, they will need to proofread only the final draft, rather than go through the laborious process of accepting or rejecting changes, deleting comments, and troubleshooting lingering inconsistencies.

The disadvantage of wikis, when compared with tracked documents, is that wikis may result in a lack of quality control. Because each group member's changes are not clearly tracked, it may be difficult for group members to keep up with the succession of changes. When using wikis, teams should set up a clear protocol outlining who may make changes to the document and when.

Online Word-Processing Systems   Online word-processing systems are another increasingly common type of Web-based collaboration tool. Applications like Google Docs, Office 365, Adobe Buzzword, and Zoho Writer allow teams of writers and editors to share and edit a variety of documents easily on the Web. Although Office 365 offers all of the features of Microsoft Office, there is a monthly fee for Office 365 whereas Google Docs is free. You might use Google Docs to write a report with a group of colleagues or a proposal about launching a new product or service. These systems are essentially online word processors that bundle file sharing, online collaboration, word processing, and document design features. Like wikis, your writing and editing are done on websites rather than within a software package like Microsoft Office. Unlike wikis, though, these systems also allow for change tracking and document design.

One of the most widely used of these various systems is Google Docs. This free application can be accessed and used through the Google site by anyone with a Gmail password to create documents, thus allowing only authorized individuals to access and revise them. Google Docs provides the following collaborative advantages:

- It allows editors and writers to create, share, revise, and comment on documents on the Web.
- It records a complete revision history of any changes made to the document.

How Google Docs Can Streamline Group Editing

- Google Docs functions like a word processor, so the user can build publishable documents with it (a feature most wikis do not include).
- It provides a chat window in the interface for real-time collaboration online.
- It safely stores documents in a secure space online, accessible from any computer.

Google Docs is increasingly used in the workplace as a collaborative tool because it streamlines the editing process. A collaborative writing project using Microsoft Word, on the one hand, requires that many separate versions of documents be sent back and forth via dozens of emails among team members. Google Docs, on the other hand, bundles all the revisions into a common, shared, and secured space. Enhancing collaboration even further, writing partners can communicate with one another in real time using a "chat" window, such as that seen in Figure 3.11. Google Docs and other online word processors are especially helpful when you need a fast turnaround on collaboratively written and edited documents.

The case study on pages 114–115 explains how an inventory manager for a lighting supply company used Google Docs to quickly create a proposal for expanding her company's warehouse space. The "problem statement" for this proposal can be seen in Figure 3.11.

Using Web-Based Collaboration Systems Safely and Efficiently   Web-based collaboration systems create powerful opportunities to write collaboratively outside the physical boundaries of the workplace, whether at off-site locations, at an employee's home, or even across continents, but they must be used carefully. All of these systems are password protected, allowing users to create their own user names and passwords. You need to follow the same security measures regarding passwords that you would with any other collaborative program: Don't share passwords with others and keep them protected from view.

Also, keep the number of your collaborators manageable. Never share a document with more than your collaborative team. Ideally, team members should have a working relationship with one another (see pages 90–91). If you share the document with others outside your group, you risk breaching the trust of your team members and the security of your document. Because groupware links people together into work partnerships, the effective use of this technology depends on commonly agreed-on rules. Establish clear guidelines for how the editing process will work using the most appropriate models for computer-supported collaboration (see pages 116–117).

Models for Computer-Supported Collaboration

Employees using groupware will find that the *integrated* model (see below) and the *sequential* model (see page 116) are the most effective models for computer-supported collaboration.

The Integrated Model

1. **All members of the team work on the entire document—developing, expanding, revising, and editing it.** The original draft is saved so that it is accessible to all team members who will work on it. When using email, the team should be careful not to lose sight of the original draft. But when employing a document tracking system, writers/revisers need to look at the original draft and continue to edit it. When using a wiki, the original draft will remain available separately as subsequent versions are uploaded.

2. **Each member of the team keeps the rest of the team in the loop when he or she makes changes to the document.** When using email, the team member making the change should copy all other members of the team on the email. But when using a document tracking system, each team member should inform the others while he or she is making changes so that two writers aren't working on the tracked document at once. With a wiki, however, only the person currently making changes will be able to access the document, since wikis automatically lock out other users when one team member has the document checked out.

3. **Each team member looks at each draft of the document as it is edited.** When using email, team members should comment on each draft when it is emailed to the entire team. Editing through a document tracking system, each subsequent reader/writer should insert comments, noting where he or she agrees or disagrees with the

Case Study

Using Google Docs as a Collaboration Tool

Allen Knutz is the inventory manager for Lightofmylife.com, a company that specializes in designer lights and in a popular line of "green" lighting fixtures, lamps, and lightbulbs. The company has grown recently because of greater customer demands for environmentally friendly products and for compact fluorescent bulbs and LEDs. As the company expanded, its warehouse space became inadequate. They needed more floor space for their inventory and additional forklifts to accommodate an accelerated delivery schedule. To address these needs, the CEO of Lightofmylife.com has asked Knutz to write a proposal about expanding their warehouse space.

Writing proposals has always been a collaborative project at Lightofmylife.com, with teams of four or five employees under the direction of a primary supervisor who manages the process. This proposal was particularly challenging because the CEO wanted it in less than a week to take advantage of available storage options. In the past, Knutz collaborated with his colleagues using the Track Changes feature in Word, but he knew that this process generates numerous email exchanges as well as multiple versions of the same document that everyone must open and read. This type of groupware collaboration works best when the team has more time.

Why Use Google Docs

Knutz decided that the group needed to use Google Docs to streamline their collaboration. Google Docs allows team members to create, share, revise, and comment on an evolving document quickly and efficiently and produce the final version—all using the same Web-based groupware application. Given the fact that some of his team would be off-site involved in sales visits and trade shows and another member was telecommuting, Knutz decided that his team would have to meet online at least three or four times during the week to participate in real-time editing through the chat room feature, rather than through a back-and-forth email exchange that would take much longer.

How to Set Up and Use Google Docs

Knutz did the initial setup for the proposal on Google Docs and then was ready to share the document to be created with his colleagues. He keyed in their Gmail addresses in the "share" window. Figure 3.11 shows what the "problem" portion of the proposal looked like in Google Docs. Before collaboration began, however, Knutz established some guidelines. Each team member could comment on the document, but Knutz was the only editor who could actually make the revisions. Using a sequential model for collaboration, he stressed that each team member would write a different section of the proposal but that everyone was responsible for suggesting changes through the comments feature for the entire document at various stages. Though his team was working at different locations, they were able to complete a successful proposal in time for the CEO's deadline. Accordingly, by the end of the next quarter, Lightofmylife.com had moved their inventory into a larger warehouse space and had purchased three new forklifts.

FIGURE 3.11 Using Google Docs to Collaborate on a Document

Google Docs records a revision history of all changes made to the document in a chronological list, allowing you to easily access different versions.

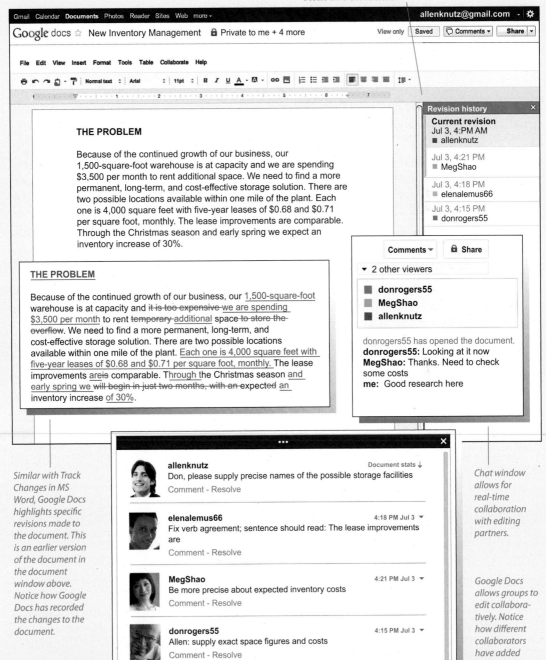

Similar with Track Changes in MS Word, Google Docs highlights specific revisions made to the document. This is an earlier version of the document in the document window above. Notice how Google Docs has recorded the changes to the document.

Chat window allows for real-time collaboration with editing partners.

Google Docs allows groups to edit collaboratively. Notice how different collaborators have added comments to this document.

previous editor's comments. With a wiki, though, each member of the team should look at each draft as it is uploaded and provide comments or express concerns to the rest of the group, which can be either fixed by the most recent editor or passed on to the next editor.

4. The team agrees on the final draft. If the steps above for email, document tracking systems, or wikis are followed, the final draft can be produced relatively smoothly. When editing via email or a document tracking system, the team must select one or more members to consolidate their various drafts. When using a wiki, the team has to agree to the second-to-last draft so that any changes to it will become the final draft version. In any instance, the final draft should be proofread one more time by all members of the team, to resolve any loose ends.

The Sequential Model

1. Each member of the team is responsible only for drafting his or her assigned section of the document. At this point, the writer is simply word processing, not using groupware. The team should choose a team leader, however, who can act as an overall editor (see step 4).

2. Each team member sends his or her completed section to everyone else in the group. Since the sections will remain separate for the time being, email should be used (it would be premature to use a document tracking system or wiki).

3. All members of the group edit the individual sections and return the revised sections to their respective authors. Email continues to be the most appropriate medium to convey these edits.

4. Each team member revises his or her section based on team comments and sends the revised version to the team leader. The team leader then combines the reviewed and edited sections into a single document, reads through the entire document for consistency of style and format within each section, and, if additional group feedback is necessary, sends the complete document to the entire group via email (in which case everyone can respond at once), as a tracked document (then each member would need to make edits and add comments in turn), or posted on a wiki (where only one member at a time will be able to check out the document and make changes).

5. The team leader finalizes the document. Now that all team members have not only written and revised their sections but also had a chance to edit and comment on the complete document, the team leader has all of the information he or she needs to finalize and submit the document.

Avoiding Problems with Online Collaboration

Regardless of the online collaborative method your team uses, it must establish ground rules by which documents are created, posted, shared, protected, and submitted. By following these guidelines, your team can avoid common problems in its online collaboration:

1. Be sure that all team members have access and authorization. Team members cannot give input if they cannot open, edit, save, or share a document. To do this, every team member must use the same software.

2. Everyone in the group must be "in the loop." Verify that you and your team members are all working on the same (and correct) version of the document at the same time. Problems result when a team member wastes time and delays a deadline by editing an earlier or otherwise incorrect version.

3. Save the original draft and subsequent ones in separate files in case the team needs to return to these earlier copies to verify that changes have been made.

4. Link each revision with the individual who made it. Set up the software's preferences or callout properties to track each contributor's changes in a unique, identifying color and/or by using initials to identify who has made any changes, as in Figure 3.10.

5. Maintain confidentiality to protect the document from unauthorized users. Do not send team emails or drafts to individuals who are not a part of the team. Team members should be cautioned not to reveal their passwords to anyone.

6. Require all team members to sign off on and agree to the complete, final document. This procedure creates a system of checks and balances to ensure quality control.

Tech Note

Virtual Meetings: Teleconferencing, Web Conferencing, and Videoconferencing

While face-to-face meetings are frequent in the workplace, technology has made virtual meetings an important option. Face-to-face meetings certainly have advantages, such as promoting group interaction, fostering close working relationships, and bypassing technical difficulties. However, sometimes one or several individuals from the group are traveling, work at a remote site, live overseas, or are unable to attend a face-to-face meeting for another reason (weather, missed flights, etc.). For these reasons, as well as for convenience and to save money on travel costs, meetings in the world of work are increasingly conducted in part or entirely over the telephone or via the Web.

Following is a brief description of three ways in which technology brings members of a group together in today's virtual office:

■ **Teleconferencing:** Teleconferencing is by far the most common way to conduct a virtual meeting. Teleconferencing has been a part of the workplace for as long as speakerphones have expanded conversations beyond two people. Teleconferencing systems allow for conference calls in which multiple participants at your office, across the country, or around the globe can speak to one another. Companies that use teleconferencing on a large scale may purchase their own advanced telephone switchboard

(Continued)

equipment to conduct conference calls. Most businesses, however, use a teleconferencing service. In either case, each individual has to be notified of the call-in telephone number and password to participate in the teleconference.

■ **Web conferencing:** Web conferencing combines the audio component of teleconferencing with the face-to-face interaction of a traditional meeting. The greatest advantage of Web conferencing over teleconferencing is that individuals attending a Web conference can view presentations and share documents electronically during a meeting. The computer's microphone and speakers provide the audio component. For the visual component, a computer-mounted video camera (webcam) feeds its images in real time to the computer network. Web conferencing is different from teleconferencing in that all of the participants attend the virtual meeting from their own computers and are connected to one another over the Internet rather than through the company's telephone switchboard. Companies hosting Web conferences can use fee-based sites such as Citrix GoToMeeting, WebEx, or Zoho. Participants meet in "real time" by logging in to the specific site, typing their password, and downloading the software. Free applications such as Skype work well for smaller Web conferences of up to 24 participants (see pages 120–121).

■ **Videoconferencing:** Videoconferencing systems allow groups to meet over distance, enabling more productive meetings and real-time decision making. Today, dedicated videoconferencing systems are primarily used for group-to-group conferences or one-way seminars in large rooms and auditoriums, bringing large groups of people together to dynamically share information without the expense and time of traveling. Dedicated video conferencing systems generally have all the required components in a single piece of equipment. Usually, this unit contains a high-quality video camera controlled remotely to pan around a room, tilt up and down, and zoom. It also includes all electrical wires and cables, the control computer, and the software. Multidirectional microphones pick up sounds throughout a room, while a large TV monitor scans the room as well.

Meetings

One of the most frequent ways to collaborate is through meetings, which can take the form of small group discussions or large, formal conferences. Whether it is regularly scheduled (a weekly staff meeting) or a special, unscheduled one, a meeting requires teamwork. Collective energy and goodwill will bear much fruit. The guidelines on collaborative writing (pages 88–91) also apply to group interactions at meetings. Basically, you need to know how to plan a meeting, create an agenda, write minutes, and take notes.

Planning a Meeting

As with a collaboratively written document, meetings have to be carefully planned in order to be successful. And both written documents and meetings need someone to organize them. A group of individuals cannot simply gather and start a conversation; they need to have some focus, some guidelines, some individual to help organize what is to be said, how, why, and when. If you have the responsibility of planning a meeting, you need to consider the following questions before you even convene the meeting:

1. What is the purpose of the meeting? Determine why the meeting is necessary, what essential topics need to be discussed, and what results or outcomes the meeting should accomplish. Jot down your ideas, which you will later use for setting the agenda.

2. Who should attend the meeting? Identify key people who need to be there, including managers, co-workers, colleagues from other departments, and any individuals outside your company (clients, vendors, etc.). Keep in mind that not every manager needs (or wants) to be present at every meeting. However, as a matter of policy, find out from your boss if management needs to be included. If someone cannot attend, solicit that person's input through email or include him or her via teleconferencing or videoconferencing (see pages 117–118 for more on these topics).

3. What specific responsibilities do individuals have? Determine who will take minutes (see pages 122–123), make an introduction, or deliver a PowerPoint presentation, for example.

4. When should the meeting take place? When planning a meeting time, have copies of each team member's schedule to help you avoid conflicts. Also consider that there are good times and bad times to hold a business meeting, as the following schedule shows:

Good Times	Bad Times
1. Mid-morning or mid-afternoon	1. Early in the morning or late in the afternoon
2. Any time during the week except Monday morning or Friday afternoon, or immediately before or after a major holiday	2. Monday morning or Friday afternoon
	3. Immediately before or after a major holiday
3. Shortly after a major company celebration when morale is high	4. Same day as a long training session or long meeting

5. Where should the meeting take place? Select an appropriate space: A meeting for eight people does not need to take place in an auditorium, while a meeting for ten people in a room with only six chairs will not work. In addition, make sure the room you choose is equipped with all the technology you, your speakers, or your off-site members plan to use, such as speakerphones and videoconferencing hook-ups (see pages 117–118).

6. What documents need to be presented at the meeting? Collect any test results, reports, client communications, statistics, models or blueprints, maps, and so on that are relevant for discussion. Make sure that electronic or hard copies of any documents you plan to discuss are available for the group to review ahead of time. Always specify when and where they can be read. If you plan to use presentation software, such as PowerPoint slides (see pages 730–733), make sure you review them ahead of time to fix any problems.

Creating an Agenda

Out of your planning will come your *agenda,* a list of the topics to be covered at the meeting. An agenda is a one- or sometimes two-page outline of the main, pertinent points. The agenda should list only those items that your group, based on its work and interaction, regards as most crucial. Prioritize your action items so that the most important ones come first. Your agenda might also include short reports or presentations for which one or two members of your group are responsible. Always distribute the agenda ahead of time (at least a day or two) so that your team will be prepared and better able to contribute.

Tech Note

Videoconferencing with Skype

Skype is a widely used software application that allows individuals to conduct videoconferences over the Internet without purchasing expensive videoconferencing systems. Skype functions like a telephone call placed from one party to another. But it uses your computer's built-in or extended webcam to send video to your Skype partner(s). You can download the free software from www.skype.com. You can easily add contacts to Skype from your email address book. A green checkmark indicates when one of your Skype contacts is online. Click on the name and click "Video Call" or "Call" and your computer will call that person's computer.

Because it is easy to use, Skype is an important tool for one-to-one meetings as well as for small group conferences (up to 24 participants). You can use it to talk to and see someone in the same office, in different locations, or even on different continents. Like other ways to videoconference, it gives you the benefits of a face-to-face meeting, allowing participants to read body language as well. Figure 3.12 shows three individuals engaged in a Skype conversation.

Like any other business meeting, you have to prepare for a Skype session. Just because it is not as elaborate as other videoconferences does not release you from your responsibility to have your notes and questions ready for the meeting. Follow these guidelines to ensure that your Skype conversations will be productive.

- **Plan the session.** Decide ahead of time on the topics, issues, and problems you want to discuss. Prepare an informal agenda or at least a list of major subjects to be covered.
- **Share information about the meeting.** To be effective, all the Skype partners need to be aware of the subject and goals of the meeting. Email this information to them, as well as the length of the session, in advance of the meeting. That way all your Skype partners will be better prepared and focused.

FIGURE 3.12 Three Individuals Engaged in a Skype Conversation

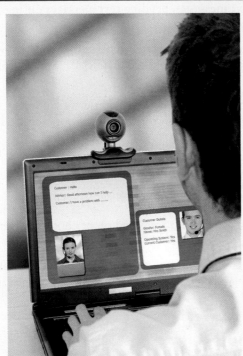

Andresr/Shutterstock.com

- ■ **Arrange a precise meeting time.** Be sure that you and your partner(s) have your Skype application window open and that you are ready to receive a call.
- ■ **Test your computer's webcam and audio.** Always test your camera to make certain it is working (and is in focus) before making a Skype call, and ask your partner(s) to do the same. In your Skype contact list, click on "Echo/Sound Test Service" to test your audio settings.
- ■ **Use file sharing to send files through Skype that are relevant to your meeting.** Click the "Share" button to share materials to be discussed at your meeting with your Skype partner(s).

Observing Courtesy at a Group Meeting

To show your team members the courtesy they deserve during a meeting, observe the following guidelines:

1. **Be on time.** Coming in late can disrupt a meeting and cause further delays if you have to be brought up-to-date.

2. Silence your cell phone. Set your phone on vibrate if you are expecting an important business call.

3. Do not send text messages. Sending text messages during a meeting is as rude as playing a video game.

4. Avoid side conversations. Talking to others around you while the meeting is in progress shows poor manners and can distract from the business and progress of the meeting.

5. Avoid interrupting. If you need to interject a comment, raise your hand or wait for the meeting planner to ask for comments from the group.

6. Be an active listener. Pay attention to what is being said at the meeting and also try to discern the message behind those words.

7. Participate; don't dominate. Give everyone a chance to speak. Contribute to the discussion, but don't monopolize it.

8. Be a focused speaker. If you are called on or choose to speak to the group, get to the point quickly, be clear, and avoid straying from your topic.

9. Do not record the meeting or take photographs. Unless you have permission, do not bring an audio recorder or video camera to a meeting, and do not take photographs.

Writing the Minutes

The *minutes* are a summary of what happened at the meeting. Transmit copies of the minutes to the team members to have them approved, to help them recall what happened at the meeting, and to help them prepare for the next one. Copies of minutes are kept on file—they are the official, permanent record of the group's deliberations and are regarded as legal documents. They may be used in courts to verify that a problem existed, that a solution was suggested, or that a specific action was or was not carried out.

Accordingly, minutes must be clear, accurate, and impartial. Keep them free from your own opinions of how well or poorly the meeting went; for example, "Saunders customarily offered the right solutions" or "Once more Hicks got off the topic" are not appropriate comments. Because the person chairing a meeting cannot take minutes and preside at the same time, another member of the group designated as the secretary should prepare the minutes. Minutes are transmitted usually within twenty-four to forty-eight hours after the meeting has adjourned.

What to Include
Minutes of a meeting should include the following information:

- date, time, and place of the meeting
- name of the group holding the meeting and why
- name of the person chairing the meeting

- names of those present and those absent; at the start of every meeting, circulate a piece of paper so those present can sign their names
- the approval or amendment of the minutes of the previous meeting
- for each major point—the action items—indicate what was done:
  - who said what
  - what was discussed, suggested, or proposed
  - what was decided and the vote, including abstentions
  - what was continued (tabled) for a subsequent study, report, or meeting
  - the time of the next meeting
  - the time the meeting officially concluded

Guidelines on Writing Minutes

To be effective, minutes must be concise and to the point. Here are a few guidelines to help you:

- Make sure of your facts; spell all names, products, and tests correctly.
- Concentrate on the major facts surrounding action items. Save the reader's time and your own by condensing lengthy discussions, debates, and reports given at the meeting.
- Do not report verbatim what everyone said; readers will be more interested in outcomes—what the group did.
- List each motion (or item voted on) exactly as it is worded and in its final form.
- Avoid words that interpret (negatively or positively) what the group or anyone in the group did or did not do.

Figure 3.13 shows how these parts fit together.

Taking Notes

Meetings are crucial to the day-to-day operation and long-range planning of a business. Some meetings are held in a company office for a small group such as the Environmental Safety Committee in Figure 3.13. Others take place online, as in a Skype conversation, as shown in Figure 3.12. Still others are held off-site where, because of time and location, you may be the only member of your team attending and will have to share your notes with co-workers.

A variety of software applications can help you to share your notes via email. Take your laptop to the meeting to help you. As you take notes, include only major issues that have top priority for your audience. Don't get bogged down in recording minor details. Inform your team or boss about important points affecting new dates, revised schedules, different prices, amended policies or regulations, changes in operations, service, key decisions, and other important business matters. Anticipate the questions your team or boss may have about the meeting, and then supply the answers through your notes. Figure 9.8 (page 441) contains a short summary from the notes two employees made of the major topics covered at a seminar they attended.

FIGURE 3.13 Minutes from a Business Meeting

NewTech, Inc.

Supplies essential information on attendance, date, and place of meeting

Minutes for Environmental Safety Committee (ESC) meeting on August 10, 2012, in Room 203 of Lab Annex Building at 10:00 a.m.

Members Present:

Thomas Baldanza, Grace Corlee (President), Virginia Downey, Victor Johnson, Roberta Koos, Kent Leviche (Secretary), Ralph Nowicki, Barbara Poe-Smith, Williard Ralston, Asah Rashid, Morgan Tachiashi, and Carlos Zandrillia

Members Absent:

Paul Gordon (sick leave); Marty Wagner

Refers to previous meeting to provide continuity

Old Business:

The minutes from the previous meeting on July 9, 2012, were approved as read.

Reports:

Concisely summarizes progress on ongoing business

(1) Morgan Tachiashi reported on the progress the Site Inspection Committee is making in getting the plant ready for the August 29 visit of the State Board of Examiners. All preparations are on schedule.

Identifies key speakers

(2) The proposal to study the use of biometric identification in place of employee ID badges is nearly complete, according to Asah Rashid, Chair of the Proposal Committee, and will be presented at next month's meeting for approval.

New Business:

Records only main points of discussion and votes

(1) Virginia Downey and Ralph Nowicki voiced concern about a computer virus that may strike the plant—Monkey. Disguised as a familiar email, the virus is contained in an attachment that destroys files. A motion was made by Barbara Poe-Smith, seconded by Virginia Downey, that management upgrade its antivirus protection software. Objecting to this expense, Williard Ralston thought the current software was sufficient. The vote carried by 9 to 3.

FIGURE 3.13 (Continued)

Page 2

(2) Thomas Baldanza believed that cross-training should be accelerated, especially in safety areas, to meet the target date of September 7 which the ESC had set in December 2010. Agreeing, Roberta Koos stressed that, without cross-training, some departments would be vulnerable to safety violations. Victor Johnson, on the other hand, found that such cross-training could not feasibly be accomplished in time since several departments could not spare employees to participate. He moved that the target date motion be amended and pushed to November 15. The vote to amend was defeated 10–2. Calling attention to the importance of the target date, Grace Corlee will ask Zandria Pickens, Plant Training Coordinator, to attend the next ESC meeting to discuss the current status of the training program and to offer suggestions for its speedy implementation.

Sticks to basic facts and results of vote

Explains what will happen at next meeting

(3) Kent Leviche calculated computer downtime in the plant during the month of July—5 outages totaling 7.5 lost working hours—and asked the ESC to address this problem. After discussion, the ESC unanimously agreed to appoint a subcommittee to investigate the outages and determine solutions. Roberta Koos and Thomas Baldanza will chair the subcommittee and then present a survey report at next month's meeting.

Includes other business to be continued

(4) Personnel in the Environmental Testing Lab were commended for their extra effort in ensuring that their department maintained the highest professional standards during the month of July. A letter of commendation was sent to Emily Lu, Lab Supervisor, and her staff.

Excellent morale builder

(5) Grace Corlee adjourned the meeting at 11:41 a.m.

Next Meeting:

Signals end of meeting and date of next one

The next meeting of the ESC will be on September 7 at 1:00 p.m. in Room 203 of the Lab Annex Building.

Conclusion

This chapter has emphasized the importance of collaboration in the world of work and explained the various collaborative models you may need to follow. It has also given you detailed guidelines on creating, editing, and revising a collaborative document, using the most current types of Web-based applications. Working successfully as part of a team, whether at face-to-face meetings or online, is one of the most valuable skills you can develop, and one that your employer will expect you to use routinely.

✓ Revision Checklist

Being a Responsible Team Member

- ☐ Succeeded in being a team player by putting the success of my group over the needs of my own ego.
- ☐ Followed the necessary steps of the writing process to take advantage of team effort and feedback.
- ☐ Attended all group meetings and understood and agreed to the responsibilities of the group and my own obligations.
- ☐ Finished the research, planning, and/or drafting expected of me as a group member.
- ☐ Conducted necessary interviews and conferences to gather, clarify, and verify information.
- ☐ Shared my research, ideas, and suggestions for revision through constructive criticism.
- ☐ Participated honestly and politely in discussions with colleagues.
- ☐ Treated members of my team with respect and courtesy.
- ☐ Was open to criticism and suggestions for change.
- ☐ Read colleagues' work and gave specific and helpful criticism and suggestions.
- ☐ Kept matters in proper perspective by not being a nitpicker and by not interrupting with extraneous points or unnecessary questions.
- ☐ Sought help when necessary from relevant subject matter experts and from co-workers.
- ☐ Secured responses and approval from management.

Using Computer-Supported Collaboration

- ☐ Took advantage of email, instant messaging, and groupware applications (e.g., document tracking systems, wikis, Google Docs) to communicate with my collaborative team.
- ☐ Investigated the research, drafting, revising, and editing benefits available with computer software.
- ☐ Answered questions and responded to requests promptly from the team leader and collaborative team members.
- ☐ Attached pertinent documents in emails to the collaborative team.
- ☐ Avoided technical problems with online collaboration by adhering to established policies.
- ☐ Respected confidentiality and used computer-assisted editing technologies responsibly and ethically.

Preparing for and Participating at a Meeting

- ☐ Prepared a clear agenda for the meeting and distributed it ahead of time.
- ☐ Wrote minutes that objectively reported what happened.
- ☐ Took notes that highlighted main points of the meeting for my collaborative team and boss.
- ☐ Participated in virtual meetings through teleconferencing or Web conferencing or videoconferencing.

Exercises

1. Assume you belong to a three- or four-person editing team that functions the way Tara Barber's does, as described in Figure 3.3. Each member of your team should bring in four copies of a paper written for this course or for another one. Exchange copies with the other members of your team so that each team member has everyone else's papers to review and revise. For each paper you receive, comment on the style, organization, tone, and discussion of ideas as Wells McCraw did in Figure 3.7.

2. With the members of your collaborative team (selected by your teacher or self-appointed) select four different brands of the same product (such as a software package, a Web browser, a smartphone, a microwave, a DVD player, or a power tool). Each member of your team should select one of the brands and prepare a two-page memo report for your instructor (see pages 133–141), evaluating the product according to the following criteria:

- convenience
- performance
- technical capabilities or capacities
- appearance
- adaptability
- price
- warranties
- weaknesses and strengths compared with competitors' models

Each team member should then submit a draft to the other members of the team to review. At a subsequent group meeting, the group should evaluate the four brands based on the team's drafts and then together prepare one final recommendation report for your instructor.

3. Your company is planning to construct a new office, and you, together with other employees from your company, have been asked to serve on a committee to make sure that plans for the new building adhere to the Americans with Disabilities Act, passed in 1990. According to that act, it is against the law to discriminate against anyone with disabilities that limit "major life activities," such as walking, seeing, speaking, hearing, or working.

   The law is expressly designed to remove architectural and physical barriers and to make sure that plans are modified to accommodate those protected by the law (for example, wider hallways to accommodate wheelchairs). Other considerations include choosing appropriate floor surfaces (reducing the danger of slipping), placing water fountains low enough for use by individuals in wheelchairs, and installing doors that require minimal pressure to open and close.

   After studying the plans for the new building, you and your team members find several problem areas. Prepare a group-written report advising management of the problems and what must be done to correct them to comply with the law. Divide your written work according to areas that need alteration—doors, floors, water fountains, restroom facilities. Each team member should bring in his or her section for the group to edit and revise. The group should then prepare the final report for management.

4. A new president will be coming to your college in the next month, and you and five other students have been asked to serve on a committee that will submit a report about campus safety problems and what should be done to solve them. You and your team must establish priorities and propose guidelines that you want the new administration to put into practice. After two very heated meetings, you realize that what you and two other students have considered solutions, the other half of your committee regards as the problems. Here is a rundown of the leading conflicts dividing your committee:

- **Speed bumps.** Half the committee likes the way they slow traffic down on campus, but the other half says they are a menace because they jar car CD players.
- **Sound pollution.** Half your team wants Campus Security to enforce a noise policy preventing students from playing loud music while driving on campus, but the other half insists that would violate students' rights.
- **Van and sport utility vehicle parking.** Half the committee demands that vans and sport utility vehicles park in specially designated places because they block the view of traffic for smaller vehicles parked next to them; the other members protest that people who drive these vehicles will be singled out for less desirable parking places on campus.

Clearly your committee has reached a deadlock and will be unproductive as long as these conflicts go unresolved. Based on the above scenario, do the following:

a. Have each student on the committee write (or email) the other five students suggesting a specific plan on how the group can resolve its conflicts. Prepare your email message and send it to the other five committee members and to your instructor. What's your plan to get the committee moving toward writing the report to the incoming president?

b. Assume that you have been asked to convince the committee to accept your views on the three areas of conflict: speed bumps, noise control, and parking. Send the three opposition students a memo or an email persuading them to your way of thinking. Keep in mind that your message must assure them that you respect their point of view and that you are considering the overall benefits for the campus community.

c. Assume that the committee members reach a compromise after seeing the plan you put forth in (a). Collaboratively draft a three-page report to the new president.

d. Collaboratively draft a letter to the editor of your student newspaper defending your recommendations to the student body and explaining how the group resolved its difficulty. This is a public statement that the group feels it is important to write; you will have to choose your words carefully to win campus-wide support.

5. You work for a hospital laboratory, and your lab manager, under pressure from management to save money, insists that you and the three other med-techs switch to a different brand of vacuum blood-drawing tubes. You and your colleagues much prefer the brand of tubes you have been using for years. Moreover, the price difference between the two brands is small. As a group project, prepare a memo to the business manager of the hospital explaining why the switch is unnecessary, unwise, and unpopular. Then prepare another collaboratively written memo to your lab manager. Be sensitive to each reader's needs as you diplomatically explain the group's position.

PART II

Correspondence

CHAPTER **4**

Writing Routine Business Correspondence

Memos, Faxes, Emails, IMs, and Blogs

Access chapter-specific interactive learning tools, including quizzes and more in your English CourseMate, accessed through www .cengagebrain.com.

Memos, faxes, emails, IMs, and blogs are the types of writing you will do most frequently on the job. These five forms of business correspondence are quick, easy, and effective ways for a company to communicate internally as well as externally. You can expect to send one or more of these routine forms of correspondence each day to co-workers in your department, to colleagues in other divisions of your company, to decision makers at all levels, and to clients and customers as well.

What Memos, Faxes, Emails, IMs, and Blog Posts Have in Common

Although memos, faxes, emails, IMs, and blog posts are very different types of correspondence, they share the following characteristics:

1. **Each of these types of correspondence is streamlined for the busy world of work**. Unlike letters, proposals, or reports, which can be long and detailed and contain formal parts and sections, these routine types of correspondence give writers and readers a particularly fast way to communicate. Even though memos do have to be formatted (see page 136), they, like blog posts, IMs, and emails, are ready-made to send and receive shorter messages.

2. **They give busy readers information quickly**. While the messages they contain can be about any topic in the world of work, most often they focus on day-to-day activities and operations at your company—sales and product information, policy and schedule changes, progress reports, orders, troubleshooting problems, and so forth.

3. **They are informal**. Compared to letters, proposals, and reports, these kinds of routine correspondence are not as formal. They emphasize a conversational, yet professional, style of writing.

4. **Even though they are routine, they still demand a great deal of thought and time**. Although memos, emails, blogs, and IMs are less formal than, for instance,

a letter to a client, they all must be written clearly and with correct grammar and punctuation, even when the correspondence is between two employees. Always plan what you are going to write.

5. They represent your company. These routine messages are reflections of your company's image and your professionalism. Be careful about what you write and how you say it. Just because the format of your message is informal does not mean you can be unprofessional, even in a brief email to a co-worker. Your success as an employee can depend as much on your writing an unbiased, ethically proper email or blog post as it does on your technical expertise.

Memos

Memorandum, usually shortened to *memo*, is a Latin word for "something to be remembered." The Latin meaning points to the memo's chief function: to record information of immediate importance and interest in the busy world of work. Memos are brief and informal but can contain official announcements that serve a variety of functions, including

- making an announcement
- providing instructions
- clarifying a policy, procedure, or issue
- changing a policy or procedure
- alerting employees to a problem or issue
- making a request
- offering suggestions or recommendations
- providing a record of an important matter
- confirming an outcome
- calling a meeting

Memos are usually written for an in-house audience, although the memo format can be used for documents sent outside a company, such as proposals or short reports (see Chapters 13 and 14) or for cover notes for longer reports (see Chapter 15).

Memos keep track of what jobs are done where, when, and by whom; they also report on any difficulties, delays, or cancellations and what your company or organization needs to do about correcting or eliminating them.

Memo Protocol and Company Politics

As with any other forms of business correspondence, memos reflect a company's image and therefore must follow the company's *protocol*—accepted ways in which in-house communications are formatted, organized, written, and routed. In fact, some companies offer protocol seminars on how employees are to prepare communications. In addition to following your company's protocol, use these commonsense guidelines when writing memos:

1. Be timely. Don't wait until the day of the meeting to announce it.

2. Be professional. Just because a memo is an informal, in-house communication does not mean you should compose a poorly organized, poorly written, or factually inaccurate document. Notice that in Figure 4.1, the informal memo between Roger and Lucy is professionally written, clearly organized, and properly spelled and punctuated.

3. Be tactful. Be polite and diplomatic, not curt and bossy. For example, in Figure 4.2, notice how Janet Hempstead adopts a firm tone regarding an important safety issue, yet she does not blame or talk down to her readers—the machine shop employees. Politeness and diplomacy count a lot at work.

FIGURE 4.1 Standard Memo Format

MEMO

Header

Memo parts

TO:	Lucy
FROM:	Roger
DATE:	November 12, 2012
SUBJECT:	Review of "Successful Website" Seminar

Introduction provides background and tells reader what memo will do

As you know, I attended the "How to Build a Successful Website" seminar on November 7 and learned the "rules and tools" we will need to redesign our own site.

Preview

Here is a review of the major topics covered by the presenter, Jackie Chen:

Numbered list in body helps readers follow information quickly

1. Keep your website content-based—identify your target audience.
2. Visualize and "map out" your site ahead of time.
3. Design your website to look the way you envision it—make it aesthetically pleasing.
4. Be sure your site is easy to navigate, especially for global readers.
5. Make sure the site is easy to find by search engines such as Google for maximum exposure.
6. Create hot links and image maps to move users from page to page.
7. Encourage visitor interaction by soliciting feedback.
8. Complete your site with appropriate sound and animation.
9. Keep your site updated.

Conclusion asks for comments

Could we meet in the next day or two to discuss recreating our website in light of these guidelines? I would really appreciate your suggestions about this project as well.
Thanks.

FIGURE 4.2 Memo on Letterhead with a Clear Introduction, Discussion, and Conclusion

Dearborn Equipment Company

To: Machine Shop Employees
From: Janet Hempstead, Shop Supervisor *JH*
Date: September 27, 2012
Subject: Cleaning Brake Machines

During the past two weeks I have received several reports that the brake machines are not being cleaned properly after each use. Through this memo I want to emphasize and explain the importance of keeping these machines clean for the safety of all employees.

When the brake machines are used, the cutter chops off small particles of metal from brake drums. These particles then settle on the machines and create a potentially hazardous situation for anyone working on or near the machines. If the machines are not cleaned routinely before being used again, these metal particles could easily fly into an individual's face or upper body when the brake drum is spinning.

To prevent accidents like this from happening, please make sure you vacuum the brake machines after each use.

You will find two vacuum cleaners for this purpose in the shop—one of them is located in work area 1-A, and the other, a reserve model, is in the storage area. Vacuuming brake machines is quick and easy: It should take you no more than a few seconds, a small amount of time to make the shop safer for all of us.

Thanks for your cooperation. If you have any questions, please call me at Extension 324, email me, or come by my office.

204 South Mill St., South Orange, NJ 02341-3420 (609) 555-9848 JHEMP@dearco.com

Writer's initals verify message

Introduction explains purpose and importance of memo

Discussion states why problem exists and how to solve it

Safety message is boldfaced for emphasis

Conclusion builds goodwill and asks for questions

4. **Send memos to the appropriate personnel.** Don't send copies of a memo to people who don't need to read them. It wastes time and energy. Moreover, don't send a memo to high-ranking company personnel in place of your immediate supervisor, who may think you are going over his or her head. For instance, in Figure 4.3, Mike Gonzalez has sent his memo only to the vice president and the public relations officer, the two people who will most likely benefit from the memo's recommendations, and in Figure 4.2, Janet Hempstead has sent her memo only to the machine shop workers, not to the upper management of the Dearborn Company.

Keep in mind, though, that memos are often sent up and down the corporate ladder. Employees send memos to their supervisors, and workers send memos to one another. Figure 4.1 shows a memo sent from one worker to another. Figure 4.2 contains a memo sent from the top down, and Figure 4.3 illustrates a memo sent from an employee to management.

Memo Format

Memos vary in format and the way they are sent. Some companies use standard, printed forms (as in Figure 4.1), while others have their names (letterhead) printed on their memos (as in Figures 4.2 and 4.3). You can also create a memo by including the necessary parts in an email, as in Figures 4.4 and 4.5, which appear later in this chapter.

As you can see from looking at Figures 4.1 through 4.3, memos look different from letters. They are also less formal. Because they are often sent to individuals within your company, memos do not need the formalities necessary in business letters, such as an inside address, a formal salutation or complimentary close, or a signature line, as discussed in Chapter 5 (see pages 171–176). But if your memo extends to a second page, you do need to carry over at least two lines and include a page notation at the top of subsequent pages (see Figures 13.6 and 14.13 as examples).

Memo Parts

Basically, the memo consists of two parts: the *header*, or the identifying information at the top, and the *message* itself. This identifying information includes four easily recognized parts: *To, From, Date*, and *Subject* lines.

```
TO:        Aileen Kelly, Chief Computer Analyst
FROM:      Stacy Kaufman, Operator, Level II
DATE:      January 30, 2012
SUBJECT:   Progress report on the fall schedule
```

You can use a memo template, as below with information filled in, in your word-processing program that lists these headings, as follows, to save time.

```
TO:        [Enter name]
FROM:      Linda Cowan
DATE:      [Enter date]
RE:        [Enter subject here.]
```

FIGURE 4.3 A Memo That Uses Headings to Highlight Organization

RAMCO TECHNOLOGIES
Where Technology Shapes Tomorrow

marketing@Ramco.com　　　　　　www.Ramco.com

TO:	Rachel Mohler, Vice President
	Harrison Fontentot, Public Relations
FROM:	Mike Gonzalez MG
SUBJECT:	Three Ways to Increase Ramco's Community Involvement
DATE:	March 2, 2012

At our planning session in early February, our division managers stressed the need to generate favorable publicity for our new Ramco facility in Mayfield. Knowing that such publicity will highlight Ramco's visibility in Mayfield, I think the company's image might be enhanced in the following three ways.

CREATE A SCHOLARSHIP FUND
Ramco would receive favorable publicity by creating a scholarship at Mayfield Community College for any student interested in a career in technology. A one-year scholarship would cost $6,800. The scholarship could be awarded by a committee composed of Ramco executives and staff. Such a scholarship would emphasize Ramco's enthusiastic support for the latest technical education at a local college.

OFFER SITE TOURS
Guided tours of the Mayfield facility would introduce the community to Ramco's innovative technology. These tours might be organized for academic, community, and civic groups. Individuals would see the care we take in protecting the environment in our production and equipment choices and the speed with which we ship our products. Of special interest to visitors would be Ramco's use of industrial robots working alongside our employees. Since these tours would be scheduled in advance, they should not conflict with our production schedules.

PROVIDE GUEST SPEAKERS
Many of our employees would be excellent guest speakers at civic and educational meetings in the Mayfield area. Possible topics include the advances Ramco has made in designing and engineering and how these advances have helped consumers and the local economy.

Thanks for giving me your comments as soon as possible. If we are going to put one or more of these suggestions into practice before the facility opens in mid-April, we'll need to act before the end of the month.

On the *To* line, write the name and job title of the individual(s) who will receive your memo or a copy of it. If you are sending your memo to more than one reader, make sure you list your readers in the order of their status in your company or agency, as Mike Gonzalez does in Figure 4.3 (according to company policy, the vice president's name appears before that of the public relations director). If you are on a first-name basis with the reader, use just his or her first name, as in Figure 4.1. Otherwise, include the reader's first and last names. Don't leave out anyone who needs the information.

On the *From* line, insert your name (use your first name only if your reader refers to you by it) and your job title (unless it is unnecessary for your reader). Some companies ask employees to handwrite their initials after their typed name to verify that the message comes from them and that they are certifying its contents, as in Figures 4.2 and 4.3. You do not have to key in your initials in a memo sent as an email.

On the *Date* line, do not simply name the day of the week. Give the full calendar date (June 1, 2012).

On the *Subject* line, key in the purpose of your memo. The subject line serves as the title of your memo; it summarizes your message. Vague subject lines, such as "New Policy," "Operating Difficulties," or "Software," do not identify your message precisely and may suggest that you have not restricted or developed it sufficiently. Note how Mike Gonzalez's subject line in Figure 4.3 is so much more precise than just saying "Ramco's Community Involvement."

Questions Your Memo Needs to Answer for Readers

Here are some key questions your audience may ask and your memo needs to answer clearly and concisely:

1. **When?** When did it happen? Is it on, ahead of, or behind schedule? When does it need to be discussed or implemented? *When* is answered in Figures 4.1 ("November 7," "in the next day or two"), 4.2 ("during the past two weeks," "after each use"), and 4.3 ("in early February," "before the end of the month").

2. **Who?** Who is involved? Who will be affected by your message? How many people are involved? *Who* is answered in Figures 4.1 (Jackie Chen), 4.2 (all machine shop employees), and 4.3 (Ramco Technologies as a whole).

3. **Where?** Where did it take place or will it take place? *Where* is answered in Figures 4.1 (the website seminar), 4.2 (the brake shop), and 4.3 (the Mayfield facility).

4. **Why?** Why is it an important topic? *Why* is clearly answered in Figures 4.1 (because the website is being redesigned), 4.2 (because it's a safety issue), and 4.3 (because favorable publicity will help the company).

5. **Costs?** How much will it cost? Will the costs be lower or higher than a competitor's costs? Not every memo will answer financial questions, but in Figure 4.3, the specific cost of an individual scholarship ($6,800) is an important issue.

6. **Technology?** What technology is involved? Why is the technology needed? Is the technology available, current, adaptable, safe for the environment? Again, not

every one of your memos will answer questions about technology, but note that Figures 4.1, 4.2, and 4.3 all especially refer to technological issues—website design, equipment safety, education, robots.

   7. What's next? What are the next steps that should be taken as a result of the issues discussed in the memo? What are the implications for the product, service, budget, staff? A good example of this is in Figure 4.3 (the company needs to decide on how to implement the suggestions before the new plant opens).

Memo Style and Tone

The audience within your company will determine your memo's style and tone (for a review of identifying audience, see Chapter 1, pages 12–17). When writing to a co-worker whom you know well, you can adopt a casual, conversational tone. You want to be seen as friendly and cooperative. In fact, to do otherwise would make you look self-important, stuffy, or hard to work with. Consider the friendly tone appropriate for one colleague writing to another as in Roger's memo to Lucy in Figure 4.1. Note how he ends in a polite but informal way.

   When writing a memo to a manager, though, you will want to use a more formal tone than you would when communicating with a co-worker or peer. Your boss will expect you to show a more respectful, even official, posture. See how formal yet conversationally persuasive Mike Gonzalez's memo to his bosses is in Figure 4.3. His tone and style are a reflection of his hard work as well as his respect for his employers. Here are two ways of expressing the same message, the first more suitable when writing to a co-worker and the second more appropriate for a memo to the boss.

> **Co-worker:** I think we should go ahead with Marisol's plan for reorganization. It seems like a safe option to me, and I don't think we can lose.
>
> **Boss:** I think that we should adopt the organizational plan developed by Marisol Vega. Her recommendations are carefully researched and persuasively answer the questions our department has about solving the problem.

When an employer writes to workers informing them about policies or procedures, as Janet Hempstead does in Figure 4.2, the tone of the memo is official and straightforward. Yet even so, Hempstead takes into account her readers' feelings (she does not blame) and safety, which are at the forefront of her rhetorical purpose.

   Finally, remember that your employer and co-workers deserve the same clear and concise writing and attention to the "you attitude" (see Chapter 5, pages 179–183) that your customers do. Memos require the same care and should follow the same rules of effective writing, outlined in Chapter 2, as letters do.

Strategies for Organizing a Memo

Don't just dash your memo off. Take a few minutes to outline and draft what you need to say and to decide in what order it needs to be presented. Organize your memos so that readers can find information quickly and act on it promptly. For longer, more complex communications, such as the memos in Figures 4.2 and 4.3, your message might be divided into three parts: (1) introduction, (2) discussion, and (3) conclusion. Regardless of how short or long your memo is, recall the three *P*'s for success: *plan* what you are going to say; *polish* your writing before you send it; and *proofread* everything.

Introduction

The introduction of your memo should do the following:

- Tell readers clearly about the problem, procedure, question, or policy that prompted you to write.
- Explain briefly any background information the reader needs to know.
- Be specific about what you are going to accomplish in your memo.

Do not hesitate to come right out and say, "This memo explains new email security procedures" or "This memo summarizes the action taken in Evansville to reduce air pollution." See how clearly this is done in Figure 4.1.

Discussion

In the discussion section (the body) of your memo, help readers in these ways:

- State why a problem or procedure is important, who will be affected by it, and what caused it and why.
- Indicate why and what changes are necessary.
- Give precise dates, times, locations, and costs.

See how Janet Hempstead's memo in Figure 4.2 carefully describes an existing problem and explains the proper procedure for cleaning the brake machines, and how Mike Gonzalez in Figure 4.3 offers carefully researched evidence on how Ramco can increase its favorable publicity in the community.

Conclusion

In your conclusion, state specifically how you want the reader to respond to your memo. To get readers to act appropriately, you can do one or more of the following:

- Ask readers to call you if they have any questions, as in Figure 4.2.
- Request a reply—in writing, over the telephone, via email, or in person—by a specific date, as in Figure 4.3.
- Provide a list of recommendations that the readers are to accept, revise, or reject, as in Figures 4.1 and 4.3.

Organizational Markers

Throughout your memo, use the following organizational markers, where appropriate:

- Headings organize your work and make information easy for readers to follow, as in Figure 4.3.
- Numbered or bulleted lists help readers see comparisons and contrasts readily and thereby comprehend your ideas more quickly, as in Figure 4.1.

■ Underlining or boldfacing emphasizes key points (see Figure 4.2). Do not overuse this technique; draw attention only to main points and those that contain summaries or draw conclusions.

Organizational markers are not limited to memos; you will find them in email, letters, reports, and proposals as well. (See Chapter 11, pages 517–522.)

Sending Memos: Email or Hard Copy?

A memo can be sent as a printed hard copy, as an email, or as a scanned email attachment. Find out your company's policy. Increasingly, email is replacing printed memos, but there are times when a hard-copy memo is preferred.

Consider the level of importance and confidentiality of your memo. If your memo is an official document, such as the policy outlined in Figure 4.2, you may not want to send it via email, because it could easily be deleted or altered. Moreover, if your memo is confidential (e.g., an evaluation of a co-worker or vendor, or a message containing sensitive financial or medical information), you may not want to send it via email, because it could easily be forwarded or fall victim to hackers. But if you are sending a routine message that must reach readers quickly, use email.

Sending Faxes: Some Guidelines

Even though business documents are commonly sent via email, faxes are still used in the world of work. A fax (facsimile) is an original document copied and transmitted over telephone or computer lines. Faxes are particularly helpful either when

Tech Note

Scanning a Document

The process of scanning a document involves taking a hard copy of text or an image and digitizing (i.e., capturing) it to a format that a computer can recognize and use. The basic principle of a scanner is to pass a beam of light over the image, analyze it, and then convert it to digital format in either color or black and white. This image and text capture (optical character recognition, or OCR) allow you to save hard-copy information to a "soft copy" file on your computer. You can then alter text or enhance the image, email it, print it out, or use it on your webpage. Scanned documents save time, money, and space. Because of the inefficiencies inherent in the time-sensitive sharing, filing, storage, and retrieval of paper documents, businesses are routinely converting all their paper files into digital information that can be edited. This is true for documents such as medical records, contracts, proposals, manuscripts, reports—anything that is now in paper form. Once a document has been scanned, it can be accessed for printing and/or emailing with the click of a button. On your computer, you need software—called a driver—that knows how to communicate with the scanner. Scanners can actually look like a printer, and many printers can both scan and print documents.

you have only hard copy to send or when you want to send an original signed letter, contract, blueprint, artwork, or other document that you could not send via an email transmission. Faxes demonstrate exactly what original documents look like and allow recipients to obtain a hard copy quickly.

Cover Page

When you send a fax, make sure your cover page includes the following information:

1. **The name of the sender and his or her fax and phone numbers**. The phone number is important because it enables the recipient to report an incomplete transmission.
2. **The name of the recipient and his or her fax and phone number**. The recipient's correct fax and phone numbers should be included to ensure delivery of the message.
3. **The total number of pages being faxed**. Note that the total number of pages includes the cover page itself.
4. **A brief explanatory note that lets the recipient know what the fax is**, what its purpose is, and how and when to respond to it.

Sending a Document

Follow these four guidelines to fax a clear and complete document:

1. **Make sure the original documents you send are clear**. An unclear faxed document will be difficult for the recipient to read. For example, penciled comments may be too faint to fax clearly.
2. **Avoid writing on the top, bottom, or edges of the documents to be faxed**. Any comments written on the outer edges may be cut off or blurred during transmittal.
3. **Do not send overly long faxes**. Be careful about sending anything longer than three or four pages because you will tie up both your own and the recipient's fax machines.
4. **Respect the recipient's confidentiality**. Because faxes may be picked up by other employees in your office, don't assume your message will be confidential unless, of course, the recipient has a private fax machine.

Email: Its Importance in the Workplace

Email continues to be one of the most common forms of communication in the workplace. It is the lifeblood of every business or organization because it expedites communication within a firm as well as outside it. On their PCs, notebooks, or mobile devices, professionals in the world of work may receive between forty and one hundred emails each day from supervisors, co-workers, clients, and vendors worldwide. Email allows you to send short messages about routine matters that make business function smoothly.

Using an email program, you can expedite workplace communication in many ways:

- Send and receive information quickly; delete it, forward it, or archive it
- Organize your correspondence in folders according to date, sender, or subject
- Identify and delete spam
- Send attachments, including documents, visuals, audio clips, tables, lists, and statistical files
- Enhance all phases of your collaborative work (see Chapter 3)
- Keep track of appointments
- Set up your business calendar and even share it with colleagues
- Communicate anytime, all the time, 24/7

Email is an informal, relaxed type of business correspondence, far more informal than a printed memo, letter, short report, or proposal, though it is more complex than instant messaging (see pages 150–152). Think of your workplace email as a polite, informative, and professional conversation. It should always be to the point and accessible, as in Figure 4.4. Yet even though business email is a way of

FIGURE 4.4 An Email Sent to a Co-worker

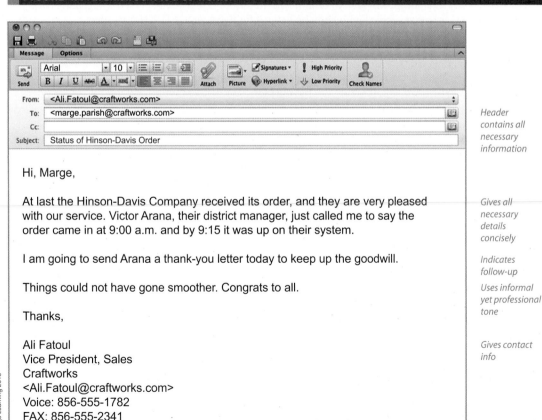

From:	\<Ali.Fatoul@craftworks.com\>	
To:	\<marge.parish@craftworks.com\>	
Cc:		
Subject:	Status of Hinson-Davis Order	

Hi, Marge,

At last the Hinson-Davis Company received its order, and they are very pleased with our service. Victor Arana, their district manager, just called me to say the order came in at 9:00 a.m. and by 9:15 it was up on their system.

I am going to send Arana a thank-you letter today to keep up the goodwill.

Things could not have gone smoother. Congrats to all.

Thanks,

Ali Fatoul
Vice President, Sales
Craftworks
\<Ali.Fatoul@craftworks.com\>
Voice: 856-555-1782
FAX: 856-555-2341
www.craftworks.com

Header contains all necessary information

Gives all necessary details concisely

Indicates follow-up

Uses informal yet professional tone

Gives contact info

FIGURE 4.5 Email Sent to a Distribution List of Co-workers

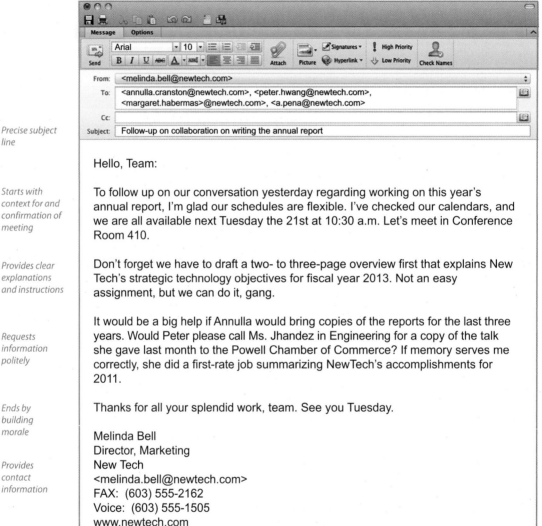

Precise subject line

Starts with context for and confirmation of meeting

Provides clear explanations and instructions

Requests information politely

Ends by building morale

Provides contact information

From: <melinda.bell@newtech.com>

To: <annulla.cranston@newtech.com>, <peter.hwang@newtech.com>, <margaret.habermas>@newtech.com>, <a.pena@newtech.com>

Cc:

Subject: Follow-up on collaboration on writing the annual report

Hello, Team:

To follow up on our conversation yesterday regarding working on this year's annual report, I'm glad our schedules are flexible. I've checked our calendars, and we are all available next Tuesday the 21st at 10:30 a.m. Let's meet in Conference Room 410.

Don't forget we have to draft a two- to three-page overview first that explains New Tech's strategic technology objectives for fiscal year 2013. Not an easy assignment, but we can do it, gang.

It would be a big help if Annulla would bring copies of the reports for the last three years. Would Peter please call Ms. Jhandez in Engineering for a copy of the talk she gave last month to the Powell Chamber of Commerce? If memory serves me correctly, she did a first-rate job summarizing NewTech's accomplishments for 2011.

Thanks for all your splendid work, team. See you Tuesday.

Melinda Bell
Director, Marketing
New Tech
<melinda.bell@newtech.com>
FAX: (603) 555-2162
Voice: (603) 555-1505
www.newtech.com

communicating, this does not mean you can forget about your responsibilities as a courteous and ethical employee, co-worker, and writer. Figure 4.5 exemplifies an email used to communicate diplomatically with a collaborative team.

Business Email versus Personal Email

The email you write on the job will require more effort than your personal email will. Don't assume you can write to your employer or a customer the way you would an old friend. In the world of work, you don't just dash off an email. You have to revise and review it before you click and send it. That means proofreading carefully and following all the rules of proper spelling (avoid text-message spellings), punctuation, capitalization, and word choice, as well as the email guidelines on pages 145–149. The tone of your business email should also be much more professional than the instant messaging you may do with friends or the e-conversations you have in chat rooms.

Unlike with your personal email, you need to consider the impact your business email will have on your company and on your career. When you send a business email, you are representing more than yourself and your preferences, as in a personal email. You are speaking on behalf of your employer. Because your email must reflect your company's best image, make sure it is businesslike, carefully researched, and polite. Sarcasm, slang, an aggressive tone, name-calling, and inappropriate clip art do not belong in a company email. As we saw, Figures 4.4 and 4.5 illustrate effectively written business email. Notice that these emails are cordial without being unprofessional.

Emails Are Legal Records

Employers own their internal email systems and thus have the right to monitor what you write and to whom (see pages 29–32). Any email written at work on the company's server can be copied, archived, forwarded, and, most significantly, intercepted. You can be fired for writing an angry or abusive email. Keep in mind that your email can easily be converted into an electronic paper trail. You never know who will receive and then forward your email—to your boss, a co-worker, a customer, an attorney, or a licensing board. Many companies include disclaimers protecting themselves from legal action because of an employee's offensive behavior in a company email. In court, an email can carry the same weight as a printed hard copy.

Here are some helpful rules to follow when using your company's email:

- Do not use it for your personal messages. Send emails only for appropriate company business, and make sure you are professional and conscientious.
- Never write an email to discuss a confidential subject—a raise, a grievance, or a complaint about a co-worker. Meet with your supervisor in person.
- Make sure of your facts before sending an email to customers. If it gives them wrong or misleading information about prices, warranties, or safety features, your company can be legally liable.
- Be careful not to expose your company server to security risks by responding to spam.

Guidelines for Using Email on the Job

When you prepare and organize your email message, always consider your reader's specific needs as well as those of your company. Following the guidelines below will help you to write effective business emails.

1. **Make sure your email is confidential and ethical.**

   - Avoid *flaming*, that is, using strong, angry language that mocks, attacks, or insults your employer, a colleague, a customer, a government agency, or a company, as in Figure 4.6 (see page 148). Abusive, obscene, or racially or culturally offensive language in an email constitutes grounds for dismissal.
   - Send nothing through email that you would not want to see on your company's website or on the front page of a local newspaper.
   - Do not forward a co-worker's or an employer's email without that person's approval.
   - Do not change the wording of a message that you are expected to read and forward.
   - Never send an objectionable photo or file.
   - Respect your company's chain of command (see page 136) by going through proper channels. There is no need to copy management (e.g., the CEO or the head of Human Resources) unless these individuals are directly involved in the communication. Always consult your immediate supervisor about who should be copied on emails.

2. **Make your email easy to read.**

   - **Provide a clear, precise subject line**. Your subject line states the purpose of your message and determines whether the reader will even look at your email. Avoid one-word subjects like "Report" or "Meeting." Instead, write "Meeting to boost declining April sales." A subject line like "Bill" leaves readers wondering if your email is about a person or an unpaid account.
   - **Try to limit your emails to one screen**. Longer messages are better sent in an attachment rather than in the body of an email.
   - **Do not send emails written in all capital or all lowercase letters**. All capital letters look as if you are shouting. Conversely, emails in all lowercase imply you do not know how to capitalize.
   - **Break your message into short paragraphs**. A screen filled with one dense block of text is intimidating. Make each paragraph no more than three to four lines long and always double-space between paragraphs. Do not indent your paragraphs.
   - **Provide URLs for all websites you reference.** Do not make your reader look them up.
   - **Use plain text**. Because different email programs can garble your message, avoid overusing typefaces like italic, script, or decorative fonts, or complex formatting (such as long numbered and bulleted lists), and symbols (monetary, accents, etc.) within the body of an email.
   - **Avoid long strings of emails**. Delete strings of previously answered emails when you reply.

3. **Observe the rules of "netiquette" (*Internet* + *etiquette*).**

   - **Respond promptly to an email**. Don't let emails pile up in your in-box. Check for new messages several times each day. If you will be offline for an extended period, use the auto-reply feature to tell those who email you when you expect to return.

- **Give your readers reasonable time to respond.** Consider time zone differences between you and your reader. It may be 2:00 a.m. when your email arrives for an international recipient.
- **Do not keep sending the same email over and over.** This is discourteous and will only antagonize your recipient.
- **Avoid unfamiliar abbreviations, jargon, and emoticons.** Don't use abbreviations common in personal emails (*BTW*, *LOL*) or that are used in text messaging. Include only those abbreviations and jargon that your recipients will understand (e.g., *FYI*). Also, do not use emoticons (smiley faces, sad faces, etc.) in your professional communications.
- **Don't use red flag words unnecessarily.** Stay away from words like "Urgent," "Crucial," or "Top Priority," along with accompanying red exclamation marks, in your subject line just to get your reader's attention. Your tactic will backfire, potentially upsetting readers or, worse yet, causing them to ignore any genuinely urgent messages you may send in the future.
- **Include a signature block.** A signature block, found at the end of your message, includes your name, title, and contact information (see Figures 4.4 and 4.5). Make it easy for others to contact you. Such information is crucial when you are part of a large organization or when you are communicating with someone outside of your company or agency.

4. **Adopt a professional business style.**

- **Use a salutation (greeting), but always follow your company's policy.** Use a comma before the party's name in a direct address.
  — to a colleague—Hi, Hello
  — to a customer—Dear Ms. Pietz, Dear Bio Tech
- **Get to the point right away.** Because readers receive a lot of email, they may look only at the first few lines you write. Start by briefly reminding readers why you are writing. Fill in the background that explains the purpose of your message.
- **Keep your message concise.** Cut wordy phrases, and send only the information your reader needs. Exclude unnecessary details and chatter.
- **Don't turn your email into a telegram.** "Send report immediately; need for meeting" is rude, as is a reply only with "Yes," "No," or "Sure." Save words like "Nope," "Yeah," and "Huh" for your personal emails.
- **End politely.** Let readers know in your last sentence that you appreciate their help or cooperation and look forward to their reply (see Figure 4.5).
- **Use a complimentary close, but always follow your company's policy.**
  — to a colleague—Thanks, Later, Take care,
  — to a customer—Sincerely yours, Sincerely, Best regards,
- Proofread and spell-check your email before you send it.

5. **Respect your international readers.**

- Use international English (see pages 9–11), which calls for short sentences, common words, and so on.
- Avoid using abbreviations, symbols, or measurements your reader may not know.

- Respect your reader's cultural traditions. For example, do not use first names unless the reader approves. Some cultures (East Asian, for instance) regard the use of abbreviations as discourteous.
- Always spell your reader's name, address, and country correctly, including the use of hyphens, accents, and capital letters.

6. **Ensure that your email is safe and secure.**

- **Use email protection services and software.** Always consult with your company's information technology (IT) department.
- **Avoid contracting email viruses** by deleting unopened, unsolicited email attachments.
- **Don't be a victim of identity theft, or "phishing."** Companies you do business with will never ask for personal information, such as your bank account or Social Security number.
- **Never provide company financial information unless you are sure that it will be relayed over a safe connection.** Always check with your boss before providing such information.
- **Create an email password that is not easy to guess.** Do not use a password such as "ABCDE" or "123456." Change your password regularly, and do not use the same password for all your accounts.

FIGURE 4.6 A Poorly Written Email Guilty of Flaming

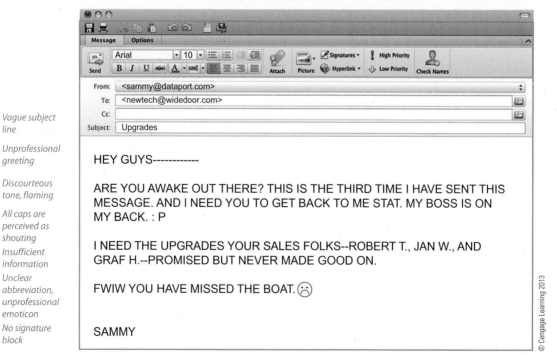

Vague subject line

Unprofessional greeting

Discourteous tone, flaming

All caps are perceived as shouting

Insufficient information

Unclear abbreviation, unprofessional emoticon

No signature block

- **Back up important files, including emails.** Save your most important and current files in case your computer contracts a virus or crashes.

Figure 4.6 shows an example of a poorly written email that violates many of the preceding guidelines. Figure 4.7 contains an effective revision that reflects the professional and courteous way the writer and his company conduct business.

FIGURE 4.7 A Revised, Effective Version of the Poorly Written Email in Figure 4.6

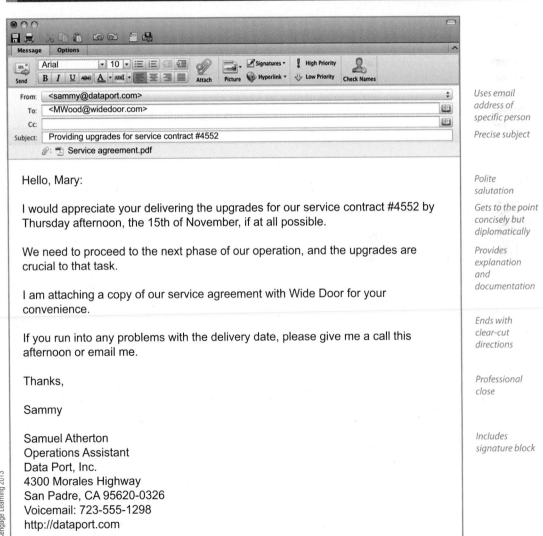

From:	<sammy@dataport.com>		*Uses email address of specific person*
To:	<MWood@widedoor.com>		
Cc:			
Subject:	Providing upgrades for service contract #4552		*Precise subject*
	📎 Service agreement.pdf		

Hello, Mary: *Polite salutation*

I would appreciate your delivering the upgrades for our service contract #4552 by Thursday afternoon, the 15th of November, if at all possible.   *Gets to the point concisely but diplomatically*

We need to proceed to the next phase of our operation, and the upgrades are crucial to that task.   *Provides explanation and documentation*

I am attaching a copy of our service agreement with Wide Door for your convenience.

If you run into any problems with the delivery date, please give me a call this afternoon or email me.   *Ends with clear-cut directions*

Thanks,   *Professional close*

Sammy

Samuel Atherton   *Includes signature block*
Operations Assistant
Data Port, Inc.
4300 Morales Highway
San Padre, CA 95620-0326
Voicemail: 723-555-1298
http://dataport.com

© Cengage Learning 2013

When Not to Use Email

Although email is convenient, easy to use, and appropriate for the routine business correspondence we have been discussing, be careful not to use it in the following situations:

- When you need a paper trail, send a printed memo or letter, whichever is most appropriate.
- Send a formal letter rather than an email when you apply for a job and for any follow-up communication.
- When you make a new business contact or welcome a new client, write a formal letter, not an email. International readers, in particular, will expect this.
- Always acknowledge a business gift or courtesy by sending a handwritten thank-you note or formal letter rather than dashing off an email.
- Never send an email in place of a letter for any type of legal notification or financial statement.

Instant Messages (IMs) for Business Use

IMs are textual conversations that take place online and in real time. They should not be confused with cell phone text messages, which occur in a different environment. Think of IMs as somewhere between a phone call and an email, or a chat with a colleague in the hallway of your office. IM conversations are almost as instantaneous as phone conversations, but at the same time they provide written records of communications like emails do. Keep in mind, though, that IMs are not just used for communication with your friends; they are also a vital part of workplace correspondence. In fact, researchers estimate that 90 percent of all businesses have used or will use IMs for routine workplace correspondence.

Exchanges through IM reflect the way people in the world of work connect and communicate with one another. IMs allow you to communicate with co-workers and managers in the same office, at remote sites, or around the globe. Crossing time zones, IMs give you access to anyone around the world who is online and connected to the same service. Figure 4.8 is an example of a workplace IM conversation.

When to Use IMs versus Emails

Like emails, IMs promote collaboration, provide a written record, and further global communication. But they are used for very different kinds of messages. Emails are more detailed than IMs. By answering the following questions, you will be better able to determine when to send an IM or an email:

1. **How quickly does my message need to be answered?** If you need information right away, use IM rather than an email because recipients will most likely reply at once if they are online.

FIGURE 4.8 An IM Exchange Between Co-workers

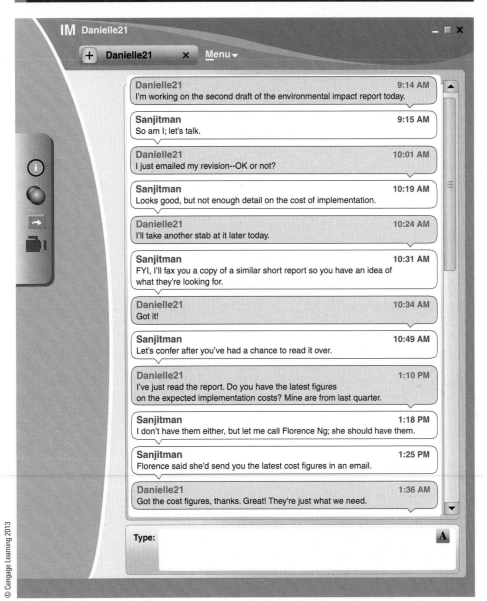

Danielle21	9:14 AM
I'm working on the second draft of the environmental impact report today.	
Sanjitman	9:15 AM
So am I; let's talk.	
Danielle21	10:01 AM
I just emailed my revision--OK or not?	
Sanjitman	10:19 AM
Looks good, but not enough detail on the cost of implementation.	
Danielle21	10:24 AM
I'll take another stab at it later today.	
Sanjitman	10:31 AM
FYI, I'll fax you a copy of a similar short report so you have an idea of what they're looking for.	
Danielle21	10:34 AM
Got it!	
Sanjitman	10:49 AM
Let's confer after you've had a chance to read it over.	
Danielle21	1:10 PM
I've just read the report. Do you have the latest figures on the expected implementation costs? Mine are from last quarter.	
Sanjitman	1:18 PM
I don't have them either, but let me call Florence Ng; she should have them.	
Sanjitman	1:25 PM
Florence said she'd send you the latest cost figures in an email.	
Danielle21	1:36 AM
Got the cost figures, thanks. Great! They're just what we need.	

IM user names are informal but appropriate

Messages are kept to 1–2 lines each

Message exchange sticks to a single topic

Clear language avoids "text speak"

Time stamp accompanies each message

Style is informal but polite

Type:

2. **How long or complex is my message?** If you need to transmit a message that is, say, more than a line or two or that contains multiple points, send an email. Using IM, you can also instantly send attachments for immediate discussion.
3. If your message requires more time than a few brief back-and-forth communications, start an email exchange that can extend over several hours or days.
4. **Do not use IMs for client or customer communications.**

Guidelines for Using IMs in the Workplace

IMs may be instantaneous and informal, but that does not mean that you can send them with little thought about their content, tone, and punctuation. Again, keep in mind that your company can monitor, trace, record, and archive your IM conversations just as it can with emails. In addition to the guidelines for writing workplace emails (pages 145–149), observe these rules for your IMs:

1. **Stay connected.** As much as possible, schedule your conversations ahead of time. Always indicate your messenger status—"Away," "Busy," "Offline. Please email me at tjones@comcast.com." If you are away, tell those on your buddy list (contact list) when you will be back ("Back in 30 minutes"), or give them alternate contact information like your email address.
2. **Keep your messages short.** Get to the point right away. A line or two at most is enough for your IMs. Don't inject unnecessary pleasantries; for example, "How was your weekend?"
3. **Write about one topic at a time.** Don't include information about two or three different topics in one IM exchange. Keep the conversation flowing in one direction, not two or three.
4. **Avoid unfamiliar abbreviations.** As in an email, commonly understood abbreviations such as "FYI" or "ASAP" are fine—even encouraged—in IMs. But avoid "textspeak" abbreviations such as "CUL8R" for "See you later," especially when writing to an international reader who may not understand them. Moreover, your boss might not appreciate a textspeak message such as "np gtg ttyl" for "No problem. Got to go. Talk to you later."
5. **Make sure the tone and style of your message are professional.** Even though IMs are the most informal business messages you can send, don't disregard professional courtesy and ethics. Never send personal messages, tell jokes, spread office gossip, or attack a co-worker or boss in an IM. Also, choose an appropriate, professional online name, not "Go-Getter Pete."
6. **Organize your contact (buddy) lists into separate groups,** such as business, co-workers, friends/family, and so on, so you do not embarrassingly send someone the wrong message.
7. **Don't prolong an IM just to chat.** When you have finished your conversation, say goodbye.
8. **Don't use IMs to send sensitive/confidential information** about personnel, financial, hiring, or medical/legal issues.
9. **Watch for viruses.** Be careful about sharing files or opening attachments via IMs.
10. **Safeguard the privacy of your contact lists** as well as the contents of any IM attachments.

Blogs

Like emails and IMs, blogs (*Web + logs*) are important correspondence tools for employees, managers, and customers. Think of a blog as an evolving website, or a daily newspaper for which managers and employees write regular short articles, or posts. Blogposts are short, conversational articles giving readers current and relevant news and commentary on a variety of issues of vital concern to your company, your organization, and your profession. Each blogpost reveals the distinctive voice of its author. Posts are generally a few paragraphs long and are often written two to three times a week, although some bloggers post their work more frequently, sometimes daily. Written in chronological order, blogs are dated, titled, and frequently archived. Figures 4.9, 4.10, and 4.11 illustrate a range of blogs.

Blogs Are Interactive

Although less formal than an article in a journal or magazine, a blog is more than just a casual entry in an online diary about the world of work. It includes more than the blogger's views. Blogs are highly interactive, allowing for a two-way or often group conversation between the author and his or her online audience. Bloggers write to express their views and to receive feedback about them. In the blogosphere, readers write comments in response to blog posts, and the blogger and other readers can reply. This interaction is the key to a blog's success. In fact, bloggers often post the number of visitors who have left comments at the site. The more posts, the better. Bloggers, therefore, need to

1. attract readers,
2. build relationships,
3. promote and market products or services, and
4. respond to suggestions.

Blog Sponsors

Blogs carry different types of messages, depending on who sponsors them and why. Individuals can host their own blogs to express their opinions on any subject, from world affairs to their interests, hobbies, careers, or families. Organizations host blogs, too. For instance, the Sierra Club, which is devoted to preserving the environment, has blogged on such topics as protecting endangered species, promoting funding of nature trails, and buying eco-friendly furniture. Almost every government agency has its own blog, or multiple blogs. The FDA, for instance, uses its blogs to keep consumers up to date about potential health risks, such as an outbreak of influenza or the spread of botulism from contaminated meat or dairy products. And every type of business, from major international corporations to small, local firms, blogs about its products, services, workforce, and commitment to customers and community. It is impossible to estimate the number of blogs in the blogosphere. But one thing is sure. Your employer will at some point ask you to blog about a work-related activity. The following sections will show you how to do that successfully.

FIGURE 4.9 An Internal Blog

HMC   Hong, McCarson, and Steinway, LLC

Blog topics clearly differentiated

Topic: New antivirus software
(comments: 6)

Topic: Importing HMS logo into memos, letters, etc.
(comments: 7)

Uses clear and concise title

Chooses an informal yet professional tone

Acknowledge responses to blog

Keeps the blog post brief and chunks text

Lets readers know IT is responsive to their needs

Tone is conversational but professional

Offers practical help for employees

Invites further responses

"Comments" link allows for further discussion

Topic: Using the new binding machines
10/5/12, 10:43 a.m. (mschwartz): Hi, this is Maxine in IT. Congratulations to those of you who have completed our training workshops on the new binding equipment, and I look forward to working with individuals in the marketing and facilities departments, who are scheduled for a session next week.

Overall, I think the sessions have gone well. However, a number of questions have come up about the new binding machines, and so I thought the best thing to do would be to post the most frequently asked ones here, with answers:

1. Why are the binding machines located only on the 6th floor? We're just waiting for three additional machines to arrive from the supplier within the next two weeks. We will put two of the new machines in the 5th floor copy room (5-204) for legal services and one in the 3rd floor copy room (3-122) for marketing and facilities.

2. Where are the supplies for the binding machines located? You can find the supplies in each of three copy rooms. IT would like to apologize about the shelf locations not being more clearly labeled. We fixed that problem. Thanks for letting us know about it. Also, to help our employees, IT has posted a detailed sheet for supplies and procedures on each copy room door.

3. Why can't I get the laminated insert pages to line up properly? Here's a tip to help you with this one: Make sure that the rounded edges of the inserts are facing out and that the squared edges are lined up with the inside edges of the paper. That should help.

4. How do I avoid damaging legal documents when using the machines? To be on the safe side, insert a few pages of scrap paper into the machine and do a test bind before inserting original documents. You'll then be sure that the machine is aligned properly.

I hope these answers help everyone out. If you have further comments or questions, please don't hesitate to email me or call me at Extension 304.
(comments: 14)

Topic: What is the company policy on maternity/paternity leaves?
(comments: 7)

Topic: PowerPoint presentations — is there a specialist in-house to help?
(comments: 3)

Two Types of Blogs

Blogs can be either internal or external, depending on their audience and use in the workplace.

Internal Blogs

Internal blogs are among the most versatile forms of workplace correspondence. They are designed exclusively for use by in-house management (up and down the corporate ladder) and by employees to post their messages and comments or questions on the company network. Many internal blogs are intended for a company-wide audience, for example, announcing an event or a new workplace policy, such as further greening the workplace. But some internal blogging is directed only to individuals in a given department or area, such as engineers tackling an energy problem or nurses in a large health care organization discussing new treatments for burn patients. Every internal blog, though, is aimed at making the workplace safer, more productive, and professionally and personally more satisfying.

Internal blogging serves many functions, including these:

- Informing employees about vital company news
- Helping employees to better understand and perform their jobs
- Conducting virtual meetings without having to make arrangements for face-to-face gatherings
- Enhancing collaboration through the interaction of blog posts and comments
- Providing a forum for workplace discussions
- Improving employee participation and morale by inviting suggestions and questions

Note how the internal blog in Figure 4.9 fulfills many of these functions.

Always follow your company's blogging policies, but be especially careful that you do not divulge information that may be confidential or sensitive. Your employer will expect you to observe the same guidelines whether posting your own blogs or commenting on others' (see pages 158–160). What you say on your own blog can reflect positively or negatively on your employer. Finally, don't let internal blogging take up so much of your time that you neglect your other duties.

External Blogs

External, or business, blogs are essential marketing and public relations tools in the world of e-commerce. They provide a fast, informal way to get information out about a company's mission and activities, as in Figures 4.10 (see pages 156–157) and 4.11 (see pages 161–162). A carefully designed and persuasively written business blog can announce and sell new products, services, or technology; share information about employee accomplishments; and describe the company's contributions to the community or the environment, as in Figure 4.10. It can provide updates about corporate changes in personnel, locations, benefits, and so on. Moreover, a business blog can be useful in announcing unpleasant news, such as environmental problems, product recalls, or the discontinuation of a service, brand, or model, as in Figure 4.11. An external blog allows companies to tell their side of the story—their interpretation of events and their clarification of the issues—quickly and publicly.

FIGURE 4.10 An External Blog on Choosing an Eco-Friendly Product

WFM HOME   STORES   PRODUCTS   RECIPES   HEALTH STARTS HERE   VALUES   COMPANY   FORUMS   SUBSCRIBE VIS RSS

WHOLE STORY
the official whole foods market® blog

SEARCH

Attention-grabbing headline

Get to Know Your Tilapia

by **Carrie Brownstein**, **January 4th, 2011** | **Permalink** | **Email this**

Convinces readers right away about the safety of the product

Unlike conventional grocers who may source tilapia from any old place as long as the price is right, Whole Foods Market sources all seafood, including tilapia, according to our Quality Standards. In the case of tilapia, we source from just three supplier partners, all of whom have passed a third-party audit to ensure that they meet our rigorous quality standards.

Provides essential jargon-free background information for readers

Our primary supplier partner, Tropical Aquaculture Inc., brings us tilapia from Santa Priscila, located in beautiful Ecuador. Santa Priscila practices polyculture by raising shrimp and tilapia together in the same ponds. This helps reduce waste and water pollution, as tilapia consume feed that the shrimp leave behind and help get rid of organic matter that otherwise could end up in the environment. The farm also recirculates its water, which further helps to protect water quality surrounding the farm.

Addresses the reader directly but sincerely; uses a friendly conversational tone

And you'll be glad to know that our Quality Standards for Aquaculture prohibit the common industry practice of using the hormone methyl testosterone to reverse the sex of tilapia. Conventional tilapia producers prefer to raise only male fish so that the fish put their energy into growth rather than reproduction and grow to a larger, more marketable size. Our farmer partners, however, grow fish the old fashioned way: they let the fish reproduce naturally. Then they separate the males and females by hand and raise them in separate ponds.

Documents the rigorous standards consumers expect from Whole Foods

And as always, Whole Foods Market prohibits slaughterhouse by-products from avian or mammalian species in feed. Fortunately, tilapia are naturally omnivorous fish that don't require a lot of fishmeal in their feed, which helps our tilapia suppliers meet our goal of reducing pressure on wild populations of fish that are used to produce animal feed, but are also important species in marine food webs. In fact, Santa Priscila's feed (as well as other supplier partners' feed), uses trimmings from other fish species processed for seafood, which also reduces wastes.

Provides hyperlink for easy reference

We launched our Quality Standards for Aquaculture in 2008 and they still remain the toughest quality standards for farmed seafood in the industry. Fish farmers who want to partner with us must complete a lengthy application detailing all of their farming practices. And it's more than just words; third-party auditors verify that the farm is meeting our standards before any of their fish makes its way to our stores. Not only that, but suppliers must continue to pass annual inspections for as long as they partner with us.

Includes logo within blog to show consumers exactly what to look for when purchasing eco-friendly tilapia

So, how do you know you're purchasing farmed seafood that meets Whole Foods Market's strict standards? Look for our aquaculture logo — Responsibly Farmed — at Whole Foods Market stores. That symbol means that the fish has been third-party verified to meet our standards.

Courtesy of Whole Foods Market. "Whole Foods Market" is a registered trademark of Whole Foods Market IP, L.P.

FIGURE 4.10 (Continued)

8 Responses to "Get to Know Your Tilapia"

Tricia: I just purchased some tilapia at WFM last night specifically for this reason. I don't shop at WFM for everything, but I think it's important to purchase items in industries that can be unsafe. The fish was reasonably priced and tasted great!
January 4th, 2011 at 3:55 pm

dining room table: I have heard of this fish before and they told me that this is really something so delicious.
January 5th, 2011 at 7:02 am

Lynda Reynolds: Tilapia are a freshwater fish, not seafood. And how did the prison system in Colorado go about getting a contract with Whole Foods to sell? . . . is this something I can start in California or do we already have the same program?
January 5th, 2011 at 1:36 pm

Sharon Miracle: I commend you for taking these steps to protect the aquaculture, and for helping protect us humans from ingesting more unnecessary hormones which may have negative consequences on our bodies over time.
January 5th, 2011 at 3:06 pm

Kat: Thank you for this info about tilapia. I just starting eating it but I did not know that hormones are added to it by certain suppliers.
January 5th, 2011 at 5:34 pm

Ellie: Most tilapia is grown in such conditions that it is gross, if not unhealthy, to eat. It is wonderful to hear yours is worth eating. Thanks!
January 5th, 2011 at 9:30 pm

Ryan: But what are they fed? Most tilapia are fed corn, resulting in an extremely high omega 6 to omega 3 ratio.
January 8th, 2011 at 10:30 pm

Bev Baker: Just checking to ensure that the tilapia are not fed GMO corn???
January 10th, 2011 at 8:44 pm

ON THE WEB

Whole Foods Market photos on Flickr

Whole Foods Market on Facebook

Whole Foods Market updates on Twitter

VIDEOS & PODCASTS

View our growing library of video content.

BE GOOD TO YOUR
WHOLE BODY

Audio podcast all about natural body care and supplements.

CATEGORIES

Back to School **(23)**

Best Meal of the Week **(15)**

Cheese **(27)**

Community – Local and Global **(20)**

Farm to Market **(67)**

South **(2)**

Field Reports **(20)**

Floral **(20)**

Food & Recipes **(357)**

Food Issues **(45)**

Food Podcasts **(68)**

Food Safety **(11)**

Grass-fed Beef **(12)**

Green Action **(115)**

Grocery **(30)**

Guidelines for Writing Business Blogs

To post a successful blog, follow all the guidelines for writing a business email or using workplace IM. Always be ethical and honest, and document what you say. Also, your writing cannot be sloppy or careless; you have to use proper spelling and punctuation. Be diplomatic, whether you are writing to co-workers, as Maxine Schwartz does in Figure 4.9, or customers, as the blogs in Figures 4.10 and 4.11 do. Avoid sounding curt, condescending, or arrogant. Keep in mind, too, that an external blog is your employer's official, and many times daily or weekly, publication, as we see in Figure 4.10. Your employer may require you to get your post approved by a blog administrator to make sure it follows corporate rules. The guidelines that follow will also help you to write appropriate business blog posts.

Use the Right Tone to Attract Readers

Essential to any business blog's success is getting information from readers—their views, concerns, and feedback. Whether you are a manager or an employee, your business blog needs to reveal the personal side of your company. Your blog needs to sound sincere and friendly, welcoming readers to your site. As we saw, every blogger's goal is to attract visitors to his or her site and to keep them coming back to read more. Using the interactive features of a blog, you can make it easy for readers to contact you. A friendly and inviting tone will let readers know you want to hear their views and will take them into account. In subsequent posts, you can address their views and concerns, as Carrie Brownstein does in Figure 4.10 and Clay Denton-Tyler does in Figure 4.11.

Follow Company Protocol

- **Project your company's best image.** Keep your company's history, mission, and reputation in mind when you prepare your blog. Be enthusiastic about its products, services, workforce, company mission, and commitment to the environment. Do not make your company look bad by undermining management or criticizing a vendor or a competitor. And never attack a boss or co-worker.
- **Respect your employer's confidentiality.** Guard your company's trade secrets. Do not blog about anything that might reveal confidential, restricted, or otherwise off-limits information. Topics to stay away from include any ongoing research and development of products and services, sales and marketing plans, financial matters including stocks, and personnel matters.
- **Avoid making your company liable for false or misleading information.** Don't make promises, offer guarantees, or provide additional warranties unless they have been approved by upper management. Be careful about using the pronoun *we,* which implies you are speaking for your employer.
- **Be professional.** Avoid posting stories and pictures unrelated to your work on the company's blog. Don't comment on company policies or make political comments, all of which may be grounds for dismissal.

Target Your Audience

- **Know what your audience cares about.** Try to reach current and potential customers by anticipating their questions and needs. Be aware of their attitudes, likes, and dislikes. Read replies to previous blog posts. Be open to suggestions, and acknowledge readers' insights. Tell readers how and why your blog will help them.
- **Write an attention-grabbing headline.** Attract readers with a title that tells them how and why they can profit from reading your blog, and encourage them to respond to your post; for example, "Getting to Know Your Tilapia" in Figure 4.10 and "A Power Tool Even Better Than the PH-450" in Figure 4.11. Avoid vague, boring headlines, such as "Important News," "Something You Need to Know," and "Any Further Ideas?"
- **Determine if your blog will attract an international audience** as well as native English speakers (see pages 8–11). To accommodate a global readership, avoid jargon and unclear abbreviations.
- **Date every blog post so readers can follow a conversation.** Update your blog to make sure the information you give readers is current and accurate.
- **Make it easy for readers to respond to your post.** Welcome feedback. Consider your blog a place where you want to listen to readers' comments. Tell them where and how to reply. Like the blogs in Figures 4.10 and 4.11, refer to customer posts to show how concerned you and your company are about readers' opinions.

Make Your Blog Persuasive

- **Structure your blog so that your first paragraph comes to the point at once and tells readers what you are blogging about and why.** For example, in Figure 4.10, Carrie Brownstein reassures readers that Whole Foods still faithfully follows its tough quality standards for aquaculture. In Figure 4.11, Clay Denton-Tyler clearly states he has information on "why this product will no longer be manufactured," a question his audience is eager to see him answer.
- **Share your story.** Draw from your own experiences. Offer your perspectives. Provide firsthand information that shows readers you are knowledgeable and sincere. Observe how, in Figure 4.11, Denton-Tyler expresses his views as an owner of the popular, but discontinued, PH-450 without in any way compromising his company's position or decision.
- **Highlight any new, improved, or special features.** Note how Denton-Tyler points to the benefits the new power tool offers customers.
- **Use facts and statistics to develop your message or point of view.** Honest numbers sell products and services. Include units sold, costs, and so on. Note how Denton-Tyler wisely cites a lower price to promote the new model SHP-1000.

Write Concisely and Sincerely

- **Keep your posts short and easy to read.** Your blog should be simple and practical. Most blogs are no more than a few paragraphs. Don't turn your blog into a report or a compilation of technical data.
- **Adopt a casual, conversational style.** Be personable and friendly. Sound authentic and upbeat. Emphasize your interest in your readers. Regard your blog as a fruitful chat between you and your readers. Don't weigh them down with long, windy paragraphs that can bore or confuse your audience.

Document Your Sources, Including Visuals

- If you use someone else's statistics, surveys, illustrations, or ideas, get permission first from the individual or the company that owns the copyright.
- Quote accurately, but do not include an extended quote without obtaining permission.
- Do not use commercial trademarks in place of generic terms, such as Jacuzzi for hot tub, Clorox for bleach cleaner, Kleenex for facial tissue. But note the appropriate use of the Whole Foods logo in the blog in Figure 4.10.

Case Study

Writing a Blog to Keep Customer Goodwill

Clay Denton-Tyler is a district manager for PowerHouse, Inc., a company that sells a large line of power tools. The company recently decided to discontinue one of its most popular models, the PH-450, which had enjoyed wide brand recognition and high consumer ratings. Customers had been blogging Powerhouse to complain about the company's decision, and Denton-Tyler faced the difficult challenge of responding to customer posts. His blog can be seen in Figure 4.11.

To respond successfully, he had to consider his audience's needs, as voiced in their posts to the PowerHouse blog. Since his readers were loyal customers, he did not want to lose their business and goodwill. But he had to acknowledge that they were understandably disappointed that a well-received product was being taken off the market. He also had to be credible, honest, and diplomatic in addressing their needs and his company's response to its customers. Most important, he had to reassure them that PowerHouse was sincerely interested in their views. He also had to convince them that the replacement model PowerHouse was offering was better and cheaper than the discontinued PH-450.

But in the interactive world of blogging, he recognized that he was also writing to potential customers, and he wanted them to view his company as being responsive to and driven by the needs of its customers. Moreover, he realized that his post would be a part of an ongoing public discussion about the new model and his company, and so he wanted to answer as many questions as he could while keeping the conversation going—all in a positive direction—and, ideally, attracting new customers around the globe.

FIGURE 4.11 An External Blog

PowerHouse, Inc.

About Us | Products | International | Jobs | Mobile | RSS

⊙ PowerBlog   ○ All of Powerhouse, Inc.   🔍 Search this site   Go

PowerBlog

| All | Company News | Product News | Distribution | Manufacturing |

<< Previous Post   >> Next Post

Today's Post (August 15, 2012):

>> A Power Tool Even Better than the PH450?
by Clay Denton-Tyler, District Manager

© iStockphoto.com/ Prill Mediendesign & Fotografie

Recent Posts:

International Dyanamics, SE CFO announces retirement.

PowerHouse announces the discontinuation of the PH-450.

PowerHouse opens new retail outlets in Sydney, Australia, and Dunedin, New Zealand.

Many of our loyal customers disagree with PowerHouse's decision to discontinue manufacturing the PH-450 All-in-One Power Tool. It is always great to hear from our customers and to receive their feedback, even when they believe we've done something wrong. To help our customers better understand our perspective, let me fill in some of the background about why the PH-450 will not be available anymore.

Discontinuing the PH-450 was not an easy decision. After all, this was the product that first brought our company to national attention and widespread customer acceptance. Also, I know from many complimentary emails and replies to earlier posts, as well as from my personal experience as a proud owner of the PH-450, that our customers have always applauded its price, compact design, durability, and all-weather usability. As one of you put it, "Why kill a popular product that has worked so well for 20 years?"

International Dyanamics, SE acquires quality Swiss Home Products company.

PowerHouse goes international.

More

All of these responses are true, but the good news is that even though the PH-450 is being discontinued, our customers now have a very similar but improved alternative. Recently, our parent company, International Dyanamics, SE, acquired a new multipurpose tool from Swiss

Search engine and navigation links help readers find information

Additional links aid site navigation

Provides a clear and concise title and date

Writes to a general audience and avoids jargon

Thanks customers for feedback

Uses a conversational but professional tone

Acknowledges customers' disappointment

Attempts to persuade customer to switch to a new model

(Continued)

FIGURE 4.11 (Continued)

Describes benefits of new model and why it is being marketed

Home Products. The SHP-1000 is not only just as compact, durable, and weather-friendly as the PH-450, but it offers several additional features, such as a nail gun attachment and lifetime limited warranty. And due to an excellent distribution deal negotiated between International Dyanamics, SE, and Swiss Home Products, we can sell it at less than 30 percent of the retail cost of the PH-450.

Sympathizes with readers but offers personal endorsement characteristic of bloggers at same time

Tone is sincere and friendly

I know it is hard to say goodbye to a reliable helper, but like many products in our increasingly technical age, the PH-450 is being replaced by a more efficient alternative model. I will miss the old PH-450, but I have found that the SHP-1000 is even more effective in my shop at home. Adapting to a new model has never come easier for me. Why not give it a try? Thanks. I would like to hear from you.

Comments link allows for further discussion

Comments: (21)
- Sign in to add a comment
- First time users, please register first, in order to add a comment

© Cengage Learning 2013

Conclusion

Writing memos, emails, IMs, and blogs is a routine yet important part of every employee's job. Although much shorter and far more informal than reports or proposals, these basic types of business correspondence keep crucial information flowing among co-workers, management, vendors, and others, so that a company can meet its day-to-day obligations. In addition, the information contained in these short messages often helps you to gather information you need to write longer documents.

By following the guidelines in this chapter, you will be better able to write clear, concise, and ethical messages for your audience. To be a successful employee, always respond promptly and courteously to memos, emails, IMs, and blogs. Your annual evaluations may in part depend on how well you research, draft, revise, and format these routine business correspondences.

✓ Revision Checklist

Memos
- ☐ Used appropriate and consistent format.
- ☐ Followed employer's policy of when to send print or email memos.
- ☐ Announced purpose of memo early and clearly.
- ☐ Organized memo according to reader's need for information, putting main ideas up front, giving documentation, and supplying conclusion.
- ☐ Wrote concise and clear memo suitable for audience.
- ☐ Included bullets, lists, boldfacing, and underscoring where necessary to reflect logic and organization of memo and make it easier to read.
- ☐ Determined when to send hard copy or an e-copy of a memo.
- ☐ Refrained from overloading reader with unnecessary details.

Faxes
- ☐ Verified reader's fax number and sent cover sheet with phone number to call in the event of transmission trouble.
- ☐ Excluded anything confidential or sensitive if reader's fax machine is not secure.

Email
- ☐ Did not send unsolicited or confidential email.
- ☐ Sent to reader's correct address.
- ☐ Formatted email with acceptable margins and spacing.
- ☐ Observed netiquette; avoided flaming.
- ☐ Wrote a separate message rather than returning sender's message with a short reply.
- ☐ Kept paragraphs short but used full—not telegraphic—sentences.
- ☐ Avoided unfamiliar abbreviations or terms that would confuse a reader.
- ☐ Received permission to repeat or incorporate another person's email.
- ☐ Observed all legal obligations in using email.
- ☐ Safeguarded employer's confidentiality and security by excluding sensitive or privileged information.
- ☐ Included enough information and documentation for reader's purpose.
- ☐ Honored reader by observing proper courtesy.
- ☐ Began with friendly greeting; ended politely.
- ☐ Considered needs of international audience.
- ☐ Used antivirus program.
- ☐ Did not forward or reply to spam.

Instant Messaging
- ☐ Used IM only for professional, job-related communications.
- ☐ Kept messages short—not over a line or two.
- ☐ Avoided "textspeak" in business IMs.
- ☐ Notified readers when IM was offline and back online.
- ☐ Did not send anything confidential through an IM exchange.

Blogs

☐ Posted nothing critical of employer or co-workers and nothing embarrassing, offensive, or confidential.

☐ Made posts conversational and informal, yet professional.

☐ Dated every blog.

☐ Targeted my audience.

☐ Included attention-grabbing headline.

☐ Posted only current and relevant information.

☐ Provided a place for readers to give feedback.

Exercises

1. Write a memo to your boss saying that you will be out of town two days next week and three days the following week for **one** of the following reasons: (a) to inspect some land your firm is thinking of buying, (b) to investigate some claims, (c) to look at some new office space for a branch your firm is thinking of opening in a city five hundred miles away, (d) to attend a conference sponsored by a professional society, or (e) to pay calls on customers. In your memo, be specific about dates, places, times, and reasons.

2. Write a memo to two or three of your co-workers on the same subject you chose for Exercise 1.

3. Send a memo informing the public relations department of your company that you are completing a degree or a certificate program. Indicate how the information could be useful for your firm's publicity campaign.

4. Write a memo notifying the human resources department that there is a mistake in an insurance claim you filed. Explain exactly what the error is, and give precise figures.

5. You are the manager of a major art museum. Write a memo to various department heads at your museum giving them the following information. Use proper memo format.

Old hours:	Mon.–Fri. 9–5; closed Sat. except during July and August, when you are open 9–12
New hours:	Mon.–Th. 8:30–4:30; Fri.–Sat. 9–9
Old rates:	Adults $12.00; senior citizens $5.00; children under 12 $3.00
New rates:	Adults $15.00; senior citizens $7.00; children under 12 $5.00
Added features:	Paintings by Thora Horne, local artist; sculpture from West Indies in display area all summer; guided tours available for parties of six or more; lounge areas will offer patrons sandwiches and soft drinks during May, June, July, and August

6. Select some change (in policy, schedule, or personnel assignment) you encountered in a job you held in the last two or three years, and write an appropriate memo describing that change. Write the memo from the perspective of your current or former employer to explain the change to employees.

7. How would the memos in Exercises 1 to 6 have to be rewritten to make them suitable as an email message? Rewrite one of them as an email.

8. Bring five or six examples of a company's or an organization's email to class. As a group activity, evaluate them for professional style, tone, layout, and preciseness of message. Write a carefully organized email to your instructor evaluating the effectiveness of these emails. Include the sample emails as an attachment with your report.

9. Send a fax to a company or an organization requesting information about the products or services it offers. Include an appropriate cover sheet.

10. Write an email requesting information from one of the following types of businesses. Submit a copy of your email request, along with the response, to your instructor.
    a. From an airline: an up-to-date schedule along a certain route and information about any bonus-mile or discount programs
    b. From a catalog order company: information about any specials for Internet users
    c. From a stock brokerage firm: free quotes or research about a particular stock
    d. From a resort: special rates for a given week
    e. From a professional organization to which you belong about any conferences to be held in your city or state

11. Write an email with one of the following messages, observing the guidelines discussed in this chapter.
    a. You have just made a big sale, and you want to inform your boss.
    b. You have just lost a big sale, and you have to inform your boss.
    c. Tell a co-worker about a union or national sales meeting.
    d. Notify a company to cancel your subscription to one of its publications because you find it to be dated and no longer useful in your profession.
    e. Request help from a listserv about research for a major report you are preparing for your employer.
    f. Advise your district manager to discontinue marketing one of the company's products because of poor customer acceptance.
    g. Write to a friend studying finance at a German, Korean, or South American university about the biggest financial news in your town or neighborhood in the last month.

12. Rewrite the following email to your boss to make it more professional.

    Hi—

    This new territory is a pain. Lots of stops; no sales. Ughhhh. People out here resistant to change. Could get hit by a boulder and still no change. Giant companies ought to be up on charges. Will sub. reports asap as long as you care rec.

The long and short of it is that market is down. No news = bad news.

13. As a collaborative venture, join with three or four classmates to prepare one or more of the email messages for Exercise 11. Send each other drafts of your messages for revision. Submit the final draft to your instructor.

14. Assume you have received permission to repost in an email all or part of the article on microwaves (pages 47–49) or virtual reality and law enforcement (pages 429–433). Prepare an email message to a listserv or Usenet group containing part of the article you have chosen.

15. Send your instructor an email message about a project you are now working on for class, outlining your progress and describing any difficulties you are having.

16. You have just missed work or a class meeting. Email your employer or your instructor explaining the reason and telling how you intend to make up the work.

17. As a group activity, instant message two or three other members of your collaborative writing team on a project you are working on. Print out your IM exchanges during this time, and submit them to your instructor.

18. As a collaborative project, write three or four external blog posts about some aspect of your current job or a previous job. Share with readers news about your company's products or services, technology you are using, professional travel, community service, work with international colleagues, and so forth. Be sure that your posts show your company, department, or agency in a good light.

19. Send a short post (200–300 words) to your company's blog administrator about a recent accomplishment you or your office, department, or section achieved. Include a link to a relevant site for readers to visit for further information.

CHAPTER **5**

Writing Letters

Some Basics for Communicating with Audiences Worldwide

Letters are among the most important writing you will do on your job. Businesses worldwide take letter writing very seriously, and employers will expect you to prepare and respond to your correspondence promptly and diplomatically. Letters will take you longer to research, draft, and revise than a typical email or memo because they are intended for readers outside your company—clients, customers—on whom your firm depends for its business. Your signature on a letter tells readers that you are accountable for everything in it. The higher up the corporate ladder you climb, the more letters you will be expected to write.

Because letter writing is so significant to your career, this chapter introduces you to the entire process and provides guidelines and problem-solving strategies. It also shows you how to write for international readers. Chapter 6 will introduce you to the most common types of letters you can expect to write on the job.

Access chapter-specific interactive learning tools, including quizzes and more in your English CourseMate, accessed through www .cengagebrain.com.

Letters in the Age of the Internet

Even in this age of the Internet, letters are still vital in the world of work. In fact, they have a special status in the workplace, where individuals receive so much of their information through electronic communications. A professional-looking letter is one of the most significant symbols in the business world for the following reasons:

1. **Letters represent your company's public image and your competence**. A firm's corporate image is on the line when it sends a letter. Carefully written letters can create goodwill; poorly written letters can anger customers, cost your company business, and project an unfavorable image of you.

2. **Letters are far more formal—in tone and structure—than any other type of business communication**. They use formal conventions not found in other types of correspondence. Memos, emails, and IMs are the least formal communications.

3. **Letters constitute an official legal record of an agreement**. They state, modify, or respond to a business commitment. When sent to a customer, a signed letter

constitutes a legally binding contract. Be absolutely sure that what you put in a letter about prices, guarantees, warranties, equipment, delivery dates, and/or other issues is accurate. Your readers can hold you and your company accountable for such written commitments.

4. Letters symbolize your professionalism. They need to be ethical and accurate, stressing that you are a well-informed correspondent who will provide up-to-date information and answer questions knowledgeably and truthfully.

5. Many businesses require letters to be routed through channels before they are sent out. Because they convey how a company looks and what it offers to customers, letters often must be approved at a variety of corporate levels. Emails generally do not need prior approval.

6. Letters are more permanent than emails. They provide a documented hard copy. Unlike emails that can be deleted, letters are often logged in, archived, and bear a written, authorized signature.

7. A letter is the official and expected medium through which important documents and attachments (contracts, specifications, proposals) are sent to readers. Sending such attachments via email or with a memo lacks the formality and respect readers deserve and expect.

8. A letter is still the most respectful and approved way to conduct business with many international audiences. These readers see a letter as more polite and honorable than an email for initial contacts and even for subsequent business communications.

9. Though it is a formal document, a letter nonetheless assures the reader of your personal attention. Written for a particular reader, your letter lets this individual know that you want to build or sustain a business relationship with him or her.

10. A hard-copy letter is confidential. It is more likely to be delivered to the proper recipient in its sealed envelope and is less likely to be forwarded to unintended readers, as an email might be.

Letter Formats

Letter format refers to the way in which you print a letter—where you indent and where you place certain kinds of information. Your company will often specify which format it prefers. Several letter formats exist. Two of the most frequently used business letter formats are full-block and modified-block, but you should also be familiar with a third format, the semi-block.

Full-Block Format

In full-block format, all information is flush against the left margin, double-spaced between paragraphs. Figure 5.1 shows a full-block letter. Many employers prefer this format when your letter is on *letterhead stationery* (specially printed paper

FIGURE 5.1 Full-Block Letter Format with Appropriate Margins

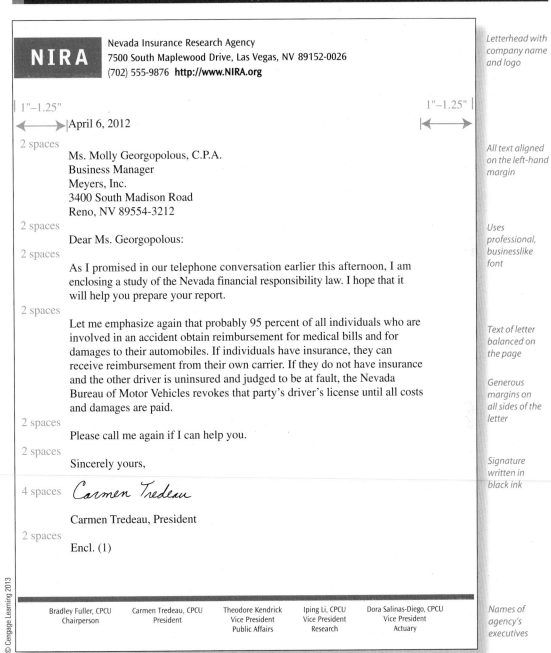

NIRA

Nevada Insurance Research Agency
7500 South Maplewood Drive, Las Vegas, NV 89152-0026
(702) 555-9876 **http://www.NIRA.org**

Letterhead with company name and logo

1"–1.25"   1"–1.25"

April 6, 2012

2 spaces

Ms. Molly Georgopolous, C.P.A.
Business Manager
Meyers, Inc.
3400 South Madison Road
Reno, NV 89554-3212

2 spaces

Dear Ms. Georgopolous:

2 spaces

As I promised in our telephone conversation earlier this afternoon, I am enclosing a study of the Nevada financial responsibility law. I hope that it will help you prepare your report.

2 spaces

Let me emphasize again that probably 95 percent of all individuals who are involved in an accident obtain reimbursement for medical bills and for damages to their automobiles. If individuals have insurance, they can receive reimbursement from their own carrier. If they do not have insurance and the other driver is uninsured and judged to be at fault, the Nevada Bureau of Motor Vehicles revokes that party's driver's license until all costs and damages are paid.

2 spaces

Please call me again if I can help you.

2 spaces

Sincerely yours,

4 spaces   *Carmen Tredeau*

Carmen Tredeau, President

2 spaces

Encl. (1)

All text aligned on the left-hand margin

Uses professional, businesslike font

Text of letter balanced on the page

Generous margins on all sides of the letter

Signature written in black ink

Bradley Fuller, CPCU	Carmen Tredeau, CPCU	Theodore Kendrick	Iping Li, CPCU	Dora Salinas-Diego, CPCU
Chairperson	President	Vice President Public Affairs	Vice President Research	Vice President Actuary

Names of agency's executives

with the company's name and logo, business and Web addresses, fax and telephone numbers, and sometimes the names of its executives).

Modified-Block Format

In modified-block format (Figure 5.2), the writer's address (if it is not imprinted on a letterhead), the date, the complimentary close, and the signature are positioned at the center point and then keyed toward the right side of the letter. The date aligns with the complimentary close. The inside address, the salutation, and the body of the letter are flush against the left margin.

FIGURE 5.2 Modified-Block Letter Format

Writer's address and date are indented

7239 East Daphne Street
Mobile, AL 36608-1012

September 28, 2012

Inside address and salutation are not indented

Mr. Travis Boykin, Manager
Scandia Gifts
703 Hardy St.
Hattiesburg, MS 39401-4633

Dear Mr. Boykin:

First lines of paragraphs can be indented or not indented

I am writing to see if you currently stock the Crescent pattern of model 5678 and how much you charge per model number. I would also like to know if you offer special prices for multiple-box orders.

Your store has been highly recommended to me by several colleagues, who have praised your service and the excellent quality of your products.

I look forward to working with you.

Sincerely yours,

Complimentary close and writer's name are indented

Arthur T. McCormack

Arthur T. McCormack

© Cengage Learning 2013

Semi-Block Format

The semi-block format (see Figure 6.1, page 211) looks just like the modified-block format in terms of aligning the date line with the complimentary close, signature, and enclosure line at the center point of the letter. But the paragraphs in the semi-block format are always indented five to seven spaces. Though not used as frequently as the full-block or modified-block formats, the semi-block format is a template an employer may ask you to use.

Continuing Pages

To indicate subsequent pages if your letter runs beyond one page, use one of the two conventions below. Always include the recipient's name, the date, and the page number. Also, make sure you continue at least two lines from the preceding page. But never put just your complimentary close and your name on a continuation page.

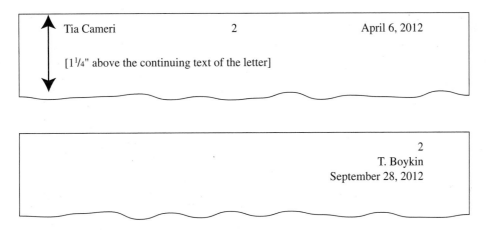

Standard Parts of a Letter

A letter contains many parts, each of which contributes to your overall message. The parts and their placement in your letter form the basic conventions of effective letter writing. Readers look for certain information in key places.

The parts of a letter discussed in the following sections will appear in every letter you write. Figure 5.3 is a sample letter containing all of the parts discussed here. Note where each part is placed in the letter.

Heading

The heading of a letter may be either your company's letterhead or your full return address. Figure 5.2 contains an example of a full return address when letterhead is not used. If you design your own letterhead, as in Figures 6.1 and 6.2 (see pages 211–212), make sure it looks professional.

FIGURE 5.3 A Sample Letter, Full-Block Format, with All Parts Labeled

Heading (letterhead)

Madison and Moore, Inc.
Professional Architects

7900 South Manheim Road
Crystal Springs, NE 71003-0092
Phone 402-555-2300 **www.mmi.com**

Date line

July 12, 2012

Inside address with correct state abbreviation and zip code

Ms. Paula Jordan
Systems Consultant
Broadacres Development Corp.
12 East River Street
Detroit, MI 48001-0422

Salutation

Dear Ms. Jordan:

Thank you for your letter of July 6, 2012. I have discussed your request with the staff in our planning department and have learned that the design modules we used for our Vestavia project are no longer available.

Body of letter

In searching through our archives, however, I came across the enclosed catalog from a California firm that might be helpful to you. This firm, California Concepts, offers plans very similar to the ones you are interested in, as you can tell from the design I flagged on page 23 of their catalog.

I hope this will help you, and I wish you every success in your project.

Complimentary close

Sincerely yours,

Company name

MADISON AND MOORE, INC.

Signature

William Newhouse

Writer's name and title

William Newhouse
Office Manager

Enclosure Copy notation

Encl.: Catalog
cc: Planning Department

Date Line

Try to leave four lines below the letterhead to the date line. Spell out the name of the month in full—"September" or "March" rather than "Sept." or "Mar." The date line is usually keyboarded this way: November 12, 2012. But different cultures express dates in different ways. Most countries, including those in Europe, list the day first, then the month, and then the year. See Figure 5.10 on page 193 for an example of how to correctly provide a date for international readers.

Inside Address

The inside address, the address of the recipient, is always placed against the left margin, two lines below the date line. It contains the name, title (if any), company, street address, city, state, and zip code of the person to whom you are writing. Use the official state abbreviations of the U.S. Postal Service found at www.usps.gov. Single-space the inside address, and do not use any punctuation at the end of the lines.

Dr. Mary Petro
Director of Research
Midwest Laboratories
1700 Oak Drive
Rapid City, SD 56213–3406

Always try to write to a specific person rather than just "Sales Manager" or "President." To find the name of the individual you need, check previous correspondence, email lists, or the company's or individual's website, or call the business. Make sure you use an abbreviated courtesy title (Ms., Mr., Dr., Prof.) before the recipient's name for the inside address (e.g., Capt. María Torres, Mr. A. T. Ricks, Rev. Siam Tau). Use *Ms.* when writing to a woman unless she has expressly asked to be called *Miss* or *Mrs.*

The last line of the inside address contains the city, state, and zip code.

Salutation

Two spaces below the inside address is your salutation, or greeting. Begin with *Dear*, and then follow with a courtesy title, the reader's last name (unless you are on a first-name basis), and a colon (Dear Mr. Brown:). *Never use a comma for a formal letter.* Avoid the sexist "Dear Sir," "Gentlemen," or "Dear Madam" and the stilted "Ladies and Gentlemen" or "Dear Sir or Madam." (For a discussion of sexist language and how to eliminate it, see Chapter 2, pages 73–76.)

Sometimes a first name does not reveal whether the reader is male or female. There are women named Stacy, Robin, and Lee, and men named Leslie, Kim, and Kelly. If you aren't certain, you can use the reader's full name: "Dear Terry Jones." Similarly, if your recipient uses just initials, write "Dear S.K. Holmes." Or if you know the person's title, you might write "Dear Credit Manager Jones."

Avoid casual salutations such as "Hello," "Hi," "Good Morning," "Greetings," or "Happy Tuesday"; these are best reserved for emails or IMs. And never begin a letter with "To Whom It May Concern," which is old-fashioned, impersonal, and trite.

Body of the Letter

The body of the letter, two spaces after the salutation, contains your message. Some of your letters will be only a few lines long, while others may extend to three or more paragraphs. Keep your sentences concise, and try to hold your paragraphs to less than seven lines. (Refer to pages 67–73 in Chapter 2.)

Complimentary Close

A close, two lines below the last line of your message, is the equivalent of a formal goodbye. For most business correspondence, use one of these standard closes:

Sincerely,

Respectfully,

Sincerely yours,

Capitalize only the first letter of the first word. The entire close is followed by a comma. If you and your reader know each other well, as in Figure 5.4, you can use

Cordially,

Best wishes,

Regards,

But avoid flowery closes, such as

Forever yours,

Devotedly yours,

Faithfully yours,

These belong in a romance novel, not in a business letter.

Signature

Allow four spaces between the complimentary close and your typed name and title so that your signature will not look squeezed in. Always sign your name in black ink. An unsigned letter indicates carelessness or, worse, indifference toward your reader. A stamped signature tells readers you could not give them personal attention.

Some firms prefer using their company name along with the employee's name in the signature section. If so, type the company name in capital letters two line spaces below the complimentary close and then sign your name. Add your title underneath your typed name. Here is an example:

Sincerely yours,

THE FINELLI COMPANY

Helen Stravopoulos

Helen Stravopoulos
Web Coordinator

FIGURE 5.4 Careful Organization of a Business Letter

Office Property Management Associates
2400 South Lincoln Highway
Livingston, NJ 07040-9990
(201) 555-3740   www.opma.com

Clear, professional letterhead with contact information

April 10, 2012

Mr. W. T. Albritton
Albritton & Sharp, CPA
Suite 400
Suburban Office Complex
Livingston, NJ 07038-2389

Accurate inside address

Dear Mr. Albritton:

Thank you for your recent suggestions on improving security at the Suburban Office Complex. You will be pleased to learn that at our March meeting OPMA has agreed to make the following improvements in services, which will go into effect within 45 days.

Introduction comes to point quickly and cordially by referencing reader's earlier request

Starting May 7, you will have an on-site manager, Thomas Vasquez, who will be happy to answer any questions you may have about the Complex and help you with any problems you may encounter. His ten years of experience in managing commercial office parks will benefit you and other businesses at the Suburban Office Complex.

Body describes changes with specific details

The new outdoor security system you asked for will be installed by May 21. It will give you and your employees greater protection through seven additional security cameras around the perimeters of the parking lot while movement sensors will monitor every outside door.

I want to reassure you that none of these changes will inconvenience the operation of your firm. We are honored to have Albritton & Sharp as residents. I welcome your comments as these changes are implemented as well as additional suggestions you may have.

Conclusion builds goodwill by promising reader what will be done and how

Best wishes,

Cheryl Hu

Cheryl Hu
Vice President

Enclosure Line

The enclosure line informs the reader that additional materials (such as a brochure, diagram, form, contract, or proposal) accompany your letter.

> Enclosure (only one item is enclosed)
>
> Enclosures (2)
>
> Encl.: Spring Quarter Sales Report

Copy Notation

The abbreviation *cc:* —two c's followed by a colon—informs your reader that a copy of your letter has been sent to one or more individuals.

> cc: Service Dept.
>
> cc: Hannah Pittman-Jarzelski
>
>   Ivor Vas

Letters are copied and sent to third parties for two reasons: (1) to document a paper trail and (2) to indicate to other readers who else needs the information contained in the letter. Unless your employer instructs you otherwise, tell your reader if others will receive a copy of your letter.

The Appearance of Your Letter

The way your letter looks can determine how readers will respond to your message. Here are some tips on how to format and print professional-looking letters:

- Use a letter-quality printer, and check ink or toner cartridge levels to avoid sending a fuzzy, faint, or messy letter.
- Stay away from fancy fonts and scripts. Use the business-like Times New Roman or Arial. (For a discussion of typography, see Chapter 11, pages 530–532.)
- Consider using letter wizards to help format and design your letters. Most word-processing programs, such as Microsoft Word or Corel Word Perfect, have them. Some organizations, however, may prefer not to use standard letter formatting. Always check with your company before using a letter wizard.
- Leave generous margins of at least 1 to 1¼ inches all around your message. For a shorter letter, as in Figure 5.5, don't expand your margins to 2 inches or increase the font size, which will only make your letter look unprofessional.
- Leave double spaces between key parts of a letter—the date line, the salutation, copy notation, and enclosure—but leave four spaces between the letterhead and date as well as the complimentary close and your typed signature.

- Single-space within each paragraph, but double-space between paragraphs. The white space enhances the look of your letter and makes it easier to read.

- Avoid crowding too much text onto one page. Squeezing too many characters on a line by using overly small fonts will make your letter look cramped and be hard to read. Also, don't cram a long letter onto one page; instead, allow your letter to flow to a second page, as in Figure 5.11 (page 197).

- Be careful about lopsided letters. Don't start a brief letter at the top of the page and then leave the lower three-fourths blank. Begin a shorter message near the center of the page.

- Use Print Preview to see an image of your letter before you print a hard copy so that you can make any necessary changes or corrections. Never print over your company's letterhead or any addresses or company logos printed across the bottom of the letter.

- Always print your letter on high-quality white bond paper (20-pound, 8½ × 11) and matching standard-size (#10) business envelopes. Avoid colored paper, which can look unprofessional.

Envelopes

The way your envelope looks also says a great deal about your message. Most companies have envelopes with their name, contact information, and logos that they will expect you to use. If you have to supply your own, use #10 envelopes, which are 9½ inches long and hold an 8½ × 11 sheet of paper. Center and single-space your recipient's name and address (which must match the inside address in your letter), and put your name and address in the upper left-hand corner of the envelope. The U.S. Postal Service recommends that all information be in all capital letters with no punctuation, to ensure that it scans effectively.

FELIX MARTIN-ROCHA
2314 SOUTH 14TH STREET
CHICAGO IL   60608-5037

                        MS. THERESA ROOKER
                        ASSISTANT MANAGER
                        PREWITT & FIDDLER
                        4800 TRUMAN AVENUE NORTH
                        SPRINGDALE MN   55439-0480

Organizing a Standard Business Letter

Like the memo in Figure 4.3 on page 137, a standard business letter can be divided into an introduction, a body, and a conclusion, each section responding to or clarifying a specific issue for your recipient. These three sections can each be one paragraph long, as in Figures 5.1, 5.2, and 5.3, or the body of your letter can be two or more paragraphs, as in Figure 5.4.

To help readers grasp your message clearly and concisely, follow this simple plan for organizing your business letters:

- In your first paragraph start with a friendly opening and explain why you are writing and why your letter is important to the recipient. Acknowledge any relevant previous meetings, correspondence, or telephone calls early in the paragraph (as in Figures 5.1 and 5.3).
- Put the most significant point of each paragraph first to make it easier for the reader to find. Never bury important ideas in the middle or at the end of a paragraph.
- In the second (or subsequent) paragraph, develop the body of your message with factual support, key details, and descriptions your reader needs. For instance, note how Figure 5.4 refers to the specific changes to improve security that the reader had requested.
- In your last paragraph, thank readers and be very clear and precise about what you want them to do or what you will do for them. Let them know what will happen next, what they can expect (Figure 5.4), or any combination of these messages. Don't leave your readers hanging. End cordially and professionally.

Making a Good Impression on Your Reader

You have just learned the basics about formatting and organizing your business letters. Now we will turn to the content of your letters—what you say and how you say it. Writing letters means communicating to influence your readers, not to alienate or antagonize them. Keep in mind that writers of effective letters are like successful diplomats in that they represent both their company and themselves. You want readers to see you as courteous, credible, and professional.

First, put yourself in the reader's position. What kinds of letters do you like to receive? Letters that are vague, impersonal, sarcastic, pushy, and condescending, or letters that are polite, businesslike, and considerate? If you have questions, you want them answered honestly, courteously, and fully. You do not want someone to waste your time with a long, puffy letter when a few sentences would suffice.

To send such effective letters, adopt the "you attitude." In other words, signal to readers that they and their needs are of utmost importance. Incorporating the "you attitude" means you should be able to answer "Yes" to these two questions:

1. Will my reader receive a positive image of me?
2. Have I chosen words that convey both my respect for the reader and my concern for his or her questions and comments?

FIGURE 5.6 A You-Centered Revision of Figure 5.5

Brown County · Office of the Tax Assessor

County Building, Room 200, Ventura, Missouri 56780-0101

712-555-3000

February 6, 2012

Mr. Ted Ladner
451 West Hawthorne Lane
Morris, MO 64507-3005

Dear Mr. Ladner:

Thank you for writing about the difficulties you encountered with your 2011 tax bill. I wish I could help you, but it is the Tax Collector's Office that issues your annual property tax bill. Our office does not prepare individual homeowners' bills.

Thanks reader and gives polite explanation

If you will kindly direct your questions to Paulette Sutton at the Brown County Tax Collector's Office, County Building, Room 100, Ventura, Missouri 56780-0100, I am sure that she will be able to assist you. Should you wish to call her, the number is 712-555-3455, extension 212. Her email address is **psutton@bctc.gov**. I hope this helps you.

Helps reader solve problem with specific information

Respectfully,

Tracey Kowalski

Tracey Kowalski
Assistant Tax Assessor

Uses appropriate close

Lists specific job title

www.browncounty.gov

I-Centered Draft

> I think that our rug shampooer is the best on the market. Our firm has invested a lot of time and money to ensure that it is the most economical and efficient shampooer available today. We have found that our customers are very satisfied with the results of our machine. We have sold thousands of these shampooers, and we are proud of our accomplishment. We hope that we can sell you one of our fantastic machines.

3. Be courteous and tactful. Refrain from turning your letter into a punch through the mail. Don't inflame your letter or email audience; review Figures 4.6 and 4.7 on pages 146 and 149. When you capture the reader's goodwill, your rewards will be great. The following negative words can leave a bad taste in the reader's mouth.

it's defective	unprofessional (job, attitude, etc.)
I demand	your failure
I insist	you contend
we reject	you allege
that's no excuse for	you should have known
totally unacceptable	your outlandish claim

Compare the following discourteous sentences with the courteous revisions.

Discourteous	Courteous
We must discontinue your service unless payment is received by the date shown.	Please send us your payment by November 4 so that your service will not be interrupted.
You are sorely mistaken about the contract.	We are sorry to learn about the difficulty you experienced over the service terms in your contract.
The new iPad you sold me is third-rate and you charged first-rate prices.	Since the iPad is still under warranty, I hope you can make the repairs easily and quickly.
It goes without saying that your suggestion is not worth considering.	It was thoughtful of you to send me your suggestion, but, unfortunately, we are unable to implement it right now.

The last discourteous example begins with a phrase that frequently sets readers on edge. Avoid using "*It goes without saying*" — it can quickly set up a hostile barrier between you and your reader.

4. Don't sound pompous or bureaucratic. Write to your reader as if you were carrying on a professional conversation. Your tone should be polite but natural and to the point. Make your letters reader-friendly and believable, not stuffy and overbearing. To do that, don't resort to using phrases that remind readers of *legalese* — language that some writers use to make themselves sound important, but that only alienates readers. It smells of contracts, deeds, and stuffy rooms.

In the following list, the words and phrases on the left are pompous expressions that have crept into letters for years; the ones on the right are contemporary equivalents.

Pompous	Contemporary
aforementioned	previously mentioned
as per your request	as you requested
at this present writing	now
I am in receipt of	I have received
contingent upon the receipt of	as soon as we receive
attached herewith	enclosed
I am cognizant of	I know
be advised that	for your information
due to the fact that	because
endeavor	try
forthwith	at once
henceforth	after this
herewith; heretofore; hereby	(drop these three *h*'s entirely)
immediate future	soon
in lieu of	instead of
pursuant	concerning
remittance	payment
under separate cover	I'm also sending you
this writer	I
we regret to inform you that	we are sorry that

International Business Correspondence

After emails, letters are the most frequent type of communication you are likely to have with international readers. Formal letter writing is a highly prized skill in the global marketplace. But as we saw in Chapter 1 (pages 8–9), you cannot assume that people in every culture write letters the way we do in the United States. The conventions of letter writing—formats, inside addresses, salutations, dates, complimentary closes, signature lines—are as diverse as international audiences are.

Case Study

Writing Reader-Friendly Letters: Two Versions

Figure 5.7 shows a letter in stilted language written by Brendan T. Mundell, the general manager of Northern Airways, responding to a complaint from Patricia Lipinski, an executive whose firm has been overcharged for airplane tickets. Responses to complaint letters must be thoughtfully planned and carefully worded so that you do not lose a customer. The language must be clear and honest. Instead, Mundell's letter overflows with flowery, old-fashioned expressions. The effect is that, even though his message offering an apology, a credit, and a promise to correct the situation brings good news, it is long-winded and pompous. Consequently, it sounds insincere and takes a condescending view of his reader. Note how Mundell's revision in Figure 5.8 is free of such stilted expressions; it is shorter, clearer, and far more personable.

FIGURE 5.7 A Letter Written in Stilted, Old-Fashioned Language

NORTHERN AIRWAYS

July 23, 2012

Ms. Patricia Lipinski
Office Manager
Lindsay Electronics
4500 South Mahoney Drive
Buffalo, NY 14214-4514

Dear Manager Lipinski:

Wordy and pompous

Please be advised that I am in receipt of yours of July 17th. We are a global air carrier with the interest of our passengers in mind. I would like to take this opportunity to say that we are cognizant of our commitment to good corporate customers like Lindsay Electronics and extend our disappointment for the problems your firm has experienced with Northern.

Legalese

Payment is due your firm, and I hasten to rectify the situation with regard to our error. Forthwith we are adjusting your account #7530, crediting it with the $706.82 you were surcharged. Inasmuch as Northern is such a wonderful company, we are going to enhance the Travel Pass we issue to preferred customers.

Condescending tone

I would also like to bring to your attention that in an endeavor to correct such billing errors in the immediate future, I have routed copies of your communication to the manager of the billing department, A. T. Padua. I am confident he will take necessary action at his earliest possible convenience to ascertain the full details of your predicament and make the necessary adjustments in our procedures.

Flowery language reeks of insincerity

Once again, I want to take the liberty to assure you that Lindsay Electronics is one of our most valued clients. Rest assured that we will take every step imaginable not to jeopardize our long-standing relationship with you. I hope that all the aforementioned problems have now been satisfactorily resolved. Thanking you, I am

Inappropriate complimentary close

Faithfully yours,

Brendan T. Mundell

Brendan T. Mundell
General Manager

3000 Airline Highway Tyler, ME 04462-3000 (207) 555-6300 Email: norair@norair.com
Fly over to our website at www.norair.com

© Cengage Learning 2013

FIGURE 5.8 A Revised Version of the Stilted, Bureaucratic Letter in Figure 5.7

Northern Airways

July 23, 2012

Ms. Patricia Lipinski
Office Manager
Lindsay Electronics
4500 South Mahoney Drive
Buffalo, NY 14214-4514

Dear Ms. Lipinski:

Thank you for your letter of July 17. I am sorry to learn about the billing problems your employees encountered while traveling on Northern earlier this month. We care very much when we have inconvenienced a loyal customer like Lindsay Electronics. Please accept my apology.

Lindsay is unquestionably entitled to compensation for our billing error. I am crediting your account #7530 with $706.82, the amount you were overcharged. Furthermore, in appreciation of Lindsay's business, I am crediting your Travel-Pass account with a bonus 3,000 miles.

To help us avoid similar incidents, I have sent a copy of your letter to the manager of our billing department, A. T. Padua. I know he will want to learn about Lindsay's experience and will use your comments constructively to revise our billing procedures.

As a Frequent Flyer account member, Lindsay is one of our most important customers, and we will continue to work hard to deserve your support. Please call me if I can help you in the future.

Sincerely yours,

Brendan T. Mundell

Brendan T. Mundell
General Manager

Opens with a sincere apology

Gives clear and concise explanations

Assures reader problem will be corrected in the future

Ends with an offer to help again

Appropriate complimentary close

3000 Airline Highway Tyler, ME 04462-3000 (207) 555-6300 Email: norair@norair.com
Fly over to our website at www.norair.com

Sometimes your international client may reside in the United States. Then you have to exercise the same diplomacy as you would when communicating with audiences living in other countries. For example, in Figure 6.3 (page 214) restaurant owner Patrice St. Jacques writes an effective sales letter by zeroing in on his reader's (Etienne Abernathy's) ethnic pride and heritage. Although Abernathy's company is located in the United States, St. Jacques persuasively sees him from a much broader cultural perspective.

It would be impossible to provide information about how to write letters to each international audience. There are at least five thousand major languages representing diverse ethnic and cultural communities around the globe. But here are some of the most important culturally sensitive questions you need to ask about writing to readers whose cultures are different from yours:

- What is your status in relationship to the reader (client, vendor, salesperson, or international colleague)?
- How should you format and address your letter?
- What is an appropriate salutation?
- How should you begin and conclude your letter?
- What types and amount of information will you have to give?
- What is the most appropriate tone to use?

To answer these and similar questions about proper letter protocol for your international readers, you need to learn about their culture by consulting the following resources:

- Country-specific websites prepared by a country's department of tourism or other English-language websites devoted to the culture of a particular country. For instance, Japan Zone (www.japanzone.com) provides information on contemporary Japanese culture, along with a forum in which you can ask questions about customs, dress, business etiquette, national holidays, and so forth.
- Travel websites such as Yahoo! Travel (http://travel.yahoo.com), which let you select the continent and country of your choice and offer "expert advice" links to the country you want to learn about.
- Business travel websites such as Cyborlink.com (www.cyborlink.com), which provides clickable maps linking to business appearance, behavior, and communication guidelines for a host of countries across the world.
- Either the print or online *World Factbook*, published by the Central Intelligence Agency. The online version is available at https://www.cia.gov/library/publications/the-world-factbook/index.html and supplies informational pull-down pages on people, government, economy, communications, and transnational issues.
- Cultural information officers at a country's embassy or consulate. EmbassyWorld.com (http://embassyworld.com) features a comprehensive database of every embassy and consulate in the world.
- A multinational colleague who is a native speaker and who can advise you about proper letter protocols in his or her country. Susan DiFusco relied on the assistance of a Korean colleague, Kim Ji, to prepare the letter in Figure 5.12.
- A foreign language instructor who is familiar with the customs of the country or countries where a particular language is spoken.

Guidelines for Communicating with International Readers

The following eleven guidelines will help you communicate more successfully with an international audience and significantly reduce the chances of readers' misunderstanding you.

1. Use common, easily understood vocabulary. Write in basic, simplified English. Choose words that are widely understood. Whenever you have a choice, use the simpler word. For example, use *stop*, not *refrain*; *prevent*, not *forestall*; *happy*, not *exultant*.

2. Avoid ambiguity. Words that have double meanings force non-native readers to wonder which one you mean. For example, "We fired the engine" would baffle your readers if they were not aware of the multiple meanings of *fire*. Unfamiliar with the context in which *fire* means "start up," a non-native speaker of English might think you're referring to "setting on fire or inflaming," which is not what you intend. Or because *fire* can also mean "dismiss" or "let go," a non-native speaker of English might even suspect the engine was replaced by another model. Such misinterpretations are likely because most bilingual dictionaries list only a few meanings.

Be especially careful of using synonyms just to vary your word choice. For example, do not write *quick* in one sentence and then, referring to the same action, describe it as *rapid*. Your reader may assume you have two different things in mind instead of just one.

3. Be careful about technical vocabulary. While a reader who is a non-native speaker may be more familiar with technical terms than with other English words, make sure the technical word or phrase you include is widely known and not a word or phrase used only at your plant or office. Double-check by consulting the most up-to-date manuals and guides in your field, but steer clear of technical terms in fields other than the one with which your reader is familiar. Be especially careful about using business words and phrases that an international reader may not know, such as *lean manufacturing*, *reverse mortgages*, *best practices*, *toxic assets*, and so forth.

4. Avoid idiomatic expressions. Idioms are the most difficult part of a language for an audience of non-native speakers to master. As with the example of *fire*, the following colorful idiomatic expressions will confuse and may even startle a non-native reader:

I'm all ears
throw cold water on it
hit the nail on the head
new blood
easy come, easy go
get a handle on it
right under your nose

think outside the box
sleep on it
give a heads-up
land in hot water
touch and go
pushed the envelope
it was a rough go

The meanings of those and similar phrases are not literal but figurative, a reflection of our culture, not necessarily your reader's. A non-native speaker of English will approach such phrases as combinations of the separate meanings of the individual words, not as a collective unit of meaning.

A non-native speaker of English—a potential customer in Asia or Africa, for example—might be shocked if you wrote about a sale concluded at a branch office this way: "Last week we made a killing in our office." Substitute the idiomatic expression with a clear, unambiguous translation easily understood in international English. "We made a big sale last week." For "Sleep on it," you might say, "Please take a week to make your decision."

5. Delete sports and gambling metaphors. These metaphors, which are often rooted in U.S. popular culture, do not translate word for word for non-native speakers and so can interfere with your communication with your readers. Here are a few examples to avoid:

out in left field	a ballpark figure
strike out	fumble the ball
drop the ball	out of bounds
down for the count	make a pass
long shot	beat the odds
be in left field	win by a nose

Use a basic English dictionary and your common sense to find nonfigurative alternatives for these and similar expressions.

6. Don't use unfamiliar abbreviations, acronyms, or contractions. While these shortened forms of words and phrases are a part of U.S. business culture, they might easily be misunderstood by a non-native speaker who is trying to make sense of them in context or by looking them up in a foreign language dictionary. Avoid abbreviations such as pharm., gov., org., pkwy., rec., hdg., hr., mfg., or w/o. The following acronyms can also cause your international reader trouble: ASAP, PDQ, p's and q's, IRA, SUV, RV, DOB, DOT, SSN. If you have to use acronyms, define them. Finally, contractions such as the following might lead readers to mistake them for the English words they look like: I've (ivy), he'll (hell), I'll (ill), we'll (well), can't (cant), won't (wont, want).

7. Watch units of measure. Do not fall into the cultural trap of assuming that your reader measures distances in miles and feet (instead of kilometers and meters as most of the world does), measures temperatures on the Fahrenheit scale (instead of Celsius), buys gallons of gasoline (instead of liters), spends dollars (rather than euros, pesos, marks, rupees, or yen), tells time by a twelve-hour clock (many countries follow a twenty-four-hour clock), and records dates by month/day/year (most countries record dates by day/month/year). See the section "International Measurements" on pages 190–191 for more detailed information.

8. Avoid culture-bound descriptions of place. For example, when you tell a reader in Hong Kong about the Sunbelt or a potential client in Africa about the Big Easy, will he or she know what you mean? When you write from California about the eastern seaboard, meaning the East Coast of the United States, the directional reference may not mean the same thing to a reader in India as it does to you. Moreover, referring to February as a winter month does not make sense to someone in New Zealand, for whom it is a summer month.

9. **Keep your sentences simple and easy to understand.** Short, direct sentences will cause a reader whose native language is not English the least amount of trouble. A good rule of thumb is that the shorter and less complicated your sentences, the easier and clearer they will be for a reader to process. Long (more than fifteen words) and complex (multiclause) sentences can be so difficult for readers to unravel that they may skip over them or simply guess at your message. Do not, however, be insultingly childish, as if you were writing to someone in kindergarten. Also, always try to avoid the passive voice. It is one of the most difficult sentence patterns for a non-native speaker to comprehend. Stick to the common subject-verb-object pattern as often as possible. See pages 68–69 and the Writer's Brief Guide to Paragraphs, Sentences, and Words (pages A-1–A-19).

10. **Be cautious about style and tone.** Culture plays a major role in how you word your message. Americans expect business letters to be concise and to the point, without flowery compliments and personal details. They want to see conclusions and recommendations up front, followed by descriptions of key items leading to the main point. Readers in other cultures, however, would find this approach offensive. For instance, German readers expect long letters that unfold slowly through a highly factual and scrupulously documented narrative of events, all of which would lead to a recommendation at the end of the letter. Japanese or Korean readers, unlike American or German readers, expect the first paragraph or two of a letter to center on the friendship and respect the writer has for the reader, as in Frank Sims's letter in Figure 5.10 (page 193) and Susan DiFusco's letter in Figure 5.12 (page 198).

11. **Use appropriate salutations, complimentary closes, and signature lines.** Find out how individuals in the recipient's culture are formally addressed in a salutation (e.g., Señor, Madame, Frau, Monsieur). Unless you are expressly asked to use a first name, always use your reader's surname and include proper titles and other honorifics (e.g., Doctor, Sir, Father). For a complimentary close, use an appropriately formal one, such as *Respectfully*, which is acceptable in almost any culture.

Modifying a Letter for an International Audience

Figures 5.9 and 5.10 on pages 192–193 show two versions of a letter to an international business executive. The letter in Figure 5.9 violates the guidelines discussed above for writing to such an audience because it

- uses the incorrect format for the date line
- misspells the name of the reader's city
- leaves out important postal information
- contains U.S. idioms (e.g., "drop you a line")
- includes troubling abbreviations
- disregards how the reader's culture records time and temperature

International Measurements

The following guidelines will help you use the proper units of measurement for your international audience.

Distance, Weight, and Volume

Most of the international community uses the metric system rather than English units (meters rather than feet, grams rather than ounces, liters rather than quarts, etc.). In fact, the United States is the only country that has not officially adopted the metric system (except in science and technology). When corresponding with an international reader, use metric units. But when writing to multiple audiences—those in the United States and overseas—display both units. For help converting pressure, temperature, length, weight, and volume measurements, consult **www.worldwidemetric.com/measurements.html**.

Dates

Different cultures express dates in different ways. Most countries, including those in Europe, list the day, the month, and then the year; for example, 1-3-12 refers to March 1, 2012. But in the United States, where the month comes first and then the day and year, the date above would be read as January 3, 2012. In Japan, a different system is used; the year is listed first, followed by the month and day: 2012.3.1 (or 12/3/1). To avoid any confusion, write out the month and use all four digits for the year.

Currency

Even though many European countries now use one currency, the euro, world currencies still vary widely. Always respect the monetary unit your reader uses and include the local symbol for it. But be careful. Don't write $ or *dollars* and think readers will automatically think of the United States. Other areas of the world—Singapore, Hong Kong, Canada, Australia, and New Zealand—also use dollars. Specify what country or region those dollars belong to: $100 (Hong Kong). To convert U.S. dollar amounts into any international currency based on the present rate of exchange, see **www.xe.com/ucc**. This site also identifies world currencies.

Numbers

In many countries, a period is used instead of a comma in large numbers (expressed in the thousands). For example, in the United States 12,001 signifies just over twelve thousand, whereas in Great Britain the number is written as 12.001. Many Latin American and European countries use spaces rather than decimals or commas to visually separate figures, for example, 12 001. In other countries, a comma is used instead of a decimal point in percentages: 82.3% in the United States stands for 82 and 3/10 percent, while in Germany or Bolivia the percentage is written as 82,3%. Be aware that conventions for numbers can change from country to country. Check the scientific guides and manuals in your discipline to find out which practice is internationally endorsed.

Time: The Twenty-Four-Hour Clock

Many countries—including most of those in Europe—use the twenty-four-hour clock, or what is sometimes referred to as "military time" in the United States. Unlike the twelve-hour clock in use in the United States, the twenty-four-hour clock does not use *a.m.* or *p.m.* In Paris, for instance, 13:30 is 1:30 p.m. in New York, and 9 p.m. in the United States translates to 21:00 in Sweden. Always specify time zones (Eastern Standard Time, Greenwich Mean Time) to help international audiences understand your message more clearly.

Temperature and Season

Most of the world, except the United States and Jamaica, records temperature using the Celsius temperature scale (represented by C) instead of the Fahrenheit scale (represented by F). For this reason, convert Fahrenheit readings to Celsius when writing to international readers. The Celsius scale has a freezing point of 0° and a boiling point of 100°, while on the Fahrenheit scale water freezes at 32° and boils at 212°. Writing to an international reader, accustomed to recording temperatures in degrees Celsius, that it is a pleasant 75 degrees, without noting that the scale is Fahrenheit, may lead to confusion: 75°C is the Fahrenheit equivalent of 120 degrees.

Be aware of seasonal differences, too. While New York is in the middle of the winter, Australia and Chile are enjoying summer. Be respectful of your readers' cultural (and physical) environment. Thanksgiving is celebrated in the United States in November, but in Canada the holiday is the second Monday in October; elsewhere around the world it may not be a holiday at all.

Even more disrespectful, the writer's overall tone is condescending ("south of the border") and inappropriately casual. At the end of the second paragraph, for instance, the writer tells his reader that his U.S. firm is superior to the Argentine company.

Note how the letter in Figure 5.10, however, respects the reader's cultural conventions because it

- spells and punctuates the reader's name and address correctly
- incorporates a clear date line

FIGURE 5.9 An Inappropriately Written Letter for an International Reader

Pro-Tech, Ltd. 452 West Main St. Concord, MA 01742   978-634-2756
www.protech.com

Misleading date line

5-9-12

Incorrect, misspelled address

Mr. Antonio Guzman
Canderas
Mercedes Ave.
Bunos Aires, ARG.

Salutation is too informal

Dear Tony,

Impolite opening disrespecting reader's status

Culturally condescending

I wanted to drop you a line before the merger hits and in doing so touch base and give you the lowdown on how our department works here in the good old U.S. of A.

None of us had a clue that Pro-Tech was going to go south of the border, but your recent meeting about the Smartboard T-C spoke volumes to the tech people who praised your operations to the hilt. So it looks like you and I both will be getting a new corp. name. Olé. I love moving from Pro-Tech, Ltd. to Pro-Tech International. We are so glad we can help you guys out.

Filled with American idioms and abbreviations

At any rate, I'm sending you an email with all the ins and outs of our department struc., layout, employees, and prod. eff. quotas. From this info, I'm hoping you'll be able to see ways for us to streamline, cooperate, and soar in the market. I understand that all of this is in the works and that you and I need to have a face-to-face and so I'd appreciate your reciprocating with all the relevant data stat.

Disregards time differences and reader's 24-hour clock

Consequently, I guess I'll be flying down your way next month. Before I take off, I would like to give you a ring. How does after lunch next Thursday (say, 1:00–1:30) sound to you? I hope this is doable.

Ignores differences between Fahrenheit and Celsius scales

We've had a spell of great weather here (can you believe it's in the low 80s today!). So, I guess I'll just sign off and wait 'til I hear from you further.

I send you felicitations and want things to go smoothly before the merger is a done deal.

Close sounds insincere

Adios,

Frank Sims

Frank Sims

FIGURE 5.10 An Appropriate Revision of Figure 5.9

Pro-Tech, Ltd. 452 West Main St. Concord, MA 01742 978-634-2756
www.protech.com

9 May 2012

Señor Antonio Mosca-Guzman
Director, Quality Assurance
Tecnología Canderas, S.A.
Av. Martin 1285, 4° P.C.
C1174AAB BUENOS AIRES
ARGENTINA

Dear Señor Mosca-Guzman:

As our two companies prepare to merge, I welcome this opportunity to write to you. I am the manager for the Quality Assurance division at Pro-Tech, a title I believe you have at Tecnología Canderas. I am looking forward to working with you both now and after our companies merge in two months.

Allow me to say that we are very honored that your company is joining ours. Tecnología Canderas has been widely praised for the research and production of your Smartboard T-C systems. I know we have much to learn from you, and we hope you will allow us to share our systems analyses with you. That way everyone in our new company, Pro-Tech International, will benefit from the merger.

Later this week, I will send you a report about our division. It describes how our division is structured and the quality inspections we make. It will also give you a brief biography of our staff so that you can learn about their qualifications and responsibilities.

The director of our new company, Dr. Suzanne Nknuma, asked me to meet with you before the merger to see how we might help each other. I would very much like to travel to Buenos Aires in the next month to visit you and take a tour of your company.

Would you please let me know by email when it may be convenient for us to talk so we might discuss the agenda for our meeting? I am in my office from 11:00 to 17:00 Buenos Aires time.

I look forward to working with and meeting you.

Respectfully,

Frank Sims

Frank Sims
Quality Assurance Manager

Clear date line

Includes appropriate title for reader

Complete and correct address

Uses courteous salutation

Clear and diplomatic opening

Respectful view of reader's company

Explains business procedures in plain English

Recognizes reader's time zone

Polite close

Gives business title

- uses an appropriate salutation and complimentary close
- is written in plain, international English
- recognizes that the reader uses a twenty-four-hour clock
- identifies the writer and makes the reader feel welcome and honored
- clearly informs the reader how the merger will affect his relationship with the writer
- strives to develop a spirit of cooperation
- is courteous throughout
- acknowledges the reader's position of authority in his company

Above all, though, the writer respects his audience's culture and role in the business world and seeks to win the reader's confidence and respect, two invaluable assets in the global marketplace.

Respecting Your Reader's Nationality and Ethnic or Racial Heritage

Do not risk offending any of your readers, whether they are native speakers of English or not, with language that demeans or stereotypes their nationality or ethnic or racial background. Here are six precautions to take.

1. **Respect your reader's cultural traditions.** Be aware of how cultures differ in terms of traditions, customs, and preferences—how people dress, communicate, and celebrate holidays. When writing a letter to someone from a different country, show respect for your reader's background and perspectives. Don't assume that your culture is the only culture. For instance, not everyone celebrates the same holidays you do, or on the same days. As we saw, Thanksgiving is celebrated in October in Canada. Take a look at the cultural calendar that Terri Ruckel incorporates into her long report on recruiting multicultural workers in Figure 15.3 (see pages 702–718).

2. **Respect your reader's nationality.** Always spell your reader's name and country properly, which may mean adding diacritical marks (e.g., accent marks) not used in English—for example, Muñoz. If your reader has a hyphenated last name (e.g., Arana-Sanchez), it would be rude to address him or her by only part of the name. In addition, be careful not to use the former name of your reader's country or city, for instance, the Soviet Union (now Russia), Malaya (now Malaysia), Czechoslovakia (now the Czech Republic), Rhodesia (now Zimbabwe), or Bombay (now Mumbai). Not only is it rude, but it also demonstrates a lack of interest about your reader's nationality.

3. **Honor your reader's place in the world economy.** Phrases like "third-world country," "emerging nation," and "undeveloped/underprivileged area" are derogatory. Using such phrases signals that you regard your reader's country as inferior. Use the name of your reader's country instead of such phrases. Saying that someone lives in the Far East implies that the United States, Canada, or Europe are the center of culture, the hub of the business community. It would be better to simply say "East Asia."

4. Avoid insulting stereotypes. Expressions such as "oil-rich Arabs," "time-relaxed Latinos," and "aggressive foreigners" unfairly characterize particular groups. Similarly, prune from your communications any stereotypical phrase that insults one group or singles it out for praise at the expense of another: "Mexican standoff," "Russian roulette," "Chinaman's chance," "Irish wake," "Dutch treat," "Indian giver." The word *Indian* refers to someone from India; use *Native American* to refer to the indigenous people of North America, who prefer to be known by their tribal affiliations (e.g., the Lakota).

Some Guides to Cultural Diversity for Businesspeople

Bridging the Culture Gap: A Practical Guide to International Business Communication, by Penny Carté and Chris Fox (Kogan Page, 2008).

Cultural Intelligence: A Guide to Working with People from Other Cultures, by Brooks Peterson (Intercultural Press, 2004).

Cultural Intelligence: People Skills for Global Business, by David C. Thomas (Berrett-Koehler, 2004).

Essential Do's and Taboos: The Complete Guide to International Business and Leisure Travel, by Roger E. Axtell (John Wiley & Sons, 2007).

International Business Etiquette Resource Page: **www.cyborlink.com/besite/ resource .htm**.

Kiss, Bow, or Shake Hands: The Bestselling Guide to Doing Business in More Than 60 Countries, by Terri Morrison and Wayne A. Conway (Adams, 2006).

Multicultural Manners: Essential Rules of Etiquette in the 21st Century, rev. ed., by Norine Dresser (John Wiley & Sons, 2005).

---

5. Be sensitive to the cultural significance of colors. Do not offend your audience by using colors in a context that would be offensive. Purple in Mexico, Brazil, and Argentina symbolizes bad luck, death, and funerals. Green and orange have a strong political context in Ireland. In Egypt and Saudi Arabia, green is the color of Islam and is considered sacred. But in China, green can symbolize infertility or adultery. Also in China, white does not symbolize purity and weddings but mourning and funerals. Similarly, in India if a married woman wears all white, she is inviting widowhood. While red symbolizes good fortune in China, it has just the opposite meaning in Korea.

6. Be careful, too, about the symbols you use for international readers. Triangles are associated with anything negative in Hong Kong, Korea, and Taiwan. Political symbols, too, may have controversial implications (e.g., a hammer and sickle, a crescent). Avoid using the flag of a country as part of your logo or letterhead for global audiences. Many countries see this as a sign of disrespect, especially in Saudi Arabia, whose flag features the name of Allah.

Writing to a Client from a Different Culture: Two Versions of a Sales Letter

Let's assume that you have to write a sales letter to an Asian business executive. You will have to employ a very different strategy in writing to this executive as opposed to an American reader. For an American audience, the best strategy is to take a direct approach—fast, hard-hitting, to the point, and stressing your product's strengths versus the opposition's weaknesses. Businesses in the United States thrive on the battle of the brands, the tactics of confrontation symbolized in the sports metaphors on page 188. A typical sales letter to an American reader would be polite but direct.

But such a strategy would be counterproductive in writing to an Asian reader. Business in East Asia is associated with courtesy and friendship and is wrapped up in a great many social courtesies. In a sales letter to an Asian reader, you have to first establish a friendship before business details are dealt with. The Asian way of doing business, including writing and receiving letters, is far more subtle, indirect, and complimentary than it is in the United States. The U.S. style of directness and forcefulness would be perceived as rude or unfair in, say, Japan, China, Malaysia, or Korea. A hard-sell letter to an Asian reader would be a sign of arrogance, suggesting inequality for the reader.

To better understand the differences between communicating with a U.S. reader and an Asian one, study the two versions of the sales letter in Figures 5.11 and 5.12. The letters, written by Susan DiFusco for Starbrook Electronics, sell the same product, but Figure 5.11 is written to a U.S. executive, while Figure 5.12 adapts the same message for a businessperson in Seoul, South Korea. See how the two letters differ not only in content but also in the way each is formatted.

Format

The full-block style of the letter to the U.S. reader signals a no-nonsense, all-business approach. Everything is lined up in neat, orderly fashion. For the Korean reader, however, DiFusco wisely chose the more varied pattern of indenting her paragraphs. For Asian readers, the visual effect suggests a much more relaxed and friendly, yet respectful, communication. Note the different typefaces, too.

Opening

Pay special attention to how the U.S. letter (Figure 5.11) starts off politely but much more directly, an opening the Korean reader would regard as blunt and discourteous. The sales letter written to the Korean audience (Figure 5.12) starts not with business talk but with a compliment to the reader and his company, praising them for trustworthiness and wishing them much prosperity in the future. DiFusco sought the advice of a Korean co-worker, Kim Ji, and from him she learned the Korean greeting that opens the letter. She knew that when addressing the Korean administrator she should not get to business right away but instead used her introductory paragraph to show respect for the company and the reader—the equivalent of East Asian business executives socializing before any mention of business is made.

The Body

Compare the second and third paragraphs of the letters in Figures 5.11 and 5.12. While the letter to the U.S. reader launches an aggressive campaign to get the reader's business, the letter

FIGURE 5.11 Sales Letter to a Native English Speaker in a U.S. Firm

Starbrook Electronic
Perry, TX 75432-3456
phone **(713) 555-2121**   Email info@starbrook.com
www.starbrook.com

April 6, 2012

Mr. Ellis Fanner, Director
Morgan General Hospital
300 Oakland Drive
Morgan, OR 97342-0091

Dear Mr. Fanner:

How many times have your physicians asked when you would have an Open MRI?
This state-of-the-art imaging equipment will help maintain your reputation as a leading
health care provider in the Morgan River Valley.

As the world leader in designing and manufacturing MRI equipment, Starbrook can
offer you the latest technology to save your patients time and improve the care you
give them. With our Open Imaging 500, Morgan General can deliver more accurate
and timely diagnoses. Within an hour you can determine whether a patient has had a
stroke rather than having to wait with less powerful scanner technology.

Thanks to our Open Imaging 500 model, Morgan General can also improve diagnoses
for orthopedic and cardiac problems. Our MRI delivers much more extensive internal
imaging than any other of our competitors' equipment.

By obtaining the Open Imaging 500, you will surpass all other health care providers
who do not offer this technology. By acting now, you will also receive Starbrook's
unsurpassed guarantee of service plus free maintenance for a year by our team of
expert technicians.

And we will even give you free upgrades to make sure your Open Imaging 500
continues to be state of the art. Since software updates change so often, no other MRI
vendor dares make such an offer. We deliver what we promise. Ask any of our recent
satisfied customers—Tennessee General, Grantsville Uptown Clinic, or Nevada
Memorial Hospital.

We are hosting a demonstration on May 4 in Portland. Please call me today to arrange
for your showing.

Sincerely yours,

Susan DiFusco

Susan DiFusco
Assistant Manager

*Begins with
an aggressive
question and
emphasis
on hospital's
competitive
role*

*Hard-hitting
appeal to sell
reader on MRI
features*

*Stresses MRI
advantages
over
competitors'
models*

*Urges reader to
act to surpass
competition*

*Praises own
company*

*Asks for
immediate
response from
reader*

FIGURE 5.12 Sales Letter to a Non-Native Speaker of English in a Foreign Firm

✴ Starbrook Electronics
Perry, TX 75432-3456
phone (713) 555-2121   Email info@starbrook.com
www.starbrook.com

Font and format more appropriate for audience— more open and inviting—than in Figure 5.11

May 4, 2012

Mr. Kim Sun-Lim
Administrator
Tangki Hospital
Nowonku 427–1
Seoul, Korea 100–175

Dear Mr. Kim:

Opens with a compliment and subtle reference to business

In your beautiful language I say "Gyuihauie bunyanggwa hangbokul kiwonhamnida" on behalf of my firm Starbrook Electronics. I am honored to introduce myself to you through this letter and send you greetings at this beautiful flower season.

Please let us know how we might be of service to you. One of the ways we may be able to serve you is by informing you about our new Open Imaging 500 MRI (magnetic resonance imaging) equipment. This new model can offer you and your patients many advantages. It is far better than conventional scanner models. Your physicians can offer quicker diagnoses for patients with strokes, heart attacks, or orthopedic injuries. MRI pictures will also give you clearer and deeper pictures than any X-ray or scanner can.

Gives important equipment information without bashing competitors; appeals to the honor of reader's company

It would be an honor to provide Tangki Hospital with one of our Open Imaging 500s. If you select the Open Imaging 500, we will be happy to give you all maintenance and software updates free for one year. Such service will provide the best in health care for the many people who come to you for help. Our firm is well known for its quality service.

Ends with courteous invitation

Kindly let me know if I might send you information about the Open Imaging 500. It would be a privilege to meet you and to give you and your staff a demonstration of our Open Imaging MRI.

Respectfully,

Susan DiFusco

Susan DiFusco
Assistant Manager

to the Korean reader avoids the hard sell of U.S. business tactics. DiFusco knew that for her Korean reader she must not promote too strenuously. The more she boasted about Starbrook's work, the less likely it was that she would make a sale. She recognized, again from discussions with other Asian businesspeople over the years, that she had to supply key information—such as the application and the advantages of her product—without overwhelming or pressuring her audience. Yet DiFusco subtly reassures Kim Sun-Lim that her company is honorable and worthy to be recommended to his friends.

View of Competitors

Observe, too, how the letter to the U.S. executive (Figure 5.11) undermines the competition by stating how much better Starbrook's offer is. For most Asian readers, it would be considered impolite to claim that your product is better than another company's or that your firm is currently doing business with other firms in the reader's country. Asian audiences prefer to avoid anything that hints of impoliteness or assertiveness.

The Conclusion

Finally, contrast the conclusions of the two letters. Writing to a U.S. audience, DiFusco strongly urges her potential customer to get in touch with her. Such a call to action is customary in a sales letter to a U.S. firm. But in her concluding paragraph to Kim Sun-Lim, DiFusco adopts a more reserved and personal tone. Her use of such appropriate phrases as "kindly let me know" and "it would be a privilege" expresses the friendly sentiments of respect and esteem that would especially appeal to her reader. Note, too, that DiFusco has chosen a complimentary close ("Respectfully") much more in keeping with her reader's cultural sensitivities than the "Sincerely yours," which was more suitable for the letter in Figure 5.11.

As these two letters show, writers need to know and respect their readers' cultures. In addressing an audience of non-native speakers of English, a writer must consider the readers' communication patterns, protocols, and cultural (or subcultural) traditions. The same guidelines apply whether you are writing to non-native speakers abroad or in this country. Your message and vocabulary should always be clear, understandable, and appropriate for the intended audience.

Sending Professional-Quality Letters: Some Final Advice to Seal Your Success

Here is some final advice to follow that will help you format, draft, and tailor the business correspondence you will be asked to write.

- Identify your reader. Are you writing to an individual, a group, a company or an agency, a new individual customer or a longtime one, a native or non-native speaker?
- Pay special attention to an international reader's needs by researching his or her cultural traditions and avoiding writing anything that may offend or confuse him or her.

- Emphasize the "you attitude." Keep the reader's needs in the forefront of your message. Consider how your reader will respond to your message.
- Organize your information. Make sure your letter has a clear introduction, body, and conclusion.
- Include essential information, such as schedules; dates; prices and expenses; personnel; explanation of services, warranties, and products; background.
- Use an appropriate style and tone. Be professional and courteous, concise and focused, and sensitive to your reader's needs, including his or her culture and traditions.
- Make sure your letter looks professional. Follow your company's format, and choose an appropriate font.

Planning your letter carefully—its purpose, organization, content, and format—will help you to make sure your message begins, continues, and ends professionally and successfully for readers in the United States and around the globe.

✓ Revision Checklist

Audience Analysis and Research
- ☐ Made sure reader's name and job title are correct.
- ☐ Found out something about my audience—interests and background, well informed or unfamiliar with topic, former client or new one.
- ☐ Determined whether audience will be friendly, hostile, or neutral about my message.
- ☐ Did necessary research—in print, through online sources, in discussions with colleagues—to give readers what they need.
- ☐ Acknowledged previous correspondence.
- ☐ Spent sufficient time drafting and revising letter before printing final copy.

Format and Appearance
- ☐ Followed one letter format (full-block, modified-block, semi-block) consistently.
- ☐ Used letter wizards with caution.
- ☐ Set left margins wide enough to make my letter look attractive and well proportioned.
- ☐ Included all the necessary parts of a letter for my purpose.
- ☐ Made sure that my letter looks neat and professional.
- ☐ Printed my letter on company letterhead or quality bond paper.
- ☐ Proofread my letter carefully and made sure each correction was made before printing final copy.
- ☐ Eliminated any grammatical and spelling errors.
- ☐ Signed my letter legibly in black ink.
- ☐ Sent copies to appropriate parties.
- ☐ Printed envelopes properly.

Content and Organization
- ☐ Clearly understood my purpose in writing.
- ☐ Put most important point first in my letter.
- ☐ Started each paragraph with the central idea of that paragraph.
- ☐ Answered all the reader's questions and concerns.
- ☐ Omitted anything offensive, irrelevant, or repetitious.
- ☐ Stated clearly what I want the reader to do.
- ☐ Used last paragraph to summarize and encourage the reader to continue cordial relations with me and my company.

Style: Words, Tone, Sentences, Paragraphs
- ☐ Emphasized the "you attitude" by seeing things from the reader's perspective.
- ☐ Avoided being too casual or colloquial.
- ☐ Chose words that are clear, precise, and friendly.
- ☐ Cut anything sounding flowery, stuffy, or bureaucratic.
- ☐ Ensured that my sentences are readable, clear, and not too long (less than fifteen to twenty words).
- ☐ Wrote paragraphs that are easy to read and that flow together.

Writing to International Readers
- ☐ Did appropriate research about the reader's culture—in print, through online sources, and with colleagues who are native speakers (as well as teachers)—especially about accepted ways of communicating.
- ☐ Adopted a respectful, not condescending, tone.
- ☐ Avoided anything offensive to my reader, especially references to politics, religion, or cultural taboos.
- ☐ Used plain and clear language that my reader would understand.
- ☐ Tested my sentences for length and active voice.
- ☐ Made sure nothing in my letter might be misinterpreted by my reader.
- ☐ Chose colors and symbols culturally appropriate for the context of my message.
- ☐ Selected the right format, salutation, and complimentary close for my reader.
- ☐ Observed the reader's units of measurement for time, temperature, currency, dates, and numbers.

Exercises

1. Find two business letters and bring them to class. Be prepared to identify and comment on the various parts of a letter discussed in this chapter.

2. Find a form letter that is addressed to "Dear Customer," "Postal Patron," or "Dear Resident," and rewrite it to make it more personal.

3. Correct the following inside addresses:

a. Dr. Ann Clark, M.D.
1730 East Jefferson
Jackson, MI. 46759

b. To: Miss Tommy Jones
Secretary to Mrs. Franks
Donlevey labs
Cleveland, O. 45362

c. Debbie Hinkle
432 Parkway
N. Y. C. 10054

d. Mr. Charles Howe, Acme Pro.
P.O. Box 675
1234 S. e. Boulevard
Gainesville, Flor. 32601

e. Alex Goings, man.
Pittfield Industries
Longview, TEXAS 76450

f. ATTENTION: G. Yancy (Mrs.)
Police Academy
1329 Tucker
N. O., La. 3410–70122

g. David and Mahenny
Lawyers
Dobbs Build.
L.A. 94756

h. Barry Fahwd
Man., Peninsular, Ltd.
Arabia

4. Write appropriate inside addresses and salutations to (a) a woman who has not specified her marital status; (b) an officer in the armed forces; (c) a professor at your school; (d) an assistant manager at your local bank; (e) a member of the clergy; (f) a government worker.

5. Rewrite the following sentences to make them more personal.

a. It becomes incumbent upon this office to cancel order #2394.

b. Management has suggested the curtailment of parking privileges.

c. ALL USERS OF HYDROPLEX: Desist from ordering replacement valves during the period of Dec. 20–30.

d. The request for a new catalog has been honored; it will be shipped to same address soon.

e. Perseverance and attention to detail have made this writer important to company in-house work.

f. The Director of Nurses hereby notifies staff that a general meeting will be held Monday afternoon at 3:00 p.m. sharp. Attendance is mandatory.

g. Reports will be filed by appropriate personnel no later than the scheduled plans allow.

6. The following sentences are discourteous, boastful, excessively humble, vague, or lacking the "you attitude." Rewrite them to correct those mistakes.

a. Something is obviously wrong in your head office. They have once more sent me the wrong model number. Can they ever get things straight?

b. My instructor wants me to do a term paper on safety regulations at a small plant. Since you are the manager of a small plant, send me all the information I need at once. My grade depends heavily on all this.

c. It is apparent that you are in business to rip off the public.

d. I was wondering if you could possibly see your way into sending me the local chapter president's name and address—if you have the time, that is.

e. I have waited for my confirmation for two weeks now. Do you expect me to wait forever, or can I get some action?

    f. It goes without saying that we cannot honor your request.

    g. May I take just a moment of your valuable time to point out that our hours for the next three weeks will change, and we trust and pray that no one in your agency will be terribly inconvenienced by this.

    h. Your application has been received and will be kept on file for six months. If we are interested in you, we will notify you. If you do not hear from us, please do not write us again.

    i. My past performance as a medical technologist has left nothing to be desired.

    j. Credit means a lot to some people. But obviously you do not care about yours. If you did, you would have sent us the $249.95 you rightfully owe us three months ago. What's wrong with you?

7. The following letter, filled with musty expressions and in old-fashioned language (legalese), buries key ideas. Rewrite and reorganize it to make it shorter, clearer, and more reader-centered.

Dear Ms. Granedi:

This is in response to your firm's letter of recent date inquiring about the types of additional services that may be available to business customers of the First National Bank of Bentonville. The question of a possible time frame for the implementation of said services was also raised in the aforementioned letter. Pursuant to these queries, the following answers, this office trusts, will prove helpful.

Please be advised that the Board of Directors at First National Bank has a continuing reputation for servicing the needs of the Bentonville community, especially the business community. For the last fifty years—half of a century—First National Bank has provided the funds necessary for the growth, success, and expansion of many local firms, yours included. This financial support has bestowed many opportunities on a multitude of business owners, residents of Bentonville, and even residents of surrounding local communities.

The Board is at this present writing currently deliberating, with its characteristic caution, over a variety of options suggested to us by our patrons, including your firm. These options, if the Board decides to act upon them, would enhance the business opportunities for financial transactions at First National Bank. Among the two options receiving attention by the Board at this point in time are the creation of a branch office in the rapidly growing north side of Bentonville. This area has many customers who rely on the services of First National Bank. The Board may also place a business loan department in the new branch.

If this office of the First National Bank of Bentonville might be of further helpful assistance, please advise. Remember, banking with First National Bank is a community privilege.

Soundly yours,

M. T. Watkins
Public Relations Director

8. Either individually or in a small group, write a business letter to one of the following individuals and submit an appropriate envelope with your letter.
    a. your mayor, asking for an appointment and explaining why you need one
    b. your college president, stressing the need for more parking spaces or for additional computers in the library
    c. the local water department, asking for information about fluoride supplements
    d. the editor of an online magazine, requesting permission to reprint an article in your school newspaper
    e. the author of an article you have read recently, telling why you agree or disagree with the views presented
    f. the director of food services on campus asking for more ethnic meals
    g. a computer retailer, inquiring about costs and availability of office software packages and explaining your company's special needs

9. Rewrite the following letters, making them appropriate for a reader whose native language is not English. Identify the intended reader's cultural heritage. As you revise the letters, pay attention to the words, measurements, and sentence constructions you employ. Be sure to consider the reader's cultural traditions and avoid cultural insensitivity.
    a. Dear Pal,

    Our stateside boss hit the ceiling earlier today when she learned that our sales quota for this quarter fell precipitously short. Ouch! Were I in her spot, I would have exploded too. Numerous missives to her underlings warned them to get off the dime and on the stick, but they were oblivious to such. These are the breaks in our business, right? We can't all bat 1000.

    Let's hope that next quarter's sales take a turn for the best by 12–1-12. If they are as disastrous, we all may be in hot water. Until then, we will have to watch our p's and q's around here. We're freezing here—20° today.

    Cheers,

b. Dear Mr. Wong,

It's not every day that you have the chance to get in on the ground floor of a deal so good you can actually taste it. But Off-Wall Street Mutual can make the difference in your financial future. Give me a moment to convince you.

By becoming a member of our international investing group for just under $250, you can just about ensure your success. We know all the ins and outs of long-term investing and can save you a bundle. Our analysts are the hot shots of the business and always look long and hard for the most propitious business deals. The stocks we select with your interests in mind are as safe as a bank and not nearly so costly for you. Unlike any of your undertrained local agents, we can save you money by investing your money. We are penny pinchers with our clients' initial investments, but we are King Midas when it comes to transforming those investments into pure gold.

I am enclosing a brochure for you to study, and I really hope you will examine it carefully. You would be foolish to let a deal like Off-Wall Street Mutual pass you by. Go for it. Call me by 3:00 today.

Hurriedly,

c. Dear Mr. Bafaloukos,

My firm is taking a survey of businesses in your part of the world to see if there is any likelihood of getting you on board our international computer network, and so I thought I would see if you might like to take the chance. In today's shaky world, business events can change overnight and without the proper scoop you could be left out in the cold. We can alleviate that mess.

Not only do we interface with major exchanges all around the globe, but we make sure that we get the facts to you pronto. We do not sit on our hands here at Intertel. Check out our website on who and how we serve and I have no doubts that you will email or ring us up to find out about joining up.

One last point: Can you really risk going out on a limb without first knowing that you have all the facts at your fingertips about worldwide business events? Intertel is there to save you.

Fondly,

10. Interview a student at your school or a co-worker who was born and raised in a non-English-speaking country about the proper etiquette in writing a business letter to someone from his or her country. Collaborate with that student to write a letter to an executive from that country—for example, a sales letter or a letter asking for information.

11. In a letter to your instructor, describe the kinds of adaptations you had to make for the international reader you wrote to in Exercise 10.

12. Assume you work for a large international corporation that has just opened a new office in one of the following cities:
   a. Dar es Salaam, Tanzania
   b. Istanbul, Turkey
   c. Caracas, Venezuela
   d. Manila, Philippines
   e. Kyiv, Ukraine
   f. Beijing, China
   g. Warsaw, Poland
   h. Mexico City, Mexico
   i. Amman, Jordan
   j. Perth, Australia
   k. Lagos, Nigeria
   l. Prague, Czech Republic

Using the resources mentioned on page 186, write a short report (1–2 pages) on the main points of letter etiquette that your boss will have to observe in communicating with the non-native speaker of English who is the manager at the new office.

13. As a collaborative project, team up with three other students in your class to write separate letters tailored to executives in each of the following cities:
   a. Riyadh, Saudi Arabia
   b. Tokyo, Japan
   c. Munich, Germany
   d. Nairobi, Kenya

Assume you are selling the same product or service to each reader, but adapt your communication to the culture represented by the reader.

Turn in the four letters, and explain to your instructor in an accompanying memo how you met the needs of those diverse cultural audiences in terms of style, tone, level of content, format, and sales tactics. Describe the research tools you used to find out about your reader's particular culture (and communication protocols) and how you benefited from using those tools.

CHAPTER **6**

Types of Business Letters

As we saw in Chapter 5, letters can be the lifeblood of any company or organization. In this chapter, you will learn to write a variety of letters for different workplace occasions. But regardless of your message, every letter you send needs to

- establish or maintain good rapport with the reader
- protect and promote your company's and your own professional image
- continue or increase business sales, relationships, and opportunities

Letters can help or hurt your company and your career. You want to write letters that lead to promotions and greater professional opportunities.

Access chapter-specific interactive learning tools, including quizzes and more in your English CourseMate, accessed through www .cengagebrain.com.

Formulating Your Message

Writing effective, diplomatic letters can be challenging. Business correspondence places a great demand on your ability to formulate and organize a suitable message for your readers. You have to identify your audience, your purpose in writing to them, and their needs in wanting to hear from you. To communicate with your audience effectively, you have to determine

- what to say
- how to say it
- where to say it

The second and third points are just as important as the first. In fact, they may be even more crucial to your success as a letter writer. As we saw in Chapter 5, the tone and organization of your letter will determine how your message is received. In the process of drafting and revising your letter, consider whether your reader will bristle at or welcome the words you use. Review pages 179–183 on the "you attitude."

You will also have to select the best communication strategy, which means taking into account where and how you give readers information. Some messages need to be direct and clear-cut; others require being indirect, holding off a bad news message until you prepare the reader for it. Consider the following two versions of extracts from a letter that send the same message:

Insensitive Version

> Your job is being outsourced next quarter. Corporate knows of your twelve years with the firm, but downsizing mandates this decision. The firm will try to place you in another job, but expect no guarantees.

Respectful Version

> Thank you for your twelve years of excellent work. We appreciate your many contributions. Unfortunately, downsizing means that your position, like several others in the department, will be outsourced. However, I will try to arrange a transfer for you. I may not succeed, but I know you will in your career because of your skills and dedication.

Clearly, the first letter has no respect for the reader's feelings or contributions, while the second is a model of respect and sincerity.

Letter Writers Play Key Roles

To be a skilled letter writer means that you will have to play several key roles, sometimes all at once. Among them are

- researcher
- problem solver
- decision maker
- honest and ethical spokesperson
- team player

These are the assets of a valued employee and the characteristics that employers want in job applicants (see pages 272–276). In fulfilling these roles through your business letters, you demonstrate that you can work well with people, are sensitive to their needs, and represent your company with integrity.

Letters and Collaboration at Work

Like other types of workplace writing, letters can be collaborative documents. For instance, you may confer with specialists in engineering to answer a complaint about one of your company's products, or you may work with individuals in the law or marketing department to draft a sales letter. Possibly you may need to get your letter approved by your boss, who may edit your letter before you send it. Or you may be asked to write a letter for another person's signature, as in Figure 3.6 (pages 99–100).

This chapter gives you guidelines and strategies to write successful business letters. It will also show you how *not* to write these letters, an equally valuable lesson worth learning about the world of work.

The Five Most Common Types of Business Letters

This chapter discusses the five most common types of correspondence that you will be expected to write on the job:

1. cover letters
2. inquiry letters
3. special request letters
4. sales letters
5. customer relations letters

These letter types involve a variety of formats, writing strategies, and techniques. Business letters can be classified as *positive*, *neutral*, or *negative*, depending on their message and the anticipated reactions of your audience. Inquiry and special request letters are examples of neutral letters that carry neither good nor bad news; they simply inform, responding to routine correspondence.

- Cover, inquiry, and special request letters introduce a document, request information about a product or service, place an order, or respond to some action or question.
- Sales letters promoting a product carry positive news, according to the companies that spend millions of dollars a year preparing them.
- Customer relations letters can be positive (responding favorably to a writer's request or complaint) or negative (refusing a request, saying "No" to an adjustment, denying credit, seeking payment, critiquing poor performance, or announcing a product recall).

Cover Letters

A cover letter accompanies a document (a proposal, a report, a survey, a portfolio) that you send to your reader. It identifies the type of document you are sending and prepares your audience to read it. Figure 5.1 (page 169) is an example of a cover letter used to send a smaller document, while Figure 15.3 (pages 702-718) contains a cover letter sent with a copy of a long report. Cover letters should do the following:

- Provide a written record that you have transmitted a document.
- Tell readers why you are sending them the document.
- Briefly summarize what the document contains—number of sections, visuals, statistics, appendices, and so on.
- Explain why the document is of interest to readers.
- Express a willingness to answer questions about the document.
- Thank readers for their time.

Inquiry Letters

Inquiry letters ask for information about a product, service, or procedure. Businesses frequently exchange such letters. As a customer, you, too, have occasion to ask in a letter about a special line of products, the price, the size, the color, and delivery arrangements. The clearer your letter, the quicker and more helpful your answers are likely to be.

Figure 6.1 illustrates a letter of inquiry from Michael Ortega to a real estate office managing a large number of apartment complexes. Note that it follows these five rules for an effective inquiry letter:

- It states exactly what information the writer wants.
- It indicates clearly why the writer must have the information.
- It keeps questions short and to the point.
- It specifies when the writer must have the information.
- It thanks the reader.

Had Ortega simply written the following very brief letter to Hillside Properties, he would not have received the information he needed about the size, location, and price of apartments: "Please send me some information on housing in Roanoke. My family and I plan to move there soon."

Special Request Letters

Special request letters make a special demand, not a routine inquiry. For example, these letters can ask a company for information that you as a student or an employee will use in a report, or they might ask an individual for a copy of a report or an agency for facts that your company needs to prepare a proposal or sell a product. The person or company being asked for help stands to gain no immediate financial reward for supplying the information; the only reward is the goodwill that a response creates. Figure 6.2, on page 212, contains a special request letter from a student asking for information to complete a report.

Because your busy reader is not obliged to respond to your letter, the way you present yourself and your request is crucial to your success. "Why should I help?" is a common question any reader might have. Therefore, try to anticipate and remove any barriers to getting his or her help. Your aim is to convince the reader to send the information you need.

To make your request clear and easy to answer, supply readers with an addressed, postage-paid envelope, an email address, and fax and telephone numbers in case they have questions. But don't expect a company to fax a long document to you. It is discourteous to ask someone else to pay the fax charges for something you need.

Guidelines for Writing Special Request Letters

When writing a special request letter, follow these nine points:

1. Address your letter to the right person.
2. In your opening paragraph, state who you are and why you are writing—a student researching a paper, a state employee compiling information for a report, and so on.
3. Indicate clearly your reason for requesting the information. Mention any individuals who may have referred you to the reader for help and information.

FIGURE 6.1 A Letter of Inquiry Written in Semi-Block Format

Michael Ortega
403 South Main Street
Kingsport, TN 37721-0217
mortega@erols.com

April 6, 2012

Mr. Fred Stonehill
Property Manager
Hillside Properties
701 South Arbor St.
Roanoke, VA 24015-1100

Dear Mr. Stonehill:

Please let me know if you will have any two-bedroom furnished apartments available for rent during the months of June, July, and August. I am willing to pay up to $750 a month plus utilities. My wife, one-year-old son, and I will be moving to Roanoke for the summer so I can take classes at Virginia Western Community College.

If possible, we would like to have an apartment that is within two or three miles of the college. We do not have any pets.

I would appreciate hearing from you by April 20. My email address is **mortega@erols.com**, or you can call me at home (606-555-8957) any evening from 6–10 p.m.

Should you have any suitable vacancies, we would be happy to drive to Roanoke to look at them and give you a deposit to hold an apartment. Thanks for your help.

Sincerely yours,

Michael Ortega

Michael Ortega

States precise request

Explains need for information

Identifies precise location

Specifies exact date when a reply is needed

Offers to confer with and then thanks the reader

FIGURE 6.2 A Special Request Letter

1505 West 19th Street
Syracuse, NY 13206
315-555-1214   jkawatsu@webnet.com

October 5, 2012

Ms. Sharonda Aimes-Worthington
Research Director
Creative Marketing Associates
198 Madison Ave.
New York, NY 10016-0092

Dear Ms. Aimes-Worthington:

Explains reason for letter and how the writer learned about the firm

I am a junior at Monroe College in Syracuse, and I am writing a report on "Internet Marketing Strategies for the Finger Lakes Region of New York" for my Marketing 340 class. Several of my professors have spoken very highly of Creative Marketing Associates, and in my own research I have learned a great deal from reading the blogs on Web designs and local economies that you posted last month.

Proves writer has done research

Acknowledges reader's expertise

Given your extensive experience in developing Internet sites to promote regional businesses and tourism, I would be grateful if you would share your responses to the following three questions with me:

1. What have been the most effective design components in websites for a regional marketplace such as the Finger Lakes?

Lists specific, numbered questions on the topic

2. How can area chambers of commerce and various municipalities help generate Web traffic to a regional marketplace website for the Finger Lakes area?

3. Which other regional area(s) do you see having the same or very similar marketing goals and challenges as the Finger Lakes?

As an incentive offers to send copy of report

Your answers to these questions would make my report much more authoritative and useful. I would be happy to send you a copy and will, of course, be honored to cite you and Creative Marketing Associates in my work.

Indicates when information is needed, and makes contact easy

Because my report is due by December 2, I would greatly appreciate having your answers within the next month so that I can include them. Would you kindly send your responses, or any questions you may have, to my email address, jkawatsu@webnet.com.

Thanks reader

Many thanks for your help.

Sincerely yours,

Julie Kawatsu

Julie Kawatsu

4. Research your topic so you can ask the most important, relevant questions.

5. Make your questions easy to answer. List, number, and separate them clearly by double spacing between them. Keep them to three or four at most.

6. Specify exactly when you need the information. Be reasonable; don't ask for the impossible. Allow sufficient time—at least three weeks.

7. As an incentive for the reader to reply, offer to forward a copy of your report, paper, or survey in thanks for the reader's help.

8. If you want to reprint or publish the materials you ask for, indicate that you will secure whatever permissions are necessary. State that you will keep the information confidential, if that is appropriate.

9. End by thanking the reader for helping.

Note how Julie Kawatsu's letter in Figure 6.2 adheres to these guidelines.

Sales Letters

Sales letters are written to persuade readers to buy a product, try a service, support some cause, or participate in some activity. A sales letter can also serve as a method of introducing yourself to potential customers. No matter what profession you have chosen or whether you are self-employed, work for a small company, or are part of a large organization or charitable project, knowing how to write a sales letter is an invaluable skill. There will always be times when you have to sell a product, a service, an idea, a point of view, or yourself—for an interview, a raise, or a promotion. Figures 6.3 and 6.4 contain two sales letters.

Preliminary Guidelines

You have undoubtedly received numerous sales letters from large companies, local merchants, charitable organizations, and campus groups. Because of the great volume of sales letters like these in the business world, the ones you write face a lot of competition. To write an effective sales letter that stands out and does its job, you have to do the following:

1. **Identify and limit your audience.** Knowing who your target audience is and how to find them is crucial to your success. You will also have to determine how many people are in your audience. A sales letter may be written to just one person (Figure 6.3) or to hundreds of readers at different companies (Figure 6.4).

2. **Use reader psychology.** Think like your reader and ask: "What am I trying to do for the customer?" Ask that question before you begin writing and you will be using effective reader psychology. Appeal to readers' health, security, convenience, comfort, or finances by focusing on the right issues (for instance, inform buyers that your product research involves no animal testing). Note that Patrice St. Jacques appeals to the reader's ethnic and community values in Figure 6.3.

3. **Don't be a bore or boast.** Save elaborate explanations about a product for after the sale. Put detailed documentation in instruction booklets, in warranties,

FIGURE 6.3 A Sales Letter That Appeals to a Specific International Audience

ISLAND JACQUES

4700 Cyprus Avenue
Philadelphia, PA 19172

www.islandjacques.netdoor.com
856-555-3295

Distinctive, functional letterhead

11 May 2012

Mr. Etienne Abernathy, President
Seagrove Enterprises
1800 S. Port Haven Road
Philadelphia, PA 19103-1800

Dear Mr. Abernathy:

Compliments reader on award

Congratulations on winning the Hanover Award for Community Service. We in the Port Haven area of Philadelphia are proud that a business with Caribbean roots has received such a distinguished honor.

Extends invitation

To celebrate your and Seagrove's success, as well as all your business entertaining needs (annual banquet, monthly meetings, etc.), I invite you to Island Jacques. We are a family-owned business that for 30 years has offered Philadelphia residents the finest Caribbean atmosphere and food west of the Islands. Our black pepper shrimp, reggae or mango chicken, and Steak St. Lucie—plus our irresistible beef and pork jerk—are the talk from here to Kingston. You and your guests can also see our original Caribbean art and enjoy our steel drum music.

Appeals to reader's senses through art, music, and food

Describes special features, such as flexible hours and dining options

Island Jacques can offer Seagrove a variety of dining options. With five separate dining rooms, we are small enough for an intimate party of 4 yet large enough to accommodate a group of 250. We can do early lunches or late dinners, depending on your schedule. And we even cater, if that's your style. Our chefs—Diana Maurier and Emile Danticat—will prepare a special calypso menu just for you. Also a benefit, our prices are competitive for the Philadelphia area.

Ties costs to benefits

Ends by giving reader incentive to act soon

Please call me soon so you can savor Island Jacques's unique hospitality. For your convenience, I am enclosing a copy of this week's menu delights. Check out our website, too, for a taste of Caribbean sound. We would love to feature Seagrove as Island Jacques's "Guest of the Week!"

Uses an appropriate close for reader

Stay Cool, Mon.

Patrice St. Jacques

Patrice St. Jacques
Manager

FIGURE 6.4 A Sales Letter Sent to a Business Reader

 Workwell Software
3700 Stewart Avenue   Chicago IL 60637-2210
Phone: (312) 555-3720   **Fax:** (312) 555-7601   **Email: sales@workwell.com**
www.workwell.com

August 13, 2012

Ali Jen, Office Manager
Circuit Systems, Inc.
7 Tyler Place
Oklahoma City, OK 73101-0761

Dear Manager Ali Jen:

Do you know how much money your company loses from repetitive
strain injury (RSI)? Each year employers spend millions of dollars on
employee insurance claims because of back pains, fatigue, eye strain,
bursitis, and carpal tunnel syndrome injuries.

Workwell can solve your problems with its easy-to-use Exercise
Program Software, which automatically monitors the time employees
spend at their computers and also measures their keyboard activity. After
each hour (or the specified number of keystrokes), **Workwell** software
will take your employees through a series of brief exercises that will help
prevent carpal tunnel syndrome and muscle strains.

Workwell's Exercise Program Software will not interfere with your busy
schedule. Each of the 27 exercises is demonstrated on screen with audio
instructions. The entire program takes less than 3 minutes and is available
for Windows XP and Windows 7. For only $1,499.00, you can provide a
networked version of this valuable software to all Circuit Systems' employees.

To help your employees stay at peak efficiency in a safe work environment,
please call us at 1-800-555-WELL or visit us at **www.workwell.com** to order
your software today.

Thank you,

Cory Soufas

Cory Soufas
Sales Manager

*Gets reader's
attention with
a question*

*Emphasizes
the product's
appeal
using precise
language*

*Shows specific
application of
the product*

*Links costs to
benefits*

*Ends with a
call for prompt
action*

or on your company's website. Further, do not turn your sales letter into a glowing commendation of your company or yourself. Figures 6.3 and 6.4 both avoid that.

4. Use words that appeal to the reader's senses. Choose concrete words instead of abstract, vague ones. Find verbs that are colorful, that put the reader in the picture, so to speak. You will have a greater chance of selling readers if they can hear, see, taste, or touch your product in their mind. That way they can visualize themselves buying or using your product or service. Note how St. Jacques fills the sales letter in Figure 6.3 with Caribbean sights, sounds, and tastes to appeal to Etienne Abernathy and his company.

5. Be ethical. Avoid untruths, exaggerations, distortions, false comparisons, and unsupported generalizations. Honesty is the best way to make a sale. Never make false claims about the cost, safety, or adaptability of your product or service. You could be prosecuted for mail fraud or sued for misrepresentation. (Review pages 40–42 in Chapter 1.) In addition, never attack a competitor and always get permission before you include an endorsement.

The Four *A*'s of Sales Letters

Successful sales letters follow a time-honored and workable plan; each sales letter follows what can be called the four A's:

1. It gets the reader's *attention.*
2. It highlights the product's *appeal.*
3. It shows the customer the product's *application.*
4. It ends with a specific request for *action.*

Those four goals can be achieved in fewer than four or five paragraphs. Look at Figure 6.4, a one-page letter in which those four parts are labeled. Holding a sales letter down to one page or less will keep the reader's attention. Television commercials, magazine ads, and Web ads provide useful models of the fourfold approach to sales. The next time you see one of these variations of the sales letter, try to identify the four *A*'s.

Getting the Reader's Attention

Your opening sentence is crucial. That first sentence is bait on a hook. If you lose readers there, you will have lost them forever. A typical television commercial or website has about two to five seconds to catch the viewers' attention. Keep your opening short, one or two sentences at most. Show how your product or service will make your reader's life or job easier, save money, or make him or her safer or happier.

The following six techniques are a few of the many interest-grabbing ways to begin a sales letter. Adapt these techniques to your product or service.

1. Ask a question. Look, for example, at the opening question in Figure 6.4. Avoid such general questions as "Are you happy?" or "Would you like to make money?"

Use more specific questions with concrete language. For instance, an ad for Air Force Reserve Nursing asks nurses, "Are you looking for something 30,000 feet out of the ordinary?"

2. Use a how-to statement. This is one of the most frequently used openers in a sales letter. Here are some effective how-to statements: "We can show you how to increase your plant growth up to 91%." "This is how to provide nourishing lunches for less than eighty cents a person." Note that the opening sentence of Cory Soufas's sales letter in Figure 6.4 combines both a how-to and a question approach.

3. Compliment your reader. Appeal to the reader's ego. But remember that readers are not naive; they will be suspicious of false praise. Patrice St. Jacques opens the sales letter in Figure 6.3 with fitting and legitimate praise.

4. Offer a gift. Often you can lure readers further into your letter by sending a coupon for a discount on their next purchase, telling them they are eligible for a rewards program, assuring them of getting a second product free or at half price, and so on. A realtor tempts customers to see lots for sale with this opening: "Enclosed is a coupon worth $50 in gas after you tour Deer Trails Estates."

5. Introduce a comparison. Compare your product or service with conventional or standard ones. For instance, a clothing firm told police officers that if they purchased a particular jacket, they were really getting three coats in one because the product had a removable lining for winter or summer use and a visibility coating for nighttime use.

6. Announce a change. Link your sales offer to a current event that will affect your prospective buyer. For instance, when the sales tax on hybrid cars was about to be increased, a dealer sent sales letters to potential buyers, alerting them to the implications of delaying their purchase: "The sales tax on hybrids will jump a WHOPPING 4 percent effective next month. You may not think that 4 percent will mean that much money, but on a new 2013 model that increase could cost you an extra $1,500." The sales letter continued: "Couldn't you use that money for something else, say, those extras you've always wanted, like a GPS system or an extended warranty?"

Highlighting the Product's Appeal

Once you have aroused your reader's attention, introduce your product or service by making it so attractive, so necessary, and so profitable that the reader will want to buy or use it. In Figure 6.3, the Island Jacques letter zeroes in on the reader's ethnic pride and heritage. In Figure 6.4, Workwell Software's mass mailing letter to office managers appeals to their desire for greater productivity and improved employee safety. Here is an appeal by the Gulf Stream Fruit Company:

> Can you, when you bite into an orange, tell where it was grown? If it tastes better than any you have ever eaten . . . full of rich, golden flavor, brimming with juice, sparkling with sunshine . . . then you know it was grown here in our famous Indian River Valley where we have handpicked it, at the very peak of its flavor, just for your order.[1]

---

[1]From "Gifts from Gulf Stream," Gulf Stream Fruit Company, Ft. Lauderdale, Fla. Reprinted by permission.

Showing the Customer the Product's or Service's Application

In the third part of your sales letter, supply evidence of the value of what you are selling. Don't overwhelm readers with facts, statistics, detailed mechanical descriptions, or elaborate arguments. Keep the emphasis on the reader's use of the product and not on the company that manufactures or sells it.

1. **Supply the right evidence.** What evidence best convinces readers about a product's or service's appeal?

- Descriptions that emphasize state-of-the-art design and construction, efficiency, convenience, usefulness, and economy.
- Special features or changes that make your product or service more attractive. For example, a greenhouse manufacturer stressed that in addition to using its structure for growing plants, customers would also find it to be a "perfect sun room enclosure for year-round 'outdoor' activities, gardening, or leisure health spa." A computer outlet store promised free IT support for a year.
- Testimonials, or endorsements, from previous customers as well as from specialists.
- Guarantees, warranties, services, or special considerations that will make your customer's life easier or happier—a loaner car, free home delivery, a ten-day trial period, a year's free Internet access, or upgrades.

2. **Do I mention costs?** You may be obligated to mention costs in your letter. But postpone them until the reader has been shown how appealing and valuable your product or service is. Of course, if price is a key selling point, mention it early in the letter. As a general rule, however, do not bluntly state the cost. See how unobtrusively St. Jacques handles price at the end of paragraph 3 in Figure 6.3.

Relate prices, charges, or fees to benefits. For example, a dealer who installs steel shutters did not tell readers the exact price of the product but indicated that individuals will save money by buying them: "Virtually maintenance free, your Reel Shutters also offer substantial savings in energy costs by reducing your loss through radiation by as much as 65% . . . and that lowers your utility bills by 35%."

Ending with a Specific Request for Action

The last section of your letter is vital. If the reader ignores your request for action, you have written your letter in vain. Tell readers exactly what you want them to do and by when. Make it easy for them to

- authorize a deduction
- visit your business (as in Figure 6.3)
- visit your website or blog (as in Figure 6.4)
- take a test drive
- participate in a meeting
- fill out a pledge card

- respond via the Internet (as in Figure 6.4)
- sign an order blank
- return a form (provide a stamped and addressed envelope)

As with price, link the benefits the customers will receive to their responses. "Respond and be rewarded" is the basic message of the last section of your letter. Note that in Figure 6.4, the call to action is made in the last paragraph; the writer urges the reader to call a toll-free number immediately in order to increase employee safety and efficiency.

Customer Relations Letters

Much business correspondence deals explicitly with establishing and maintaining friendly working relations. Such correspondence, known as customer relations letters, sends readers good news or bad news, acceptances or refusals. Good news customer relations letters tell customers that you

- have the product or service they want at a reasonable price
- agree with them about a problem they brought to your attention
- will solve their problem exactly the way they want
- are approving their loan
- are grateful to them for their business

Thank-you letters, congratulations letters, and adjustment letters saying "Yes" with the above messages are all examples of good news messages.

Bad news messages, however, inform readers that you

- do not like their work or the equipment they sold you
- do not have what they want or cannot provide it at the price they want to pay
- are rejecting a proposal they offered
- are denying them further use of a facility
- cannot refund their purchase price or perform a service again
- want them to pay what they owe you now
- are downsizing or moving and cannot continue to provide a service

Bad news messages often come to readers through complaint letters, adjustment letters that say "No," and collection letters.

Diplomacy and Reader Psychology

Writing effective customer relations letters requires skill in human relations and reader psychology. Regardless of the news—good or bad—you need to be a diplomatic and persuasive writer. In fact, customer relations letters, like other types of letters, call upon your most effective skills in persuasion (see pages 27–29). To be at your best persuasively, do some research about your readers—their business needs, their schedules, their areas of authority in the chain of command, and even their grievances (if applicable).

Customer relations letters show how you and your company regard the people with whom you do business. The letters should reveal your sensitivity to their needs. The first lesson to learn is that you cannot look at your letter only from your perspective as the writer. You have to see the letter from the reader's perspective and anticipate the reader's needs and reactions.

The Customers Always Write

As you read this section on customer relations letters, keep in mind the two basic principles captured in the pun "the customers always write."

1. Customers will write about how they would like to be or have been treated—to thank, to complain, to request an explanation.
2. Customers have certain rights that you must respect in your correspondence with them. They deserve a prompt and courteous reply, whether or not they are correct. If you refuse their request, they deserve to know why; if they owe you money, you should give them an opportunity to explain and a chance, up to a point, to set up a payment schedule. Always be ethical in responding—be fair, honest, undeniably legal, and professional. (Review pages 29–42.)

Being Direct or Indirect

Your message, tone, and knowledge of your reader are essential ingredients in a successful customer relations letter. But success also involves knowing where and how to start, and, especially, where to present your main point. Not every customer relations letter starts by giving the reader the writer's main point, judgment, conclusion, or reaction. *Where you place your main idea is determined by the type of letter you are writing.* Good news messages require one tactic; bad news, another.

Good News Messages
If you are writing a good news letter, use the direct approach. Start your letter with the welcome, pleasant news that the reader wants to hear. Don't postpone the opportunity to put your reader in the right frame of mind. Then provide any relevant supporting details, explanations, or commentary. Being direct is advantageous when you have good news to convey.

Bad News Messages
If you have bad news to report, do *not* open your letter with it. Be indirect. Prepare your reader for the bad news; keep the tension level down. If you throw the bad

news at your reader right away, you jeopardize the goodwill you want to create and sustain. Consider how you would react to a letter that begins with these slaps:

- Your account is in error.
- Your order cannot be filled.
- Your application for a loan has been denied.
- It is our unfortunate duty to report. . . .

Having been denied, disappointed, or even offended in the first sentence or paragraph, the reader is not likely to give you his or her attentive cooperation thereafter.

Two Versions of a Bad News Message

Figures 6.5 and 6.6 illustrate two versions of a letter written by A. J. Griffin, the manager of a large mall, to one of her tenants, Daniel Sobol, notifying him about an increase in rent. Notice how Griffin's bad news letter in Figure 6.5 curtly starts off with the bad news of a rent increase. Receiving such a letter, the owner of Flowers by Dan certainly could not be blamed for looking for a new place of business. Or if Daniel Sobol did pay the increase, Griffin's letter would hardly ensure that Sobol would remain a happy tenant. Griffin was too direct when he should have been diplomatically indirect. He did not consider his reader's reaction; all he was concerned about was delivering his message.

Compare the curt version of Griffin's letter in Figure 6.5 with her revised message in Figure 6.6. In the revised version, Griffin begins tactfully with pleasant, positive words designed to put her reader in a good frame of mind. Then Griffin gives some background information that the owner of Flowers by Dan can relate to. A businessperson himself, Sobol doubtless has experienced some recent increases in his own costs. Griffin makes one more attempt to encourage Sobol to recall his good feelings about the mall—last year they did not raise rents—before introducing the bad news of a rent increase.

Even after giving the bad news, Griffin softens the blow by saying that the River Road Mall knows it is bad news. Griffin's tactic here is to defuse some of the anger that Sobol will inevitably feel. In fact, Griffin words the bad news so that the tenant sees the mall as acting in the best interest of his flower shop. The mall will not lower or compromise on the services that the tenant has enjoyed and profited from in the past. Griffin then ends on a positive, upbeat note: a prosperous future for Flowers by Dan.

You will learn more strategies about conveying bad news messages on pages 234–239.

FIGURE 6.5 An Ineffective Bad News Letter

River Road Mall

December 3, 2012

Mr. Daniel Sobol
Flowers by Dan
Lower Level 107
River Road Mall

Dear Mr. Sobol:

Blunt opening disregards audience's needs and feelings

This is to inform you of a rent increase. Starting next month your new rent will be $3,500.00, resulting in a 15 percent increase.

Ends with a demand

Please make sure that your January rent check includes this increase.

Sincerely,

No attempt to help audience understand or accept message

A. J. Griffin
Manager
ajg@rrmall.com

300 First Street
Canton, Ohio 44701
(216) 555-6700
www.RRMall.com

FIGURE 6.6 A Diplomatic Revision of the Bad News Letter in Figure 6.5

River Road Mall

December 3, 2012

Mr. Daniel Sobol
Flowers by Dan
Lower Level 107
River Road Mall

Dear Mr. Sobol:

It has been a pleasure to have you as a tenant at the Mall for the past two years, and we look forward to serving you in the future.

Over these last two years we have experienced a dramatic increase in costs at River Road Mall for security, maintenance, landscaping, pest control, utilities, insurance, and taxes. Last year we absorbed those increases and so did not have to raise your rent. We wish we could do it again, but, unfortunately, we must increase your rent by 15 percent, to $3,500.00 a month, effective January 1.

Although no one likes a rent increase, we know that you do not want us to compromise on the quality of service that you and your customers expect and deserve from River Road Mall.

Please let us know how we can assist you in the future. We wish you a very successful and profitable 2013. If you have any questions, please call or visit my office.

Cordially,

A. J. Griffin

A. J. Griffin
Manager
ajg@rrmall.com

300 First Street
Canton, Ohio 44701
(216) 555-6700
www.RRMall.com

Opens with positive association

Prepares reader for bad news to follow

States bad news in most concise, upbeat way

Links bad news to reader benefits

Does not apologize but ends with respectful tone

Friendly complimentary close

© Cengage Learning 2013

Follow-Up Letters

A follow-up letter is sent by a company after a sale to thank the customer for buying a product or using a service and to encourage the customer to buy more products and services in the future. A follow-up letter is a combination thank-you note and sales letter that resells the reader. The letter in Figure 6.7 shows how an income tax preparation service attempts to obtain repeat business by doing the following:

1. beginning with a brief and sincere expression of gratitude
2. discussing the benefits already known to the customer and then drawing the customer's attention to a new sales area
3. ending with a specific request for future business

Occasionally, a follow-up letter is sent to a good customer who, for some reason, has stopped doing business with the company. Such a follow-up letter should try to find out why the customer has stopped doing business and to persuade that customer to resume business. Study the letter in Figure 6.8 (page 226), in which Jim Margolis first politely inquires whether Edward Janeck has experienced a problem and then urges him to return to the store.

Complaint Letters

All of us, either as customers or businesspeople, at some time have been frustrated by a defective product, a delayed or wrong shipment, unavailable products or parts, inadequate or rude service, or incorrect billing. When we get no satisfaction from calling an 800 number and are routed through a series of menu options, our frustration level goes up. Usually our first response is to write a letter in all capital letters (LIKE THIS), dripping with juicy insults. But an angry letter, like a piece of flaming email (see Figure 4.6), rarely gets positive results and can hurt your company's image. In the world of business, be careful about offending readers.

Establishing the Right Tone

A complaint, or claims, letter is a delicate one to write. First, avoid the following:

- name-calling
- sarcasm
- insults
- threats
- unflattering clip art
- all capital letters

The key thing to keep in mind is that you can disagree without being disagreeable. Be rational, not hostile. Just to let off steam, you might want to write an angry letter but then tear it up and write a new letter with courteous and diplomatic language.

A complaint letter is written for more reasons than just blowing off steam. You want some specific action taken. The "you attitude" is especially important here to

FIGURE 6.7 A Follow-Up Letter to Encourage Repeat Business

Taylor Tax Service
Highway 10, North Jennings, TX 78326
(888) 555-9681 email taylor@aol.com
www.taylor.com

December 3, 2012

Ms. Laurie Pavlovich
345 Jefferson St.
Jennings, TX 78326

Dear Ms. Pavlovich:

Thank you for using our services in February of this year. We were
pleased to help you prepare your 2011 federal income tax return. Our
goal is to save you every tax dollar to which you are entitled. If you ever
have questions about your return, we are open all year long to help you.

*Links business
goal to
customer
advantage*

We are looking forward to serving you again next year. Several new
federal tax laws enacted this year will change the types of deductions
you can declare. These changes might appreciably increase your refund.
Our consultants know the new laws and are ready to apply them to your
return.

*Stresses reasons
for customer
to return for
service*

Another important tax matter influencing your 2012 returns will be any
losses you may have suffered because of the hailstorms and tornadoes
that hit our area five months ago. Our consultants are specially trained to
assist you in filing proper damage claims with your federal return.

*Makes it easy
and profitable
for customer to
act soon*

To make using our services even more convenient, we file your tax return
electronically to speed up any refund. Please call us at (888) 555-9681 or
email us at **taylor@aol.com** as soon as you have received all your forms
to set up an appointment. We are waiting to serve you seven days a
week from 9:00 a.m. to 9:00 p.m.

*Ends with
commitment
to customer
convenience*

Sincerely yours,

TAYLOR TAX SERVICE

Demetria Taylor

Demetria Taylor

Photo, Yuri Arcurs/Shutterstock.com

FIGURE 6.8 A Follow-Up Letter to Maintain Customer Goodwill

BROADWAY CLEANERS

May 9, 2012

Mr. Edward Janeck
34 Brompton Lane, Apt. 13
Baltimore, MD 21227-0102

Dear Mr. Janeck:

Thanks customer

Thank you for allowing us to take care of your cleaning needs for more than three years. It has been our pleasure to see you in the store each week and to clean your shirts, slacks, and coats to your satisfaction. Since you have not come in during the last month, we are concerned that in some way we may have disappointed you. We hope not, because you are a valuable customer whose goodwill we do not want to lose.

Expresses concern and respect for customer

Asks for feedback and shows goodwill

If there is something wrong, please tell us about it. We welcome any suggestions on how we can serve you better. Our goal is to have a spotless reputation in the eyes of our customers.

Offers incentive to resume business dealings

The next time you need your garments cleaned, won't you please bring them to us, along with the enclosed coupon worth $10 off your next bill? We look forward to seeing you again—soon.

Cordially,

Jim Margolis

Jim Margolis, Manager

Encl. 1 coupon

Broadway at Davis Drive   Baltimore, Maryland 21228-6210
(443) 555-1962 broadclean@metdoor.com

maintain the reader's goodwill. An angry or defensive letter does not get the best results. Complaint letters that are professional and considerate are more likely to receive positive attention than letters bristling with angry words. An effective complaint letter can be written by an individual consumer or by a company. Figure 6.9 (page 228) shows Michael Trigg's complaint about a defective fishing reel. Figure 6.10 (page 229) expresses a restaurant's dissatisfaction with an industrial dishwasher.

Writing an Effective Complaint Letter

To increase your chances of receiving a speedy settlement, follow these seven steps in writing your letter of complaint. They will help you build your case.

   1. Send your letter to the right person to further ensure your success. As we saw, never address it "To Whom It May Concern." Do your homework—search the company's website or go to Hoover's business directory (www.hoovers.com) to get contact information. But don't send your letter to the CEO. Find the appropriate person or office that responds to customer problems.

   2. Be concise. Keep your letter to one page. Your reader wants essential details, not a saga of your troubles.

   3. Begin with a detailed description of the product or service. Give the appropriate model and serial numbers, size, quantity, color, and cost. Specify check and invoice numbers. Indicate when, where (specific address), and how (through a vendor, the Internet, at a store) you purchased it and also the remaining warranty. If you are complaining about a service, give the name of the company, the date of the service, the personnel providing it, and their exact duties.

   4. State exactly what is wrong with the product or service. Be factual. Precise information will enable the reader to understand and act on your complaint.

- How many times did the product work before it stopped?
- What parts were malfunctioning?
- What parts of a job were not done or were done poorly?
- When did all this happen? How many times?
- Where and how were you inconvenienced?
- Was the service late, incomplete, rude?

Stating that "the brake shoes were defective" tells very little about how long they were on your car, how effectively they may have been installed, or what condition they were in when they ceased functioning safely.

   5. Briefly describe the inconvenience you experienced. Show that your problems were directly caused by the defective product or service. To build your case, give precise details about the time and money you lost. Don't just say you had "numerous difficulties." Did you have to pay a mechanic to fix your car when it was stalled on the road? Did you have to buy a new printer or Blu-Ray DVD player? Where appropriate, refer to any previous telephone calls, emails, or letters. Give the names of the people you have written to or spoken with and the dates.

FIGURE 6.9 A Complaint Letter from a Consumer

17 Westwood Drive

Magnolia, MA 02171

mtrigg@roof.com

October 10, 2012

*Identifies
appropriate
persons to
resolve problem*

Mr. Ralph Montoya
Customer Relations Department
Smith Sports Equipment
P.O. Box 1014
Tulsa, OK 74109-1014

Dear Mr. Montoya:

*Documents all
relevant details
about the
product*

On September 21, 2012, I purchased a Smith reel, model 191, at the Uni-Mart Store on Marsh Avenue in Magnolia. The reel sold for $94.95 plus tax. The reel is not working effectively, and I am returning it to you under separate cover by first-class mail.

*Explains
politely what
is wrong*

I had made no more than five casts with the reel when it began to malfunction. The button that releases the spool and allows the line to cast would not spring back into position after casting. In addition, the gears made a grinding noise whenever I tried to retrieve the line. Because of these problems, I was unable to continue my participation in the Gloucester Fishing Tournament last week.

*Clearly states
what should be
done*

I request that a new reel be sent to me free of charge in place of the defective one I returned. I would also like to know what was wrong with the defective reel.

*Specifies an
acceptable
time frame*

Thank you for processing my claim within the next two weeks.

Sincerely yours,

Michael Trigg

Michael Trigg

FIGURE 6.10 A Complaint Letter from a Business

The Loft

Camerson and Dale
Sunnyside, California 91793-4116
213-555-7500

June 21, 2012

Ms. Priscilla Dubrow
Customer Relations Department
Superflex Products
San Diego, CA 93141-0808

Dear Ms. Dubrow:

On September 15, 2011, we purchased a Superflex industrial dishwasher, model
3203876, at the Hillcrest store at 3400 Broadway Drive in Sunnyside, for $5,000. In the
last three weeks, our restaurant has had serious and repeated problems with this
machine. Three more months of warranty remain on the unit.

The machine does not complete a full cycle; it stops before the final rinsing and thus
leaves the dishes dirty. It appears that the cycle regulators are not working properly
because they refuse to shift into the next necessary gear. Attempts to repair the
machine by the Hillcrest service team on June 4, 11, and 15 have been unsuccessful.

The Loft has been greatly inconvenienced. Our kitchen team has been forced
to sort, clean, and sanitize utensils, dishes, pans, and pots by hand, resulting in
additional overtime. Moreover, our expenses for proper detergents have increased.

We want your main office to send another repair crew to fix this machine. If your crew is
unable to do this, we want a discount worth the amount of the warranty life on this
model to be applied to the purchase of a new Superflex dishwasher. This amount
would come to $1,000, or 20 percent of the original purchase price.

So that our business is not further disrupted, we would appreciate your resolving
this problem promptly within the next four to five business days.

Sincerely yours,

Emily Rashon

Emily Rashon
Co-owner

▶ Browse our menu, which changes daily, at www.theloft.com

Writes to
specific reader

Gives all
product's facts
and warranty
information

Describes what
happened and
when

Documents
problem and
reason for
adjustment

Provides clear
description of
how problem
should be solved

Concludes
politely with
justification for
prompt action

6. Indicate precisely what you want done. Be realistic. Don't inflate costs or damages. And do not simply write that you "want something done" or that you "want to be compensated." State precisely that you want one or more of the following:

- your purchase price refunded in full
- a credit made to your account (give precise amount)
- a credit toward the purchase of another model (list precise amount)
- your exact model repaired or replaced
- a new repair crew assigned to the job
- an apology from the company for discourteous or late service

If you are asking for damages, state your request in dollars and cents and always be prepared to supply any sales receipts, canceled checks, or previous correspondence. Note that if Emily Rashon in Figure 6.10 were asking for reimbursement for her extra expenses because of a defective dishwasher, she would have to present such documentation. Other expenses you or your company may incur include renting a car, paying for cleanup services, or obtaining replacement equipment at a higher rate because the company did not make its deliveries on time.

7. Ask for prompt handling of your claim. In your concluding paragraph, ask the reader to answer any questions you may have (such as finding out where calls came from that you were billed for but did not make). Also specify a reasonable time by which you want to hear from the reader or need the problem fixed. Note how the last paragraphs in Figures 6.9 and 6.10 do this effectively.

Adjustment Letters

Adjustment letters respond to complaint letters by telling customers dissatisfied with a product or service how their claim will be settled. Adjustment letters should reconcile the differences that exist between a customer and a firm and restore the customer's confidence in that firm.

How to (and Not to) Write an Adjustment Letter

An effective adjustment letter requires diplomacy. Be prompt, courteous, and decisive; do not brush the complaint aside in hopes that it will be forgotten. Investigate the complaint quickly, and determine its validity by checking previous correspondence, warranty statements, guarantees, and your firm's adjustment policies on merchandise and service. In some cases you may even have to send returned damaged merchandise to your company's laboratory to determine who is at fault.

A noncommittal letter signals to the customer that you have failed to investigate the claim or are stalling for time. Do not resort to vague statements such as the following:

- We will do what we can to solve your problems as soon as possible.
- A company policy prohibits our returning your purchase price in full.
- Your request, while legitimate, will take time to process.
- While we cannot now determine the extent of an adjustment, we will be back in touch with you.

Customers want to be told that they are right; if they cannot get what they request, they will demand to know why. When you comply with a request, a begrudging tone will destroy the goodwill that your refund or replacement would have created. At the other extreme, do not overdo an apology by agreeing that the company is "completely at fault" or that "such shoddy merchandise is inexcusable." If you make your company look too bad, you risk losing the customer permanently, and saying your company was negligent or that a product was defective may result in a lawsuit. Before you reply, always check your company's policies about what should and should not be said in an adjustment letter.

Adjustment Letters That Tell the Customer "Yes"

It is easy to write a "Yes" letter if you remember a few useful suggestions. As with other good news messages, start with the favorable news the customer wants to hear; that will put him or her in a positive frame of mind to read the rest of your letter. Let the customer know that you sincerely agree with him or her—don't sound as if you are reluctantly honoring the request. For example, if your airline lost or misplaced luggage, apologize before you offer a settlement.

Study the two examples of adjustment letters saying "Yes" in this chapter to see how to write this kind of correspondence. The first example, Figure 6.11 (page 232), says "Yes" to Michael Trigg's letter in Figure 6.9. You might want to reread the Trigg complaint letter to see what problems Ralph Montoya faced when he had to respond to him. The second example of an adjustment letter that says "Yes" is in Figure 6.12 (page 233), responding to a customer who complained about an incorrect billing.

You might also want to review the letter in Figure 5.8 (page 185), which is an adjustment letter saying "Yes."

Guidelines for Writing a "Yes" Letter

The following four steps will help you write a "Yes" adjustment letter.

1. **Admit immediately that the customer's complaint is justified and apologize.** Briefly state that you are sorry, and thank the customer for writing to inform you. Let the customer know that someone in your company is paying attention to his or her letter.

2. **State precisely what you are going to do to correct the problem.** Let the customer know that you will

- extend warranty coverage
- credit the account with funds, more air miles, as in Figure 5.8 (page 185)
- offer a discount on the next purchase
- cancel a bill or give credit toward another purchase
- repair damaged equipment
- enclose a free pass, coupon, or waiver
- upgrade a product or service

FIGURE 6.11 An Adjustment Letter Saying "Yes" to the Complaint Letter in Figure 6.9

Smith Sports Equipment
P.O. Box 1014   Tulsa, Oklahoma 74109-1014
(918) 555-0164 ■ www.smithsport.com

Responds within time frame specified in complaint letter

October 19, 2012

Mr. Michael Trigg
17 Westwood Drive
Magnolia, MA 02171

Dear Mr. Trigg:

Apologizes and announces good news

Thank you for alerting us in your letter of October 10 to your problems with one of our model 191 spincast reels. I am sorry for the inconvenience the reel caused you. A new Smith reel is on its way to you.

Explains what happened and why problem will not recur

We have examined your reel and found the difficulty. It seems that a retaining pin on the button spring was improperly installed by one of our new soldering machines on the assembly line. We have thoroughly inspected, repaired, and cleaned this machine to eliminate the problem from happening again.

Expresses respect for customer

Since we began making quality reels in 1955, we have taken pride in helping loyal customers like you who rely on a Smith reel. We hope that your new Smith reel brings you years of pleasure and many good catches, especially next year at the Gloucester Fishing Tournament.

Closes with friendly offer to help again

Thank you for your business. Please let me know if I can assist you again.

Respectfully,

SMITH SPORTS EQUIPMENT

Ralph Montoya

Ralph Montoya, Manager
Customer Relations Department

FIGURE 6.12 An Adjustment Letter Saying "Yes"

Brunelli Motors

- -

Route 3A, Giddings, Kansas 62034-8100 (913) 555-1521

August 6, 2012

Ms. Kathryn Brumfield
34 East Main
Giddings, KS 62034-1123

Dear Ms. Brumfield:

We appreciate your notifying us, in your letter of July 30, about the problem you experienced with the warranty coverage on your new Phantom Hawk GT. The bills sent to you were incorrect, and I have canceled them. Please accept my apologies. You should not have been charged for a shroud or for repairs to the damaged fan and hose, since all those parts, and labor on them, are fully covered by your warranty.

The problem was the result of an error in the way the charges were listed. Our firm has begun using new billing software to give customers better service, and the technician apparently entered the wrong code for your account. We have since programmed our system to flag any bills for vehicles still under warranty. We hope that this new procedure will help us serve you and our other customers more efficiently.

Thank you for taking the time to write to us. We value you as a customer at Brunelli Motors. When you are ready for another Phantom, I hope that you will once again visit our dealership. Happy motoring!

Sincerely yours,

Susan Chee-Saafir

Susan Chee-Saafir
Service Manager

- -

Experience virtual reality: Drive a new Phantom at
http://www.brunelli.com

Responds promptly

Thanks customer and complies with request

Explains why problem occurred and how it has been resolved

Ends courteously and leaves reader with good feeling about the dealership

© Cengage Learning 2013

Do not postpone the good news the customer wants to hear. The rest of your letter will be much more appreciated and convincing if the customer is told the good news right away. In Figure 6.11, Michael Trigg is told that he will receive a new reel; in Figure 6.12, Kathryn Brumfield learns that she will not be charged for parts or service.

3. Tell customers exactly what happened. They deserve an explanation for the inconvenience they suffered. Note that the explanations in Figures 6.11 and 6.12 give only the essential details; they do not bother the reader with side issues, excessive apologies, or petty remarks about who was to blame. Assure customers that the mishap is not typical of your company's operations. Strive to keep the reader one of your satisfied customers.

4. End on a friendly—and positive—note. Do not remind customers about their trouble. Leave customers with a positive feeling about your company. You want them to purchase your product or service again.

Adjustment Letters That Tell the Customer "No"

Writing to tell customers "No" is obviously more difficult than agreeing with them. You are faced with the sensitive task of conveying bad news, while at the same time convincing the reader that your position is fair, logical, and consistent. Your tone is crucial. You have to be friendly but firm; you cannot be critical or condescending. Do not bluntly start off with a "No." Do not accuse or argue. Avoid remarks, such as the following, that blame, scold, or remind customers of a wrongdoing:

- You obviously did not read your instruction manual.
- Our records show that you purchased the equipment after the policy went into effect.
- The company policy plainly states that such refunds are not allowed.
- You were negligent in running the machine.
- You claim that our scanner was poorly constructed.
- Your complaint is unjustified.

Guidelines for Saying "No" Diplomatically

The following six suggestions will help you say "No" diplomatically. Practical applications of these suggestions can be found in Figures 6.13 and 6.14 (pages 235 and 236). Contrast the refusal of Michael Trigg's complaint in Figure 6.13 with the favorable response to it in Figure 6.11.

1. Thank customers for writing. Open with a polite, respectful comment, called a *buffer*, to soften your reader's response before he or she sees your "No." Don't put your reader on the defensive by beginning with "We regret to inform you." The letter writers in Figures 6.13 and 6.14 use buffers to thank the customers for bringing the matter to their attention and to sympathize with them about their inconvenience. As with other bad news letters, never begin with a refusal. Telling readers

FIGURE 6.13 An Adjustment Letter Saying "No" to the Complaint Letter in Figure 6.9

Smith Sports Equipment

P.O. Box 1014   Tulsa, Oklahoma 74109-1014
(918) 555-0164 ▪ www.smithsport.com

October 19, 2012

Mr. Michael Trigg
17 Westwood Drive
Magnolia, MA 02171

Dear Mr. Trigg:

Thank you for writing to us on October 10 about the trouble you experienced with our model 191 spincast reel. We are sorry to hear about the difficulties you had with the release button and gears.

We have examined your reel and found the difficulty. It seems that a retaining pin in the button spring was pushed into the side of the reel casing, thereby making the gears inoperable. The retaining pin is a vital yet delicate part of your reel. In order to function properly, it has to be pushed gently. Since our warranty does not cover the use of the pin in this way, we cannot send you a replacement.

However, we want you to have many more hours of fishing pleasure, and so we would be happy to repair your reel for $19.98 and return it to you within 5–7 days. Please let us know your decision.

I look forward to hearing from you. Thank you for writing to us.

Respectfully,

SMITH SPORTS EQUIPMENT

Ralph Montoya

Ralph Montoya, Manager
Customer Relations Department

Buffer— thanks and sympathizes with reader

Explains problem without directly blaming the reader; gives firm decision

Turns a "No" into a "Yes" for customer

Ends politely without any reference to the problem

© Cengage Learning 2013

FIGURE 6.14 Another Adjustment Letter Saying "No"

Health**AIR**

4300 Marshall Drive
Salt Lake City, Utah 84113-1521
(801) 555-6028
www.healthair.com

August 28, 2012

*Inside address
and salutation
list reader's title*

Ms. Denise Southby, Director
Bradley General Hospital
Bradley, IL 60610-4615

Dear Director Southby:

*Professional
you-centered
opening with
buffer*

Thank you for your letter of August 20 explaining the problems you encountered with our Puritan MAII ventilator. We were sorry to learn that you were unable to get the high-volume PAO_2 alarm circuit to work.

*Justifies firm
decision by
explaining
causes of
problem and
conditions of
sale*

Our ventilator is a high-volume, low-frequency unit that can deliver up to 40 ml. of water pressure. The ventilator runs with a center of gravity attachment on the right side of the diode. The trouble you had with the high oxygen alarm system is due to an overload on your piped-in oxygen. Our laboratory inspection of the ventilator you returned indicated that the high-pressure system had blown a vital adapter in the MAII. An overload in an oxygen system is not covered by the warranty on the ventilator, and so we cannot replace it free of charge.

*Provides
practical
alternative
with financial
incentive to
keep customer's
business*

*Ends with
goodwill*

We would, however, be pleased to send you another model of the adapter, which would be more compatible with your system, as soon as we receive your order. The price of the adapter is $600, but because you are a valued customer, our service representative will install it at no charge to you.

Please let me know your decision. I look forward to hearing from you.

Sincerely yours,

R. P. Gifford

R. P. Gifford
Customer Service Department

"No" in the first sentence or two will negatively color their reactions to the rest of your letter. Use the indirect approach discussed earlier in this chapter (pages 220–223), and avoid these reader-hostile openings:

- I was surprised to learn that you found our product unsatisfactory.
- We have been in business for years, and nothing like this has ever happened.
- There is no way we could give you what you demand.

2. State the problem so that customers realize that you understand their complaint. Let customers know that you have read their letter carefully and that you are responding to it fairly. Reassure them that you are aware of the facts of the situation. You thereby prove that you are not trying to misrepresent or distort what the customer told you. You want to keep the reader as a customer.

3. Explain what happened with the product or service before you give the customer a decision. Provide a factual, respectful explanation to show customers they are being treated fairly. Rather than focusing on the customer's misunderstanding of instructions or failure to observe a service agreement or contract, explain the proper use of the product or the terms of the agreement.

> Poor: By reading the instructions on the side of the paint can, you would have avoided the streaking condition that you claim resulted.
> Revised: Hi-Gloss Paint requires two applications, four hours apart, for a clear smooth finish.

Note how the explanations in Figures 6.13 and 6.14 emphasize the appropriate ways to use the products without blaming the reader and thereby losing his or her business for good.

4. Give your decision without hedging. It might seem easier to dangle some hope before readers that you might say "Yes." But this fence-sitting is not a good strategy for establishing a business relationship. Don't say, "Perhaps some type of restitution could be made later" or "Further proof would have been helpful." Indecision will only infuriate your customers. Never apologize for your decision. You could be leaving yourself and your company open to a lawsuit. Arrive at a fair and firm decision, but don't dwell on it. Keep your bad news short, and anchor it to an honest explanation, as in Figures 6.13 and 6.14. Never apologize for your decision, and avoid using words such as *reject, claim,* or *grant. Reject* is harsh and impersonal. *Claim* implies you distrust the customer's complaint. *Grant* signals that you have the power to respond favorably but decline to do so. Instead, use words that reconcile or mend relationships.

5. Turn your "No" into a benefit for readers. End by convincing readers of your goodwill. Move from bad news to good news. Link your firm "No" to an attractive alternative for readers. Your attitude needs to be "Here's what we can do to solve your problem." Find some way to make your "No" more acceptable by supplying a workable alternative to what customers have asked for but did not receive. Surround the bad news with benefits.

- Offer to repair a part for them (Figure 6.13).
- Sell them a replacement and install it free of charge or for a lower fee (Figure 6.14).
- Encourage readers to see the "No" of a refusal of credit as good for the writer's business (as in Figure 6.15).

Never promise to do the impossible or go against company policy, but do continue to convince readers you have their needs in mind.

6. Leave the door open for better and continued business. Close with a diplomatic sentence or two showing respect for your readers and expressing your desire to do business with them again. Figure 6.13 ends by letting the reader know that the writer is looking forward to hearing from him, and Figure 6.14 graciously thanks the reader for being a valued and loyal customer.

Refusal-of-Credit Letters

A special type of bad news letters deals with a company's refusing credit to an individual or another company. Writing such a letter requires a great deal of sensitivity. You want to be clear and firm about your decision; at the same time, you do not want to alienate the reader and risk losing his or her business in the future.

How to Say "No"

Follow these guidelines when saying "No" in a refusal-of-credit letter:

1. Begin on a positive—not a negative—note. Find something to thank the reader for; make the bad news easier to take. Compliment the reader's company or previous good credit achievements (if known), and certainly express gratitude to the individual for wanting to do business with your company.

2. In a second paragraph, provide a clear-cut explanation of why you must refuse the request for credit, but base your explanation on facts, not personal shortcomings or liabilities. Appropriate reasons to cite for a refusal of credit include

- a lack of business experience or prior credit
- being "overextended" or needing more time to pay off existing obligations (Figure 6.15)
- current unfavorable or unstable financial conditions
- an order that is too large to process without prepayment
- a lack of equipment or personnel for the company to do the business for which they are seeking credit

3. End on a positive note. Encourage the reader to reapply when business conditions have improved or when the reader's firm is in a better financial position. Try to keep the reader as a potential customer, as in Figure 6.15.

FIGURE 6.15 An Effective Letter Refusing Credit

WEST COAST CREDIT INC.

4800 Ridge Road
Los Angeles, CA 91666
Phone (714) 555-3500
FAX (714) 555-4323

March 19, 2012

Mr. Otto L. King
Sunshine Interiors
8235 Mimosa Highway
Vinedale, CA 92004-0318

Dear Mr. King:

We appreciate your interest in doing business with West Coast Credit. It is always gratifying to see a store like yours open in an expanding community like Vinedale.

In reviewing your credit application, we checked into the business history and credit references you supplied. We also called your local credit bureau and while there was nothing negative in your credit profile we did determine that for a business of your size you have already reached a maximum level of indebtedness. For that reason, we believe that this would not be the best time to extend your credit line.

We would, however, encourage you to visit our website and fill out the credit survey. This site is periodically reviewed and can give you up-to-date information on your credit availability. In the meantime, we wish you every success in your new business.

Cordially,

B. Rimes-Assante

B. Rimes-Assante
Supervisor

www.wccredit.com

Begins on a positive note

Denies credit but explains precisely why in factual terms

Ends encouragingly by urging reader to reapply

Collection Letters

Collection letters require the same tact and fairness as do complaint and adjustment letters. Each nonpayment case needs to be evaluated separately. A nasty collection letter sent to a customer who is a good credit risk after only one month's nonpayment can send that customer elsewhere. Three easygoing letters to a customer who is a poor credit risk may encourage that individual to postpone payment, perhaps indefinitely.

Types of Collection Letters

Many businesses send several letters to customers before turning matters over to a collection agency. Each letter in the series employs a different technique, ranging from giving compliments and offering flexible credit terms to issuing demands for immediate payment and threatening legal consequences. One hospital uses the collection letters illustrated in Figures 6.18 and 6.19 (pages 243 and 244) to encourage

Case Study

Writing About Credit to a Non-Native Speaker

Jan Buwalda, who is the operations manager for the Dutch firm Mendson SA, wants to open an account with Consolidated Plastics in New Hampshire in order to purchase their highly respected plastic casings. Thinking that his firm's esteemed reputation in the Netherlands carried over to the United States, he wrote a letter to Emma Corson, Consolidated's Accounts Executive, to open a credit account with Consolidated. Compare Corson's inappropriate refusal letter in Figure 6.16 with her more diplomatic and acceptably worded revision in Figure 6.17.

The letter in Figure 6.16 is rude, uses words a non-native speaker of English may not understand ("herewith," "expedited"), and does not encourage future business dealings with Consolidated. Corson's revision in Figure 6.17 follows the guidelines for effective communication with non-native speakers of English and adheres to the suggestions for denying credit. In fact, Corson diplomatically compliments her reader and his firm, expresses an interest in doing business with Mendson SA, and helps Buwalda understand how Consolidated's credit policy might help Mendson in the future.

As these two examples show, writing a letter refusing credit to a non-native speaker of English requires double tact. As we saw in Chapter 5, you have to consider the cultural expectations of your audience as well as use easily understood international English.

FIGURE 6.16 An Inappropriate Letter Refusing Credit to a Non-Native Speaker of English

Consolidated Plastics

May 25, 2012

Mr. Jan Buwalda
Operations Manager
Mendson SA
Hoofdstraat 23
Dokkum, The Netherlands 1324 XK

Dear Mr. Buwalda:

I have received herewith your request of the 12th and news about your company.
Thanks.

Curt, begrudging tone; uses legalese

Regarding that request to open a credit account with us, it just cannot be done. I don't
know how things are expedited in your country, but in the United States giving credit
to a first-time foreign customer is just not standard business practice. As you must
understand, your credit rating could be unacceptable as far as we are concerned. You
will be expected to pay in cash for your first transaction with us. We'll evaluate the
situation thereafter.

Insensitive to reader's needs and cultural heritage

Let me know how you anticipate proceeding.

Disrespectful close—no attempt to encourage future business

Sincerely,

Emma Corson

Emma Corson
Accounts Executive

900 Technology Blvd., Bambrake, NH 03243
Phone (603) 555-7000 ▪ Fax (603) 555-4321 ▪ conplastics@compuserv.com
www.conplastics.com

FIGURE 6.17   A Diplomatic Revision of Figure 6.16, a Letter Refusing Credit to a Non-Native Speaker of English

Consolidated Plastics

May 25, 2012

Mr. Jan Buwalda
Operations Manager
Mendson SA
Hoofdstraat 23
Dokkum, The Netherlands 1324 XK

Dear Mr. Buwalda:

Expresses gratitude to and interest in reader's company

Thank you very much for your letter asking about opening a credit account with our firm. It is always a pleasure to hear from potential customers in Holland. I was most interested to learn about Mendson's diverse activities.

Welcomes future business contact

We understand and share your company's wish to have a U.S. supplier. Having Mendson as a customer would be beneficial for Consolidated Plastics, too. Working with you would allow us to enter a new market.

Politely explains procedure to ensure reader's goodwill

However, I am sorry that we cannot open any new account on credit. If you would kindly send us your check for your first month's supplies, we will rush your shipment to you. Doing this, you can establish an account with us and you will be able to charge your second month's supplies on that account.

Closes with friendly offer to help

Please write or call me if you have any questions. I look forward to serving you and Mendson in the future.

Cordially,

Emma Corson

Emma Corson
Accounts Executive

900 Technology Blvd., Bambrake, NH 03243
Phone (603) 555-7000 ■ Fax (603) 555-4321 ■ conplastics@compuserv.com
www.conplastics.com

FIGURE 6.18 A First, or Early, Collection Letter

SABINE MEMORIAL HOSPITAL

7200 Medical Blvd.
Sabine, TX 77231-0011
512-555-6734   www.sabinememorial.org

May 15, 2012

Mr. Cal Smith
24 Mulberry Street
Valley, TX 77212-3160

Re: Inpatient Services
Date of Hospitalization: March 11–12, 2012
Balance Due: $4,725.48

Dear Mr. Smith:

We are honored that we were able to serve your health care needs during your recent stay at Sabine County Hospital. It is our continuing goal to provide the best possible care for residents of Sabine County and its vicinity. To do so, we must keep our finances up-to-date.

Our records indicate that your account is now overdue and that we have not received a payment from you for two months. If you have recently sent one in, kindly disregard this letter and accept our thanks.

If for any reason you are unable to pay the full amount at this time, we would be happy to set up a convenient payment schedule. Just fill in the appropriate blanks below, and return this letter to us. That will enable us to avoid billing you on a "Past Due" basis. Thank you for your cooperation.

Please call me if you have any questions about our billing options.

Sincerely,

Morris T. Jukes
Accounts Department

( ) I will pay $ _____ ( ) monthly ( ) quarterly on my account.

( ) Enclosed is a check for full payment in the amount of $ _____.

Signature

_____

Links hospital mission to patient's payment

Diplomatic reminder to pay now

Offers options to maintain goodwill

Closes with offer of assistance if customer cannot pay now

Makes it easy for reader to respond

© Cengage Learning 2013

Photo, Thomas Robbin/Photolibrary

FIGURE 6.19 A Final Collection Letter

SABINE MEMORIAL HOSPITAL

7200 Medical Blvd.
Sabine, TX 77231-0011
512-555-6734   www.sabinememorial.org

September 21, 2012

Mr. Cal Smith
24 Mulberry Street
Valley, TX 77212-3160

Re: Inpatient Services
Date of Hospitalization: March 11–12, 2012
Balance Due: $4,725.48

Dear Mr. Smith:

Direct opening about history and current status of the account

During the past few months we have written to you several times about your balance of $4,725.48 for hospital services you received on March 11 and 12. Your account is more than 190 days overdue, and we cannot allow any further extensions in receiving a payment from you.

Reminds patient of goodwill and insists on payment

As you will recall, we have tried to help you meet your obligations by offering several options for paying your bill. You could have arranged for installment payments that would be due each month, or even each quarter, whichever would be more convenient. Because you have not replied, we must ask for full payment now.

States final option in a respectful yet firm tone

If we do not hear from you within the next ten days, we will have no alternative but to turn your account over to our collection agency, which will seriously hurt your credit rating. Neither of us would find this a welcome alternative.

Sincerely,

Morris T. Jukes

Morris T. Jukes
Accounts Department

© Cengage Learning 2013

Photo, Thomas Robbin/Photolibrary

patients to pay their bills. Figure 6.18 is a letter sent early in the collection process when a client is only a month or two late. The collection letter in Figure 6.19, however, is sent much later to a client who has ignored earlier notices.

The tone of Figure 6.18 is cordial and sincere—now is not the time to say "Pay up or else." Instead, it stresses how valuable the patient is and underscores how pleased the hospital is to have provided the care he needed. The second-to-last paragraph makes a request for payment, offering: (1) a flexible payment schedule and (2) an escape from the inconvenience (or embarrassment) of receiving past due notices. The bottom of the letter conveniently lists payment options available to the patient. The last paragraph leaves the door open for communication with the hospital.

The late collection letter in Figure 6.19, however, points out that the time for concessions is over and reminds the patient of all the efforts that the hospital has expended to collect its bills. Appealing to the reader's need to maintain his good credit record, the last paragraph then announces what unfortunate consequences will result if he still does not pay.

Writing Business Letters That Matter: A Summary

As we saw, letters are an important part of workplace writing. Your employer will expect you to respond to and send a variety of messages. This chapter has introduced you to five different types of letters you can expect to write on the job. Although you will send many more memos, emails, and IMs than letters, every job will require you to write effective, professional letters.

Regardless of the type of letter you have to write, follow these six guidelines:

1. **Analyze your audience and their needs.** Anticipate the types of information readers are expecting to receive from you and how they will use it.
2. **Determine your reason for writing.** Are you writing to make a routine or special request, offer an explanation, file a complaint, apologize, sell a product or service, build goodwill, express thanks, refuse credit, or collect a debt?
3. **Organize your information.** Would a direct or an indirect approach be best? Begin with good news or a buffer, and save negative messages for the middle of the letter.
4. **Draft your letter carefully.** Be clear, concise, and diplomatic. Select the most appropriate language for your reader. Letters sent to international readers should respect their cultural traditions.
5. **Revise your letter.** Double-check all facts; make sure you left nothing out.
6. **Proofread, proofread, proofread.**

✓ # Revision Checklist

Planning Correspondence

- ☐ Made sure reader's name, job title, and address are correct.
- ☐ Wrote to a specific individual.
- ☐ Determined if audience will be friendly, hostile, or neutral about my message.
- ☐ Did sufficient research—in print, through online sources, in discussions with colleagues and, when necessary, with my boss—to meet my reader's needs.
- ☐ Proved to reader that I am knowledgeable, professional, and easy to work with.
- ☐ Followed my company protocol in choosing the format, organization, style, tone, and content of my letter.
- ☐ Ensured that my correspondence was timely. Answered all correspondence both from people in my company and from customers promptly and reasonably.

Cover Letters

- ☐ Told reader what I am transmitting.
- ☐ Concisely summarized contents of document.
- ☐ Explained why document is important for reader.

Inquiry Letters

- ☐ Explained why I am writing and what information I need and why.
- ☐ Researched and formulated specific questions that are brief and to the point; put them in easy-to-read format.
- ☐ Indicated when information is needed.
- ☐ Thanked reader for reply.
- ☐ Offered to share a copy of my report with reader.

Special Request Letters

- ☐ Stated why I am writing.
- ☐ Explained reasons for writing and how I will use the requested information.
- ☐ Numbered questions to make them easy to answer.
- ☐ Allowed reader enough time to reply; indicated by what date information is needed; supplied stamped, addressed envelope or email address.
- ☐ Offered to cite reader's help and share a copy of my work.

Sales Letters

- ☐ Identified and convinced my targeted audience.
- ☐ Got reader's *attention* with question or with attention-capturing statement.
- ☐ *Appealed* to the reader's senses.
- ☐ Emphasized how product/service *applies* to solving reader's problem or ensures competitor cannot offer benefits.
- ☐ Asked reader to take *action*.
- ☐ Was honest and ethical and did not attack the competition.

Customer Relations Letters

☐ Began my correspondence with reader-effective strategies. If reporting good news, told the reader right away. If reporting bad news, was diplomatically indirect and considerate of my reader's reactions.

Follow-Up Letters

☐ Thanked reader for his or her patronage.
☐ Courteously stressed benefits to reader for his or her continuing business.
☐ Promised to resolve any problems or answer any questions.
☐ Requested future business.

Complaint Letters

☐ Wrote promptly using a courteous, professional tone free from name-calling, insults, or threats.
☐ Described the problem with product or service in detail, including necessary documentation (invoices, model numbers, dealer's name and address).
☐ Specified exactly what I wanted done and by when; was reasonable and objective.
☐ Ended politely.

Adjustment Letters Saying "Yes"

☐ Started off with good news reader wants to hear.
☐ Apologized sincerely for problem.
☐ Told reader what was wrong and how problem will be corrected; did not blame or make excuses.
☐ Ended on a friendly note without referring to the problem but encouraging further business.

Adjustment Letters Saying "No"

☐ Did not begin with a "No." Instead, followed the indirect approach and used a neutral buffer.
☐ Acknowledged reader's point of view but provided clear and direct explanation for denial of claim.
☐ Concluded by giving or suggesting an alternative to keep customer's goodwill.

Refusal of Credit Letters

☐ Began positively by thanking reader for letter.
☐ Gave appropriate factual reasons for refusal that did not lay personal blame on reader.
☐ Encouraged further inquiries or applications.
☐ Took special care to meet the needs of non-native speakers of English in both tone and message.

Collection Letters
- ☐ Tailored letter according to audience, time, and circumstances of overdue account.
- ☐ Tried to maintain goodwill.
- ☐ Followed all legal guidelines in informing reader about past due account and payment policies.

Exercises

1. Write a letter of inquiry to a utility company, a safety or health care agency, or a company in your town, and ask for a brochure or an annual report describing its services to the community. Be specific about your reasons for requesting the information.

2. In which courses are you or will you be writing a report? Write to an agency or company that could supply you with helpful information, and request its aid. Indicate why you are writing, precisely what information you need, and why you need it. Offer to share your report with the company.

3. Choose one of the following, and write a sales letter addressed to an appropriate reader on why he or she should
   a. work for the same company you do
   b. move to your neighborhood
   c. take a vacation where you did last year
   d. dine at a particular restaurant
   e. use a particular cell phone service
   f. have a car repaired at a specific garage
   g. use the services of a particular real estate agency
   h. use your company's new website when ordering replacement parts

4. As a collaborative project, rewrite the following sales letter to make it more effective. Add any details you think are relevant.

   Dear Pizza Lovers:

   Allow me to introduce myself. My name is Rudy Moore and I am the new manager of Tasty Pizza Parlor in town. The Parlor is located at the intersection of North Miller Parkway and 95th Street. We are open from 10 a.m. to 11 p.m., except on the weekends, when we are open later.

   I think you will be as happy as I am to learn that Tasty's will now offer free delivery to an extended service area. As a result, you can get your Tasty Pizza hot when you want it.

Please see your weekly newspapers for our ad. We also are offering customers a coupon. It is a real deal for you.

I know you will enjoy Tasty's pizza and I hope to see you. I am always interested in hearing from you about our service and our fine product. We want to take your order soon. Please come in.

Hungry for your business,

Rudy

5. Send a follow-up letter to one of the following individuals:
   a. a customer who informs you that she will no longer do business with your company because your prices are too high
   b. a family of four who stayed at your motel for a week last summer
   c. a wedding party or professional organization that used your catering services last month
   d. a customer who returned a coat for the purchase price
   e. a customer who purchased a used car from you and who has not been happy with your service
   f. a company that bought software from you nine months ago, alerting them about updates

6. Write a bad news letter based on an experience at your workplace—rejecting an applicant for a job, notifying tenants of an increase in parking rates, or denying a request for funding, for example.

7. Write a bad news letter to an appropriate reader about one of the following situations:
   a. Your company will be discontinuing Saturday deliveries because of rising labor and fuel costs.
   b. You are the manager of an insurance company writing to tell one of your customers that, because of reckless driving, his or her rates will increase.
   c. You have to refuse to send a bonus gift to a customer who sent in an order after the promotion period ended.
   d. You have discontinued a model that a business customer wants to reorder.
   e. You have to notify residents of a community that a bus route or hours of operation are being discontinued.
   f. You represent the water department and have to tell residents of a community that they cannot water their lawns for the next month because of a serious water shortage in your town.
   g. You cannot send customers a catalog—which your company formerly sent free of charge—unless they first send $10 for the cost of that catalog.
   h. You cannot repair a particular piece of equipment because the customer still owes your company for three previous service visits.

8. Write a good news letter about the opposite of one of the situations listed in Exercise 7.

9. Write a complaint letter about one of the following:
   a. an error in your utility, telephone, credit card, or Internet service bill
   b. discourteous service you received on an airplane, train, or bus
   c. a frozen food product of poor quality
   d. a shipment that arrived late and damaged
   e. an insurance payment to you that is $357.00 less than it should be
   f. a public television station's decision to discontinue a particular series
   g. junk mail or spam that you have been receiving
   h. equipment that arrived with missing parts
   i. misleading representation by a salesperson
   j. incorrect or misleading information given on a website

10. This exercise might be done as a collaborative project. You are a section manager at e-Tech. Your company has a service contract with Professional Office Cleaners (POC). However, each morning when you arrive at work you are disappointed with what they've done. POC has overlooked some essential tasks and done a poor job on others. Your staff is also disappointed and has emailed or spoken to you about problems with POC. Write the following:
   a. a memo or email to your boss, the vice president, about POC's shoddy work
   b. a complaint letter to POC that the vice president has asked you to write and to sign his name to
   c. a letter to the vice president from the manager of POC, apologizing for the problem and offering a solution
   d. a letter from POC to the vice president taking issue with the complaint made against the cleaning company and offering proof that the work was done according to contract specifications
   e. an email you send to your staff about what's happened with POC

11. Write the complaint letter that prompted the adjustment letter in Figure 6.12.

12. Write the complaint letter that prompted the adjustment letter in Figure 6.14.

13. Rewrite the following complaint letter to make it more precise, less emotional, and more persuasive.

Dear Sir:

We recently purchased a machine from your Albany store and paid a great deal of money for it. This machine, according to your website, is supposedly the best model in your line and has caused us nothing but trouble each time we use it. Really, can't you do any better with your technology?

We expect you to stand by your products. The warranties you give with them should make you accountable for shoddy workmanship. Let us know at once what you intend to do about our problem. If you cannot or are unwilling to correct the situation, we will take our business elsewhere, and then you will be sorry.

Sincerely yours,

14. The following story appeared in a local newspaper.

Residents Concerned About Relocation of Pet Food Plant

OCEAN SPRINGS (AP) Finicky Pet Food is moving its processing plant from Pascagoula to Ocean Springs, a decision that has some residents concerned about possible odor and other problems.

The plant is moving to an industrial area bordering a subdivision of expensive homes.

"The wind doesn't discriminate," said Jo Souers, who lives in the Bienville Place subdivision. "I don't want this in our neighborhood."

City officials said the plant is moving to an area zoned to accommodate it.

"We don't have a lot of control over it," said city planner Donovan Scruggs. "It is a permitted use for this property."

Scruggs said the property was zoned industrial before the subdivision was built. A body shop, cabinet shop, and boat business are located nearby.

The plant will be built in the small industrial area on U.S. 90, directly across the highway from the Super Wal-Mart.

It is moving into a vacant building, the interior of which has been renovated for its new purpose, city officials said.

The plant will process frozen fish and fish parts for bait and pet food. It will employ 10 workers, with that number doubling during fishing season.

Plant manager Dean Niemann said in a statement that the company no longer needed its Pascagoula location near deep water, which was rented from the county.

Scruggs said the city has investigated the possibility that the plant will emit odors.

"We've told Dean (Niemann) from day one, 'You're locating next to a residential area. If you start stinking, action will be taken,'" Scruggs said.

He said the city has a nuisance ordinance that should handle anything that might arise.

Source: "Residents Concerned About Relocation of Pet Food Plant," Associated Press, September 27, 2005. Reprinted by permission.

Based upon information in this story, which you may want to supplement, write the following complaint and adjustment letters:

  a. a complaint letter to the city from resident Jo Souers
  b. a complaint letter to Finicky Pet Food from city officials warning about dangers of pollution to the residential area
  c. a letter from plant manager Dean Niemann to the residents of Bienville Place subdivision
  d. a letter from city officials to the residents of Bienville Place subdivision

15. Write an adjustment letter saying "Yes" to the manager of The Loft, whose letter is in Figure 6.10.

16. Write an adjustment letter saying "No" to the customer who received the "Yes" adjustment letter shown in Figure 6.12.

17. Rewrite the following ineffective adjustment letter saying "Yes."

Dear Mr. Smith:

We are extremely sorry to learn that you found the suit you purchased from us unsatisfactory. The problem obviously stems from the fact that you selected it from the rack marked "Factory Seconds." In all honesty, we have had a lot of problems because of this rack. I guess we should know better than to try to feature inferior merchandise along with the name-brand clothing that we sell. But we originally thought that our customers would accept poorer quality merchandise if it saved them some money. That was our mistake.

Please accept our apologies. If you will bring your "Factory Second" suit to us, we will see what we can do about honoring your request.

Sincerely yours,

18. Rewrite the following ineffective adjustment letter saying "No."

Dear Customer:

Our company is unwilling to give you a new toaster or to refund your purchase price. After examining the toaster you sent to us, we found that the fault was not ours, as you insist, but yours.

Let me explain. Our toaster is made to take a lot of punishment. But being dropped on the floor or poked inside with a knife, as you probably did, exceeds all decent treatment. You must be careful if you expect your appliances to last. Your negligence in this case is so bad that the toaster could not be repaired.

In the future, consider using your appliances according to the guidelines set down in their warranties. That's why they are written.

Since you are now in the market for a new toaster, let me suggest that you purchase our new heavy-duty model, number 67342, called the Counter-Whiz. I am taking the liberty of sending you some information about this model. I do hope you at least go to see one at your local appliance center.

Sincerely,

19. You are the manager of a computer software company, and one of your salespeople has just sold a large order to a new customer whose business you have tried to obtain for years. Unfortunately, the salesperson made a mistake writing out the invoice, undercharging the customer $429. At that price, your company would not break even. You must write a letter explaining the problem so the customer will not assume all future business dealings with your firm will be offered at such "below market" rates. You have to decide whether you should ask for the $429 or just write it off in the interest of keeping a valuable new customer.

a. Write a letter to the new customer, asking for the $429 and explaining the problem while still projecting an image of your company as accurate, professional, and very competitive.

b. Write a letter to the new customer, not asking for the $429 but explaining the mistake and emphasizing that your company is both competitive and professional.

c. Write a letter to your boss explaining why you wrote the letter in (a).

d. Write a letter to your boss explaining why you wrote the letter in (b).

e. Write a letter to the salesperson who made the mistake, asking him or her to take appropriate action with regard to the new customer.

20. You just found out that a business that applied for credit has missed its mortgage payment. You have to refuse credit to this local firm, which has been in business successfully for eight years. Write a refusal letter without jeopardizing future business dealings.

21. Write an appropriate collection letter to one of the following:

a. a loyal customer who has not responded to a first notice letter

b. a new business customer who placed a large order with you last quarter and paid for it promptly but who has ignored two notices you have already sent about an order filled this quarter

c. a customer who has just placed an order over the Internet but has not responded so far to any of your notices for payment for previous purchases

d. a customer who has been continually late but has always paid eventually

e. an international customer who has sent in only partial payment

CHAPTER 7

How to Get a Job

Searches, Networking, Dossiers, Portfolios/ Webfolios, Résumés, Letters, and Interviews

Access chapter-specific interactive learning tools, including quizzes and more in your English CourseMate, accessed through www .cengagebrain.com.

Obtaining a job in today's tough market involves a lot of hard work. Before your name is added to a company's payroll, you will have to do more than simply walk into the human resources office and fill out an application form or send a résumé online. Finding the *right* job takes time in this highly competitive job market. And finding the right person to fill that job also takes time for the employer.

Steps the Employer Takes to Hire

From the employer's viewpoint, the stages in the search for a valuable employee include the following:

1. Deciding what duties and responsibilities go with the job and determining the qualifications the future employee should possess
2. Advertising the job on the company website, on online job-posting sites, in newspapers, and in professional publications
3. Reviewing and evaluating résumés and letters of application
4. Having candidates complete application forms
5. Requesting further proof of candidates' skills (letters of recommendation, transcripts, portfolios/webfolios)
6. Interviewing selected candidates
7. Doing further follow-ups and selecting those to be interviewed again
8. Offering the job to the best-qualified individual

Sometimes the steps are interchangeable, especially steps 4 and 5, but generally speaking, employers go through a long and detailed process to select employees. Step 3, for example, is among the most important for employers (and the most crucial for job candidates). At that stage employers often classify job seekers into one of three groups: those they definitely want to interview, those they may want to interview, and those in whom they have no interest.

Steps to Follow to Get Hired

As a job seeker, you will have to know how and when to give prospective employers the kinds of information the preceding eight steps require. You will also have to follow a definite schedule in your search for a job. Expect to go through the following eight procedures:

1. analyzing your strengths and restricting your job search
2. enhancing your image
3. looking in the right places for a job
4. assembling a dossier and a portfolio
5. preparing a résumé
6. writing a letter of application and filling out a job application
7. going to an interview
8. accepting or declining a job offer

Your timetable should match that of your prospective employer. This chapter shows you how to begin your job search, design an effective portfolio, prepare an appropriate résumé, write a persuasive letter of application, and prepare for an interview.

Analyzing Your Strengths and Restricting Your Job Search

Job counselors advise students to start planning for their careers several years before they graduate. The more you find out about what career path you want to take ahead of graduation, the better you will be able to target the jobs that are right for you. But formulating a clear job objective takes lots of preparation. Learn as much as you can about the type of work that characterizes your chosen area and the preparation you need to be successful in it.

Before you apply for jobs, analyze your job skills, career goals, and interests. Here are some points to consider:

1. Make an inventory of your most significant accomplishments in your major or on the job. What are your greatest strengths—writing and speaking, working with people, organizing and problem solving, managing money, speaking a second language, developing software, designing websites?
2. Decide which specialty within your chosen career appeals to you the most. If you are in a nursing program, do you want to work in a large teaching hospital, for a home health or hospice agency, or in a physician's office? What kinds of patients do you prefer to care for—pediatric, geriatric, psychiatric?
3. What types of working conditions most appeal to you—small groups, traveling, telecommuting, relocating overseas?
4. What are the most rewarding prospects of a job in your profession? What most interests you about a position—travel, technology, international contacts, on-the-job training, helping people, being creative?

5. What are some of the greatest challenges you face in your career today—or will face in five years?
6. Which specific companies or organizations have the best track record in hiring and promoting individuals in your field? What qualifications will such firms insist on from prospective employees?

Once you answer these questions, you can avoid applying for positions for which you are either overqualified or underqualified. If a position requires ten years of related work experience and you are just starting out, you will only waste the employer's time and your own by applying. However, if a job requires a certificate or license and you are in the process of obtaining one, go ahead and apply.

Enhancing Your Professional Image

Whether you are looking for your first job in your career field, returning to the job market, or changing careers, there are several steps you can take to help improve your chances of getting hired. Here are some suggestions that will help you develop your professional career plans:

- Attend job fairs and interviewing workshops on campus as well as those sponsored by municipal, state, and federal agencies.
- Go to trade shows to learn about the latest products, services, and technologies in your profession and to meet contacts and even potential employers.
- Join student and professional organizations and societies in your area of interest. Membership rates for students are often reduced.
- Apply for relevant internships and training programs to gain real-world experience and increase your networking contacts.
- If available, take a temporary job in your profession to gain some experience.
- Ask your instructors to critique your work in light of your career plans.
- Confer with your academic adviser regularly, not just once a semester.
- Find a mentor—someone in the field you might want to join.
- Consider developing a competency in a second (or third) language.
- Learn more about Web design and other interactive media.
- Do volunteer work to gain or enhance experience working in a group setting, preparing documents, and so on.
- Write a blog to increase your networking potential and to demonstrate your interest in and knowledge about your profession.

Looking in the Right Places for a Job

One way to search for a job is simply to send out a batch of letters and résumés to companies you want to work for. But how do you know what jobs, if any, those companies have available, what qualifications they are looking for, and what deadlines they might want you to meet? You can avoid these uncertainties by knowing where to look for a job and knowing what responsibilities a specific job entails.

Consult the following resources for a wealth of job-related information. Use as many of them as you can.

1. **Personal (face-to-face) networking.** One of the most successful ways to land a job is through networking. In fact, most jobs come through consulting with other people. John D. Erdlen and Donald H. Sweet, experts on job searching, cite the following as a primary rule of job hunting: "Don't do anything yourself you can get someone with influence to do for you." Let your professors, co-workers, friends, classmates, neighbors, relatives, and even your clergy know you are looking for a job. They may hear of something and can notify you or, better yet, recommend you for the position. See how the job seekers in Figures 7.16 (page 304) and 7.17 (page 305) have successfully networked with people they know. Attend professional and organization meetings related to your field as well as community and civic functions to increase your chances of meeting the right contact people to ask for advice and also for possible follow-up help and even recommendations.

2. **Your campus placement office.** Counselors keep an online file of current available positions, and they also make your résumé available to recruiters when they come on campus to conduct interviews. Placement offices also have recruiting databases, allowing students access to a broad range of contacts and interview information. Counselors can help you locate summer and part-time work as well, both on and off campus, positions that might lead to full-time jobs. Most important, they will give you sound advice on your job search, including strategies for finding the right job, salary ranges, and interview tips. Many placement offices also sponsor career fairs to bring job seekers and employers together in specific professional fields. Finally, your placement office will help you set up and archive your dossier, or credentials (see pages 265–268).

3. **Online job-posting sites.** A majority of posted jobs can be found on the Internet. You can learn about jobs at a specific company or organization by visiting its website to see what vacancies it has and what the qualifications are for them. Also consult the *Riley Guide: Employment Opportunities and Job Resources* on the Internet (www.rileyguide.com). This invaluable resource surveys and classifies job openings on the Web by field, location, and category (private or public) and provides you with links for direct access. In addition, you might want to explore many of the job-posting sites, such as those in Table 7.1, which list positions and sometimes give you advice about applying for them.

4. **Newspapers.** Look at local newspapers as well as the Sunday editions of large city papers with a wide circulation, such as *The New York Times* (http://jobmarket .nytimes.com). The *National Business Employment Weekly* (www.careerjournal .com), published by *The Wall Street Journal*, also lists jobs in different areas, including technical and managerial positions.

5. **Federal and state employment offices.** The U.S. government is one of the biggest employers in the country. During 2010 and 2011, for instance, the most active career site on the Web was operated by the federal government, with 2 million new hires. Counselors at federal and state employment centers also help

TABLE 7.1   Job-Posting Sites on the Web

Website	URL	Description
After College	www.aftercollege.com	Lists more than 200,000 entry-level jobs and internships; connects students, alumni and employers through faculty and career networks across the country
Career One Stop	www.careeronestop.com	Sponsored by the Department of Labor; offers employment services, job search sites, vocational trends ("What's Hot," "Green Jobs"), and help for military transitions
Career Builder	www.careerbuilder.com	Hosts the career sites for more than 9,000 websites, including 140 newspapers and broadband portals such as MSN and AOL
College Recruiter	http://collegerecruiter.com	Leading job board for college students searching for internships and recent graduates hunting for entry-level jobs and other career opportunities
College Grad	www.collegegrad.com	Targets college students and recent grads exclusively. Provides more entry-level job search content to job seekers and linked to more colleges and universities than any other career site
Diversity Employers	www.diversityemployers.com	Largest database of equal-opportunity employers committed to workplace diversity. Dedicated to providing career- and self-development information on careers, job opportunities, graduate/professional schools, internships/co-ops, and study-abroad programs
Indeed	www.indeed.com	In addition to allowing you to search for jobs on your mobile device, this site offers job-search tips, including a tutorial on how to get started, how to either narrow or broaden your search, and provides media postings covering the current job market
Monster	www.monster.com	Largest commercial online job board; lists hundreds of thousands of openings; includes global postings; offers advice on the job-search process
Monster College	http://college.monster.com	Jobs posted for college graduates and internships
Net Temps	www.net-temps.com	Postings for temporary, temp-to-perm, and full-time employment through the staffing industry

job seekers find career opportunities. Consult the following websites for listings of government jobs:

- USAJOBS—www.usajobs.gov
- Federal Jobs—www.federaljobs.net
- Student Jobs—www.studentjobs.gov

Figure 7.1 shows the home page of USAJOBS, a U.S. government site.

   6. **Professional and trade journals and associations in your major.** Identify the most respected periodicals (print and online) in your field and search their ads. Each issue of *Food Technology*, for example, features a section called "Professional Placement," a listing of jobs all over the country. Similarly, *CIO Magazine— Information Technology Professional Research Center* (http://itjobs.cio.com) can help you find jobs in the computer industry, engineering, and

FIGURE 7.1 USAJOBS Website

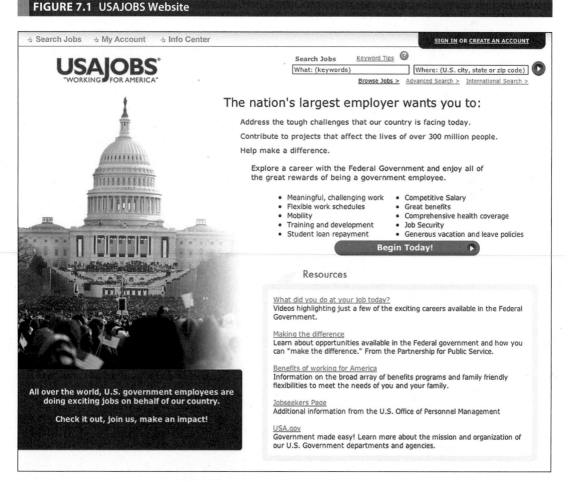

Courtesy of the United States Office of Personnel Management

technology. Consulting the *Encyclopedia of Associations* (http://library.dialog.com/bluesheets/html/bl0114.html) is the quickest way to find out about professional organizations and the journals they publish in your field.

7. The human resources department of a company or an agency you would like to work for. Often you will be able to fill out an application even if there is not a current opening. But do not call employers asking about openings; a visit shows a much more serious interest.

8. A résumé database service. A number of online services will put your résumé in a database and make it available to prospective employers, who scan the database regularly to find suitable job candidates. Check to see if a professional society to which you belong (or might join) offers a similar service.

But be careful about posting personal information, such as your phone number or social security number. You never know who can gain access to this information. If your current boss finds out you are looking for another job, you risk being fired. (See page 268.)

9. Professional employment agencies. Some agencies list jobs you can apply for free of charge (because the employer pays the fee), while others charge a stiff fee, usually a percentage of your first year's salary. Be sure to ask who pays the fee for this service. Because employment agencies often find out about jobs through channels already available to you, speak to someone at your campus placement center first.

Using Online Social and Professional Networking Sites in Your Job Search

Social networking sites (Facebook, Twitter, MySpace, LinkedIn, Google+) are essential tools to help you find a job and advance your career. Do not think of them only as personal media sites where you exchange news and photos with friends and family. These sites contain valuable information about the companies that might hire you. Millions of companies maintain pages, or detailed profiles, on social networking sites to showcase their products, to publicize their service to the community and their respect for the environment, and to provide information about their workforce. Social networking sites are also excellent tools for building a professional network.

Finding Jobs Through Networking Sites

Consulting these networks will alert you to job possibilities in your area. Facebook, for example, contains a "marketplace" where you can search for jobs by location and title in your network. Twit Job Search (www.twitjobsearch.com) is another site you may find useful as you begin your job search. If you follow businesses and organizations on Twitter, you can receive notifications from them about job openings. Facebook and Twitter also allow you to post your profile or a link to your blog, to your portfolio, or to both, to attract potential employers. These networks help you not only to locate a job but also to advertise your expertise for it. The best way to promote yourself is to post your qualifications online. As we will see, though, the most important media site in your job search is LinkedIn (www.linkedin.com), which is designed exclusively to promote professional networking. Each of these networking sites provides powerful search engines to help you contact the right people and to be contacted by them.

Using Facebook to Start Your Network

Facebook can help you start your professional networking. Begin with former employers or co-workers on Facebook, for example, who know your work and who will say good things about your education, skills, and previous work experience. As you prepare to enter the job market, contact these people through Facebook and ask if they would write a reference for you. Also, be sure to "friend" through Facebook former professors, especially in your area of study and who liked your work. Your teachers are important professional contacts because many of them are plugged into professional networks themselves. Contact them to ask if they have heard about any jobs in your field. Moreover, with their permission, your instructors can become your professional references. And, finally, don't forget to connect with friends or classmates who may be employed, even part-time, by companies or organizations you would like to join.

Another way to profit from social networking sites such as Facebook is to search for "friends" who are also on LinkedIn. By connecting with Facebook friends on LinkedIn, you can tap into their professional networks and can begin a professional network quickly by leveraging connections you already have. Facebook is also a powerful self-promotion tool because it allows layers of access. But be careful not to create a profile meant to be viewed only by your friends. Instead, create a public profile on Facebook that is professional and employer-savvy.

Promoting Your Best Image—Some Do's and Don'ts

Used effectively, networking sites allow you to publicize your qualifications through text, audio, video images, and even podcasts. Hiring managers and recruiters often check these social networks for potential job candidates as well as to screen applicants to interview and eventually hire. In fact, more than a million hiring managers, recruiters, and human resource professionals look for qualified candidates on LinkedIn alone. These individuals scout other social networking sites, too, and ask colleagues and other professionals to alert them to qualified job candidates.

But when you use these networks, make sure you enhance and don't jeopardize your chances with prospective employers. Success is all about how you look online, including your social networking profile and what you post on your page and in your portfolio (see pages 268–272), in a blog, or on a website, whether it is your own or someone else's. To project your best professional image, employ the same care as you would in preparing other job-related documents, such as memos, letters, reports, and proposals.

Here are some do's and don'ts to follow when you create your profile on a social networking site.

Do's	Don'ts
Supply a current, professional picture showing how you want employers to see you at an interview.	Never use an unprofessional-looking photo, such as those taken at a party, a sports event or on your vacation. Exclude photos with revealing clothing or compromising poses or gestures.
Choose appropriate "likes" and "activities" relating to your professional, community service, or charitable work. Mention the titles of current books related to your major or articles in *Time, US News & World Report*, or *BusinessWeek*.	Make sure that the hobbies or activities you list do not detract from your professional profile (e.g., playing computer games, gambling, etc.).
Highlight your strongest career accomplishments to demonstrate your knowledge of the industry where you want to work.	Be careful not to quote someone or give out personal information, such as email addresses, phone numbers, or names, without first obtaining that person's permission.
Ensure that your tone, words, and comments are ethical.	Never use sexist, racist, or obscene language online. Steer clear of sensitive or inflammatory topics, such as politics and religion.
Respect all the ethical guidelines of your current or former employer (see pages 31–32).	Refrain from giving out any privileged company information (e.g., names of clients, financial details, marketing plans, etc.). Do not use your company email address to frequent social networking sites via your computer at the office.
Check all other sites (YouTube, a friend's website, etc.) where you may appear or may be quoted to make sure you look and sound professional. Google yourself to find these and other links.	Delete anything that makes you look unprofessional in dress, actions, or words. Adjust privacy settings to protect anything personal.
Be careful when blogging, even if you host your own site; exclude anything embarrassing or damaging to your job search.	Avoid criticizing a former or current co-worker, employer, client, vendor, competitor, or government official or agency.
Keep all information up-to-date. Be consistent, too, in your online postings so that one site does not contradict another.	Do not bombard a potential employer with repeated small posts, queries, or updates.
Verify that all images and video clips are clear and professional.	Make sure you eliminate background noise and crop unnecessary background images.
Be discreet. Check your security settings to ensure that information about you is available only to those you want to have it.	Your profile is never as private as you think. Your current network of friends may include someone who has some type of connection to a prospective employer.

LinkedIn

LinkedIn is the most important social network for your job search. As the name implies, its purpose is to link, or connect, you to people who can help you professionally. As job counselors repeatedly advise, the best way to land a job is through networking—that is, one person helping another. LinkedIn is all about networking, making contacts who have inside information at a company and who may introduce (and maybe even recommend) you to the individual who makes the decision to interview applicants and eventually hire. Through its immense networks and special-interest groups, LinkedIn can make your search for a job more effective and successful. More than 90 million companies and individuals belong to LinkedIn, including Fortune 500 companies such as Microsoft, McDonald's, Walmart, IBM, GM, and Pfizer, plus a host of charitable organizations. LinkedIn also includes special occupational groups such as Medical LinkedIn, which has more than two million members who work in a variety of health care settings, from the pharmaceutical industry to hospitals to fitness centers. Other specific groups include Women 2.0 and the Dallas Business Network.

Benefits of Joining LinkedIn

Joining LinkedIn is easy and free. Not only will belonging to LinkedIn help you find a job, it will enhance your professional image. Being a member of LinkedIn signals to prospective employers that you are serious about pursuing your professional goals, that you possess the qualifications to enter and advance in the world of work, and that you want to participate in a forum to discuss and share professional information about your major/industry. It also signals that you value teamwork. But networking entails more than someone helping you. You need to be prepared to help others through your contacts. Joining LinkedIn tells prospective network contacts and employers you are ready to do that.

Five Ways LinkedIn Can Help Your Job Search

LinkedIn can help you find a position in five key ways.

1. **LinkedIn allows you to search for a specific type of job by targeting a particular specialized field, company, job title, or even zip code.** LinkedIn includes thousands of classified posts helping you to search the wider Internet more efficiently. But keep in mind that some jobs will be listed only on LinkedIn.

2. **LinkedIn gives you valuable information about a company, such as its mission statement, the names of the CEO and directors, its products and services, its locations, its awards, and even its competitors.** Company profiles also provide relevant job statistics—the number of current employees, new hires, and internships, as well as possible future job needs. You will even find the names of individuals who once worked for the company and left for another position. Most important of all, company profiles on LinkedIn provide the names of and, many times, contact information for hiring managers and human resource professionals, the very individuals you want to reach and have reach you. LinkedIn entries also often have links to company websites with job postings, making researching a company or an

organization easier and helping you to prepare a stronger letter of application and make sure you send it to the right person.

3. LinkedIn helps you to become part of a network and expand your list of contacts. The working principle behind LinkedIn is to assist you in joining a network and then finding out if someone in your network knows someone at the company you want to work for. As a LinkedIn member, you can advertise that you are looking for a job in a given area or career field. LinkedIn will then identify individuals in your network, or those who may know someone in another network, who can give you inside information about a potential employer and, ideally, put in a good word for you. You want to join groups that help you to add influential contacts. Even if you have a small network—five to ten people—LinkedIn can still improve your chances of getting hired. Through the "contact settings" for your profile, you can indicate what areas you are interested in, such as job inquiries, career opportunities, or reference requests.

4. LinkedIn improves your search by including short recommendations from people in your network. Solicit as many recommendations as you can, and offer to write them for people you know through LinkedIn (see the profile in Figure 7.2 on pages 266–267). Recommendations praising your work will enhance your profile on LinkedIn, providing potential employers with evidence that you are qualified, conscientious, and respected.

5. LinkedIn allows you to create a public profile that shows you are a professional. Like Facebook, LinkedIn lets you craft a public profile that you can share with potential employers. This profile will be one of the first pages that come up in a Google search when someone looks for you by name. Be sure that your profile is appropriate for the kinds of jobs you are applying for.

Establishing Your LinkedIn Network

When establishing your LinkedIn network, the first thing you need to do is identify as many individuals as possible who might be able to help you in your job search. Look for people who know you and like your work. As with Facebook, your goal on LinkedIn is to find individuals who might be part of a larger, broader network of contacts—people you can benefit from knowing. Here are some of the individuals you might consider asking to be possible members of your network:

- Current and former instructors who have complimented your work
- Former bosses, managers, or co-workers who are willing to write a recommendation on your behalf—for example, "I have worked with Agnes Delancy for two years and always found her cooperative and efficient. She is a great hire."
- Individuals who belong to a professional association or organization that you have joined, such as the National Society of Black Engineers, the National Student Nurses Association, or the Society of Marketing Specialists
- Community leaders you have worked for and who value what you have done
- Individuals in the military you served with or under who can comment on your technical skills, cooperation, leadership skills, and so on

- People you do business with on a regular basis and who may have a wide circle of connections (e.g., current and former customers, sales representatives, bankers, insurance agents, etc.)
- Alumni and parents of classmates
- Neighbors and friends, but use caution here; select individuals who are members of a professional organization that fits with your career goals

A Sample LinkedIn Profile

Figure 7.2 (pages 266–267) contains the LinkedIn profile of Daniel Ricks Solter. The profile lists such job-relevant categories as current and past positions held, education, and a brief summary of accomplishments. It also indicates that Solter is interested in such LinkedIn categories as job inquiries, expertise requests, reference requests, and getting back in touch. And it includes important recommendations for him.

Dossiers and Letters of Recommendation

Dossiers play a major role in the job search. A *dossier*, French for "bundle of documents," provides a file of information about you and your work—recommendations and so on—that others have supplied.

Basically, your dossier contains the following documents:

- letters of recommendation
- letters that awarded you a scholarship, gave you an academic honor, or acknowledged your community service
- letters that praised your work on the job, notified you of a merit raise, promotion, or recognition ("Employee of the Month")
- your academic transcript(s)

A dossier collects important information about you that prospective employers will want to see to decide whether to interview you. You may ask your placement office to send your dossier to an employer, or employers may request it themselves if you have listed the placement office address on your résumé.

Obtaining Letters of Recommendation

Should You See Your Letters?

You have a legal right to see your recommendation letters, but some employers believe that if candidates will see what is written about them, their references may be overly complimentary and more inclined to withhold information. Also, some of your references may refuse to write a letter of recommendation that they know you will see. However, you may feel more comfortable knowing what your recommendation letters contain. But before you make any decisions about seeing your recommendation letters, get the advice of your instructors and placement counselors.

Whom Should You Ask?

Be careful about whom you ask. Whether your recommendation letters are confidential or not, they can sell you or sink your chances, so select your references

FIGURE 7.2 LinkedIn Profile

Includes recent, professional-looking photo

Supplies information about work, military, and educational background

People in Solter's network, many of whom are in a position to help his career

Secured recommendations from current and former supervisors and co-workers praising his expertise, team spirit, and cooperative attitude, all qualities that hiring managers prize

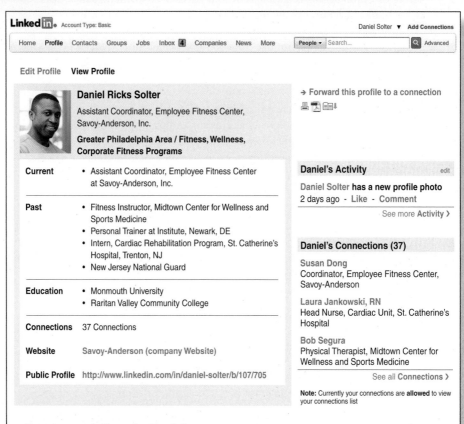

Dan Vallero/LinkedIn. Photo, © iStockphoto.com/Randy Plett

FIGURE 7.2 (Continued)

"Daniel is dynamic and talented. He can build rapport and relationships quickly and is always available for individual counseling about fitness needs."

 Buddy Rirerson, **Client Relations Director, Savoy-Anderson, Inc.**

Fitness Instructor
Midtown Center for Wellness and Sports Medicine

"When he worked for MCWSM (Midtown Center for Wellness and Sports Medicine), Daniel Solter was a supportive and articulate spokesperson for our facility. His group classes were among the most popular ones we offered, and he knows fitness programs extremely well."

 Joyce Hwang, **Assistant Director, Midtown Center for WSM**

Summary

Excel at providing fitness evaluations and exercise prescriptions and at teaching diverse classes—customized, group, and Web-based—in corporate settings. Help oversee in-house fitness center (1,600+ employees), including promoting and scheduling health/wellness events, offering health screenings, purchasing supplies and equipment, and supervising a staff of three full-time and two part-time workers. Solid motivational/leadership skills in implementing corporate policy aimed at improving employee fitness to decrease health care cots, absenteeism, and stress while improving morale and workplace safety. Adept at organizational communication with employees and management. Wrote fitness manual for Savoy-Anderson, Inc., and collaborated on drafting company safety manual. Passionate about improving employee fitness.

Specialties
Employee Fitness, Managing Corporate Fitness Facilities

Additional Information

Volunteered for the last 3 years at Trenton Veterans Hospital in Dept. of Cardiac Rehabilitation; coached 9th grade soccer at Boys and Girls Club, Trenton; and worked as physical fitness trainer for high school football players at South Trenton High School.

Groups and Associations

ASEP (American Society of Exercise Physiologists)
Young Professionals of Trenton Fitness

Personal Information

Email   dsolterfitness@yahoo.com
Birthday   **February 19**
Marital status   **Married**

Contact Settings
Interested in

- job inquiries
- reference requests
- expertise requests
- getting back in touch

Concisely describes job responsibilities with an emphasis on technical, leadership, and communication skills

Uses incomplete sentences, the standard style of LinkedIn profiles; chooses strong active verbs

Emphasizes volunteer works to show professional commitment and strong link between job goals and community service

Groups from which Solter has built his network

Identifies areas in which Solter wants to exchange information to further expand his network of professional contacts

Dan Vallero/LinkedIn

carefully. Ask the following individuals to be your references and to write enthusiastically about your work qualifications and skills:

- previous employers (even for summer jobs or internships) who commended you and your work
- two or three of your professors who know and like your work, have graded your papers, or have supervised you in fieldwork or laboratory activities
- supervisors who evaluated and praised your work in the military
- community leaders or officials with whom you have worked successfully on civic projects

Recommendations from such individuals will be regarded as more objective—and more relevant—than letters from friends, neighbors, or members of the clergy.

Whoever you ask, make sure he or she is a strong supporter of yours, someone who has sincerely and consistently complimented your work and encouraged you in your career. Find out by asking if this person is willing to write a strong, enthusiastic letter on your behalf. Stay clear of individuals who are lukewarm about your work or who might be reluctant to recommend you for another reason. Figure 7.3 shows a letter requesting a letter of recommendation.

Always Ask for Permission

Always ask for permission before you list an individual as a reference. You could jeopardize your chances for a job if a prospective employer called one of your references and that person did not even know you were looking for a job or, worse yet, reveals that you did not have the courtesy to ask to use his or her name.

Should You Ask Your Current Boss?

Asking your current boss can be tricky. If your present employer is already aware that you are looking for work elsewhere (for instance, if your job is temporary or if your contract is about to run out) or you are working at a part-time job, by all means ask for a letter of recommendation. However, if you are employed full time and are looking for professional advancement or for a better salary elsewhere, you may not want your current employer to know that you are searching for another job. You may want to speak to a job counselor.

Career Portfolios/Webfolios

Like dossiers, career portfolios/webfolios play a major role in the job search. The documents they contain work together to support and supplement your résumé. Increasingly, job seekers are providing both a dossier and portfolio/webfolio to prospective employers. Unlike a dossier, which provides information about you and your work that others have written, a portfolio (or webfolio if it is submitted electronically) contains documents or multimedia items you have created or produced yourself.

The word *portfolio* historically refers to a collection of work that an artist would show to a prospective client. For example, a portfolio might contain reproductions

FIGURE 7.3 Request for a Letter of Recommendation

TADEUS MAJESKI • 5432 South Kenneth Avenue • Chicago, IL 60651

312-555-7733   tmajeski@gatenet.com

March 30, 2012

Mr. Sonny Butler, Manager
Empire Supermarket
4000 West 79th Street
Chicago, IL 66052-4300

Dear Mr. Butler:

I was employed at your store from September 2010 through August 2011. During my employment, I worked part time as a stock clerk and relief cashier, and during the summer I was a full-time employee in the produce department, helping to fill in while Bill Dirksen and Vivian Ho were on vacation.

Reviews employment history

I enjoyed my work at Empire, and I learned a great deal about the latest inventory tracking systems, ordering stock, calculating and helping to prevent merchandise shrinkage, and assisting customers.

Emphasizes skills learned on the job

This May, I will receive my A.A. degree from Moraine Valley Community College in retail merchandising and I have already begun preparing for my job search for a position in retail sales. Would you be willing to write an enthusiastic letter of recommendation for me describing what you regard as my greatest strengths as one of your employees? Having your endorsement would be a great help to me.

Diplomatically requests strong letter of recommendation

To assist you, I can send you a letter of recommendation form from the Placement Office at Moraine Valley. Your letter would then become part of my permanent placement file.

Explains how letter will be used

I look forward to hearing from you. I thought you might like to see the enclosed résumé, which shows what I have been doing since I left Empire. Thank you for the opportunity to work at your supermarket.

Encloses copy of résumé and thanks reader

Sincerely yours,

Tadeus Majeski

Tadeus Majeski

Encl.: Résumé

or slides of the artist's paintings or illustrations or samples of photography. Similarly, your career portfolio would contain a collection of samples of your work—written or visual—to show to prospective employers.

Keep in mind that your portfolio/webfolio must represent your best work. Try to see it from the employer's point of view. A portfolio/webfolio is not a personal/family scrapbook but a record of your professional accomplishments. Your portfolio should say, "This is a collection of my most important and highly praised work. The documents here illustrate the high professional standards and skills that I will bring with me as an employee of your company."

What to Include in a Career Portfolio/Webfolio

What you include in your career portfolio/webfolio and how you organize it says as much about your qualifications for the job as the documents themselves. It is never too early to start assembling relevant documents for your portfolio.

Following is a sampling of the kinds of documents you might include in your career portfolio/webfolio, as illustrated in Figure 7.4.

- a mission statement (two or three paragraphs) that outlines your career goals and work skills
- an additional copy of your résumé
- copies or scans of diplomas, certificates, licenses, internships, papers, unofficial transcripts
- copies of awards (academic and job-related), promotion letters, or commendations (e.g., for protecting the environment)
- impressive examples of written work you did for college courses, such as reports or proposals (include any positive comments provided by instructors)
- newspaper or newsletter stories about your academic, community, or on-the-job successes
- pertinent examples of media presentations or other graphic work you have done, PowerPoint presentations you have created, a CD of a website you designed, layout work you completed, career-related photographs you have taken, or graphics you have created
- a list of your references with contact information

What Not to Include in a Career Portfolio/Webfolio

Be highly selective about what you include. Never include anything that would contradict or call into question information in your résumé or letter of application. Exclude the following types of documents from your career portfolio/webfolio.

- documents or scans of documents that show your memberships in clubs, fraternities/sororities, sports teams, and so on, unless directly relevant to the job (e.g., applying for a job at the national office for Sigma Sigma Kappa or with the Professional Golfers Association)
- links to or printouts from personal Web pages, including Facebook, LinkedIn, and MySpace pages (these may contain inappropriate personal information or portray you as unprofessional; see pages 261–262)

FIGURE 7.4 A Sample Webfolio with Annotated Artifacts/Documents Page Open

Mission Statement

Resume

College Transcript

Artifacts/Documents

Presentations

References

Sample Business Documents

To demonstrate my clear and concise writing style, my ability to design a variety of complex workplace documents, and my understanding of international communication, I have provided links to various business documents I created and wrote. None of these documents contains proprietary or confidential information.

View Sample Sales Proposal on Saving Energy

View Sample Research Report (Table of Contents, Abstract, First 3 Pages)

View Sample Sales Letter to International Client

View Sample Survey Questionnaire and Analysis

Company Newsletter Article

At the end of 2011, I was named "Employee of the Year" at my firm's Cincinnati office. An article about the award appeared in the company's online newsletter, outlining the selection criteria and describing the job accomplishments that earned me this award.

View Newsletter Article

PowerPoint Presentations

I have excelled at making frequent presentations at company meetings and those before clients. My colleagues and supervisors have praised my abilities as a speaker who can simplify complex ideas for audiences and zero in on the main ideas. To give you a sense of my ability to organize information and provide visual support for my ideas, I provide a link to a sample PowerPoint presentation that accompanied one of my talks.

View Powerpoint Presentation

Website

To illustrate both my design skills and my keen understanding of the Web environment, I have provided several screen captures from a company website I created. The construction of the site was handled by my company's Web group, which I supervised for 14 months.

View Website Screen Captures

Photo, kropic1/Shutterstock.com

- pictures of your family, friends, pets, and the like
- scans of newspaper or newsletter stories about you that are not directly related to your job search, such as your winning a cruise or playing on a bowling team

Career Portfolio/Webfolio Formats

When you provide a prospective employer with your career portfolio, you can either mail it or send it electronically as a webfolio. If you submit a hard-copy portfolio, always make high-quality copies of each document. Never include originals. If you submit a webfolio, make sure that all of your scans are clean and clear. In addition, you can hyperlink parts of your career portfolio to your résumé, as Anthony Jones does in Figure 7.5, making it easy for a job recruiter or prospective employer to find evidence of your qualifications.

Whether you submit your career portfolio in hard copy or electronically, provide short annotations explaining what each document is and why it is important, as the job applicant did in Figure 7.4.

Preparing a Résumé

The résumé, sometimes called a *curriculum vitae (CV),* may be the most important document you prepare for your job search. It merits doing some careful homework. A résumé is not your life history or your emotional autobiography, nor is it a transcript of your college work. It is a factual and concise summary of your qualifications, convincing a prospective employer that you have the education and experience to do the job you are applying for. Regard your résumé as a persuasive ad for your professional qualifications. It is a billboard advertising you.

What you include—your key details, the wording, the ordering of information, and the formatting—are all vital to your campaign to sell yourself and land an interview. Employers want to see the most crucial and current details about your qualifications quickly. Accordingly, keep your print résumé short (preferably one page, never longer than two) and hard-hitting. The same thing goes for your online résumés. See Figures 7.11–7.14 for examples of online, digital résumés. Everything on your résumé needs to convince an employer you have the exact skills and background he or she is looking for.

What Employers Like to See in a Résumé

Prospective employers will judge you and your work by your résumé; it is their first view of you and your qualifications. They will expect an applicant's résumé to be

- **Honest.** Be truthful about your qualifications—your education, experience, and skills. Distorting, exaggerating, or falsifying information about yourself in your résumé is unethical and could cost you the job. If you were a clerical assistant to an attorney, don't describe yourself as a paralegal. Employers demand truthfulness.

- **Attractive.** The document should be pleasing to the eye, with generous margins and appropriate spacing, typeface, and use of boldface; it shows you have a sense of proportion and document design and that you are visually smart. The print should be clear and dark, not faded, and on high-quality paper. Do not use gimmicks like clip art or excessive capitalization. (See pages 528–535 on document design.)

- **Carefully organized.** Arrange information so that it is easy to follow, logical, and consistent; the way you organize information shows you have the ability to process information and to summarize. Employers prize analytical thinking. Include plenty of white space to separate major sections, and use bullets to list and highlight key facts within each section, as in Figures 7.5 and 7.6 (pages 274–276).

- **Concise.** Make sure your résumé is to the point. Generally, keep your résumé to one page, as in Figure 7.5. However, depending on your education or job experience, you may want to include a second page. Résumés are written in short sentences that omit "I" and that use action-packed verbs, such as those listed in Table 7.2 on page 274.

- **Accurate.** Make sure your grammar, spelling, dates, names, titles, and programs are correct; typos, inconsistencies, and math errors say you didn't check your facts and figures.

- **Current.** All information needs to be up-to-date and documented, with no gaps or sketchy areas about previous jobs or education. Missing or incorrect dates or leaving key information out are red flags to employers to reject your résumé.

- **Relevant.** The information on your résumé must be appropriate for the job description and level. It must show that you have the necessary education and experience and must confirm that you can be an effective team player.

- **Quantifiable.** For instance, include specifics about how much revenue you generated for an employer (or how much money you saved, or how many times you did a complex job).

Your goal is to prepare a résumé that shows the employer you possess the sought-after job skills. A résumé that is unattractive, difficult to follow, poorly written, filled with typos and spelling mistakes or that is sketchy, vague, boastful, or not relevant for the prospective employer's needs will not make the first cut.

Create Several Versions of Your Résumé

It might be to your advantage to prepare several versions of your résumé and then adapt each one you send out to the specific job skills a prospective employer is looking for. It pays to customize your résumé based on information you may have gathered about the job from researching the company (see pages 260–265), through networking on sites such as LinkedIn, and from analyzing the keywords the employer used to describe the position. Following the process detailed in the next section will help you prepare any résumé.

TABLE 7.2 Action Verbs to Use in Your Résumé

accommodated	conducted	generated	organized	settled
accomplished	converted	guided	oversaw	sold
achieved	coordinated	handled	performed	solved
adapted	created	headed	planned	spearheaded
adjusted	customized	hired	posted	streamlined
administered	dealt in	implemented	prepared	supervised
analyzed	delivered	improved	programmed	surveyed
arranged	designed	increased	provided	taught
assembled	determined	informed	purchased	teamed up
assisted	developed	initiated	ranked	tested
attended	devised	installed	reappraised	tracked
awarded	directed	instituted	received	trained
bridged	discovered	instructed	reconciled	transcribed
budgeted	drafted	interned	recorded	translated
built	earned	interpreted	reduced	tutored
calculated	economized	judged	re-evaluated	updated
chaired	edited	launched	reported	upgraded
coached	elected	maintained	researched	verified
collaborated	established	managed	reviewed	volunteered
collected	estimated	mapped	saved	weighed
communicated	evaluated	monitored	scheduled	wired
compiled	expanded	motivated	searched	won
completed	expedited	navigated	secured	worked
composed	figured	negotiated	selected	wrote
computed	fulfilled	operated	served	

The Process of Writing Your Résumé

To write an effective résumé, ask the following important questions:

1. What classes did you excel in?
2. What papers, reports, surveys, or presentations earned you your highest grades?
3. What computer skills have you mastered—languages, software, Internet, e-commerce, blog or website design, collaborative online editing?
4. What other technical skills have you acquired?
5. What jobs have you had? For how long and where? What were your primary duties? Did you supervise other employees?
6. How did you open or expand a business market? Increase a customer base?
7. What did you do to earn a raise or a promotion in a previous or current job?
8. Do you work well with people? What skills do you possess as a member of a team working toward a common job goal (e.g., finishing a report)?
9. Can you organize complicated tasks or identify and solve problems quickly?
10. Have you had experiences or responsibilities managing money—collecting fees or receipts, preparing payrolls, conducting nightly audits, and so on?
11. Have you won any awards or scholarships or received a commendation or other recognition at work?

Pay special attention to your four or five most significant, job-worthy strengths, and work especially hard on listing them concisely and persuasively.

Although not everything you have done relates directly to a particular job, indicate how your achievements are relevant to the employer's overall needs. For example, supervising staff in a convenience store points to your ability to perform the same duties in another business context.

Balancing Education and Experience

If you have years of experience, don't flood your prospective employer with too many details. You cannot possibly include every detail of your jobs for the last ten or twenty years.

- Emphasize only those skills and positions most likely to earn you the job.
- Eliminate early jobs that do not relate to your present employment search.
- Combine and condense skills acquired over many years and jobs.
- Include relevant military schools or service.

Figures 7.7 and 7.10 (page 279 and 287) show the résumés of individuals who have a great deal of experience to offer prospective employers.

Many job candidates who have spent most of their lives in school are faced with the other extreme: not having much job experience to list. The worst thing to do is to write "None" for experience. Any part-time, summer, or other seasonal jobs, as well as volunteer work, apprenticeships, and internships, show an employer that you are responsible and knowledgeable about the obligations of being an employee. Figure 7.5 contains a résumé from Anthony Jones, a student with little job experience; Figure 7.6 shows the résumé of María López, a student with a few years of experience.

What to Exclude from a Résumé

Knowing what to exclude from a résumé is as important as knowing what to include. Because federal employment laws prohibit discrimination on the basis of age, sex, race, national origin, religion, marital status, or disability, do not include such information on your résumé. Here are some other details best left off your résumé:

- salary demands, expectations, or ranges
- preferences for work schedules, days off, or overtime
- comments about fringe benefits
- travel restrictions
- reasons for leaving your previous job
- your photograph (unless you are applying for a modeling or acting job)
- your Social Security number
- information about your family, spouse, or children
- height, weight, hair or eye color
- sexual orientation, religious and political affiliations
- hobbies, interests (unless relevant to the job you are seeking)

Save comments about salary and schedules for your interview (see pages 307–312). The résumé should be written to earn you that interview.

FIGURE 7.5 Résumé from a Student with Little Job Experience

Headlines major achievements and gives contact details

73 Allenwood Boulevard
Santa Rosa, California 95401-1074
707-555-6390
ajdesigner@plat.com
www.plat.com/users/ajones/resume.html

Anthony H. Jones
• WEBSITE DEVELOPER
• DESIGNER
• GRAPHIC ARTIST

CAREER OBJECTIVE

Offers precise, convincing objective

Full-time position as a layout artist with a commercial publishing house using my training in state-of-the-art design technology.

EDUCATION

Starts with most important qualification—education

Santa Rosa Junior College, 2010–2012, A.S. degree to be awarded in June 2012
  Dean's List in 2011; GPA 3.45
  Major: Commercial Graphics Illustration, with specialty in design layout
  Related courses included:
  • Digital Photography
  • Graphics Programs: Illustrator, Photoshop, Dreamweaver
  • Desktop Publishing: Adobe InDesign, QuarkXPress
Internship, 2011–2012, McAdam Publishers
  Major projects included:

Stresses job-related activities of internship

  • Assisting layout editors with page composition and photo archiving
  • Writing detailed assessment reports on digital photography, designs, and artwork used in *Living in Sonoma County* (www.sonomacounty.com) and *Real Estate in Sonoma County* (www.resc.net) magazines

EXPERIENCE

Includes part-time work experience

Salesperson (part-time), 2009–2011, Buchman's Department Store, Santa Rosa
  Duties included assisting customers in sporting goods and appliance departments, designing custom window displays each month for the main entrance, and helping to manage the inventory database

COMPUTER SKILLS

Demonstrates skills in Web design software

InDesign, QuarkXPress, Illustrator, Photoshop, Dreamweaver, FinalCut Pro

RELATED ACTIVITIES

Designed the website and a three-fold brochure for the Santa Rosa Humane Society's 2011 fund drive; helped raise $5,600

REFERENCES / WEBFOLIO

Credentials and portfolio document accomplishments

References and webfolio containing designs, photographs, and graphics are available at **www.plat.com/users/ajones/resume.html**

FIGURE 7.6 Résumé from a Student with Some Job Experience

María López

1725 Brooke Street   Miami, Florida 32701-2121   (305) 555-3429   mlopez@eagle.com

CAREER OBJECTIVE

Position assisting dentist in providing dental care, counseling, and preventive dental treatments, especially in pediatric dentistry.

EDUCATION

A.S. in Dental Hygiene, Miami-Dade Community College
August 2010–May 2012
GPA: 3.38 (Ranked in the top 10 percent)

Major courses:

- Oral pathology
- Dental materials and specialties
- Periodontics
- Community dental health

- Experienced with procedures and instruments used with oral prophylaxis techniques.
- Subject of major project was proper nutrition and dental health for preschoolers.
- Will take American Dental Assisting National Board Exams for licensure on June 2.

Minor: Psychology (twelve hours in child and adolescent psychology)

EXPERIENCE

St. Francis Hospital
(Miami Beach, Florida) April 2008–July 2010
Unit secretary on pediatric unit. Maintained medical supply levels using inventory tracking software, assisted with the computerization of all medical records into hospital-wide database, transcribed medical orders and surgical notes, greeted and assisted visitors.

Murphy Construction Company
(Miami, Florida) June 2007–April 2008
Office assistant-receptionist. Did data entry and filing, and assisted with billing and creating project schedules in a small office (5 employees).

City of Hialeah, Florida
Summers 2005–2006
Lifeguard. Established safety procedures and tested pool chlorine levels.

COMPUTER SKILLS

PowerPoint, DentiMax, Microsoft Office, FileMaker Pro

LANGUAGE SKILLS

Fluent in Spanish

Provides easy-to-find contact information

Lists course and clinical work required for licensure and job

Calls attention to counseling skills that benefit employer and patients

Emphasizes professional licensure qualifications

Highlights previous job responsibilities in health care setting

Includes career-relevant computer skills

Calls attention to bilingual skills of value to employer and patients

© Cengage Learning 2013

(Continued)

FIGURE 7.6 (Continued)

Lopéz 2

<u>REFERENCES</u>

The following individuals have written letters of recommendation for my dossier available from the Placement Center, Miami-Dade Community College, Medical Center Campus, Miami, FL 33127-2225.

Sister Mary Pela, R.N. Pediatric Unit St. Francis Hospital 10003 Collins Avenue Miami Beach, FL 33141 (305) 555-5113	Professor Mitchell Pelbourne Department of Dental Hygiene Miami-Dade Community College Medical Center Campus Miami, FL 33127 (305) 555-3872
Tia Gutiérrez, D.D.S. 9800 Exchange Avenue Miami, FL 33167 (305) 555-1039	Mr. Jack Murphy 1203 Francis Street Miami, FL 33157 (305) 555-6767

Parts of a Résumé

As with memos, letters, and reports, résumés consist of specific parts shown in boldface headings on the following pages. These parts—contact information, career objective, credentials (education and experience), related skills and achievements, and references/portfolios—need to be included in any résumé.

Contact Information

At the top of the page, provide your full name (do not use a nickname), address including your zip code, telephone number (use a cell phone number if you always have your phone with you), and email address. If your academic address is different from your home address, list and identify both. The contact information can either be centered on the page or flush left or right. Avoid unprofessional email addresses such as toughguy@netfield.com or sassygirl@techscape.com. Make sure your voice mail message is straightforward and professional, as well. Also include a URL for your website and a fax number if you have one.

Career Objective

One of the first things a prospective employer will read is your career objective statement, which specifies the exact type of job you are looking for and in what ways you are qualified to hold it. Create an objective that precisely dovetails with the prospective employer's requirements. Such a statement should be the result of your focused self-evaluation and your evaluation of the job market. It will influence everything else you include on your résumé. Depending on your background and the types of jobs you are qualified for, you might formulate two or three different

FIGURE 7.7 Résumé from an Individual with Significant Job Experience

ANNA C. CASSETTI

6457 Blackstone Avenue	MacMurray Real Estate
Fort Worth, TX 76321-6733	1700 Ross Boulevard
(817) 555-5657	Haltom City, TX 77320-1700
acassetti@netdor.com	(817) 555-7211

Provides home and work contact information

Objective
Full-time sales position with large real estate office in the Phoenix or Tucson area with opportunities to use proven skills in real estate appraisal and tax counseling.

Experience
MacMurray Real Estate, Haltom City, Texas, 2010–present
Real estate agent. Excelled in small suburban office (five salespersons plus broker) with limited listings; **sold individually over $3 million in residential property**; appraised both residential and commercial listings.

Begins with the most important job qualification first— experience

Dallman Federal Savings and Loan, Inc., Fort Worth, Texas 2003–2009
Chief Teller. Responsible for supervising, training, and coordinating activities of six full-time and two part-time tellers. Promoted to Chief Teller, March 2000, with bonus.

Job history focuses on sales achievements, promotion, and responsibilities

H&R Block, Westover Hills, Texas, 2003 (Jan.–Apr.)
Tax Consultant. Prepared personal and business returns.

Cruckshank's Hardware Store, Fort Worth, Texas, 2000–2002
Salesperson.

U.S. Navy, 1993–1999
Honorably discharged with rank of Petty Officer, Third Class.
Served as stores manager; earned three commendations.

Gives related military experience

Education
Texas Christian University, Fort Worth, Texas, 2002–2007
Awarded B.S. degree in Real Estate Management. Completed thirty-three hours in business and real estate courses with a concentration in finance, appraising, and property management. Also took twelve hours in programming and Web design. Wrote reports on appraisal procedures as part of supervised training program.

Documents education; cites specific skills

H&R Block, Westover Hills, Texas, 2002 (Sept.–Dec.)
Earned diploma in Basic Income Tax Preparation after completing ten-week course.

U.S. Navy, 1993–2000
Attended U.S. Navy's Supply Management School
Applied principles of stores management at Newport Naval Base

Includes computer and communication skills

Skills and Activities
Licensed Texas Realtor—756a2737
Chair, Financial Committee, Grace Presbyterian Church, Fort Worth
Adviser, Junior Achievement

Provides necessary license information

References
Available upon request.

career or employment objectives to use with different versions of your résumé as you apply for various positions. Always try to incorporate keywords an employer lists in the job announcement.

To write an effective career objective statement, ask yourself four basic questions:

1. What kind of job do I want?
2. What kind of job am I qualified for?
3. What capabilities do I possess?
4. What kinds of skills do I want to learn?

Avoid trite, vague, or self-centered goals, such as "Looking for professional advancement," "Want to join a progressive company," "Seeking high-paying job that brings personal satisfaction," or "Job where I can use my proven leadership abilities." Compare the vague objectives on the left with the more precise ones on the right.

Unfocused	**Focused**
Job in sales to use my aggressive skills in expanding markets.	Regional sales representative using my proven skills in e-commerce and communication to develop and expand a customer base.
Full-time position as staff nurse.	Full-time position as staff nurse on cardiac step-down unit to offer excellent primary care nursing and patient/family teaching.
Position in cable industry.	Position as part of a service team to provide efficient cable repair service.

A career objective can help you when you are applying for a specific job opening. You need to tailor it to meet the needs of a prospective employer.

Credentials

The order of the next two categories—Education and Experience—can vary. Generally, if you have lots of work experience, list it first, as Anna Cassetti does in Figure 7.7. However, if you are a recent graduate short on job experience, list education first, as Anthony Jones does in Figure 7.5. María López (Figure 7.6) also decided to place her education before her job experience because the job she was applying for required the formal training she recently received at Miami-Dade Community College.

Education   Begin with your most recent education first, then list everything significant since high school. For each school, give the name, the dates you attended, and the degree, diploma, or certificate you earned. Don't overlook relevant military experience or major training programs (EMT, court reporter), institutes, internships, or workshops you have completed.

Remember, however, that a résumé is not a transcript. Simply listing a series of required courses will not set you apart from hundreds of other applicants taking similar courses across the country. Focus on courses that have specifically prepared you for the job. Avoid vague titles such as Science 203 or Nursing IV. Instead, concentrate on describing the kinds of skills you learned.

> 30 hours in planning and development courses specializing in transportation, land use, and community facilities; 12 hours in field methods of gathering, interpreting, and describing survey data in reports.

> Completed 28 hours in major courses in business marketing, management, and materials in addition to 12 hours in information science, including HTML/Web publishing.

Mention any special projects, experiments, or reports that bear directly on the job you are seeking. Note how María López (Figure 7.6) briefly references her major project on preschoolers' nutrition and dental health, a topic sure to interest her potential employer.

List your grade point average (GPA) only if it is 3.0 or above; otherwise, indicate your GPA in just your major or during your last year or term, again if it is above 3.0.

Experience   Your job history is the key category for many employers. It shows them that you have held jobs before and that you are responsible. Here are some guidelines about listing your experience.

1. Begin with your most recent position and work backward—in reverse chronological order. List the company or agency name, location (city and state), your job title, and dates of employment. Do not mention why you left a job.

2. For each job or activity, provide a short description (one or two lines) of your duties and achievements. If you were a work-study student, don't say that you helped an instructor. Emphasize your responsibilities; for example, you helped to set up a chemistry laboratory, ordering supplies and keeping an inventory of them. Rather than saying you were an administrative assistant, indicate that you wrote business letters and used various software programs, maintained records, designed a company website, prepared schedules for part-time help in an office of twenty-five people, or assisted the manager in preparing minutes, accounts, and presentations.

3. In describing your position(s), emphasize any responsibilities that involved handling money (for example, assisting customers, filing insurance claims, or preparing payrolls); managing other employees; working with customer accounts, services, and programs; or writing letters and reports. Prospective employers are interested in your leadership abilities, teamwork, financial shrewdness (especially if you saved your company money), tact in dealing with the public, and communications skills. They will also be favorably impressed by commendations "Earned Highest Sales Record"; recognized for "Exceptional Clinical Skill" in serving geriatric patients) and promotions you have earned.

4. Include any relevant volunteer work you have done, as Anthony Jones did for an animal shelter in Figure 7.5. Note, too, how Dora Cooper Bolger's volunteer work translates into marketing skills an employer wants to see in a prospective employee's résumé in Figure 7.8.

5. If you have been a full-time parent for ten years or a caregiver for a family member or friend, indicate the management skills you developed while running a household and any community or civic service, as Dora Cooper Bolger does in her résumé. She skillfully relates her family and community accomplishments to the specific job she seeks.

Related Skills and Achievements

Not every résumé will have this section, but the following are all employer-friendly things to include:

- second or third languages you speak or write
- extensive travel
- certificates or licenses you hold
- memberships in professional associations (e.g., American Society of Safety Engineers, National Black Law Students Association, National Hispanic Business Association, Texas Executive Women, National Association for the Education of Young Children)
- memberships in community service groups (e.g., Habitat for Humanity, Salvation Army, Big Brothers/Big Sisters); list any offices you held—recorder, secretary, fund drive chairperson, etc.

Computer Skills   Knowledge of computer hardware, software, word-processing programs, Web design, social media, and search engines is extremely valuable in the job market. Note how Anthony Jones and María López inform prospective employers about their relevant technical competencies in Figures 7.5 and 7.6. See also how Donald Kitto-Klein describes his knowledge of and experience in IT in Figure 7.10.

Honors/Awards   List any civic honors (mayor's award, community service award, cultural harmony award) and academic honors (dean's list, department awards, scholarships, grants, honorable mentions) you have won. Memberships in honor societies in your major and technical/business associations also demonstrate that you are professionally accomplished and active.

References

As a rule, do not list references with personal contact information. Simply say they are available on request. The only exception would be when you are applying for a specific job, as María López did in Figure 7.6, and your references are well known in the community or belong to the same profession in which you are seeking employment; that way, you profit from your association with a recognizable name or title. Here is where networking can help you. Ask your instructors, previous employer, or individuals who have supervised your work. Always give the person providing the reference a copy of your current résumé, as the job seeker did in Figure 7.3.

FIGURE 7.8 Dora Cooper Bolger's Résumé Organized by Skills Areas

DORA COOPER BOLGER

1215 Lakeview Avenue
Westhampton, MI 46532
Cell: 616-555-4773
dcbplanner@aol.com

Objective Seek full-time position as public affairs officer to promote the goals of a health care, educational, or charitable organization

Skills **Organizational Communication**
- Delivered 25 presentations to civic groups on educational issues
- Recorded minutes and helped formulate agenda as president of large, local PTA (800 members) for past 6½ years
- Possess excellent computer skills in Microsoft Word, PeopleSoft, PowerPoint, and BusinessPax
- Updated and maintained computerized mailing lists for Teens in Trouble and Foster Parents' Association

Financial
- Spearheaded 3 major fund-raising drives (total of $200,000 collected)
- Prepared and implemented large family budget (3 children, 8 foster children)
- Budget planner, Foster Parents' Association
- Served as financial secretary, Faith United Methodist Church, for 4 years

Administrative
- Organized volunteers for American Kidney Fund (last 5 years)
- Established and oversaw neighborhood carpool (17 drivers; more than 60 children) for 7 years
- Coordinated after-school tutoring program for Teens in Trouble; president since 1998
- Vice-president, Foster Parents' Association, 2010

Honors "Volunteer of the Year" (2011), Michigan Child Placement Agency

Education Metropolitan Community College, A.A., 2008
Mid-Michigan College, B.S., expected 2012
Major: Public Administration; Minor: Psychology
GPA: 3.55 | Dean's List: 2010–2012

Work Experience Secretary, 1995–2007 (full- and part-time): Merrymount Plastics; Foley and Wasson; Westhampton Health Dept.; G & K Electric

References Furnished on request

Restricted objective

Aptly features skill areas before education

Links achievements from volunteer and home-based activities most important to employer

Chooses strong, active verbs to convey image of a results-oriented professional

Places education after skills; includes major and related minor plus strong GPA

Excludes details about least recent jobs

© Cengage Learning 2013

Organizing Your Résumé

There are two primary ways to organize your résumé: chronologically and by function or skill area. You may want to prepare two versions of your résumé—one chronological and one by function or skill area—to see which sells your talents better. Don't hesitate to seek the advice of a placement counselor or instructor about which may work best for you.

Chronologically

The résumés in Figures 7.6 and 7.7 are organized chronologically, with most recent education and experience listed first. This is the traditional way to organize a résumé. It is straightforward and easy to read, and employers find it acceptable. The chronological sequence works especially well when you can show a clear continuity toward progress in your career through your employment and schoolwork or when you want to apply for a similar job with another company. A chronological résumé is also appropriate for students who want to emphasize recent educational achievements.

By Function or Skill Area

Depending on your experiences and accomplishments, you might organize your résumé according to function or skill area. According to this plan, you would *not* list your information chronologically in the categories "Experience" and "Education." Instead, you would sort your achievements and abilities—whether from course work, jobs, extracurricular activities, or technical skills—into two to four key skill areas, such as

- Sales
- Public Relations
- Training/Teaching
- Management
- Research
- Technical Capabilities
- Counseling
- Leadership
- Communications

- Network Operations
- Customer Service
- People Skills
- Teamwork
- Troubleshooting
- Opening New Markets
- Multicultural Experiences
- Information Technology
- Problem-Solving Skills

Under each area you would list three to five points illustrating your achievements in that area. Functional and skills résumés are often called *bullet résumés* because they itemize the candidate's main strengths in bulleted lists. Some employers prefer the bullet résumé because they can skim the candidate's list of qualifications in a few seconds.

Note Dora Cooper Bolger's profitable use of a skills résumé format in Figure 7.8. She delayed school for several years because of family commitments, yet she uses the experiences she acquired during those years to her advantage in her résumé organized by "Skills." No gap of ten years interrupts her valuable marketable skills.

Preparing a Functional or Skills Résumé

When you prepare a functional or skills résumé, start with your name, address, telephone number, and career objective, just as in a chronological résumé. To find the best two or three functional areas to include, use the prewriting strategies (especially clustering and brainstorming) discussed in Chapter 2.

The following individuals would probably benefit from organizing their résumés by function or skill area instead of chronologically:

- Nontraditional students who have had diverse job experiences
- People who are changing professions
- Individuals who have changed jobs frequently
- Ex-military personnel reentering the civilian marketplace

Note how Donald Kitto-Klein (Figure 7.10, page 287) was able to pull together a series of related, marketable skills from the many different jobs he had held over several years.

After you discover and suitably revise the information to be included in your categories, briefly list your educational and work experiences, as Dora Cooper Bolger (Figure 7.8), Anna Cassetti (Figure 7.9), and Donald Kitto-Klein (Figure 7.10) do.

Tech Note

Developing Your Own Website for Your Job Search

Most employers do a simple Google search on potential employees, even before they interview them. Having your own website gives you instant credibility and affords a prospective employer an insight into you and your accomplishments that is more extensive than your basic written résumé. A website also allows you to update information quickly and give employers easy access to it.

But be careful about what you put on your website. You do not want to make personal information available to a worldwide audience and risk becoming the victim of identity theft. Moreover, don't equate a professional website with a personal one. Include only professional details that will increase your chances of getting an interview without compromising your privacy. You might safely include a link to your webfolio (see pages 268–272) or other examples of work you've done.

And always make sure your website is clear, readable, user-friendly, and easy to navigate. The Internet is full of information that can assist you in building a website. You might want to start with the following sites to get some help.

- www.web.com
- www.wix.com
- www.weebly.com
- www.godaddy.com

Also see Chapter 11 (pages 538–546) for specific advice on writing for a Web audience.

FIGURE 7.9 Anna Cassetti's Résumé Organized by Function or Skill Areas

ANNA C. CASSETTI   www.acas.yahoo.com

6457 Blackstone Avenue   MacMurray Real Estate
Fort Worth, TX 76321-6733   1700 Ross Boulevard
(817) 555-5657   Haltom City, TX 77320-1700
acassetti@netdor.com   (817) 555-7211

Begins with a clear and focused objective

Objective
Sales position with real estate office in the Phoenix or Tucson area with opportunities to use proven skills in property appraisals and tax counseling

Groups accomplishments into three relevant and marketable skills areas for prospective employer

Sales/Financial
- Sold over $3 million of residential property (2010–2011)
- Served as a tax consultant with special interest in real estate sales/market conditions
- Performed general banking procedures as chief teller
- Ordered and maintained ship's stores, U.S. Navy

Public Relations
- Helped clients select appropriate property for their needs and income
- Counseled commercial and individual clients about taxes/benefits/liabilities
- Supervised, trained, and coordinated the activities of six bank tellers
- Commended for rapport in assisting customers with their banking needs

Uses bullets, strong verbs, and specific examples

Communication
- Prepared detailed real estate appraisals
- Wrote in-depth business reports on appraisal procedures, property management problems, and banking policies affecting real estate transactions
- Achieved proficiency in CorpSheet and other spreadsheet programs
- Conducted small group training and sales sessions

Emphasizes written and oral skills

Stresses educational preparation after skills

Education
B.S. in Real Estate Management, 2008, Texas Christian University, Fort Worth, Texas
Advanced course work in finance and real estate; minor in statistics

Diploma, Basic Income Tax Preparation, 2002, H&R Block, Westover Hills, Texas

Employment
MacMurray Real Estate, Haltom City, Texas
2010–present; sales agent

Dallman Federal Savings and Loan, Inc., Fort Worth, Texas
2003–2009; Chief Teller

Demonstrates professional commitment and successes

H & R Block, Westover Hills, Texas
2003 (January–April); consultant, tax preparer

Cruckshank's Hardware Store, Fort Worth, Texas
2000–2002; salesperson

U.S. Navy, 1993–2000
stores manager, 1996–1999; honorably discharged with rank of Petty Officer, Third Class

References
Available on request.

FIGURE 7.10 Functional Résumé by an Applicant Who Has Held a Variety of Jobs

DONALD KITTO-KLEIN
kitto@gar.com   www.dkk.opengate.com

56 South Ardmore Way   Garland.com
Petersburg, NY 15438   Grand Banks, NY 15532
(716) 555-9032   (716) 555-4800, Ext. 5398

Objective   Interested in supervisory position in IT and service to provide network continuity to staff and clients.

Computer Languages   Java, JavaScript, UNIX, HTML

Computer Platforms   UNIX (Solaris), Windows, Mac OS, Warp

Networking   DECnet, Ethernet, Secure ID, Novell, PGH

IT Maintenance/Service Skills
- Serviced computers and workstations on a regular basis for 3½ years
- Worked extensively on spreadsheets/database software
- Modified software billing program and saved company $400,000
- Coordinated maintenance/service activities

People Skills
- Promoted to Support Services Team Leader
- Supervised IT staff of three
- Worked closely with computer manufacturers and vendors to minimize hardware downtime
- Elected to employee benefits committee

Communication Skills
- Collaboratively wrote safety manual for power company road crews
- Taught in-house training sessions on computer maintenance, networking, data security
- Devised routing systems to expedite work orders
- Coordinated small group meetings in systems analysis

Employment   Garland.com, 2008–present
Business Graphics and Computers Store, Salesperson, 2003–2008
U.S. Army, Specialist, 4/E, 1997–2003

Education   B.S., Grand Valley Technical Institute, 2008
U.S. Army, four schools in IT, 2001–2004

References/Webfolio   Available on request.

Provides professional website

Connects writer's technical and administrative skills

Establishes technical background/ qualifications for job

Emphasizes the management skills and experiences prospective employer wants

Demonstrates ability to be a problem solver, a leader, and a spokesperson for company

Includes relevant military background

The Digital Résumé

In addition to drafting a hard copy of your résumé, expect to prepare multiple digital versions of it, including (1) creating and posting it to the Web, (2) formatting a scannable text, or (3) emailing it. In today's highly competitive job market, where employers have differing requirements, it is to your advantage to use these various formats to attract the interest of prospective employers.

Five Ways to Post, Email, Scan, or Videotape Your Résumé

In this digital age, your electronic résumé may be the most important document in your job search since employers determine who will make their interview list based on what they see in a résumé. There are five basic ways to create and post your digital résumé to the web or to scan, email, or videotape it. But whether you post, scan, or email it, you must remember to make any changes in all versions of your résumé to be consistent. Use the following guidelines to prepare different versions of your résumé:

1. **Post your résumé on the Web.** The Internet offers a variety of sites to disseminate your résumé. But regardless of where you post your résumé, the main advantage is that you will be reaching a large pool of potential employers. Here are some tips to help you post your résumé correctly for a potential employer.

   - Send it to one of the large job posting sites listed in Table 7.1 to reach the maximum number of employers. Make sure you carefully follow all the directions listed on the site, and always format and save your résumé before posting it. Keep a log, too, of where you have posted your résumé.
   - Post your résumé directly on an employer's website where it can be indexed and stored. Follow the employer's instructions precisely or your application may be automatically deleted.
   - Include hyperlinks (e.g., Education, Experience, Honors) to your own website, to appropriate blogs you wrote, or to reviews of your professional work. If you include a webfolio, as in Figures 7.11 or 7.13, hyperlinks will help an employer to access examples of your work quickly.
   - Since employers often print out the résumés of job seekers they are interested in, be sure your résumé can be downloaded easily and quickly. But keep in mind that reading a résumé on a screen is different from seeing it on the printed page.
   - If you post your résumé on your own website, do not provide more information than a prospective employer needs, and do not give out personal information. Protect your privacy by following the guidelines on pages 296–298. Remember, though, that posting your résumé on your own website will not attract employers the way a large job board will.

2. **Send your résumé via an email attachment.** Many ads will ask you to email your résumé. To do so, prepare your résumé as a Word file and then save it as a

FIGURE 7.11 A Scannable, Electronic Version of Anthony Jones's Hard-Copy Résumé in Figure 7.5

Anthony H. Jones
ajdesigner@plat.com
Phone: (707) 555-6390

KEYWORDS
Web designer, graphic designer, Illustrator, Photoshop, QuarkXPress, InDesign, fundraiser, budgets, sales, virus protection, team player

OBJECTIVE
Position as layout artist with small commercial publisher using my training in state-of-the-art technology

EDUCATION
Santa Rosa Junior College, A.S. degree to be awarded in June 2012. Commercial Graphics Illustration major. GPA 3.45

COMPUTER SKILLS
Excellent knowledge of computer graphics and design software: QuarkXPress, InDesign, Illustrator, Dreamweaver, Photoshop, and FinalCut Pro

EXPERIENCE
* Intern in layout and design department. Preparing page composition, importing visuals, manipulating images, McAdam Publishers, 8 Parkway Heights, Santa Rosa
* Salesperson; display merchandise coordinator, Buchman's Department Store, Greenview Mall, Santa Rosa
* Volunteer: designed website, brochures, and other artwork for successful fund drive, Santa Rosa Humane Society
* Web designer, graphic artist, proofreader, student magazine

REFERENCES
Available on request.

WEBFOLIO
www.plat.com/users/ajones/resume.html

All lines aligned flush with the left margin

Keywords (for search engine readiness) go at the top of the online résumé

Objective appears in the body of the online resume

All caps rather than bold, italics, or fancy fonts used to highlight categories

Uses terminology appropriate for position

Chooses nouns rather than action verbs to increase employer matches

Asterisks rather than bullets mark beginning of lines

Webfolio is hyperlinked to make it easily accessible

FIGURE 7.12   A Scannable, Electronic Version of Dora Cooper Bolger's Hard-Copy Résumé in Figure 7.8

Avoids giving personal information

Dora Cooper Bolger
P.O. Box 3216
Westhampton, MI 46532
dcbplanner@aol.com

Uses keywords from job description

KEYWORDS
Activity planner, budget coordinator, fund-raising, community service campaign, child advocacy, grant writer, public affairs, public speaking, interpersonal communication, management internal and external groups, team builder, strategic planner

OBJECTIVE
Position as public affairs coordinator for health care, educational, or charitable organization

Uses nouns to list accomplishments

RELEVANT EXPERIENCE
Presenter at 25 civic group functions
Fund raiser for 3 major campaigns, over $200,000 collected
President and coordinator, Teens in Trouble, a tutoring and mentoring program
Budget planner, Foster Parents' Association; large urban church

Summarizes experience concisely

VOLUNTEER WORK/AWARDS
Volunteer of the Year, Michigan Child Placement Agency, 2011
Volunteer Coordinator/Leader, American Kidney Fund
President and Secretary, local PTA, 2003–2010
Vice-president, Foster Parents' Association, 2010

EDUCATION
Metropolitan Community College, A.A., 2008
Mid-Michigan College, B.S., 2012; Major: Public Administration
Minor: Psychology
GPA: 3.45. Dean's List: 2010–2012

Avoids any symbols, boldfacing, or italics that could garble text

COMPUTER SKILLS
Microsoft Word; PeopleSoft; PowerPoint; BusinessPax

REFERENCES
Available on request.

read-only PDF file. Be aware, though, that a Word file may not print your résumé exactly as it appears on your computer screen, whereas a résumé sent as a PDF file will automatically retain the format, fonts, and graphics in your document, ensuring that it will look as you want it to no matter where and when it is printed. If you are given the option, email your résumé as a PDF document. Here are some practical guidelines to follow when emailing your résumé.

- Create your résumé as an MS Word file (the most common word-processing program in the world of work) and, unless directed otherwise, send it as a PDF attachment.
- Include a short cover email, but also attach a longer application letter.
- If either Word or PDF poses problems, you always have the option to save your Word file as a rich text format (rtf) document. But make sure you use a plain and simple design. Avoid underlining, boldface, italics, or shadowing, which can garble the text of your résumé, making it almost impossible to read. Use all capital letters instead of bold or italics for emphasis and insert an asterisk (*) or a plus sign (+) in place of bullets.
- Stay away from hard-to-read, nonstandard, or fancy fonts like script. Instead, choose a font like Arial, Times New Roman, or Verdana that is easy to scan and does not mask letters. Avoid Courier, which looks like a typewriter font.

3. **Create a scannable résumé.** Many companies scan hard copy résumés (such as those in Figures 7.11 and 7.12) into their databases so they can search for keywords (see pages 288 and 289). An increasing number of employers are now asking job candidates to paste the text of their résumé into a special submission window. To be successful, format your résumé so that key information is easy to locate and stands out clearly in the thirty-to-forty-second review of your credentials. Adhere to the following points when creating a scannable résumé:

- Follow all of the employer's instructions on formatting and submitting a résumé online. Otherwise, your application will be rejected.
- Where possible, make your scannable résumé longer than your hard copy version to increase the number of keywords or matches with the employer's job description.
- Use ample white space, which a scanner recognizes as separating one heading or section from another.
- If asked to submit a scannable hard copy résumé, use a high-quality laser printer and put your résumé on white or off-white paper.
- Avoid a script font. Use Times New Roman or Arial instead.
- Use at least 10- to 12-point type and allow a maximum width of 6½ inches to make your résumé easier to scan and read.
- Do not surround your résumé with a frame or border since this can lead to formatting difficulties.
- Do not staple or fold the pages.

4. **Create an HTML version of your résumé.** You can post your résumé on your own website, and you can link to an HTML version of your résumé from other

websites or from an email you send to a potential employer. HTML is the computer language used to create much of the content on the Internet. Since employers are always impressed by job candidates with some HTML knowledge or Web design experience, be sure to list this skill on your résumé. It may be one of your most powerful assets in your job search. Here are some options for creating an HTML version of your résumé.

- In using Word, you need to click "Save As" and then select the HTML format.
- Always view your résumé on different browsers before you send it to an employer.
- If you have a personal website, as Anthony Jones (see Figure 7.5) and Donald Kitto-Klein do (see Figure 7.10), add the HTML version of your résumé in a separate page on your website.
- Consider creating a website that functions as an online résumé with separate pages about your education, experience, skills, professional interests, letters of reference, and so on. An online résumé can contain a portfolio with samples of your work. This type of online résumé can be used as a model of your writing and technical skills.
- Keep in mind that complex images will not be preserved in the HTML view of your résumé.

5. Send a video résumé. Video résumés are growing in popularity with prospective employers. A video résumé allows them to see and hear you, thus displaying your communication skills and dedication to your professional goals. You can post a video résumé on YouTube and other sites, as well on the Web. You can also include it as part of your LinkedIn profile (see pages 263–265). But make sure you prepare your presentation carefully before you record it and ask an instructor or someone at your job placement center to critique it.

Developing an Electronic Résumé and Building a Network

Beth Pryor has just earned her B.S. in marketing at Southern Ohio University. She realizes that the competition for jobs is fierce and that she has to prepare a persuasive résumé as well as a cover letter to go with it. Taking advantage of the resources on campus, she met several times with an adviser in the career placement office who gave her some excellent advice on the strategies she might use to locate suitable positions. He also urged her to join LinkedIn and to list her profile on her résumé.

As Beth followed her adviser's advice about looking in many places for job openings, she quickly discovered that she would have to create more than one type of résumé. Some companies wanted job applicants to send their résumé as a Word or PDF document attached to an email. Others required applicants to paste a text résumé into a submission form on their human resources website. And many employers asked candidates to submit online portfolios of their work. Beth's roommate, Stella, an industrial design major with expertise in creating online portfolios, helped Beth develop and format her portfolio.

Beth's former instructor, Professor Lisa Brady, also offered to evaluate Beth's job-search materials. Beth knew that her feedback would be essential. Perhaps the most useful advice she received from her former teacher was: "When you prepare your résumé, focus on the content first, and then prepare multiple versions for different job searches." Professor Brady also cautioned Beth about what she needed to include and how: "Make certain your résumé emphasizes the skills that a marketing manager will want to have—teamwork, leadership, enthusiasm, creativity, and most of all your strong sense of visual design and thinking." Professor Brady added, "Make sure your résumé demonstrates the visual and technical qualities that will distinguish you from other applicants."

After consulting with the adviser at the career office, her tech-savvy roommate, Stella, and Professor Brady, and after searching the various sources for job leads, Beth created different versions of her résumé: a professionally formatted Word document, a PDF version of this document (Figure 7.13 on page 294), and a scannable résumé—a simplified text version of her Word résumé with the formatting removed (Figure 7.14 on page 295). She created the Word version of her résumé first, and then saved it as a PDF file. To prepare a scannable résumé, she removed the formatting from her Word résumé. Note how Beth's PDF résumé in Figure 7.13 is clearly formatted so an employer can read it quickly. Unlike her PDF résumé, Beth's scannable résumé does not contain any bold, italics, or indentations because these elements might prevent a scanner from capturing her information. To accompany her résumé, Beth developed a webfolio to display logo designs and advertising work from her college marketing courses and an internship at Archer Media. In her webfolio, she also included downloadable Word and PDF versions of her résumé.

Convinced that she needed to join LinkedIn, she asked Professor Brady and Stella to be part of her network. Both were members of groups that would help her in her job search. After establishing a network and creating different types of résumés, Beth was ready to enter a highly competitive job market with a professional Web presence that persuasively demonstrated her qualifications for a marketing position.

FIGURE 7.13 A PDF File of Beth Pryor's Résumé

Phone number not given because résumé posted on Web

PDF file retains all formatting— boldface, bullets, etc.

Information chunked into logically divided sections

Information is easy to access for employer through bulleted lists

Hyperlinks allow employer to access further relevant information about candidate

Beth Pryor
PO Box 5112, Oxford, Ohio 45056
bethpryor@hotmail.com

Objective A position with marketing firm emphasizing analysis, management, and leadership skills

Education **Southern Ohio University**, Oxford, OH
B.S. in Marketing; Minor: Spanish
Graduation: May 2012 — GPA in major: 3.36

Relevant Areas of Study
- Buyer Behavior
- Ethics of e-Business
- Management: Leadership and Learning
- Marketing Analysis
- Collaborating in the Workplace
- Promotional Strategies
- International Business

Study Abroad, Summer 2011, Southern Ohio University School of Business Administration Program in Santiago, Chile
- Completed 7-week program
- Developed second-language skills in Spanish

Business Internship **Archer Media Associates**, Marketing Assistant, Spring 2010
- Wrote marketing copy for one of Archer's largest clients, Techsure, Ltd.
- Led a team that created a marketing plan for 3 Amazon.com clients
- Prepared 8 major press releases for clients in health care, food service management, IT
- Tracked media coverage for clients
- Participated in corporate training seminars

Retail Experience **The Boutique**, Sales Representative, 2008–2010
- Promoted to "key holder" (opened and closed store; supervised store short-term, Summer 2010)
- Helped train staff of 4 in weekly meetings
- Chaired 2-3 weekly meetings per quarter
- Implemented goal-oriented management strategies

Software Skills • Lotus Notes, Access, InDesign, Illustrator, Photoshop, Excel

Profiles
- LinkedIn: www.linkedin.com/in/bethpryor
- MarketingEdge: www.ms.marketingedge.com/profiles/bpryor

Webfolio www.bethpryor.com/portfolio

Professional Memberships Student Marketing Association
- Treasurer (senior year)
- Public Relations Committee (2 years)

References Provided upon request.

© Cengage Learning 2013

BETH PRYOR
PO Box 5112
Oxford, Ohio 45056
bethpryor@hotmail.com

KEYWORDS
Marketing analyst, marketing management, e-business, business ethics, international clients, marketing plans, team spirit, media coverage writer, motivator

EDUCATION
Southern Ohio University
B.S. in Marketing; Minor: Spanish
Graduation: May 2012 — GPA in major: 3.36

RELEVANT AREAS OF STUDY
Buyer Behavior
Finance: Ethics of e-Business
International Business
Management: Leadership and Learning
Marketing Analysis
Collaborating in the Workplace
Promotional Strategies

STUDY ABROAD, Summer 2011
Southern Ohio University, School of Business Administration Program in Santiago, Chile
Graduate of 7-week program
Reader, speaker, intermediate Spanish

BUSINESS INTERNSHIP, Archer Media Associates, Spring 2010
Team marketing planner, amazon.com client
Writer, press releases and copy for clients in health care, food service, IT
Researcher, media coverage and visual designs
Participant, corporate training seminars

SALES MANAGEMENT EXPERIENCE, The Boutique, 2008-2010
Sales Representative
Key holder, 2009-2010
Staff trainer, 4 employees
Chair, 2-3 weekly meetings per quarter
Developer, goal-oriented management strategies

SOFTWARE
Lotus Notes, Access, InDesign, Illustrator, Photoshop, Excel

PROFILES
LinkedIn: www.linkedin.com/in/bethpryor
MarketingEdge: www.ms.marketingedge.com/profiles/bpryor

WEBFOLIO
www.bethpryor.com/portfolio

PROFESSIONAL MEMBERSHIP
Student Marketing Association

REFERENCES
Provided on request

Omits personal details

Simple text easily scanned into company's HR database, which will search for keywords

Does not use boldfacing, italics, etc.

Key sections are separated with extra spacing and all caps

Quantifies achievements

Supplies hyperlinks to profiles and individual website

Making Your Online Résumé Search-Engine Ready

The most important section of your online résumé contains the keywords you use. Prospective employers scan résumés to find the keywords they most want to see in the job seeker's description of his or her experience, education, and interpersonal skills. The more matches, or hits, they find between appropriate keywords in your résumé and those on their list, the better your chances are of being interviewed. List keywords throughout your résumé in appropriate places. Keywords should highlight your technical expertise, training and education, knowledge of a field, leadership ability, teamwork, writing/speaking skills, sales experience, and so on. Keywords, including degrees and licenses appropriate for an online résumé appear in Table 7.3.

Here are a few tips to help you select and use appropriate keywords:

1. Provide a keyword section at the top of your résumé to give employers an immediate snapshot of your skills, as in Figures 7.11 and 7.12.
2. Include the descriptive keywords found in the employer's ad and website in sections on education and experience to increase your chances of landing an interview.
3. Do not be afraid of using the shoptalk (or jargon) of your profession. An employer will expect you to be familiar with current terminology.
4. Use keywords to connect sections or categories of your résumé. Keyword headers will help you emphasize your job strengths and make it easy for employers to scroll back to an appropriate section of your résumé.
5. Replace the action verbs found in conventional résumés (on the left in the following list) with keyword nouns. Here are some examples:

Conventional Résumé	Online Résumé
Wrote business report	Business report writer
Performed laboratory tests	Laboratory technician
Solved consumer complaints	Consumer advocate
Responsible for managing accounts	Accounts manager
Won three awards	Award winner
Edited company newsletter	Newsletter editor
Solved software problem	Software specialist

Making Your Résumé Cybersafe

Whether you use a database service or post your résumé on your own website, protect your identity and your current job. Be careful about revealing personal information.

- Post your résumé only on legitimate sites. Avoid those that say they will flood the market. You don't know where your résumé will end up.
- Do not put personal information in your résumé—home address, phone number, Social Security number, birthday, health status, or photograph.
- You may want to use an anonymous email address rather than your personal one if you are concerned about sharing personal information. Consider a

TABLE 7.3   Sample Keywords for an Online Résumé

Job Title	Area of Expertise	Computer Skills	Degrees and Licenses	Job Skills and Personal Qualities
Accountant	Accounting	Access	AA	Analytical
Consumer advocate	Automotive repair	Adobe Illustrator	AS	Bilingual
Editor	Budget	Dreamweaver	ASID	Competency based
Environmentalist	Computer support	E-commerce	BA	Competitive
Exercise physiologist	Construction	Excel	BS	Cooperative
Fashion designer	Consumer affairs	Filemaker Pro	BSN	Critical thinking
Health care manager	Counseling	HTML	CPA	Customer oriented
Hospitality management	Customer service	InDesign	EMT	Dynamic
Intern	Engineering	JavaScript	LAT	Ethical
Law enforcement officer	Financial affairs	Lotus Notes	LTC	Experienced
Licensed practical nurse	Graphics	Lotus 1-2-3	LPC	Flexible
Maintenance expert	Health care	Mac OS	LPN	Fund raising
Musician	Human resources	Microsoft Word	MAT	Goal oriented
Officer	Information technology	Outlook	MBA	High energy
Paralegal	Law enforcement	PageMaker	MT	International
Pharmacy technician	Legal	PeopleSoft	NAVT	Leadership
Planner	Management	Photoshop	OT	Motivated
Programmer	Marketing	PowerPoint	PT	Multitasking
Public relations	Media	QuarkXPress	PTA	Research oriented
Registered nurse	Networking systems	QuickBooks	RN	Results oriented
Report writer	Office support	SPSS	RT	Risk taking
Resident manager	Public relations	UNIX		Safety conscious
Sales associate	Purchasing	Warp		Self-motivated
Teacher	Recruitment	WordPerfect		Speaking/writing
Technician	Safety/security	XML		Strong work ethic
Trainer	Sales			Studied abroad
Tutor	Technical support			Team player
Underwriter				
Veterinary technician				
Waitperson				
Web designer				

generic email address that includes a word or phrase that identifies your area of expertise, for example, JosieProgrammer@aol.com. Or include an email hyperlink with a built-in "mailto" on your résumé.

- Never put the names of your references or their contact information online. Simply say, "References available on request."
- Block certain readers from searching your résumé, such as your current employer or firms that you know send out spam.
- Never use your present employer's company name or business email address.

Testing, Proofreading, and Sending Your Online Résumé

Never underestimate the negative impact of errors in your résumé, email, or application letter. Many prospective employers will discard a résumé if they spot a typo.

1. Test your formatting. Send your résumé to a friend to be sure your file is readable and formatted correctly.
2. Print out your résumé and proofread the hard copy carefully. Do not rely on spell-check alone. It is easy to overlook mistakes if you only proofread what is on your computer screen.
3. Don't just put "résumé" as the subject of your email when sending your résumé to an employer. List the title, number, or code of the position for which you are applying.
4. Simply posting your résumé online is not enough. Also send a scannable hard copy and a letter of application (discussed below) to prospective employers. Do not fold or staple your résumé. Send it, along with your letter, in a large envelope ($8\frac{1}{2} \times 11$ inches).
5. Always keep a log of where you have posted your résumé online.

Letters of Application

Along with your résumé, you must send your prospective employer a letter of application, one of the most important pieces of correspondence you may ever write. Its goal is to get you an interview and ultimately the job. Letters you write in applying for jobs should be *personable*, *professional*, and *persuasive*—the three *P*'s. Knowing how the letter of application and résumé work together and how they differ can give you a better idea of how to compose your letter.

How Application Letters and Résumés Differ

The résumé is a persuasive record of dates, important achievements, skills, names, places, addresses, and jobs. As noted earlier, you may prepare several different résumés, depending on your experience and the job market.

Your letter of application, however, is much more personal. It introduces you to a prospective employer. Because you must write a new, original letter to each prospective employer, you may write (or adapt) many different letters. Each letter of application should be tailored to a specific job. It should respond precisely to the qualifications the employer seeks.

The letter of application is a sales letter that emphasizes and applies the most relevant details (of education, experience, and talents) in your résumé. In short, the résumé contains the raw material that the letter of application transforms into a finished and highly marketable product—you.

Résumé Facts to Exclude from Letters of Application

The letter of application should not simply repeat the details listed in your résumé. In fact, the following details that you would include in your résumé should *not* be restated in the letter:

- personal data, including license or certificate numbers
- specific course numbers
- names and addresses of all your references

Writing the Letter of Application

The letter of application, such as those in Figures 7.15–7.18, can make the difference between your getting an interview and your being eliminated early from consideration. It should convince a prospective employer that you will use the experience and education listed on your résumé in the job he or she is hoping to fill. You want your letter to be placed in the "definitely interview" category. As you prepare your letter, use the following general guidelines.

1. **Follow the standard conventions of letter writing.** Print your letter on good-quality, white 8½ × 11 inch paper. Proofread meticulously; a spelling error, typo, or grammatical mistake will make you look careless. As with your résumés, don't rely only on your spell-checker. (See pages 176–177.)

2. **Supply all contact information as part of your heading.** Include home address, phone numbers, email address, and your website, if you have one. (See page 171.)

3. **Make sure your letter looks attractive.** Use wide margins, and don't crowd your page. Keep your paragraphs short and readable—no more than four or five sentences each.

4. **Send your letter to a specific person.** Never address an application letter "To Whom It May Concern," "Dear Sir or Madam," or "Dear Director of Human Resources." Get an individual's name from the company's website or by calling the company's main office, and be sure to verify the spelling of the person's name and his or her title.

5. **Don't send a form letter to every potential employer.** Stay away from generic application letter templates. Customize your letter to make sure you address the employer's specific needs.

6. **Be concise.** A one-page letter is standard in today's job market unless you have years of experience.

7. Emphasize the "you attitude." (See pages 179–183.) See yourself as an employer sees you. Focus on how your qualifications meet the employer's needs, not the other way around. Employers are not impressed by vain boasts ("I am the most efficient and effective safety engineer"). Convince prospective employers that you will be a valuable addition to their organization—a team player, a problem solver, an energetic representative, a skilled professional. (See pages 178–183.)

8. Don't be tempted to send out your first draft. Write and rewrite your letter of application until you are convinced it presents you in the best possible light. Getting the job may depend on it. A first or even second draft rarely sells your abilities as well as a third, fourth, or even fifth revision does.

9. Tell the truth. Don't exaggerate the importance of any previous job experience or academic work. Never mislead a potential employer by lying about anything relating to continuing education units (CEUs), licenses, certificates, permits, or other professional qualifications.

The sections that follow give you some suggestions on how to prepare the various parts of an application letter successfully.

Your Opening Paragraph

The first paragraph of your letter of application is your introduction. It must get your reader's attention by answering four questions:

1. Why are you writing?
2. Where or how did you learn of the vacancy, the company, or the job?
3. What is the specific job title for which you are applying?
4. What is your most important qualification for the job?

Begin your letter by stating directly that you are writing to apply for a job. Don't say that you "want to apply for the job"; such an opening raises the question, "Why don't you, then?"

Avoid an unconventional or arrogant opening: "Are you looking for a dynamic, young, and talented accountant?" Do not begin with a question; be more positive and professional.

If you learned about the job through a newspaper or journal, make sure you italicize or underscore its title.

> I am applying for the food service manager position you advertised in the May 10 edition of the *Los Angeles Times* online.

Because many companies announce positions on the Internet, check there first to see if their position is listed online, as Anthony Jones did in Figure 7.15.

If you learned of the job from a professor, a friend, or an employee at the firm, indicate that. Take advantage of a personal (networking) contact who is confident that you are qualified for and interested in the position, as María López (Figure 7.16, page 304) and Dora Cooper Bolger (Figure 7.17, page 305) did. But first confirm that your contact gives you permission to use his or her name.

FIGURE 7.15 Letter of Application from Anthony Jones, a Recent Graduate with Little Job Experience

73 Allenwood Boulevard
Santa Rosa, California 95401-1074
707-555-6390
ajdesigner@plat.com
www.plat.com/users/ajones/resume.html

Anthony H. Jones
- WEBSITE DEVELOPER
- DESIGNER
- GRAPHIC ARTIST

May 24, 2012

Ms. Jocelyn Nogasaki
Human Resources Manager
Megalith Publishing Company
1001 Heathcliff Row
San Francisco, CA 94123-7707

Dear Ms. Nogasaki:

I am applying for the layout editor position advertised on your website, which I accessed on May 14. Early next month, I will receive an A.S. degree in commercial graphics illustration from Santa Rosa Junior College.

With a special interest in publishing, I have successfully completed more than 40 credit hours in courses directly related to layout design and gained experience using Adobe InDesign, QuarkXPress, Illustrator, and Photoshop. You might like to know that many Megalith publications were used as models of design and layout in my graphics communications and digital photography courses.

My studies have also given me practical experience at McAdam Publishers as part of my Santa Rosa internship program. While working at McAdam, I was responsible for assisting the design department in page composition and archiving photos. Other related experiences I have had include creating a website and brochure for the Santa Rosa Humane Society and designing and executing custom window displays at Buchman's Department Store. As the enclosed résumé indicates, my references and a webfolio are available at **www.plat.com/users.**

I would appreciate the opportunity to discuss my qualifications in graphic design with you. My phone number, email address, and website are listed above. After June 12, I will be available for an interview at any time that is convenient for you. Thank you for your consideration.

Sincerely yours,

Anthony H. Jones

Anthony H. Jones
Encl.: Résumé

Clear and professional–looking letterhead

Writes to a specific person

Identifies position and source of ad

Applies education directly to employer's business

Convincingly cites related job experience

Refers to résumé/ webfolio

Asks for an interview and thanks employer

You have to attract the reader's attention quickly and persuasively. In a sentence or two, tell the reader how your education and experience qualify you for the job. Whenever possible, use keywords from the job announcement.

The Body of Your Letter

The body of your letter, comprising one or two paragraphs, cites evidence from your résumé to prove you are qualified for the job. You might want to spend one paragraph on your education and one on your experience or combine your accomplishments into one paragraph.

Follow these guidelines for the body of your letter:

1. **Keep your paragraphs short and readable—four or five sentences.** Avoid long, complex sentences. Use the active voice to emphasize yourself as a doer. Review the action verbs in Table 7.2 and, again, use keywords found in the employer's ad.

2. **Don't begin each sentence with "I."** Vary your sentence structure. Write reader-centered sentences, even those beginning with "I."

3. **Concentrate on seeing yourself as a potential employer sees you.** Prove that you can help an employer's sales and service, promote an organization's mission and goals, and be a reliable team player.

4. **Highlight your qualifications by citing specific accomplishments.** Tell your reader exactly how your schoolwork and job experience qualify you to perform and advance in the job advertised. Show how you can make a positive contribution to the employer's company. Don't simply say you are a great salesperson. Demonstrate your accomplishments by stressing that you increased the sales volume in your department by 15 percent within six months, you won an award or received a promotion for customer service, or you reduced costs by 10 percent. Prove you saved a lab money by pricing and ordering technical equipment. Employers are not impressed by boasting or arrogance. They want hard facts to prove you are the right person for the job.

5. **Mention you are enclosing your résumé.** Put an "Encl." notation at the bottom of your letter.

Education   Recent graduates with little work experience, such as Anthony Jones in Figure 7.15, will, of course, spend more time discussing their education. Emphasize why and how your most significant educational accomplishments—course work, degrees, certificates, licenses, training—are relevant for the particular job. Mention significant extracurricular activities if they relate to the job description. Employers want to know which specific skills from your education translate into benefits for their company.

Simply saying you will graduate with a degree in criminal justice does not explain how you, unlike all the other graduates of such programs, are best suited for a particular job. Ask yourself which classes you took are most relevant for the employer. Consider grouping classes to show how and why you are the best qualified applicant for the job. For example, when you indicate that you have completed 36 credit hours in software security and have another 12 credit hours in global business, you prove you have an expertise other job candidates may not have. Note how Anthony Jones in Figure 7.15 and María López in Figure 7.16 establish their educational qualifications with specific details about their training. Be sure to also mention internships or clinical training, as Jones and López do.

Experience   After you discuss your educational qualifications, turn to your job experience. But if your experience is your most valuable and extensive qualification for the job, put it before education and stress any previous experience that is similar to what a new position calls for. This is what Donald Kitto-Klein did in his letter in Figure 7.18. Be sure to highlight any promotions or other leadership roles you have had. If you are switching careers or returning to a career after years away from the workplace, start the body of your letter with your experience or your community and civic service, as Dora Cooper Bolger does in Figure 7.17. Her volunteer work convincingly demonstrates she has the organizational and communication skills her prospective employer seeks. Never minimize such contributions.

Relate Your Education and Experience to the Job   Link your education and experience as benefits to the particular job you apply for. Persuasively show a prospective employer how your previous accomplishments, especially teamwork and responsibility, have prepared you for future success on the job. Relate your course work in computer science to being an efficient programmer. Indicate how your summer work for a local park district reinforced your exemplary skills in customer service. Connect your background to the prospective employer's company. Any homework you can do about the company's history, goals, or structure (see pages 307–308) will pay off.

- By citing Megalith publications as a model in his courses, Anthony Jones stresses he is ready to start successfully from the first day on the job (Figure 7.15).
- Note how María López links her major school project and her work on a hospital pediatric floor to Dr. Henrady's specialty (Figure 7.16).
- Dora Cooper Bolger likewise shows her prospective employer that she is familiar with and can contribute to Tanselle's programs in community mental health though her volunteer work and public speaking experience (Figure 7.17).
- Stressing his responsibilities in maintaining equipment and training staff, Donald Kitto-Klein shows he is qualified for the IT position at the Patterson Corporation (Figure 7.18).

Closing
The purpose of your last paragraph is clear-cut—to convince the reader to call or email you for an interview. Keep your closing paragraph short—about two or three sentences—but be sure it fulfills the following four important functions:

1. briefly emphasizes once again your major qualifications
2. asks for an interview or a phone call
3. indicates when you are available for an interview
4. thanks the reader

FIGURE 7.16 Letter of Application from María López, a Recent Graduate with Some Job Experience

Uses professional-looking letterhead with contact information

1725 Brooke Street
Miami, Florida 32701-2121
(305) 555-3429 • mlopez@eagle.com

May 15, 2012

Dr. Marvin Henrady
Suite 34
Medical/Dental Plaza
839 Causeway Drive
Miami, FL 32706-2468

Dear Dr. Henrady:

Begins with personal contact

Verifies she will have necessary licensure

Mr. Mitchell Pelbourne, my clinical instructor at Miami-Dade Community College, informs me you are looking for a dental hygienist to work in your northside office. My education and experience qualify me for that position. This month I will graduate with an A.S. degree in the dental hygienist program, and I will take the American Dental Assisting National Board Exams in early June.

Links training to job responsibilities; demonstrates knowledge of employer's office

I have successfully completed all course work and clinical programs in oral hygiene, anatomy, and prophylaxis techniques. During my clinical training, I received intensive practical instruction from several local dentists, including Dr. Tia Gutiérrez. Since your northside office specializes in pediatric dental care, you might find the subject of my major project—proper nutrition and dental care for preschoolers—especially relevant.

Relates previous experience to employer's needs; refers to résumé

I also have related job experience in working with children in a health care setting. For over two years, I was a unit clerk on the pediatric floor at St. Francis Hospital, and my experience in greeting patients, transcribing medical orders and surgical notes, and assisting the nursing staff would be valuable to you in your office. An additional job strength I would bring to your office is my bilingual (Spanish/English) communication skills. You will find more detailed information about my accomplishments in the enclosed résumé.

Ends with a polite request for an interview and thanks reader

I welcome the opportunity to talk with you about the position and my interest in pediatric dental care. I am available for an interview any time after 2:00 pm until June 11, but after that date I could come to your office any time at your convenience. Thank you for considering my application.

Sincerely yours,

María López

María López

Encl. Résumé

FIGURE 7.17 Letter of Application from Dora Cooper Bolger, a Job Candidate with Years of Community and Civic Experience

DORA COOPER BOLGER

1215 Lakeview Avenue
Westhampton, MI 46532
616-555-4773
dcbplanner@aol.com

February 9, 2012

Dr. Lindsay Bafaloukos, Director
Tanselle Mental Health Agency
4400 West Gallagher Drive
Tanselle, MI 46932-3106

Dear Dr. Bafaloukos:

At a recent meeting of the County Services Council, a member of your staff, Homer Steen, told me that you will soon be hiring a public affairs coordinator. Because of my extensive experience in and commitment to community affairs, I would appreciate your considering me for this opening. I expect to receive my B.S. in Public Administration from Mid-Michigan College later this year.

Begins with contact made at professional meeting, highlighting her qualifications

For the past ten years, I have organized community groups with outreach programs similar to Tanselle's. I have held administrative positions in the PTA and the Foster Parents' Association and served as president of Teens in Trouble, a volunteer group providing assistance to dysfunctional teens. My responsibilities with Teens have included coordinating counseling activities with various school programs, scheduling tutorials, and representing the organization before local and state government agencies. I have been commended for my organizational and communication skills. My twenty-five presentations on foster home care and Teens in Trouble also demonstrate that I am an effective speaker, a skill I could put to work for Tanselle immediately.

Relates proven past successes to employer's needs; gives concrete examples of her skills

Because of my work at Mid-Michigan and in Teens and Foster Parents, I have the practical experience in communication and psychology to successfully promote Tanselle's goals. The enclosed résumé provides details about my experience, education, and awards.

Encourages reader to see her as best-prepared candidate; includes résumé

I would appreciate an opportunity to discuss my work with Teens and other organizations I represented and how I might help Tanselle with its programs. I am available for an interview at your convenience. Thanks for considering my interest in and qualifications for your public affairs coordinator position.

Requests interview and thanks reader

Sincerely yours,

Dora Cooper Bolger

Dora Cooper Bolger

Encl. Résumé

FIGURE 7.18 Letter of Application from Donald Kitto-Klein, Who Has Strong Work Experience

Uses a professional, distinctive letterhead

DONALD KITTO-KLEIN

56 South Ardmore Way   (716) 555-9032
Petersburg, NY 15438   kitto@gar.com
716-555-3177   http://www.dkk.opengate.com

November 5, 2012

Ms. Michelle Washington
Vice President, Operations
Patterson Corporation
Sun Valley, CA 94356

Dear Ms. Washington:

Identifies specific position and his major qualification

I am interested in the position of Director of IT Services posted on your website on October 30 . My knowledge of IT services and my proven ability to work with people in a large organization such as Patterson would qualify me to be a productive member of your management team.

Documents accomplishments in technology, training, and leadership pertinent to the job

For the last four years, I have been responsible for all phases of IT maintenance and service at Garland.com. As a regular part of my duties I have serviced and repaired 20 to 25 PCs a month as well as supervised a mainframe. I have successfully coordinated the activities of my department with Garland's other offices and worked closely with vendors and manufacturers, relationships I would like to continue for Patterson. The training programs and in-house conferences I have conducted have repeatedly received high marks from Garland's management. Because I have worked so effectively with management and staff, I was promoted to support services team leader.

Puts education after experience

My educational achievements include a degree in IT from Grand Valley and certificates from four U.S. Army schools in IT systems and security. The enclosed résumé will give you information about these and my other accomplishments.

Emphatically closes by offering to put his career successes to work for employer and requests an interview

I would like to put my ability to motivate personnel and to manage computer services to work for you and Patterson Corporation. I am available for an interview at your convenience. I look forward to hearing from you.

Sincerely yours,

Donald Kitto-Klein

Donald Kitto-Klein

Encl. Résumé

End gracefully and professionally. Be straightforward. Don't leave the reader with a single weak, vague sentence: "I would like to have an interview at your convenience." That does nothing to sell you. Say that you would appreciate talking with the employer further to discuss your qualifications. Then mention your chief talent. If you are applying for an international job or one far from home, you might request a phone call instead of an in-person interview. You might also express your willingness to relocate if the job requires it.

After indicating your interest in the job, give the times you are available for an interview and specifically tell the reader where you can be reached. If you are going to a professional meeting that the employer might also attend, or if you are visiting the employer's city soon, say so.

The following samples show how *not* to close your letter and explain why.

Pushy:	I would like to set up an interview with you. Please phone me to arrange a convenient time. [That's the employer's prerogative, not yours.]
Too Informal:	I do not live far from your office. Let's meet for coffee sometime next week. [Say instead that since you live nearby, you will be available for an interview.]
Introduces New Subject:	I would like to discuss other qualifications you have in mind for the job. [How do you know what the interviewer might have in mind?]

Note that the closing paragraphs in Figures 7.15 through 7.18 avoid these errors.

Going to an Interview

If an employer is impressed with your résumé and letter of application, he or she will invite you to an interview. There are various ways for a prospective employer to conduct an interview. It might be a one-on-one meeting—you and the interviewer—or you may visit with a group of individuals or even with several groups from different divisions in the company to decide if you would fit in. You could even have an interview over the telephone or through a videoconference or via Skype (see the Tech Note on page 308). An interview can last thirty minutes or take all day. It can be a positive, uplifting experience or intentionally hard-hitting if the employer wants to see how you behave under pressure.

Preparing for an Interview

Before you go to an interview, prepare by doing the following:

   1. Do your homework about the company. Show you are interested in the company by learning as much as you can about it. Imagine how embarrassing it would be if an employer asked you what you know about the firm and you could not reply with a relevant answer. Click on "About Us" and other links on the corporate or agency website to find out who founded the company, who the current CEO is, if it is a local firm or a subsidiary, its chief products or services, how many years it

Skype Interviews

Prospective employers are increasingly conducting interviews over Skype (see pages 120–121). Much more effective than a telephone interview and less expensive than an on-site one, a Skype interview allows an employer to see you and assess your professional image. To have a successful Skype interview, be sure your webcam works and that your Internet connection is effective. Most webcams come with their own software and instructions that are usually fairly easy to follow. The faster your Internet speed is, the better the quality of the picture. Look directly into the webcam and do not stare at your computer screen. Eye contact is as important via Skype as in an on-site interview.

Be prepared. Keep a set of notes handy with the most crucial information you need to give to an employer. Dress professionally, too, even though the interview may be conducted from your dorm room or home via Skype. Look your best. Also, position your webcam so that the background for your interview looks neat and business-like. Don't sit or stand in front of posters, a pile of dirty clothes, or anything that detracts from your professional image. Turn off all radios, televisions, and cell phones and make sure you will not be interrupted, which can sink your chances of getting hired. Use Skype as a tool of job success.

has been in business, how many employees it has, where its main office and plants are, and who its major clients and competitors are. Find the company's profile on LinkedIn. Read company websites, blogs, and brochures to get a sense of the corporate culture. Also look for recent stories about the company in leading business publications, such as the following:

- *Wall Street Journal*—http://online.wsj.com
- *New York Times*—www.nytimes.com
- *USA Today*—www.usatoday.com
- *Businessweek*—www.businessweek.com
- *Fast Company*—www.fastcompany.com
- *Fortune*—www.fortune.com
- *Forbes*—www.forbes.com

2. Review the job description carefully. Bring a copy of the job description with you to review just prior to the interview so that you are clear about what the job entails.

3. Prepare a one- or two-minute summary of your chief qualifications. You will most likely be asked to summarize your education, experience, and professional goals during the job interview. In doing so, identify how specific classes, course projects, jobs you have held, or community service have equipped you for the position the company wants to fill.

4. Take your portfolio, including three or four extra copies of your résumé, with you. Also bring a notepad and a pen to jot down essential details.

5. **Practice your interview skills with a friend or job counselor.** Be sure that this person asks tough questions about your education and experience so that you will get practice answering these questions realistically and convincingly.

6. **Brush up on business etiquette.** Silence your cell phone before your interview. Remember the name(s) of the interviewer(s) and others you may meet. Always be polite and respectful, saying "Thank you," "You're welcome," and so on. Pay special attention to acceptable ways of communicating with international audiences if your interview is with a multinational company or with a non-native speaker of English.

7. **Bring your photo ID and Social Security card.** Bring any licenses or certificates you may be asked to present to a human resources office. If you are not a U.S. citizen, bring your work visa.

Questions to Expect at Your Interview

The following questions are typical of those you can expect from interviewers, with advice on how to answer them.

- **Tell us something about yourself.** Emphasize achievements that show you are responsible (e.g., working to pay for your tuition), conscientious (participating in a community or service activity), and eager to contribute to and learn more about your profession and potential employer.
- **Why do you want to work for us?** Recall any job goals you have and apply them specifically to the job under discussion.
- **What qualifications do you have for the job?** Point to educational achievements and relevant work experience, especially IT skills.
- **What could you offer us that other candidates do not have? Why should we hire you?** Say enthusiasm, being a team player, problem-solving skills, ability to meet deadlines. Stress that you are diplomatic yet goal oriented.
- **Why did you attend this school?** Be honest—location, costs, programs.
- **Why did you major in "X"?** Do not simply say financial benefits; concentrate on professional goals and interests.
- **Why did you get a grade of C in a course?** Don't say that you could have done better if you'd tried. Explain what the trouble was, and mention that you corrected it in a course in which you earned a B or an A.
- **What extracurricular activities did you participate in while in high school or college?** Indicate any responsibilities you had—managing money, preparing minutes, coordinating events. If you were unable to participate in such activities, tell the interviewer that a part-time job or community or church activities prevented you from participating. Such answers sound better than saying that you did not like sports or clubs in school.
- **Did you learn as much as you wanted from your course work?** This is a loaded question. Indicate that you learned a great deal but now look forward to the opportunity to gain more practical skills, to put into practice the principles and procedures you have learned.

- **What is your greatest strength?** Say being a team player, planning and organizing tasks efficiently, being sensitive to others' needs and concerned about the environment, being cooperative and willing to learn, having the ability to grasp difficult concepts easily, wanting to find a more efficient or economical way of doing something, being proficient at managing time or money, taking criticism easily, and profiting from it.
- **What is your greatest shortcoming?** Be honest here and mention it, but then turn to ways in which you are improving. Don't say something deadly like, "I can never seem to finish what I start" or "I hate being criticized." You should neither dwell on your weaknesses nor keep silent about them. Saying "None" to this kind of question is as inadvisable as rattling off a list of faults.
- **How do you handle conflict with a co-worker, supervisor, or customer?** Stress your ability to be courteous and honest and to work toward a productive resolution. State that you avoid language, tone of voice, or gestures that interfere with healthy dialogue. Describe a specific situation where you resolved a problem with maturity and grace.
- **Why did you leave your last job?** Say "I returned to school full time" or "I moved from Jackson to Springfield," or say that you changed professions. *Never attack your previous employer.* That only makes you look bad.
- **Why would you leave your current job?** Again, never attack an individual or an organization. Say your current job has prepared you for the position you are now applying for. Emphasize your desire to work for a specific company because of its goals, work environment, and opportunities. Say you want a challenge.
- **What are your career goals over the next three to five years?** State your career objectives in terms of what you would like to accomplish for the company, your profession, the community, and yourself. Be confident, not cocky.

What Do I Say About Salary?

Find out what the salary range is for your professional level in your area. Consult the U.S. Bureau of Labor Statistics' *Occupational Outlook Handbook* at www.bls.gov/oco as well as www.salary.com. You can also ask your instructors or individuals you know who work for the company or call your professional organization for information. If the issue of salary comes up, ask if the company has established a salary range for the position and where you stand in relationship to that range. However, because many companies set fixed salaries for entry-level positions, it may be unwise to try to negotiate.

If you are asked what salary you expect for the job, do not give an exact figure. You may undercut yourself if the employer has a higher figure in mind. By doing your homework on salary ranges, you will have a better feel for the market when the employer does mention salary.

Factor other things into your salary calculations—health insurance, day care, housing, uniform/clothing allowances, product or service discounts, opportunities for travel and foreign language instruction, and tuition reimbursement.

Questions You May Ask the Interviewer(s)

You will have a chance to ask the interviewer(s) questions. Watch for appropriate cues, and be prepared to say more than "No, I don't have any questions," which suggests either indifference or unpreparedness on your part. Here are some legitimate questions you can ask interviewers:

1. Will there be any safety, security, or proficiency requirements I will need to meet?
2. When is the starting date?
3. Is there a probationary period? If so, how long?
4. How often will my work be evaluated (monthly, quarterly, semiannually) and by whom (immediate superior, committee)?
5. What types of on-the-job training are required or offered?
6. Are there any mentoring programs in place?
7. Is there any support for continuing my education to improve my job performance?

Also, ask questions about the company's products and services, including a dedication to greening the environment.

What Interviewer(s) Can't Ask You

There are some questions an interviewer may not legally ask you. Questions about your age, marital status, ethnic background, religion, race, or physical disabilities violate equal opportunity employment laws. Even so, some employers may disguise their interest in those subjects by asking you indirect questions about them. A question such as "Will your husband care if you have to work overtime?" or "How many children do you have?" could probe into your personal life. Confronted with such questions, it is best to answer them positively ("My home life will not interfere with my job," "My family understands that overtime may be required") rather than bristling defensively, "It's none of your business if I'm married."

Ten Interview Do's and Don'ts

Keep in mind these other interview do's and don'ts:

1. Be on time. In fact, show up about fifteen minutes early in case the interviewer or human resources office wants you to complete some forms.
2. Turn off your cell phone and any other portable media device! The last thing you want is for your phone to ring or receive a text message during your interview. Never text during an interview.
3. Dress appropriately for the occasion and be well groomed. Avoid using strong perfume or cologne. Men: Wear a dark solid-color or pinstripe suit and tie. Women: Wear a suit (pants or skirt) or other equally business-like attire.
4. Be careful about tattoos. Job counselors warn that visible tattoos can hurt a job seeker's chance for success.

5. Greet the interviewer with a friendly and firm, but not vicelike, handshake. Don't be a wimp, either, with a limp, fishy handshake. Thank the interviewer for inviting you.

6. Don't sit down before the interviewer does. Wait for the interviewer to invite you to sit and to indicate where.

7. Speak slowly and distinctly; do not nervously hurry to finish your sentences, and never interrupt or finish an interviewer's sentences. Avoid one- or two-word answers, which sound unfriendly or unprepared. Do not use slang (e.g., "Awesome!" or "Chill") or overly casual language ("Like…" "You know?"). Don't monopolize the discussion by talking too much and always about yourself. And don't act arrogantly as if the job is yours already. Show a keen interest in the company and its products, services, employees, and organization.

8. Do not chew gum, click a ballpoint pen, fidget; twirl your hair, or tap your foot against the floor, a chair, or a desk.

9. Maintain appropriate eye contact with the interviewer; do not sheepishly stare at the floor or the desk. If you are interviewed by a group of individuals, make eye contact with each one of them. Body language is equally important. Don't fold your arms—a signal that you are closed to the interviewer's suggestions and comments. Sit up straight; do not slouch. Smile; it shows you are confident.

10. When the interview is over, thank the interviewer(s) for considering you for the job, and say you look forward to hearing from him or her.

The Follow-Up Letter

Within a week after the interview, it is wise to send a follow-up letter, not an email, thanking the interviewer for his or her time and interest in you. In your letter, reemphasize your qualifications for the job by showing how they apply to the requirements described by the interviewer. You might also ask for further information to show your interest in the job and the employer. The sample follow-up letter in Figure 7.19 accomplishes all of these things.

Accepting or Declining a Job Offer

If you accept a job, send the employer a letter within a week of the offer. Accepting verbally on the phone is not enough. Your letter will make your acceptance official and will probably be included in your permanent personnel file. Accepting a job is easy. Make the communication with your new employer a model of clarity and diplomacy.

A sample acceptance letter appears in Figure 7.20 (page 314). In the first sentence, tell the employer that you are accepting the job and refer to the date of the letter offering you the position, the specific job title, and the salary. Indicate when you can begin working. Then mention any pleasant associations from your interview or any specific challenges you are anticipating. That should take no more than a paragraph.

In a second paragraph, express your plans to fulfill any further requirements for the job—going to the human resources office, taking a physical examination,

FIGURE 7.19 A Follow-Up Letter

2739 EAST STREET
LATROBE, PA 17042-0312

610-555-6373
mlb@springboard.com

September 20, 2012

Mr. Jack Fukurai, Director
Human Resources
Global Tech
1334 Ridge Road N.E.
Pittsburgh, PA 17122-3107

Dear Mr. Fukurai:

I enjoyed talking with you last Wednesday and learning more about the security officer position available at Global Tech. It was especially helpful to take a tour of the plant's north gate section to see the challenges it presents for the security officer stationed there.

Expresses gratitude for an interview and singles out main company feature

As you noted at the interview, my training in surveillance electronics has prepared me to operate the sophisticated equipment Global Tech has installed at the north gate. I was grateful to Ms. Turner for taking time to demonstrate this technology.

Reemphasizes qualifications

I am looking forward to receiving the handbook about Global Tech's employee services. Would you also kindly email me a copy of the newsletter from last quarter that introduced the new security equipment to employees?

Asks for newsletter to express future interest

Thank you, again, for considering me for the position and for the hospitality you showed me. Please let me know if you have any other questions. I look forward to hearing from you.

Ends politely by thanking interviewer

Sincerely yours,

Marcia Le Borde

Marcia Le Borde

FIGURE 7.20 Letter Accepting a Job

• KEVIN DUBINSKI •

73 Park Street
Evansville, WI 53536-1016

Home: 608-555-3173 • Cell: 608-898-4291
KDubinski@global.net

June 27, 2012

Ms. Melinda Haas, Manager
Weise's Department Store
Janesville Mall
Janesville, WI 53545-1014

Dear Ms. Haas:

Accepts job and shows willingness to be team member

I am pleased to accept the position of assistant controller that you offered me at a salary of $31,250 in your letter of June 22. Starting on July 18 will be no problem. I look forward to helping Ms. Meyers in the business office. In the next few months I know that I will learn a great deal about Weise's.

Agrees to conditions

As you asked, I will make an appointment early next week with the Human Resources Department to discuss travel policies, salary payment schedules, and insurance coverage.

Looks forward to starting

I am eager to start working for Weise's.

Cordially,

Kevin Dubinski

Kevin Dubinski

FIGURE 7.21 Letter Refusing a Job

George Alexander

March 8, 2012

Ms. Gail Buckholtz-Adderley
Assistant Editor
The Everett News
Everett, WA 98421-1016

Dear Ms. Buckholtz-Adderley:

I enjoyed meeting you and the staff photographers at the recent interview for the graphics position at the *News*. Your plans for the special online magazine are exciting, and I know that I would have enjoyed my assignments greatly.

However, because I have decided to continue my education part-time at Western Washington University in Bellingham, I have accepted a position with the *Bellingham American*. Not having to commute to Everett will give me more time for my studies and also for my freelance work.

Thank you for your generous offer and for the time you and the staff spent explaining your plans to me. I wish you much success with the new online magazine.

Sincerely yours,

George Alexander

George Alexander

345 Melba Lane ▪ Bellingham, WA 98225-4912 ▪ galexander@micro.com

Does not begin with refusal

Gives creditable reason for not taking job

Thanks reader and ends on a friendly note

having a copy of a certificate or license forwarded, sending a final transcript of your college work. A final one-sentence paragraph might state that you look forward to starting your new job.

Refusing a job requires tact. You are obligated to inform an employer why you are not taking the job. See the sample refusal letter in Figure 7.21.

Do not bluntly begin with the refusal. Instead, prepare the reader for bad news by starting with a complimentary remark about the job, the interview, or the company. Then move to your refusal and supply an honest but not elaborate explanation of why you are not taking the job. Many students cite educational opportunities, work schedules, geographic preference, or more relevant professional opportunities. End on a friendly note, because you may be interested in working for the company in the future and do not want to leave any bad feelings.

Searching for the Right Job Pays

As we saw, finding the right job takes a lot of hard work (researching, organizing, and writing). But all your efforts will pay off with your first and subsequent checks. May all your letters, résumés, portfolios/webfolios, and applications be models of successful writing at work.

✔ **Revision Checklist**

☐ Enhanced my professional image to apply for jobs for which I am qualified.
☐ Looked for relevant jobs on social and professional networking sites.
☐ Created a professional profile for Facebook and LinkedIn.
☐ Joined groups on LinkedIn to increase my network.
☐ Did not put personal information on a job board/the Internet.
☐ Prepared a dossier at school placement office, including supporting letters from professors, employers, and community officials.
☐ Identified places where relevant jobs are advertised.
☐ Networked with instructors, friends, relatives, and individuals who work for the companies I want to join; notified them that I am looking for a job.
☐ Researched the companies I am interested in—on the Internet, through printed sources, and by networking with current employees.
☐ Inventoried my strengths carefully to prepare résumé.
☐ Wrote a focused and persuasive career-objective statement.
☐ Determined the most beneficial format of résumé to use—chronological, functional, or both.
☐ Investigated creating a website for my job search–related documents.
☐ Prepared a portfolio/webfolio that includes documents demonstrating my professional skills and achievements relevant for the job.

☐ Made résumé attractive and easy to read, with logical and persuasive headings and descriptive keywords.

☐ Made sure résumé contains neither too much nor too little information.

☐ Proofread résumé to ensure everything is correct, consistent, and accurate.

☐ Created properly formatted online résumé to send to prospective employers.

☐ Wrote a letter of application that shows how my specific skills and background meet an employer's exact needs.

☐ Prepared a short oral presentation about myself and my accomplishments for an interview.

☐ Researched prospective employer's company or organization and salary range.

☐ Sent prospective employer a follow-up letter within a few days after interview to thank the interviewer and show interest in position.

Exercises

1. Using at least four different sources, including LinkedIn, compile a list of ten employers for whom you would like to work. Get their names, street and email addresses, phone numbers, and the names of the managers or human resources officers. Then select one company and profile it—locations, services, kinds of products or services offered, number of employees, clients served, awards, contributions to the community or environment, and any other pertinent facts.

2. Create an appropriate professional profile for Facebook and LinkedIn.

3. Write a letter to a former employer, a community leader, or an instructor requesting a letter of recommendation.

4. Which of the following would belong on your résumé? Which would not belong? Why?
   a. your student ID number
   b. your driver's license number
   c. the zip codes of your references
   d. a list of all your English courses in college
   e. the section numbers of the courses in your major
   f. a statement that you are recently divorced
   g. subscriptions to journals in your field
   h. the titles of stories or poems you published in a high school literary magazine or newspaper
   i. your GPA for each year you were in college
   j. foreign languages you studied

    k. years you attended college
    l. the date you were discharged from the service
    m. names of the neighbors you are using as references
    n. your religion
    o. job titles you held
    p. your summer job waiting tables
    q. your telephone number
    r. the reason you changed schools
    s. your current status with the National Guard
    t. the URL of your website or blog
    u. your volunteer work for the Red Cross
    v. the number of hours per week you spend reading science fiction
    w. the title of your last term paper in your major
    x. the name of the agency or business where you worked last

5. Indicate what is wrong with the following career objectives, and rewrite them to make them more precise and professional.
    a. Job in a lawyer's office.
    b. Position with a safety emphasis.
    c. Desire growth position in a large department store.
    d. Am looking for entry position in health sciences with an emphasis on caring for older people.
    e. Position in sales with fast promotion rate.
    f. Want a job working with semiconductor circuits.
    g. I would like a position in fashion, especially one working with modern fashion.
    h. Desire a good-paying job, hours: 8–4:30, with time and a half for overtime. Would like to stay in the Omaha area.
    i. Insurance work.
    j. Working with media.
    k. Personal secretary.
    l. Job with preschoolers.
    m. Full-time position with hospitality chain.
    n. I want a career in nursing.
    o. Police work, particularly in a suburb of a large city.
    p. A job that lets me be me.
    q. Desire fun job selling cosmetics.
    r. Any position for a qualified dietitian.
    s. Although I have not made up my mind about which area of forestry I shall go into, I am looking for a job that offers me training and rewards based upon my potential.

6. As part of a team or on your own, revise the following poor résumé to make it more precise and persuasive. Include additional details where necessary and exclude any details that would hurt the job seeker's chances. Also correct any inconsistencies.

RÉSUMÉ OF

Powell T. Harrison
8604 So. Kirkpatrick St.
Ardville, Ohio
345 37 8760
614 234 4587
harrison@gem.com

<u>PERSONAL</u>	Confidential
<u>CAREER</u> <u>OBJECTIVE</u>	Seek good paying position with progressive Sunbelt company.
<u>EDUCATION</u>	
2010–2012	Will receive degree from Central Tech. Institute in Arch. St. Earned high average last semester. Took necessary courses for major; interested in systems, plans, and design development.
2008–2010	Attended Ardville High School, Ardville, OH; took all courses required. Served on several student committees.
<u>EXPERIENCE</u>	None, except for numerous part-time jobs and student apprenticeship in the Ardville area. As part of student app. worked with local firm for two months.
<u>HOBBIES</u>	Surfing the Net, playing Nintendo DSi, Member of Junior Achievement.
<u>REFERENCES</u>	Please write for names and addresses.

---

7. Determine what is wrong with the following sentences in a letter of application. Rewrite them to eliminate any mistakes, to focus on the "you attitude," or to make them more precise.
   a. Even though I have very little actual job experience, I can make up for it in enthusiasm.
   b. My qualifications will prove that I am the best person for your job.
   c. I would enjoy working with your other employees.
   d. This email résumé is my application for any job you now have open or expect to fill in the near future.

e. Next month, my family and I will be moving to Detroit, and I must get a job in the area. Will you have anything open?

f. If you are interested in me, then I hope that we make some type of arrangements to interview each other soon.

g. I have not included a résumé since all pertinent information about me is in this letter.

h. My GPA is only 2.5, but I did make two B's in my last term.

i. I hope to take state boards soon.

j. Your company, or so I have heard through the grapevine, has excellent fringe benefits. That is what I care about most, so I am applying for any position that you may advertise.

k. I am writing to ask you to kindly consider whether I would be a qualified person for the position you announced in the newspaper.

l. I have made plans to further my education.

m. My résumé speaks for itself.

n. I could not possibly accept a position that required weekend work, and night work is out, too.

o. In my own estimation, I am a go-getter—an eager beaver, so to speak.

p. My last employer was dead wrong when he let me go. I think he regrets it now.

q. When you want to arrange an interview time, give me a call. I am home every afternoon after 4:00.

8. Explain why the following letter of application is ineffective. Rewrite it to make it more precise and appropriate.

Apartment 32

Jeggler Drive

Talcott, Arizona

Monday

Grandt Corporation

Production Supervisor

Capital City, Arizona

Dear Sir:

I am writing to ask you if your company will consider me for the position you announced online recently. I believe that with my education (I have an associate degree) and experience (I have worked four years as a freight supervisor), I could fill your job.

My schoolwork was done at two junior colleges, and I took more than enough courses in business management and information technology. In fact, here is a list of some of my courses: Supervision, Materials Management, Work Experience in Management, E-commerce, Safety Tactics, Introduction to Software Analysis, Art Design, Contemporary Business Principles, and Small Business Management. In addition, I have worked as a loading dock supervisor for the last two years, and before that I worked in the military in the Quartermaster Corps.

Please let me know if you are interested in me. I would like to have an interview with you at the earliest possible date, since there are some other firms also interested in me, too.

Eagerly yours,

George D. Milhous

9. In the Sunday edition of your local newspaper or in one of the other sources discussed on pages 257–260, find notices for two or three jobs you believe you are qualified to fill, and then write a letter of application for one of them.

10. Write a chronological résumé to accompany the letter you wrote for Exercise 9.

11. Write a functional résumé to accompany your application letter in Exercise 9.

12. Bring the two résumés you prepared for Exercises 10 and 11 to class to be critiqued by a collaborative writing team. After your résumés are reviewed, revise them. Write an email to your instructor about the revisions you made, and explain why they will help you in your job search. Attach your résumé to the email.

13. Prepare an online version of the résumé you prepared in either Exercise 10 or Exercise 11. Use persuasive keywords, such as those in Table 7.3.

14. Write an appropriate job application letter to accompany Anna Cassetti's résumés in Figures 7.7 and 7.9.

15. Prepare a hard-copy portfolio. Collect appropriate documents, and arrange them in the most relevant order. Provide an annotation for each.

16. Write a letter to a local business inquiring about summer employment. Indicate that you can work only for three months because you will be returning to school by September 1. Include an appropriate résumé.

PART III

Gathering and Summarizing Information

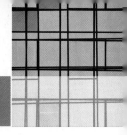

CHAPTER 8

Doing Research and Documentation on the Job

Access chapter-specific interactive learning tools, including quizzes and more in your English CourseMate, accessed through www.cengagebrain.com.

Being able to do research is crucial for success on the job, whatever company or department you work for and whatever your job title. Research is the lifeblood of a company. You can expect to spend as much as 25 to 30 percent of your time at work doing research. Companies use research to make major decisions that affect production, sales, service, hiring, promotions, and locations, as the research report at the end of this chapter illustrates (pages 397–413). In the world of work, research helps companies

- stay within budget
- remain up-to-date with market trends and new technologies
- meet the needs of their customers worldwide
- avoid problems with equipment, procedures, and the environment
- explore and open new markets
- increase profits
- report to stockholders, community groups, and government agencies

Research follows a process. You have to gather, summarize, and organize information before you can interpret it. Then, in interpreting it, you must be able to answer questions and solve problems.

Skills Necessary to Do Research

To do effective research, you need to know how to

- network with people in your department, within your company, outside of your company, and potentially across the globe to gather relevant data
- read a host of print and online sources at your company or a library to find the most relevant studies/opinions on your topic
- do direct observations, perform tests, and make site visits
- interview one person or a carefully selected group of people
- prepare and send out surveys and analyze the results

- organize information into clear and accurate reports that answer questions and solve problems
- carefully and completely document your sources to give proper credit and to help readers find your sources

Understanding Research Strategies

Although you may already have written reports for school assignments, this chapter focuses specifically on the types of research you will be expected to do on the job. Basically, it guides you through the following seven fundamental stages of doing research:

1. Understanding the differences between research done for school and research done in the workplace (pages 325–326)
2. Going through the various and interlinked stages involved in the research process (pages 328–331)
3. Distinguishing between primary and secondary research and understanding how they differ and how they relate to one another (pages 331–336)
4. Knowing how to do primary research, including making observations, doing tests, going on site visits, conducting interviews and focus groups, and conducting surveys (pages 337–353)
5. Being able to do secondary research, including using libraries and the Internet (pages 353–378)
6. Taking effective notes as you research (pages 378–383)
7. Documenting your sources and knowing how to cite them correctly in both MLA and APA style (pages 383–397)

At the end of this chapter (pages 397–413) you will find a sample business report that illustrates the strategies and tools of primary and secondary research that you will need to become a successful researcher on the job.

The Differences Between School and Workplace Research

The research tools and methods you use in school for a course paper or report are far different from what you can expect to find in the workplace. Understanding the differences in goals, audience, resources, and format/presentation, as well as how credit for your work may be assigned, will help you do your workplace research more effectively.

1. Goals. In school, you do research for educational reasons: to learn about a valuable topic and to prepare for a career. In the world of work, research is aimed at basically one goal: to provide bottom-line information for your employer. Workplace research is practical, almost always addressing IT, financial, operational, or personnel issues. It affects the way your company does business, how successful it is, and what the future of its employees, including yourself, will be.

2. Audience. In school, your instructor and possibly your classmates constitute the audience for your research paper. Your instructor will ask you to write your paper following his or her specific guidelines, may help you locate sources, and will grade you on how well you present your findings to him/her and/or your class. The audience for your work-related research, however, can include numerous and diverse individuals and groups, ranging from stockholders and upper management to your immediate supervisor, co-workers, customers, and vendors worldwide. Your boss may give you a question to answer or a problem to solve and may even suggest some sources to consult, but usually you cannot expect to receive the guidance you would when writing a paper for a class assignment.

3. Resources. Most of the resources you use for a school paper or report focus on academic publications—books, journals, encyclopedias, and websites. On the job, however, your research will be far more diverse and less print-centered. In business, research is often based on networking, interviewing individuals (customers, experts in the field, vendors, etc.), visiting sites, examining and "shopping" competitors, consulting government and market reports (for population and business statistics), and using in-house and trade literature. Of course, you will also use libraries and online resources for statistical data, as well as trade journals to keep up on the job, but the central tools of business research involve much more direct contact with individuals.

4. Format/presentation. Your school research report needs to follow the guidelines and format your instructor specifies. Your boss may also ask you to follow specific formats and methods of documentation, but they may be different from those used in school. In the workplace, you will be expected, alone or with your team, to discover the best and most persuasive ways to present and adapt your research. Depending on your audience's needs and the scope of your assignment, you may present your research as a formal report, in an email or a memo, as a thorough PowerPoint presentation, or as an informal oral presentation at a meeting or a Skype conference.

5. Credit for your work. Your name (as well as the names of any collaborative team members) would certainly be listed on any research report or paper you write in school. But in the world of work, you may not always see your name on the research reports you do. Instead, your manager's name or just the name of your department, division, or company may appear on the document. See the long report at the end of Chapter 15, where just the name of the manager appears on a report written by a team.

Some Research Scenarios

Here are a few scenarios that will help you understand the different kinds of research you can expect to do on the job:

- Your division is launching a new product or service, and the manager wants you to find out exactly what customers expect and want.

- You have been assigned to find out how a product or service your company provides can successfully compete against another company's product or service.
- Your company is lagging behind technologically, and you have to write a report recommending the most appropriate and cost-efficient way to upgrade equipment or procedures, taking into account purchasing and installing the new technology and training employees on how to use it.
- Your department is in danger of exceeding its annual budget, and your boss instructs you to write a report showing how the department can cut costs but still get the job done on time.
- A new market trend that directly affects your company has recently received a lot of media attention, and the head of your division wants you to find out if this trend is likely to be long lasting or just a fad.
- Your co-workers have elected you to lead a five-person group researching ways to attract new customers. You have to write a report on the group's recommendations to your employer.
- Your company or organization wants to open a branch in a country where it has not done business before, and you must investigate this potential market for your firm, focusing on the culture and buying preferences of its people.
- The Bureau of Labor Statistics projects that your city will experience an acute shortage of skilled workers in the next five to ten years. Your supervisor asks you to interpret those statistics in light of your company's future and to make recommendations on the types of positions your company will need to fill.

Characteristics of Effective Workplace Research

The research you do on the job needs to follow the highest professional and ethical standards. Businesses leave little margin for error and often do not give employees a second chance to get it right. To make sure your research meets your employer's expectations, it must be

  1. **Relevant.** Job-related research must focus directly on providing specific answers and solutions to the key questions and problems affecting your company. Because research in the business world is so costly, your employer will not allow you to waste time and money on side issues, unproductive leads, or personal interests.

  2. **Current.** Your information must be up-to-date. Markets and technologies change rapidly, and employers will insist that your research is on the cutting edge of your profession. By keeping up with the most recent trends and ideas, you can give your employer a bigger picture of the market and help your firm to stay competitive.

  3. **Accurate.** Double- and triple-check all of the facts and figures, dates, addresses, names, regulations, URLs, and so on, used in your research. Don't substitute guesswork and unsupported estimates for hard facts. Make sure you record all information accurately. Your boss will expect your reports to be based on quantifiable, measured data, which are essential to precise research.

4. Original. Employers want well-thought-out and supported conclusions. Don't just repeat what you have read in a report or in an article or have heard at a conference. Evaluate all data. Question and test what you have read, heard, or observed. That way, you can persuasively modify or expand it, reject it, or use it to project and predict. This is the way to arrive at original answers and solutions for your on-the-job research.

5. Thorough. Look at a question or problem from all sides. Network with colleagues to look for any gaps or inconsistencies, as well as business opportunities. Confirm all options and opinions. Never omit important data.

6. Realistic. Base your research on realistic, profitable conclusions. Unsubstantiated recommendations that fly in the face of a company's protocol (e.g., drop a product line, hire or fire 25 people, or move a plant) may not be logical, profitable, or acceptable. Be sure that your research is consistent with your company's policies.

7. Ethical and legal. Obtain your findings ethically and lawfully so that you do not infringe on the rights of others. Raiding someone's unpublished research, sharing confidential or privileged information with a third party, or skewing the results of a survey are all unethical acts. So, too, is discounting or minimizing facts that don't agree with the findings you want or expect. Plagiarism (see page 384) and omitting evidence are serious violations in the world of work, just as they are in the classroom. Be sure, too, that all of your recommendations are environmentally sound. Businesses today endorse a strong green philosophy.

The Research Process

Just as there is a process for writing—brainstorming, drafting, revising, editing—there is a process involved in doing research. In research you go through the steps of finding, assessing, and incorporating information into your written work.

As in the writing process, in doing research you may find yourself repeating certain steps. Say, for example, you are writing a business proposal and are incorporating information from several sources you've researched. At this stage of the process, you might think you've gathered enough information. However, as you work on the proposal, you may realize that it raises new questions. That may lead you back to repeating previous steps.

Let's look at the process in more detail:

Step 1. Confirm the purpose and audience of your report. Know who your audience is and why you are writing to them. Will your readers be management, co-workers, clients, government agencies, or a combination of these groups? What are their expectations about length, sources of your information, level of technical detail, conclusions/recommendations, and so on? What is the bottom line—what do you have to answer, argue, propose, solve, uncover, change, describe, or compare—and why?

Step 2. Consult a variety of resources. Not only should you consult different sources, but they should also be in different media formats: print, online, and possibly audio and video. If you rely on only a single source—one website or one trade journal—your research findings could be inaccurate and misleading. Your company could miss out on a sale, or a costly error could result.

Library resources typically include encyclopedias, almanacs, and other reference works; journals; bibliographies; and collections of books and other media. But also consult industry publications; databases; information services (e.g., LexisNexis; see also Table 8.2, pages 359–360); and in-house publications, blogs, newsletters, reports, customer files, vendor contracts, and institutional archives (see pages 353–378). You may also have to do fieldwork, such as observing an activity firsthand or interviewing a variety of people.

Step 3. Identify and retrieve information. Once you've identified valuable sources of information, you may need to master different techniques for retrieving and using those sources. Information can be stored in a variety of formats and systems in the world of work. Just as the navigation of two websites differs, information from different media is organized according to different technical criteria.

Step 4. Evaluate sources, both in print and online. Given the explosion of information online and in print, you have to be able to evaluate the content of what you read. Be prepared to read tests, surveys, interviews, websites, and printed sources critically to see if the writers have a particular agenda or bias that might slant their opinion on the topic. Whether the source of information is one person's blog, a local newspaper story, or an online journal, here are some questions your boss will expect you to ask:

- Is the information accurate and reliable?
- Who authored this work, and what are this person's qualifications in the field?
- Is the information current? If the information appears on a website, when was the site last updated?
- Is the information complete? What is missing? If the source uses statistics, for example, is the amount of data sufficient to draw the conclusions the author reached?
- Is the information valid? If the source uses statistics, how were those statistics gathered and analyzed?
- Have you located another independent source to verify or corroborate the information?

See pages 375–378 on how to evaluate websites.

Step 5. Confer with appropriate resource people and experts at work, in your profession, and in your community. Doing research in the world of work involves asking interested knowledgeable individuals for information and their perspectives on the topic you are studying. These can be people from different divisions of your company (IT, human resources, finance) or co-workers and members of your collaborative team. You might also consult various specialists who work for the local, state, or federal government. Your research could be as simple as sending an email to a customer to ask a brief question or as complex as analyzing the plans to launch a new product line.

Step 6. Continue to ask questions. Research on the job involves more than gathering information. It also means asking the right questions at each stage of your investigation. As you read, conduct an interview, make a site visit, send emails, or search databases, you may encounter dead ends, contradictions, and even new sources or leads you need to investigate. Research does not always go as smoothly as you might expect it to. Don't get discouraged. Understand that such hurdles are temporary, and see them as opportunities to make sure your work is accurate, complete, and relevant. Investigating your original problem may reveal several others you weren't aware of, or you may discover that there is more than one way to solve a problem. Compare and contrast what you find. Be careful to describe and define the information you gather in terms of your company's purpose. In fact, ask these key questions at various times during your research to keep on track:

- Which of my sources will be of the most value to my audience and my purpose?
- What is the best way to incorporate my research within my report?
- Does the information I have found raise more questions? If so, what is the best way to research them? How do they relate to my company's needs?
- Do I need to find more information? If so, where and why?
- Do I agree with the conclusions presented in my sources? Are there major differences of opinion among my sources? How do I determine which, if any, is more appropriate, effective, and economical?

Step 7. Document your sources. One of the most important steps in the research process is documenting—citing the various sources of information (online, in print, from personal interviews, etc.) on which your report or presentation is based. A later section of this chapter (pages 383–397) will give you specific guidelines on how to document your sources. Remember, to claim another's works as your own is plagiarism. It will undermine you on the job as it does in school. Avoid the mistake of "borrowing" information from an authoritative source or omitting the names of co-workers who assisted you with your report. Readers will inevitably discover the truth, and then you will be considered unreliable if not downright dishonest.

Step 8. Submit specific recommendations based on your research. Depending on the directions your employer gave you, you may be required to give recommendations. These can be informed projections, predictions, alternative measures, specific solutions, or a plan of action. See, for instance, the business report on pages 397–413.

Step 9. Adhere to a schedule. Research for a business report cannot go on indefinitely; Your employer will demand that you meet deadlines. Businesses are run by deadlines; for example, weekly summaries, monthly surveys, quarterly reports. If you delay, you will jeopardize not only other ongoing business projects that may depend on your research, but also your company's chance for acquiring new business. Budget your time wisely. Plan ahead, confer with supervisors on a regular basis, and network with colleagues and others in your field through emails, IMs, blogs, professional networking sites such as LinkedIn, and, when possible, in person.

Two Types of Research: Primary and Secondary

As we saw, you can expect to use many sources of information during the research process. But essentially your research will fall into two categories: *primary* and *secondary*. Both kinds of research are important to help you obtain a better understanding of your topic and provide your supervisor or customers with the careful and complete answers and recommendations they expect. You will often do both types of research, as the marketing report at the end of this chapter (pages 397–413) illustrates. In fact, one type of research sheds light on the other.

Conducting Primary Research

Doing primary research means consulting sources of information not found in printed documents or on the Web. It involves interacting directly with people, places, and things, and it is often done in the office, in the field, or in a laboratory. This type of research often requires gathering information from customers, clients, or other individuals who rely on your company's products or services. Here are some examples:

- An Internet provider may want to know what customers think about a plan to redesign the company's website, a business problem that cannot be solved by reading existing print and online documents. To find out whether or how to redesign its website, the company has to conduct primary research through a customer survey or focus groups (group interviews).
- A city planner may want to find out what residents think about constructing a new park in their neighborhood. To determine their views, the city planner needs to conduct primary research by interviewing residents and/or making a site visit to the neighborhood to observe the best place to locate the park.

Doing Secondary Research

Secondary research involves consulting existing print and online sources. When you conduct secondary research, you work with materials that someone else—an expert in your field, a government agency, even a competitor—has published, posted, or distributed. Consider the following:

- Your boss might ask you to write a report on the latest techniques in occupational therapy given to employees injured on the job so you can develop a similar program for your company. This assignment requires you to locate, read, and summarize the literature in this field.

■ Your advertising department might need to learn more about how and why a competitor's product is winning a larger share of the market. To find out, you might have to study the competitor's website and product literature as well as articles in newspapers and trade journals written about the product. You might also need to gather statistical information on customer buying habits (often provided by professional associations or the government), look at relevant websites in which customers provide feedback about the competitor's products, and consult your company's own archived consumer and competitive studies.

Methods of Primary versus Secondary Research

Here are some examples of the different methods of doing primary and secondary research on the job:

Primary	Secondary
making direct observations	evaluating websites
performing tests	searching databases
going on site visits/inspections	reading books, journals, and magazines
conducting interviews	consulting manuals and reference works
coordinating focus groups	examining product reviews
sending and analyzing surveys	using government documents

Using Primary and Secondary Research at B & L Stores

The following sections of Chapter 8 describe the various primary and secondary research materials and how you use them in your workplace research. But before you look at these sections, read through Figure 8.1, an explanation of how and why research is done in the workplace, written from the perspective of Shay Melka, a vice president for research and marketing at B & L Stores, a large retail chain with locations in North America and Asia.

The overview demonstrates how essential workplace research is to the day-to-day operations as well as the long-range goals of a company. Melka describes many types of both primary and secondary research and often a wide range of examples and suggestions that apply to almost any business.

As you read the discussion of the tools and strategies of primary and secondary research later in this chapter (and the report at the end of this chapter), refer to Figure 8.1 to see how Melka's company has successfully employed them to meet customer needs, increase and improve business, and troubleshoot problems.

FIGURE 8.1 Advice from a Pro: Why and How Research Is Used in the Workplace

How Research Is Conducted at B & L Stores
Shay Melka

I have worked for twelve years with B & L Stores, first as a salesperson, then as an assistant manager, and now as a vice president in our Research and Marketing Division. B & L has over 300 stores in the United States, Canada, the Caribbean, and Asia. In today's highly competitive world of retail sales, we must track the buying habits of our customer base so that we are prepared to act quickly when buying trends change or when demand is no longer ahead of supply. We need to understand who our customers are by finding out about the following:

- What they purchase
- How they pay for their purchases (cash, credit card, check, debit card, etc.)
- Why they shop with us
- Why they purchase from the competition
- How much they purchase
- Whether they purchase their products in the store or online
- What the in-store experience is like for our customers
- How many times during the year our customers visit our locations
- How advertising campaigns affect current promotions

To identify the buying decisions of our current as well as future clientele, we use some basic strategic research tools. Here is a rundown of these tools, or what we call our "outreach programs in customer logistics," with explanations of why, when, and how they work.

Power Think Sessions

This is the name B & L gives to in-house networking, in person and electronically, to do market research, gather vital statistics, and identify key performance indicators. Our weekly Power Think sessions help us obtain essential, current information from sales staff and store managers to highlight the accomplishments of the week, identify problems, and welcome proposals for change. (At one such Power Think session, a B & L part-time employee gave us the idea for our distinctive gift wrap.) We also have larger Power Think sessions several times each year, when B & L brings the company's top performers together to brainstorm at the home office. These sessions are usually composed of a cross section of sales staff and managers within our chain, representing urban and rural stores, stores within higher and lower

(Continued)

FIGURE 8.1 (Continued)

per capita income areas, and stores in large mall and strip mall locations. Each Power Think team submits marketing ideas and business plans. We have also had Power Think sessions with consultants from marketing firms outside the company with experience in information gathering.

Monthly Business Reviews

Sales results are, of course, the driving force for every aspect of the B & L Stores' P & L (that is, our profits and losses). Each store manager sends the stats (sales, returns, losses, inventory, etc.) to the home office daily and discusses the store's performance at weekly meetings. While we review numbers on a daily and weekly basis at the home office, B & L, like most corporations, also performs monthly business reviews (MBRs) to answer the following key questions:

- Are sales up, down, at the same level for the last month, quarter, year?
- What is selling and what is not?
- What is our gross margin of profit (what we paid for the item vs. what we sell it for)?
- What are our additional revenues (from warranties, installation, deliveries)?
- Are inventory dollars being wasted on product A when product B is the hot seller?
- What merchandise is selling better in various markets?
- Do we need to negotiate new lease terms for a store with a landlord?
- Should older stores be remodeled or closed?
- Is one store strong in cash sales and another in credit sales?
- Does the traffic in one store demand more payroll dollars?
- Do store hours have to be revised, e.g., stay open later, close earlier?
- Are there loss prevention issues at a store (e.g., cash, merchandise)?

The numbers we receive, and then analyze in our monthly business reviews, help us to answer the questions above. We also gather statistics on individual performance at all levels—from our part-time sales staff to our district managers and vice presidents.

Customer Surveys

We never make a business decision based just on in-house networking and weekly, monthly, and quarterly sales data. In-house networking and statistics are, of course, key sources of information and help us plan. But before we can put any plan into action we need to know what our customer base thinks of our stores, our Web presence, our merchandise, our employees, our service,

FIGURE 8.1 (Continued)

and our corporate image. To ensure a positive shopping experience in-store or online, we rely on feedback from both our current and prospective customers.

To ensure customer satisfaction, we have developed and tested a variety of survey questionnaires available at our in-store locations and online. For example, several years ago we created a generic survey used across our chain to find out customer opinions about the physical appearance of our stores (Are they clean, attractive, and easy to navigate? Is there enough parking?) and the professionalism of our sales associates (Are they courteous, knowledgeable, and attentive?).

In addition, because so many of our customers have been concerned about identity theft and other transaction issues, we modified our survey to include five basic questions on this key issue. Based on the survey and our marketing research, we prepared a policy statement displayed in every store and on our website that lists the steps B & L takes to prevent identity theft. Here is a brief rundown of what B & L promised to do for each customer based upon the feedback we received from our survey:

- Signatures on a credit card are always matched with those on valid IDs.
- Checks are always processed through the Safeguard Approval System.
- Large purchases are always reviewed by district managers.
- Large bills are always verified using a counterfeit pen.
- If a customer uses a private label credit card (Walmart, Kmart), the credit department is always contacted to verify address, phone number, and payment account.

We have also constructed other distinctive surveys for different markets and regions (e.g., those dealing with new store openings, mergers, overseas stores, new Web links, etc.). Customer surveys provide crucial information about how we can provide better customer service. In fact, we have a Customer Support department where any problems or questions raised by our surveys are reported and corrected or answered.

Focus Groups

Focus groups also provide us with informed opinions. We select participants for our focus groups very carefully at B & L. They come from three different areas: (1) our most loyal current customers as well as prospective customers who have filled out a form, online or in print, indicating that they would be willing to participate, (2) individuals who have agreed to be in focus groups for other organizations and may therefore be willing to participate in our groups, and (3) paid consultants. Recommendations gathered from these various focus groups are digested, analyzed, and implemented as needed.

(Continued)

FIGURE 8.1 (Continued)

Many organizations convene focus groups annually, but at B & L we like to conduct focus groups four or five times a year to help us more confidently forecast trends and to test new products and merchandise in a given market. Our investors (shareholders) and board of directors love focus groups because they allow us to sample the opinions of the ideal consumer we want to reach. Beyond doubt, our focus groups help B & L gain a better feel for the market and for our customer base by supplying us with a reality check on existing policies and merchandise as well as new directions our company will take.

Focus group meetings are recorded, whether the group is held in person or via teleconferencing. These meetings are presided over by a moderator who tries to keep the participants from falling into "groupthink," that is, being influenced by a majority of the group rather than voicing an individual opinion that might, in the long run, be very valuable for B & L's future business. Focus groups do B & L a disservice if all they do is give us "yes, great" answers. We want to hear negative criticism in order to remedy a bad situation or solve any problems. The information from a focus group is used to improve customer relations, advertising, online and in-store shopping experiences, merchandise/ordering/stocking, and pricing issues.

Trade Publications, Message Boards, and Blogs

B & L employees and managers also must read the trade journals and professional magazines that are distributed throughout the B & L chain. These publications release sales and real estate news, merchandise trends, new regulations, vendor information, and the like. Message boards and blogs also help our employees stay current. For instance, individuals in our Retail Architecture department consult online sources and print trade publications to stay up-to-date on trends in displaying merchandise, whether that involves the arrangement of items, mood lighting, or traffic patterns.

Conferences, Seminars, and Workshops

B & L routinely sends staff and managers in Human Resources, Loss Prevention, Diversity, and other departments to conferences, seminars, and workshops. Various departments also attend conventions throughout the year to confer with experts in the field and then make formal reports to the home office, district managers, and relevant store sites. Employees in Human Resources, for instance, regularly attend conferences to find out about the latest federal rulings and guidelines dealing with such matters as overtime, the Americans with Disabilities Act, discrimination, family medical leave, and military leave.

Research is intensive, ongoing, and essential at B & L Stores. Gathering, analyzing, and applying information are in a very real sense our most important business.

Primary Research

The key to primary research is planning. You can't set off to do primary research hoping interesting and relevant information will just pop up. Instead, you have to follow procedures just as scientists do, formulating hypotheses and deciding how best to research them, systematically conducting the research, and then compiling and interpreting the information you have gathered.

There are several methods of doing primary research, including the following:

- direct observation, site visits, and tests
- interviews and focus groups
- surveys

Direct Observation, Site Visits, and Tests

Direct observation, site visits, and tests are widely used methods of primary research that involve actively observing people, places, and things. Your company or organization will expect your findings and conclusions to be based on keen powers of observation.

Direct observation is seeing what is right in front of you—for instance, watching how an individual performs a task, determining how a piece of equipment works, or studying how a procedure is performed. The key to conducting effective research is observing actively, not passively. For example, if you work in the sales department and have been asked to write a report on how your company can improve its relationships with its customers via telephone sales calls, you need to

- observe how reps interact with customers
- listen to the reps' tone of voice
- record how clearly they explain product or service benefits

Site visits require you to use the same keen attention to detail that you use in direct observation, except you will need to go to an off-site location to report what you find there. A site visit could take you to another department in your company, a prospective customer's office, the scene of an incident or accident, or an agricultural or manufacturing location relevant to your business report. See Figure 14.13 (see pages 677–678) for an example of an incident report based on visiting the site where a railroad accident occurred. Regardless of the location, you will have to describe for your boss precisely what you witnessed firsthand. For instance, if your company—say, a chain of dry cleaners—wanted to open a new store and asked you to investigate the most profitable locations, you would need to conduct a great deal of primary research based on site visits. To select the best location, you would have to first identify relevant places for a new store and then find out such things as

- volume of traffic
- availability of parking
- location and size of competing dry cleaners in the area
- zoning ordinances and restrictions

Figure 14.9 (pages 667–668) contains an example of a trip report about opening a new restaurant based on information obtained from a site visit.

Conducting *tests* is another productive way to do primary research involving the observation of people, places, conditions, and things. A test can be as simple as examining two pieces of comparable office equipment side-by-side and noting how they compare, or trying out a new email marketing strategy. Or it can be as scientifically demanding as conducting a laboratory test. For example, let's say you work for a company that manufactures waterproof sunscreen lotion and you have been asked to find out why a competitor's product is outselling yours. To answer the question, you will have to arrange for laboratory tests of your competitor's product to determine what ingredients it contains, how effective and safe it is compared to your product, and how its advertising and packaging may affect customer decisions. Figures 14.11 (page 673) and 14.12 (pages 674–676) are examples of reports based on laboratory tests conducted in the world of work.

Sometimes you may have to use all three types of research based on observation when preparing your report, as Kirk Smith did for his water-quality study in Figure 8.2. Not only did he observe and record the data-collection methods used at the three reservoirs, but he also visited these sites and conducted his own tests.

Guidelines for Research Based on Observation

Regardless of the type of observation(s) you make for your research, follow these six guidelines:

- **Always plan ahead.** Determine what problems you have to solve, what questions you have to answer, and what data you have to gather—procedural guidelines, maps to locations, sample products—to answer those questions and solve those problems.
- **Obtain necessary permissions.** Permission may involve verbal or written consent, permits, or visas. By obtaining approval to make observations, you respect the time, privacy, and working conditions of others.
- **Observe company protocol and ethical standards.** Follow your company's or agency's guidelines and standards for observation. In addition, follow the standards and codes of ethical conduct of your profession. Note how Kirk Smith refers to the U.S. Geological Survey's standards in Figure 8.2.
- **Remain objective and impartial.** Although observations can often be influenced by your subjective point of view, make sure you are open-minded as you record your observations.
- **Keep careful notes and records.** Take a laptop or tablet PC, a pen and paper, a digital recorder, a camera, or whatever other device you may need to preserve your observations effectively.
- **Write down precise details.** Be sure to indicate the complete and accurate dates and times you made your observations, any relevant environmental conditions that had an impact on your observations, and any other factors that may have influenced what you observed. See how Kirk Smith recorded the precise methods of water flow/quality measurement in Figure 8.2.

Interviews and Focus Groups

Two other important sources of primary information come from interviews and focus groups. You can do a one-on-one *interview* with an expert in the field, a co-worker, a client, or another resource person. Or you can hold a *focus group*,

FIGURE 8.2 A Report Based on Direct Observation, Site Visits, and Tests

Water Flow and Quality Evaluation of the Cambridge, Massachusetts, Drinking Water Source Area

Kirk P. Smith

The drinking water source for Cambridge, Massachusetts, consists of three primary storage reservoirs (Hobbs Brook Reservoir, Stony Brook Reservoir, and Fresh Pond), two principal streams (Hobbs Brook and Stony Brook), and nine small tributaries. Because previous investigations identified specific areas as potentially important sources of contaminants, several sites were selected for continuous monitoring to address the water supply regulations followed by the Cambridge Water Department (CWD). The purpose of this report is to evaluate the measurement methods used by the CWD.

Reservoir altitude and meteorological measurement were recorded by monitoring stations installed at each reservoir. Water quality measurements of reservoir water were also recorded at USGS stations 01104880 and 42233020. These data were recorded at a frequency of 15 minutes, were uploaded to a U.S. Geological Survey (USGS) database on an hourly basis by phone modem, and were put on the Web at **http://ma.water.usgs.gov**. Stream-stage measurements were also recorded by monitoring stations on each principal stream and at the outlet of the Stony Brook Reservoir. These data were recorded every 15 minutes and were uploaded to a USGS database on an hourly basis by phone modem.

In addition to measurements made on the principal streams, stream-stage and water-quality data were recorded by monitoring stations on 4 of the 9 small tributaries. My visits to these sites and independent water samplings confirm that CWD's measurements comply with USGS standards.

Since the drainage areas of these sites are small and have large percentages of impervious surface, the hydrologic responses, and often the water-quality responses, change rapidly. To document these responses effectively, the monitoring stations have recorded stream-stage and water-quality measurements at variable frequencies as high as 1 minute. These data were uploaded to a USGS database on an hourly basis and are available through **http://ma.water.usgs.gov**. I have found through visits and water sampling that CWD is not only compliant with, but exceeds, USGS standards in measuring drainage area water quality.

Sidenotes:
Gives key background information

States purpose

Explains methods using specific techniques to record accurate measurements

Records data objectively

Identifies variables

Gives conclusion based on tests and observations

Source: Adapted from *Hydrologic, Water-Quality, Bed-Sediment, Soil-Chemistry, and Statistical Summaries of Data for the Cambridge, Massachusetts, Drinking-Water Source Areas, Water Year 2004,* by Kirk P. Smith. U.S. Department of the Interior/U.S. Geological Survey. Open-File Report 2005–1383.

a question-and-answer session with multiple people—both company representatives and customers—in attendance. See how in Figure 8.1 Shay Melka emphasizes the importance of interviews with individual employees as well as with focus groups to gather essential information from and about a variety of customers.

Interviews

Interviews can be conducted in person, over the telephone, or through email, although Skype conversations and face-to-face meetings are the most productive way to generate relevant information. Figure 8.3 contains an excerpt from an

---

FIGURE 8.3 An Excerpt from an Interview Transcript

Asks pertinent background question

Q: Did you have any experience working in another country before going to Hongzhou?

A: Not much really, except for a summer-long internship in Lima, Peru, I had as part of my undergraduate education.

Key question on interviewee's responsibilities

Q: What precisely were your duties when I-Systems transferred you to Hongzhou?

A: I was responsible for setting up a local operation for I-Systems and managing a staff of about eighty Chinese employees. As part of all that, I worked closely with local suppliers and developed different e-markets for our new office.

Logical question after asking about background

Q: How did you prepare for your transfer to China?

A: Before I left for my eighteen-month stay, I profited most from participating in teleconferences with our other Chinese offices and attending China trade fairs in the United States and Canada. I also immersed myself in intensive, but admittedly very basic, conversational Chinese. And, of course, I partnered with several of I-Systems Chinese employees and managers here in Pittsburgh.

Turns to problems in new job

Q: What would you say was the biggest obstacle an American manager might face when working in China?

A: Seeing China through Western eyes.

Asks for clarification

Q: When you say "seeing China," what do you mean?

A: By that I mean looking at China from an American business perspective. We tend to think in U.S. terms about expanding and opening markets, that is,

FIGURE 8.3 (Continued)

what we can do for China. But my Chinese colleagues reminded me about China's impact on American markets. While the United States accounts for only about 5 percent of the world's population, China has about 20 to 25 percent of it and can powerfully influence our company's decisions. Accordingly, we needed to shift our thinking about what China could do for us. To do this, we must have an appreciation of the Chinese way of doing business.

Q: What characterizes the Chinese way of doing business?

A: For example, the Chinese pharmaceutical industry is one of the most powerful in the world. I think there must be over thirty or forty companies that are developing, manufacturing, and marketing product lines both in China and here. Did you know that China makes nearly 10 percent of the world's aspirin? I found that relevant in light of the western perception of Chinese health care—alternative procedures such as acupuncture and homeopathic remedies.

Relevant follow-up question

Q: That is fascinating and a topic for a separate interview sometime. But getting back to characterizing Chinese business customs—

A: Oh, yes. Let me illustrate a couple of major differences I had to get used to. Americans have no problems mixing business and pleasure. In fact, we are famous for the business lunch or dinner. Banquets are great occasions to talk shop, to sell our products, services, and websites. But in China a dinner is strictly a social event, one for entertaining and not marketing. It is considered rude in China to inject talk about sales, quotas, operations, or e-markets at a dinner. If you want to give a banquet as a sign of respect for a Chinese executive, leave your specs, stats, briefcases, notebooks, and cell phones at home.

Directs interviewee back to subject of interview

Keeps interviewer focused on topic

Q: Do you have any other advice for American workers whose companies relocate them to China?

A: Be careful about gestures and gifts.

Asks for further information

Q: Why do you link the two?

A: To illustrate a major blunder, one of my colleagues kept patting a Chinese executive on the back, a sign in America of friendship and approval. Not so in China. It is seen as discourteous.

Good follow-up question

Q: And the gifts?

A: While some business gifts are appropriate, never give a Chinese executive a clock or stopwatch. It signals doom or death.

interview with a U.S. manager whose company transferred her to the company's Hongzhou, China, location for eighteen months. Note how the interviewer researched and structured his questions to help other employees who might be transferred to China.

Follow the process below when you do an interview related to your workplace research.

1. Set Up the Interview

- Determine whether you need to speak with an expert in the field, with a co-worker or manager, or with a client or customer to obtain the relevant information.
- Ask your supervisor or co-workers to help you identify experts or relevant customers you should interview, or consult other sources, such as business directories, client or customer lists, professional organizations, or relevant websites.
- Politely request an interview with the individual at his or her convenience. Be flexible. Your interviewee is giving you his or her time. Always let the individual know ahead of time exactly what you would like to discuss and why you are conducting the interview.
- Whether for a personal visit or a telephone call, specify how much time you will need for the interview. Be realistic—fifteen minutes may be too short; two hours much too long.

2. Prepare for the Interview

- Gather background information about the person you are interviewing as well as about the organization or professional group that he or she represents, if applicable.
- Continue to research your topic so that you have sufficient background information and do not waste time asking the obvious or requesting information that is readily available on the Web or from another source.
- Determine what information you need from the interview to help you solve the problem or answer the questions essential to your report. Be sure to prioritize getting the essential information you need. Be aware, too, that the interviewee may raise some relevant questions you didn't expect but need to follow up on.

3. Draft Your Questions
Prepare your questions ahead of time, and take them to the interview. Never try to wing it. Your questions should be

- focused on the topic you want to find out about to avoid vague answers
- open-ended and designed to prompt thoughtful responses, not just yes or no answers
- objectively worded so that the interviewee is not forced to respond to loaded questions

Here are some examples of poorly written questions with effective revisions:

Vague Question	Restricted Question
How can a website help customers?	In what ways can we improve the navigational signals on our website to help customers find information quicker?

Yes or No Question	Open-Ended Question
Do you think big business is opposed to a healthy environment?	Would you identify two or three ways we could green our office space?

Loaded Question	Objectively Worded Question
Isn't the future of real estate security investments doomed to a bleak future?	What are your thoughts about the future of real estate security investments?

4. Conduct the Interview

- Show up for the interview on time, and be dressed appropriately.
- Always ask if you have permission to record the interview or to take photographs.
- Start off with a few minutes of small talk to give the interviewee time to get comfortable and to talk about his or her interests and accomplishments.
- Never begin with your toughest or most controversial question. Save it for last.
- Stay focused. Don't stray from the topic or delve into personal matters. Should the individual get off the topic, politely return to your questions.
- Be an attentive and appreciative listener. Let the interviewee do most of the talking.
- If you discuss anything confidential, politely remind the interviewee about the sensitivity of the topic.
- If the interviewee does not want to answer a question or has no further information to add, don't press the point. Move to the next question.
- If the interviewee says that something is "off the record," respect his or her request and do not include it in your transcript or notes, or on tape.
- At the end of the interview, allow time for your interviewee to clarify any of his or her responses.
- Thank your interviewee and ask if you might call back with any follow-up questions or further clarifications.

5. Follow Up After the Interview

- Don't wait too long after the interview to review your notes or start your transcription. It's best to read through your notes immediately after the interview, while the conversation is still fresh in your mind. That way you will be better able to understand your notes when you return to them later.

- Thank the interviewee by letter or email within a day or two following the interview.
- If the interviewee requested a transcript of the interview, send it to him or her for approval.
- Determine how much of the interviewee's comments, ideas, statistics, or recommendations you can incorporate into your ongoing research. You cannot put the entire interview into your report. Your reader wants only the bottom-line essentials.
- Always request permission to quote anything from the interview in your report or presentation to your company or clients.

Focus Groups

Focus groups are typically made up of loyal or prospective customers who have been invited to give a company their opinions about a specific product, service, or future project. A company might also include paid consultants and even individuals selected from competitors' lists, as in Figure 8.1. Focus groups are used to obtain a wider variety of opinions than individual interviews may give and they are more personal and interactive than surveys. Businesses rely heavily on these groups to get honest, well-considered feedback from interested individuals and to incorporate that feedback into their research. Focus groups are usually conducted in face-to-face meetings, but virtual meeting technologies (pages 117–118) allow people outside of the area, even globally, to participate.

Follow the guidelines below to conduct a successful focus group:

1. Set Up the Focus Group

- Determine the location, time, length, topic, and agenda of the focus group prior to contacting potential participants.
- Identify who should be invited to the focus group and how many individuals should make up that group. Effective focus groups usually consist of six to twelve participants to get a diversity of opinions but keep the group from being too crowded and unmanageable. Consult with your supervisor, the company's sales and marketing departments, and customer lists to locate the most helpful participants or consultants.
- Screen potential participants for appropriateness, e.g., knowledge of the product or service, willingness to provide input in a group, and so on.
- Once you decide on the participants, provide them with all of the details they need about the location, payment or reimbursement, and topics to be discussed.

2. Prepare for the Focus Group

- Determine the specific questions you will need to ask the focus group. As in a one-on-one interview, prepare your questions ahead of time, avoiding vague, yes or no, or loaded questions. Limit the number of questions to allow for ample discussion time.

- Plan to record the focus group and to bring in a co-leader or moderator to take notes. Unlike in a one-on-one interview, you will not be able to take effective notes while leading a focus group.
- Create an information form for the participants to fill out at the beginning of the meeting, asking for contact information and which of your company's products or services they use or are familiar with.

3. Conduct the Focus Group

- At the beginning of the meeting, establish reasonable ground rules, such as the importance of staying on topic, speaking in turn, and meeting the goals of the group (see "Sources of Conflict in Collaborative Groups and How to Solve Them" on pages 92–95).
- Politely remind the group about confidentiality. Many companies have participants fill out confidentiality agreements before the group meets.
- Stick closely to the agenda. Don't stray off topic yourself or allow participants to do so.
- Allow sufficient time for each question, and ask participants for their recommendations. Then quickly summarize those conclusions for the group to see if participants agree with your assessment.

4. Follow Up After the Focus Group Meets

- Read through your notes soon after the meeting, while the group experience is fresh in your mind. Also, write down any observations about the group dynamic as a whole and about the individual participants, since this information may affect your results.
- If the meeting was recorded, transcribe the recording.
- Thank the participants again by letter or email within a day or two after the group meets.

Use of Social Networking Sites as a Recruiting Tool   Many researchers find social networking sites such as Facebook and Twitter to be a useful tool for recruiting participants for research projects and for studying consumer trends. Organizers of focus groups, for example, often use Facebook as a recruiting tool. By creating a Facebook "event" for the research project, researchers can advertise the focus group session, distribute information about the kind and importance of research, and even begin to collect data about participants, all from the same site. Facebook is also useful for market researchers who need to find information about new products, trends, fashions, or technology. This type of research is especially useful when your goal is to gather facts and figures from within specific communities. Because Facebook allows users to create groups around virtually any topic, researchers can find a ready-to-use data set by simply joining the group and following the posts already there.

Surveys

Surveys are among the most frequently used instruments for gathering primary research in the world of work. As Shay Melka points out in Figure 8.1, B & L Stores relies heavily on surveys to conduct its business. Think of a survey as an interview with a relatively large number of people. The goal of a survey is simple—to collect and then quantify information about people's attitudes, habits, beliefs, product loyalty, knowledge, or opinions using a survey questionnaire. You can conduct a survey over the phone, online, or by mail.

Case Study

The WH eComm Survey

Many online vendors ask customers, after a purchase, to rate their online shopping experience. Online customer feedback not only helps e-commerce companies learn about the level of their customers' satisfaction, but it also helps them find out about customer preferences to make crucial business decisions.

Note how the WH eComm survey in Figure 8.4 asks both types of questions. Some questions ask about customer preferences (questions 1, 2, 5, 6, 8, and 9 fall into this category), while others ask about customer satisfaction (questions 3, 4, 7, and 10). The questions about customer preferences can be used to help the company decide where to advertise ("How did you hear about our website?") and to determine which products to promote ("What types of products have you purchased from WH eComm?"). The questions about customer satisfaction, meanwhile, elicit information to help the company improve its service by assessing such things as the usefulness of its website ("Is the website easy to navigate?") and the quality of its customer service ("How helpful did you find our customer service?").

In order to ensure that a meaningful number of customers replied to the survey, WH eComm made the questions easy to answer by simplifying the options as well as the format of the questions. Moreover, asking only ten questions and providing an incentive (a 5% discount on the next purchase) also encouraged customers to respond.

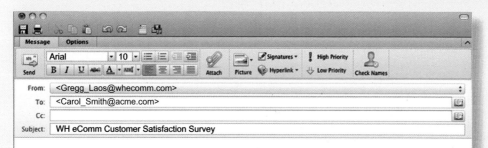

FIGURE 8.4 An Example of an Online Survey

From: <Gregg_Laos@whecomm.com>
To: <Carol_Smith@acme.com>
Cc:
Subject: WH eComm Customer Satisfaction Survey

Dear Valued WH eComm Customer,

At WH eComm, we are committed to providing our customers with high-quality e-commerce software through an efficient and user-friendly website. Customer feedback is extremely important in helping us continue to improve our web presence. So that we may best meet your needs, please answer a short survey about your experience at WH eComm. You will find the survey on our website by clicking here. Your answers will help us continue to improve WH eComm online and to offer you the efficient service you deserve.

To say thank you for filling out this short questionnaire, we want to offer you a 5 percent discount on your next purchase. When you have completed the survey, your discount will automatically be recorded in your WH eComm online account.

Many thanks for your time and your confidence in us,

Gregg Laos
Manager
WH eComm

Cover email explains why survey is important to customers

Provides incentive to reply

WH eComm
revolutionizing e-commerce

1. How many times have you visited our website?
○ First visit   ○ 2–4 times   ○ 5–7 times   ○ More than 7

2. How did you hear about our website?
○ Colleague
○ Advertisement in business journal
○ Another website
○ Search engine
○ Other (please specify) _____

3. Is the website easy to navigate?
○ Very easy   ○ Easy   ○ Somewhat easy   ○ Not easy

Easy-to-read format

Fill-in option does not require a lengthy answer

Question is not phrased in a leading manner

(Continued)

FIGURE 8.4 (Continued)

One question can elicit a great deal of consumer information

4. How effective did you find the following sections of WH eComm?

	Extremely effective	Effective	Could be improved	Ineffective
Web features	○	○	○	○
Search	○	○	○	○
FAQ	○	○	○	○
Online checkout	○	○	○	○

5. How many times have you purchased our products?
○ 1–2 times   ○ 3–4 times   ○ 5–7 times   ○ 8–9 times
○ More than 9 (please specify) [   ]

Multiple-choice items clearly differentiated

6. What types of products have you purchased from WH eComm? (Check as many as apply.)
○ E-commerce software   ○ Web design software
○ Networking software   ○ E-conferencing software

7. How helpful did you find our customer service?
○ Extremely helpful
○ Helpful
○ Could be improved
○ Not helpful
○ No response

Provides non-overlapping choices

8. How have you most often contacted our customer service center?
○ Phone   ○ Email   ○ Fax   ○ Web

9. How soon was your query answered?
○ Same day   ○ Next day   ○ Within 3 days   ○ Within a week   ○ Longer

10. How satisfied were you with the speed and efficiency of our customer service center?

Limits options

○ Very satisfied   ○ Somewhat satisfied   ○ Dissatisfied

11. Please rank, in order of importance, which factors most influence your online purchases.

	Price	Shipping options/time	Returns policy	Website quality
Most important	○	○	○	○
	○	○	○	○
	○	○	○	○
	○	○	○	○
	○	○	○	○
Least important	○	○	○	○

Ranking question supplies all necessary options

12. What would you most like to see changed or improved on our website?

Provides opportunity for respondent to elaborate

[▲▼]

Thank you for taking the time to answer our questions.

<u>Home</u>

Five Steps for Using a Survey

There are five basic steps you need to follow when using a survey as a part of your research on the job:

1. Determine the best way to deliver the survey.
2. Create the survey questionnaire.
3. Choose the survey recipients.
4. Reach the survey recipients.
5. Compile and analyze the survey results.

Determine the Best Way to Deliver the Survey   Surveys can be conducted over the phone, online, or by mail. Each medium entails various advantages and disadvantages, as outlined in Table 8.1. Decide which medium you think will yield the best results and will work within your time frame and budget. For instance, if your top priority is to reach a very large audience, conduct an online or mail survey rather than

TABLE 8.1   Advantages and Disadvantages of Different Ways to Conduct a Survey

Method	Advantages	Disadvantages
Telephone	• Instant results • Ability to ask respondents for clarification • Low incidence of skipped or partial responses • Ability to screen out inappropriate respondents • May be longer than online or mail surveys	• Inability to reach a large population • Time-consuming • Expensive • Possibility of hang-ups (risk of incomplete responses) • Lack of anonymity can discourage objective, candid responses • Respondents do not have time to carefully consider their responses
Online	• Ability to reach a large population • Fast results • Not time-consuming • Inexpensive • Anonymity encourages objective, candid responses • Respondents have time to carefully consider their responses	• Low response rate • Inability to probe respondents for clarification • High incidence of skipped or partial responses • Inability to screen out inappropriate respondents • Must be kept brief
Mail	• Ability to reach a large population • Less time-consuming than telephone surveys • Inexpensive • Anonymity encourages objective, candid responses • Respondents have time to carefully consider their responses	• Slow results • Low response rate • Inability to probe respondents for clarification • High incidence of skipped or partial responses • Inability to screen out inappropriate respondents

one over the telephone. If you need to receive detailed answers about a restricted or highly sensitive subject, conduct a telephone survey, which gives you an opportunity to talk directly to the respondents and allows them to clarify their answers. But if your goal is to obtain results quickly and inexpensively, do an online survey.

Create the Survey Questionnaire   The most crucial part about writing a questionnaire is crafting valid and accurate questions. There are many types of questions you can ask, as illustrated in Figure 8.4, including yes/no, ranking, rating, multiple-choice, and open-ended questions. Researchers advise asking questions that require the least amount of effort on the part of the respondents (yes/no, multiple-choice) to increase their chances of answering your questionnaire. Accordingly, ask open-ended and rating or ranking questions sparingly. In addition, keep your survey to ten to fifteen questions, both to encourage your recipients to reply and to stay focused on those questions that matter most to your company or organization. Again, note how the WH eComm questionnaire in Figure 8.4 includes only twelve key questions, which the company needs to have answered to help it make important decisions. Finally, design your questionnaire to look inviting and streamlined. (See pages 528–535 in Chapter 11 for guidelines on document design.)

Here are some guidelines for writing specific questions to help you get the results you want—whether you are writing a mail or an online questionnaire or preparing a script for a telephone survey.

1. Phrase questions precisely. Vague questions only elicit answers that you cannot use or will be unable to analyze. Use valid, quantifiable questions.

> Ineffective: Are we open enough hours on Saturdays?
> Yes____ No____
>
> Better: How many hours would you like us to be open on Saturdays?
> 4____ 5____ 6____ 7____ 8____

2. Ask only one question at a time. Avoid multiple questions within the same question, since you will not know the exact answer to each question.

> Ineffective: What is your overall impression of our customer support and delivery services?
> poor___ fair____ good____ very good____ excellent____
>
> Better: (Turn the two questions above into two separate queries as follows.)
> What is your overall impression of our customer support service?
> poor___ fair____ good____ very good____ excellent____
> How would you rate our delivery service?
> poor___ fair____ good____ very good____ excellent____

3. Clearly differentiate each option in multiple-choice questions. If respondents are not sure of the differences among options, they may answer inappropriately because of question overlap, or they may skip the question altogether.

> Ineffective: When is the best time to call you?
> Daytime___ Afternoon___ Weekday___ After work___ Evening___ Night___
>
> Better: When is the best time to call you?
> Morning (8:00 a.m.–noon)___ Afternoon (noon–5:00 p.m.)___
> Evening (5:00 p.m.–10:00 p.m.)___

4. Supply all of the necessary options in multiple-choice questions. If you omit an important option, respondents may choose a misleading answer or not answer at all.

> Ineffective: Which types of nonalcoholic beverages would you like us to offer?
> soda___ juice___ coffee/tea___ milk___
>
> Better: Which types of nonalcoholic beverages would you like us to offer?
> soda___ juice___ coffee/tea___ milk___ bottled water___ other (please specify)___
> (The "bottled water" and "other" options give respondents a fuller range of answers.)

5. Do not use unfamiliar jargon or abbreviations. Don't assume that respondents will understand the jargon your company or profession uses.

> Ineffective: What was your overall impression of the CGI in this film?
> poor___ fair___ good___ very good___ excellent___
>
> Better: What was your overall impression of the computer-generated imagery used in this film to create the global village scene?
> poor___ fair___ good___ very good___ excellent___

6. Do not ask inappropriate questions. Refrain from asking questions about income, education level, or other personal matters such as age, ethnicity/race, gender, disability, religion, or sexual orientation unless these questions give you essential demographic information directly relevant to the topic of your survey.

7. Avoid leading or biased questions. Do not give your respondents slanted questions that bias their answer and thus the results of your survey.

> Ineffective: Were you impressed by this award-winning product?
> Yes___ No___
>
> Better: Did you think this was an award-quality product?
> Yes___ No___

8. Limit multiple-choice and ranking items to five items. The more complicated your list of multiple-choice or ranked items, the more difficult it will be for your respondents to give a clear and helpful answer and for you to analyze the survey results.

9. Limit rating ranges to a scale of 1 to 5. As with item 8 above, do not complicate your survey by providing a scale with such a wide range of options that respondents are unclear about how they differ or overlap.

Choose the Survey Recipients The success of a survey, of course, depends on targeting the right audience and in the right numbers. Sometimes that audience is small and easy to reach. For example, you might survey people within your own company, agency, or department (all of the nurses in ICU). But more often, the group you want to survey—all of your California customers or vendors, for example—is so large that you could not possibly survey the opinions of every member of that group. In that case, you need to gather information from enough people to make reliable and relevant judgments about the larger population, and you have to select a representative cross section of individuals from the larger group (by age, gender, background, experience, education, etc.).

For example, let's say you want to do a survey on the quality of food in a school cafeteria in order to write a report to the superintendent. You would need to obtain the opinions of a representative sample, or cross section, of people in that school. A *cross section* is defined as a preselected group that is representative of the population as a whole. Nielsen Media Research, **www.nielsenmedia.com/nc/portal/site/ Public/**, which rates the popularity of television shows in the United States, conducts its surveys in this way. Obviously, Nielsen cannot survey millions of television viewers, so it selects a representative sample of 5,000 households to determine TV ratings.

Reach the Survey Recipients   Don't expect all of your respondents, or even 40 or 50 percent, for that matter, to reply to your questionnaire. Researchers find that a response rate of 12 percent from a statistically chosen sample group is still valid. But to increase the chances of receiving replies from as many respondents as possible, follow these time-tested procedures:

- Provide a cover letter or email, as in Figure 8.4, asking recipients to reply and thanking them in advance for doing so.
- Offer respondents some incentive to answer the survey, such as the discount that WH eComm promises in Figure 8.4.
- Indicate whether respondents should identify themselves or remain anonymous.
- Clearly specify how the respondents are to answer the questions—using a check mark, circling the correct response, writing in a number, or just pointing and clicking.
- If you mail your questionnaire, provide each recipient with a stamped, return-addressed envelope to increase the chance of a reply.

Compile and Analyze the Survey Results   The final step in conducting a survey is compiling and analyzing the results. The task of analyzing survey data in the workplace is often left to experts, who employ complex statistical methods (such as those listed at **www.spss.com**) to arrive at significant conclusions. However, if you are asked to analyze survey data, especially if your boss wants you to survey a manageable (small) number of responses, here are some helpful tips to follow:

1. **Keep your completed surveys organized, and save them.** Assign each completed survey a number or code so that you can easily differentiate similar-looking responses from one another. Don't throw away the completed surveys; you may need to refer to specific answers later, or your company or department may need to archive all surveys.

2. **Create a data sheet.** Originate a data sheet—for instance, an Excel spreadsheet— so that you can record all of the survey responses in one central document. Break the data sheet into logical categories, for instance, separate rows for each survey question and separate columns for each possible answer.

3. Record responses completely and accurately. Make sure you include the responses to all of the survey questionnaires on your data sheet. Always record responses exactly as you receive them. If the response to a question is blank or illegible (or gives several answers to the same question), discard that response rather than making something up or guessing what the respondent meant. As you record each response to each question, be careful that you don't record the same answer or survey twice.

4. Present your findings clearly and effectively. To help your boss or other readers understand your findings, create one or more simple tables in which you present key information in an easy-to-read format. Also supply a blank sample questionnaire for reference.

Secondary Research

As we saw earlier (pages 331–336), secondary research requires you to consult sources that are already available (books, periodicals, reference works, websites, blogs, etc.), as opposed to interacting directly with people, places, and things via direct observation, site visits, tests, interviews, focus groups, and surveys for primary research. Secondary research involves gathering documents and reading, summarizing, and incorporating them into your report. As with primary research, your secondary research involves planning and organizing your time to concentrate directly on the problem or issue your company expects you to investigate.

Libraries

Libraries are essential to conducting secondary research. To get started in your workplace research, you need to know about the different types of libraries you can use and the various services they offer, such as online catalogs and periodical databases. You will also need to learn how to use effectively the vast materials available on the Web, such as government documents, sites posted by professional organizations, and a host of company websites and blogs. Let's begin with an overview of different types of libraries and how and why they can help you in your research.

As part of your workplace research, you can expect to use one or more of the following types of libraries:

- corporate libraries
- public and academic libraries
- e-libraries

Corporate Libraries

One of the fastest, easiest, and most profitable ways to locate and collect crucial research data about your company is to consult your company or agency library. Depending on the size of your organization, your corporate library may be housed in a small room with file cabinets and a computer (with or without a full-time librarian) or a large, fully staffed library, such as those maintained by multinational pharmaceutical companies, law firms, brokerage houses, and other megacorporations.

Why Corporate Libraries Are Valuable for Research   Corporate libraries contain a vital history of a company's activities, past and present. Using your company's library, you can obtain much of the sales, historical, competitive, and other data you may need for your business report. A company library also houses business-specific and confidential documents not found in a public or academic library. Moreover, corporate libraries include documents often referred to as grey literature (see the Tech Note on page 368). If your company or agency does not provide or sponsor its own physical in-house library, most likely it will at least have a password-protected archive that you can access through the company's intranet (see the Tech Note on page 355).

Types of Research Materials in a Corporate Library   Regardless of the size of your company, the following documents and information are likely to be found in its corporate library, in its intranet archive, or in departmental files (e.g., legal, marketing, etc.):

- client and customer records
- corporate reports, studies, surveys, and proposals
- corporate newsletters—from inception to the present
- maps, diagrams, blueprints/specs
- legal records, including patents and contracts
- financial and operational information organized by month, year, or longer cycles
- books and trade periodicals directly related to your company's research and business
- product and service literature (catalogs, descriptions, technical specifications, training manuals, warranties, etc.) (See Tech Note on grey literature, page 368.)
- competitor information, including comparative analyses, competitor catalogs, and sales information

As Shay Melka points out in Figure 8.1, B & L Stores uses many materials from its corporate library or available through numerous divisions of that company. While company libraries and archives offer a wealth of valuable research information, keep in mind the following guidelines when you use them:

1. **Don't expect to find everything in one central location.** As efficient as a corporate library may be, you won't necessarily find every company document you need there. You may have to go to individual departments or branches to locate materials, particularly sensitive legal documents that for ethical reasons are available only through permission.

2. **Don't assume the library will be organized like a public or academic library.** A corporate library may use the Dewey decimal system, or it may not. Become familiar with your company library's location, checkout policies, references, and system of coding and classification.

3. **Use corporate library materials with discretion.** Recognize that much of your company's financial, statistical, or competitive information may now be irrelevant to your purpose, out of date, or overly optimistic or pessimistic, depending on the business climate in which it was written.

4. Supplement information from your corporate library with information from other sources and locations. Materials found in a corporate library may take you only part of the way toward what you need to learn. Expect to supplement what you find in your corporate library with primary research, Web research, and trips to public or academic libraries.

Public and Academic Libraries

In the world of work, do not neglect the vast resources and assistance provided at public and academic libraries. While public libraries are available to everyone, you may need to be a resident to access particular resources. To use an academic library, you will have to be enrolled at or work for the institution that owns it.

When you use public or academic libraries, you benefit from the expertise of professionally trained librarians who have selected, organized, and recommended authoritative resources. In person or online, a librarian can help you with your research by directing you to appropriate sources (page 360) and assist you in formulating your online searches (pages 369–375). A few minutes of conversation with a librarian before you begin your search can save you a great deal of time and effort. Also, a librarian can identify the most relevant databases, subject directories, professional journals, and trade magazines, in print and online, to help you accomplish your workplace research.

To start your library research, access the library's home page for a full range of its services and for directions on how to conduct a search. Figure 8.5 shows the home page for one academic library, including links for reaching a librarian,

Tech Note

Intranets

To share and archive information, diversified companies, government agencies, and many large corporations have created intranets. As the name suggests, intranets are internal communication networks modeled after the Internet. They use some of the tools and procedures found on the Web—passwords, directories, search engines, and multimedia content. Like the Internet, companies use intranets to post documents, coordinate calendars, hold virtual meetings, conduct training sessions, make announcements, and post newsletters. (See Figure 4.9, page 154.)

Unlike the Internet, which is available to everyone, an intranet is designed for internal use by the management and employees of the organization or company. Coordinating the content of various groups and departments, network administrators and other editors oversee the company's intranet. From a centralized directory, information is sent to, from, and within various divisions within the company—management, engineering, sales, human resources, environmental safety, public relations, and so forth. Documents might be designated as available to all employees or restricted to certain individuals only, depending on the audience for and content of the document. Some files may, therefore, be closed to you because of confidentiality. Intranets are protected behind firewalls so that unauthorized individuals cannot gain access to them.

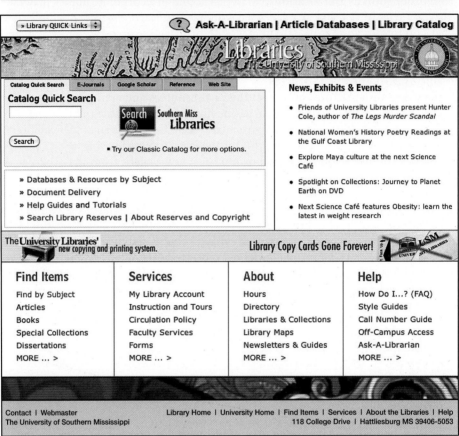

FIGURE 8.5 An Academic Library's Home Page

Courtesy of the University of Southern Mississippi

accessing electronic resources, and locating relevant documents. Note how the library has made it easy for patrons to connect with the right department for assistance with their research. To make searching even more efficient and convenient, many libraries belong to a network, regional or global, of participating libraries, enabling patrons to access the catalogs of member libraries in the group. For instance, WorldCat, a universal catalog of resources, lets you know what public and academic libraries in your area or the world over own a particular book.

To conduct your research, you'll need to know how the library's online catalog works. At public and academic libraries, the online catalog lists all of the materials the library owns, subscribes to, or has access to. See the library home page in Figure 8.5, which lists reference works, databases, websites, audio and video recordings, visuals, and the individual journals and magazines (electronic and print) to which the library subscribes. Starting with a library's online catalog will make your research easier because of the powerful search options available, as well as the various databases the library subscribes to.

Here are some guidelines to help you access an online catalog:

1. Know your topic. To get background information on your topic, start with an encyclopedia or other general reference work so that you are aware of the various issues and subtopics involved.

2. Restrict your subject. When you search a public or academic library's catalog, you need to narrow your topic to find a manageable amount of information. Suppose you are researching the use of lasers in plastic surgery. You might start your search with "medical uses of lasers." To narrow your subject, you might specify "use in cosmetic surgery." You could then further refine it by specifying "in ophthalmic cosmetic surgery."

3. Take advantage of links. An online catalog usually provides links to related materials that may be even more helpful to your search. Use the "Previous" and "Next" links to navigate through sources on the same topic.

E-Libraries

Unlike public, academic, and corporate libraries, which include print and online materials, e-libraries are exclusively Web-based. But do not confuse e-libraries with periodical databases (pages 358–362), which provide access to online articles only. E-libraries are designed to duplicate the experience of going to a library, as much as that is possible in a digital environment. They provide links to librarian-approved websites in a variety of subject areas, offer links to complete texts of books available online, and connect you with general resources online (dictionaries, almanacs, encyclopedias, and more). They also make it easy to reach librarians who are available to answer reference questions online. Most public and academic libraries have duplicated their library resources online, including access to librarians (see the Tech Note on page 360). These sites offer a variety of helpful links as well.

The most complete e-library is the Internet Public Library (www.ipl.org), or ipl2. It is the most well-known, comprehensive, and academically reliable e-library currently available. Figure 8.6 shows the "Ask a Librarian" page from the ipl2. This nonprofit e-library not only provides the resources listed above, but it also includes

- links to online periodical databases
- links to online newspapers around the world
- links to reliable blogs, exhibits of images, and much more

When you look for an e-library for your workplace research, keep these helpful tips in mind:

1. Use a reliable e-library. Consult e-libraries that have clear navigation, provide links to noncommercial websites, and verify links as current. You will lose valuable time if you don't use a high-quality, up-to-date e-library.

2. Consult a nonprofit e-library. Use e-libraries whose Web addresses end in .org or .edu, rather than commercial (.com) e-libraries. Nonprofit e-libraries are designed to help you, while commercial e-libraries are designed to sell something.

3. Realize that e-libraries do have limits. Despite the convenience of having a wealth of information at your fingertips, not every document found in a public or academic library will be available for online use. Use e-libraries in conjunction with your primary research and visits to public or academic libraries.

FIGURE 8.6 The "Ask a Librarian" Page from the Internet Public Library

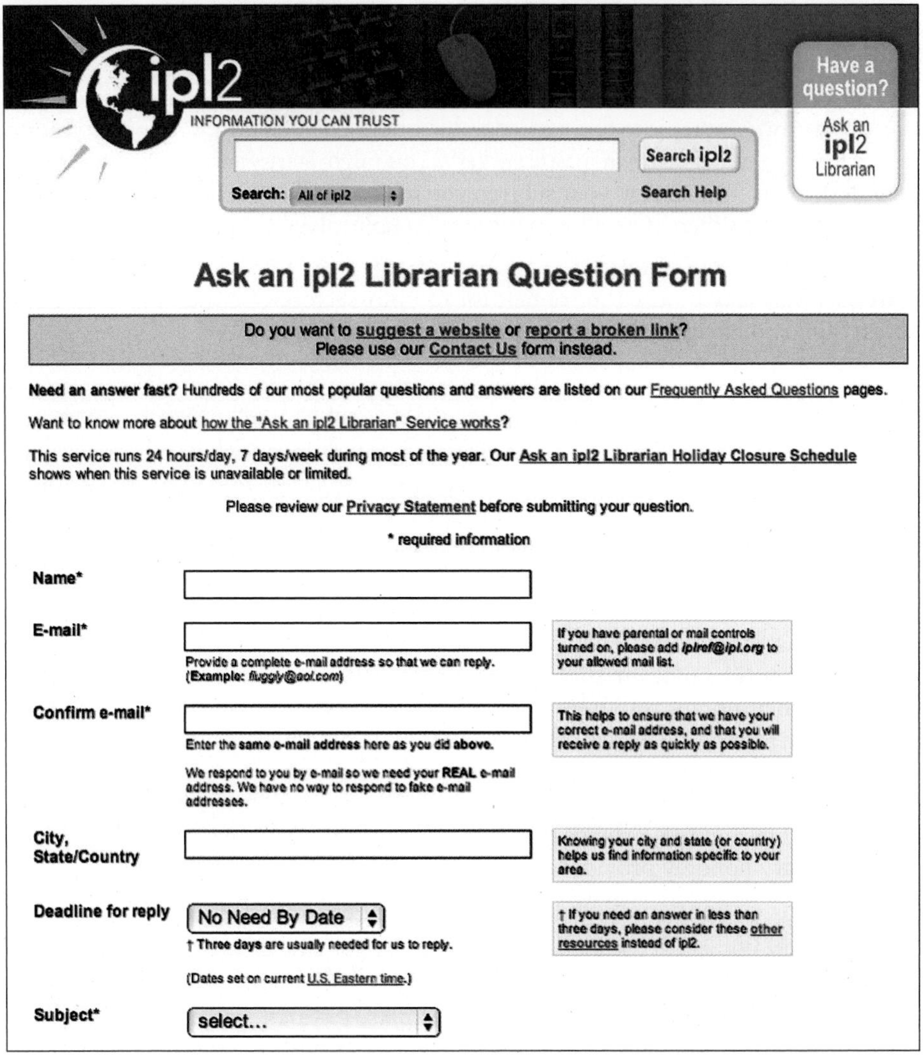
ipl2
INFORMATION YOU CAN TRUST

Search ipl2

Search: All of ipl2

Search Help

Have a question?
Ask an ipl2 Librarian

Ask an ipl2 Librarian Question Form

Do you want to suggest a website or report a broken link?
Please use our Contact Us form instead.

Need an answer fast? Hundreds of our most popular questions and answers are listed on our Frequently Asked Questions pages.

Want to know more about how the "Ask an ipl2 Librarian" Service works?

This service runs 24 hours/day, 7 days/week during most of the year. Our Ask an ipl2 Librarian Holiday Closure Schedule shows when this service is unavailable or limited.

Please review our Privacy Statement before submitting your question.

* required information

Name*

E-mail*
Provide a complete e-mail address so that we can reply. (Example: fluggly@aol.com)
If you have parental or mail controls turned on, please add iplref@ipl.org to your allowed mail list.

Confirm e-mail*
Enter the same e-mail address here as you did above.
This helps to ensure that we have your correct e-mail address, and that you will receive a reply as quickly as possible.
We respond to you by e-mail so we need your REAL e-mail address. We have no way to respond to fake e-mail addresses.

City, State/Country
Knowing your city and state (or country) helps us find information specific to your area.

Deadline for reply
No Need By Date
† Three days are usually needed for us to reply.
(Dates set on current U.S. Eastern time.)
† If you need an answer in less than three days, please consider these other resources instead of ipl2.

Subject*
select...

Courtesy of IPL2.org

Periodical Databases

Periodical databases are among the most helpful resources for doing research in your library. These online indexes allow you to search for and retrieve a wealth of magazine, journal, and newspaper articles cataloged and classified by various search engines. These databases can be located through a library's online catalog or through a periodical databases link at an e-library such as ipl2. Some libraries allow you to access their databases from a remote location, so you can use them when you are traveling.

Because most periodical databases are available only by subscription, you need a library card to access them at a public library or an access code at a corporate library. To use an academic library, you have to be a registered student or alumnus. With these databases, you can search through thousands of articles in a few minutes. Many databases are updated often—daily, weekly, or monthly. Table 8.2 lists periodical and other databases useful for research in the world of work.

TABLE 8.2   **Useful Business and Specialty Databases**

Database	Contents
ABI/INFORM	Business management articles from U.S. and international sources
AGRICOLA	Research materials relating to agriculture
Alt Healthwatch	Full-text articles from 180 international journals devoted to wellness and health care
Animal Science	Information on animal production and veterinary medicine
Business & Company Resource Center	Global business topics, company profiles, case studies, and investment information
Business Source® Complete	Full text, index, and abstracts of business journals and magazines
Computer Science Index	Abstracts and indexes of over 500 academic journals, professional publications, and other sources on computer science
Consumer Health Complete	Surveys, books, journals, popular sources, reports, and videos
Criminal Justice Abstracts	Journals, books, and reports on criminology and related disciplines
EconLit	Bibliographic citations, abstracts and journal articles, books, collected essays, and full-text book reviews dealing with economics
E-journals Service	Thousands of e-publications with direct links
eMarketer	Market information related to digital marketing and media
Environmental Sciences & Pollution Management	Journal articles on environmental sciences and pollution management
FactSearch	Over 1,000 newspapers, periodicals, newsletters, and government documents dealing with current social, economic, political, environmental, and health issues
Facts On File	Worldwide news stories from newspapers, periodicals, journals, and online government sources
General BusinessFile ASAP	Articles on business and management from over 3,000 publications
GreenFILE	Journal articles and government reports, abstracted and indexed, that "cover all aspects of human impact to the environment," including global warming, pollution, and recycling
Homeland Security Information Center	Government resource covering national security concerns, biological and chemical warfare, preparedness and response, and safety training
Hospitality and Tourism	Full text of over 700 publications, including books, journals, and reports

(Continued)

TABLE 8.2 (Continued)

Database	Contents
Insurance Periodicals Index	Over 200 journals about the insurance industry, abstracted and indexed
Internet and Personal Computing Abstracts	Popular magazines and professional journals on personal computing products and development in business, industry, education, and home use
LexisNexis®	Full range of sources for business information, financial news, international company information, market research, industry reports
National Criminal Justice Reference Center	Information, documents, and reports from the U.S. Department of Justice
Regional Business News	Full-text articles from nearly 100 regional U.S. and Canadian business publications
USA.gov	Official database of the U.S. federal government housing files, documents, and reports from all government agencies
U.S. Bureau of Consular Affairs	Government resource for international travel information, advice, and warnings
U.S. Department of the Treasury	Government information on accounting and budget, currency, financial markets, small businesses, and taxes
Wall Street Journal	Full-text articles from *The Wall Street Journal* from 1984 to the present

Information Found in Periodical Databases

Thousands of periodical databases are available from many different information services. They vary in terms of how they list information and what they offer. (See Figure 8.7, page 362 for an example of an article page from the periodical database EBSCOhost.) But most of them provide the following information, crucial for on-the-job research, when you do a keyword search to locate a particular article:

- Bibliographic citation, including author, title, publication date, source information, and perhaps subject categories.
- Keywords, which guide users to the main subjects that the article covers to help them determine if it is useful in their research.
- The full text of the article, either retyped for the database or scanned in as a PDF version of the original document. You may need to click on a further link to get past the informational page and open the complete article.
- Additional factual information that may be useful, such as the article's International Standard Serial Number (ISSN), word count, and the number and type of graphics included.

Frequently Used Periodical Databases

You need to find out which databases are most relevant to your job-related research. Here are a few full-text databases that your corporate or community library will likely subscribe to:

Library Chat Rooms

Library chat rooms are available at no cost to patrons through many public library websites and through the Library of Congress (a U.S. government library that collects almost every publication produced in the United States). Visiting these chat rooms, you can go beyond simply telephoning or emailing a question to a librarian and waiting for a response: You can actually carry on a conversation with a librarian for more detailed assistance. Most chat rooms are open during library hours, but some are open twenty-four hours a day, seven days a week.

Library chat rooms work like instant messaging (pages 150–152): You access the chat rooms using your Internet browser, key in your questions for the librarian, and receive responses online. Also, like instant messaging, library chat rooms are convenient because they allow for the ongoing communication of an email conversation combined with the immediate response of a telephone call.

In addition to communicating with you in real time, the librarian can send you attachments and URLs, providing you with the kind of assistance you would get if you went to an academic or public library in person. If you need further help, an e-librarian may refer your question to another librarian in the network who will follow up your request for information. When your chat is finished, the librarian usually sends you a transcript of your communications, which you can file for your records and refer to as you prepare your report.

1. **EBSCOhost.** This database provides full-text articles from thousands of popular magazines, professional journals, and newspapers and claims to offer the largest full-text collection of professional and academic articles in the world.
2. **Lexis/Nexis Academic.** This database is helpful for locating articles, as far back as the 1940s, in business and law.
3. **InfoTrac.** One of the largest periodical databases, InfoTrac contains a variety of articles from both academic journals and general-interest magazines.
4. **NewsBank.** NewsBank's database supplies over 70,000 news articles annually — from over 500 U.S. and Canadian newspapers.
5. **ProQuest.** Indexing more than 7,400 publications, this database includes newspapers and scholarly and general-interest sources in business, news, medicine, humanities, social sciences, hard sciences, and technology.

Business and Other Specialized Databases

Many libraries, including corporate ones, subscribe not only to the most popular databases above, but also to several business and specialized databases. Almost every professional discipline has its own database. Table 8.2, for example, lists both general business databases and highly specialized subject-specific databases. In addition to these databases, international companies frequently develop their own databases to assist employees in their research. GlaxoSmithKline, the international pharmaceutical manufacturer,

FIGURE 8.7 Page from a Periodical Database

Courtesy of Ebsco Publishing

for instance, prepared and archived a password-protected periodical database of its research studies for use by its 120,000 employees in labs and offices worldwide.

Reference Materials

In addition to finding articles in periodical databases in your library or through an e-library, expect to use reference materials available in print or online for your workplace research. Reference works include encyclopedias, dictionaries, almanacs, and atlases, as well as government documents, industry directories, handbooks and manuals, and statistics. These works will give you up-to-date information when you just need a quick and accurate overview of a topic or specific statistical, historical, or financial data.

But use general reference works such as encyclopedias cautiously. While they supply basic information and are easily accessible, do not confine your research to them, because they are limited in scope. A wise rule of thumb is to always double-check the facts you garner from a general reference work against those you find in more specialized works.

Encyclopedias

The word *encyclopedia* comes from a Greek phrase meaning "general education." An encyclopedia is simply a collection of overviews of a wide range of topics organized alphabetically, either in book form or online. A general print encyclopedia, such as the *Columbia Encyclopedia* or the *Encyclopaedia Britannica*, is a useful starting point for research because it contains knowledgeable introductions to topics, summarizes events or processes, explains key terms, and includes recent updates, along with lists of further readings. Online encyclopedias include *Wikipedia* (www.wikipedia.org), an encyclopedia written and updated by volunteers that contains over 2 million entries, and the *Encyclopaedia Britannica Online* (www .britannica.com).

Be especially careful when using *Wikipedia*. Unlike the more traditional encyclopedias above, which are authored by scholars and overseen by professional editors, *Wikipedia* articles are written by general readers and are not checked for accuracy or for bias by experts in the area.

For your workplace research, you can start with a general encyclopedia, but you may need to consult a more specialized business or technology encyclopedia, such as one of the following:

- *Goliath*, the Cengage-Gale Encyclopedia of Business—http://goliath .ecnext.com/
- *Blackwell Encyclopedia of Management*—www.managementencyclopedia .com/
- *The Encyclopedia of Banking and Finance*—www.eagletraders.com/books/ int_fin_encyclopedia.htm
- *Webopedia*—www.webopedia.com/

Dictionaries

Rather than relying exclusively on general dictionaries, such as *The American Heritage Dictionary of the English Language* (available online at www.dictionary .com), you will have to use business and industry-specific dictionaries that define the words and phrases (jargon) of your profession. Most general and specialized dictionaries are available in print and online. Use only those specialized dictionaries that are officially endorsed by your profession. The following are just a few reliable online business dictionaries:

- *The New York Times Glossary of Financial and Business Terms*—www .nytimes.com/library/financial/glossary/bfglosa.htm
- *Deardorff's Glossary of International Economics*—www-personal.umich .edu/~alandear/glossary/
- *The Washington Post Business Glossary*—www.washingtonpost.com /wp-dyn/business/specials/glossary/index.html

Almanacs

An almanac is a collection of statistical data—charts, tables, graphs, and lists—published annually and carefully organized by general topics such as geography, awards/prizes, and science and technology. Here are some useful almanacs:

- *Infoplease*—www.infoplease.com
- *Plunkett Research*—http://igpweb.igpublish.com/igp/plunkett
- *New York Times Almanac*

In addition, you may want to refer to an industry-specific almanac, such as one of Plunkett's various specialist almanacs (e.g., advertising, investing, retail, e-commerce, information technology, travel, biotech, insurance, etc.).

Atlases

Atlases are collections of worldwide maps. Many kinds of workplace research involve consulting an atlas for locations as well as for statistical information about specific geographical areas. While print atlases are limited in what they can show because of space considerations, online atlases provide much more detail, including three-dimensional images, shading, and elevations. Among the most widely used print atlases are the *Times Atlas of the World* and the *National Geographic Atlas*. Here are some other handy online atlases:

- *Google Earth*—www.google.com/earth
- *Mapquest*—http://atlas.mapquest.com
- *National Geographic Map Machine*—http://maps.nationalgeographic .com/maps/map-machine

Government Documents

The U.S. government is the largest publisher in the country. Virtually every federal agency and department conducts primary and secondary research and then publishes its findings. These publications, collectively referred to as *government documents*, include books, reports, articles, manuals, speeches, census figures, maps, and photos. Government documents are valuable for the statistical information your company may need.

Many government documents are available through the website of the department or agency that created them. But you can access all of these websites via www.USA.gov, the official government Web portal. You can use this link to take advantage of the millions of federal Web resources. Figure 8.8 shows the USA.gov Reference Center and General Government page, listing by category a variety of sites. Other major sites that provide access to government documents include

- *American Fact Finder*—http://factfinder2.census.gov
- *Library of Congress*—www.loc.gov

Directories

If in your job-related research you need to find information about specific companies and their decision makers, turn to one of the many industry-specific directories. They include a wide range of companies along with links to those companies'

FIGURE 8.8 USA.gov Reference Center and General Government Page

USA.gov

websites, or, more helpfully, staff-written overviews of companies, contact information, lists of key people, and financial statistics. Directories are invaluable in helping you find and build email lists and lists of prospective clients. Figure 8.9 shows the home page for Vault Career Intelligence (http://www.vault.com/wps/portal/usa), a Web-based resource for career management and job search information, including insider intelligence on salaries, hiring practices, and company cultures. Figure 8.10 shows the home page for Hoover's (http://www.hoovers.com) which provides industry overviews, company contact information, business statistics, and lists of competing companies. Other useful online directories include these:

- Corporate Information—www.corporateinformation.com. Similar to Hoover's, Corporate Information provides snapshots of over 31,000 companies in 59 countries.
- Zacks—www.zacks.com. Intended for investment purposes, Zacks supplies financial information about many companies.

Handbooks and Manuals

Similar to industry-specific almanacs, handbooks and manuals collect relevant and reliable information that you can use to write your report. Unlike almanacs, though, handbooks and manuals are less focused on statistical data. Instead, they include explanations of procedures, definitions of terms and concepts, descriptions of industry standards, and overviews of professional issues within the field. The *Occupational Outlook Handbook* (www.bls.gov/oco), produced by the U.S.

FIGURE 8.9 Home Page for Vault Career Intelligence Business Information

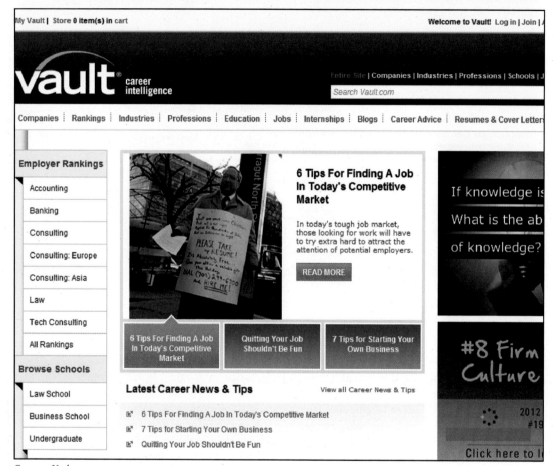

Courtesy Vault.com

Bureau of Labor Statistics, for instance, "describes what workers do on the job, working conditions, the training and education needed, earnings, and expected job prospects in a wide range of occupations."

Among the most frequently consulted handbooks and manuals for workplace research are *Moody's Manuals* and *Standard and Poor's Corporate Descriptions*, both of which give information on the history of a company, descriptions of products and services, and basic financial details (stocks, earnings, and mergers).

Statistics
Statistical reference works give you valuable numerical data on a wide range of business-related subjects—employment, housing, immigration, population,

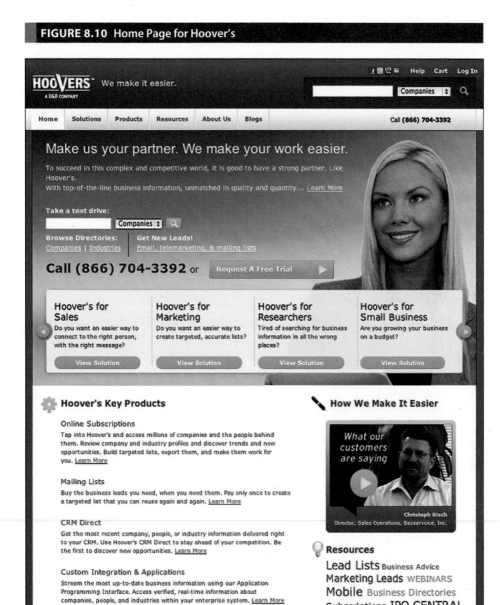

FIGURE 8.10 Home Page for Hoover's

Courtesy of Hoover's, Inc.

pollution, technology, and overseas markets, among others. Statistical data are generated from numerous sources, mostly from the U.S. government, private agencies, international organizations, and colleges and universities. Note how Shay Melka's overview in Figure 8.1 stresses the ways B & L Stores uses statistical data to formulate its major plans. Here are several U.S. government websites that provide valuable statistical references to assist you in your workplace research:

- Federal Trade Commission—www.ftc.gov
- Fedstats—http://fedstats.gov
- Minority Business Development Agency—www.mbda.gov
- Small Business Administration—www.sba.gov
- U.S. Bureau of Economic Analysis—www.bea.gov
- U.S. Bureau of Labor Statistics—www.bls.gov
- U.S. Census Bureau—www.census.gov
- U.S. Data and Statistics—www.usa.gov/Topics/Reference_Shelf/Data.shtml
- U.S. Department of Education—www.eddataexpress.ed.gov
- U.S. Department of Housing and Urban Development—www.huduser.org/portal

Tech Note

Grey Literature

Grey literature is the name given to documents that are prepared by a specific company, industry, or agency but not readily available through databases or online catalogs. Examples of grey literature include internal reports, newsletters, fact sheets, conference proceedings, and instruction booklets. Government agencies, special interest groups, universities, research centers, businesses, and professional associations all create or produce grey literature.

Grey literature is an important resource for workplace research because it often provides information unavailable in commercially published books, journals, or newspapers or in scholarly publications such as monographs and academic journals. Grey literature resources include public health information leaflets; appliance repair manuals; consumer product ratings; and business and industry reports on such topics as annual stockholder meetings, promotions, and new markets. You would be wise to consult these resources to compare your product or service with a competitor's, to assess the corporate health of a firm, or to locate the technical information you need.

Although grey literature can be a rich and diverse source of information, it also poses challenges. Grey literature is typically not indexed by libraries, making it difficult to learn about and locate via bibliographies and databases. In addition, basic information, such as author, date, and source, are often missing from these materials when you do find them. Moreover, you might want to double-check information you find in, say, a company report, with another source.

The Internet, of course, has transformed the production, distribution, and availability of grey literature. The American Library Association has provided a helpful list of places to locate grey literature on the Internet at **www.ala.org/ala/mgrps/divs/acrl/publications/crlnews/2004/mar/graylit.cfm**. But when you use grey literature for your business report, be cautious. Try to find out as much as you can about the company or organization from other sources and double-check and verify information included in a piece of grey literature. Keep in mind that grey literature reflects the viewpoints and preferences of the company or industry that produced it and the financial climate in which it was written.

Source: Sherry Laughlin, Librarian, William Carey University.

Internet Searches

The enormous amount of information available on the Internet grows larger every day. Acting as a "network of networks," the Internet gives you access to millions of websites, databases, libraries, newsgroups, chat rooms, blogs, and other online sources. As a result, it is one of the quickest, easiest, and most effective ways to do secondary research. It is also one that requires caution.

Searching, Not Surfing, the Web

When you surf the Web, you wander from website to website without any particular goal in mind. However, surfing accomplishes very little when you are researching. Just like conducting library research, your Internet research needs to be carefully planned and focused. You wouldn't wander into a library and browse the stacks until you stumbled across something useful if you were there to do research. Instead, you would consult a database or catalog to locate relevant materials.

Searching the Web, however, can be more difficult than finding a book in your online library catalog. The Web is much larger and contains much more information. Moreover, there is no single catalog of all the information it holds. To conduct effective research online, you need to use subject directories and search engines, which index webpages like a database, acting as gateways to vast amounts of information.

Subject Directories

Many businesses and groups have created Internet subject directories that allow users to easily locate appropriate websites (as opposed to articles, which are located via periodical databases). Subject directories are helpfully organized into logical categories (e.g., Business) and subcategories (e.g., International Business and Trade) for easy navigation. Typically, each subcategory is annotated. These directories are also searchable, but the search area is limited to what the directory creators have selected. Think critically when using subject directories, as some subject directories provide more reliable information than others. Here are a few examples of reliable subject directories:

- Academic Info—www.academicinfo.net. As the name indicates, this is an academic subject directory aimed at both undergraduate and graduate students and is an excellent way to conduct research.
- Business.com—www.business.com. This site provides an easy way to locate business websites for news and information on companies, products, services, people, and jobs.
- Yahoo! Directory—http://dir.yahoo.com. The Yahoo! Directory provides over 1 million annotated links to various sites.

Search Engines

When you begin your research, start by using a search engine. A search engine scans webpages to find the keywords most relevant for your research. It then indexes the information it finds to create a frequently updated database of results. Search engines rank the results of your search using a computer algorithm that takes into

TABLE 8.3   Different Types of Internet Search Engines

Search Engines	Metasearch Engines
AltaVista	Dogpile
Ask.com	Excite
Bing	Mamma
Google	MetaCrawler
Yahoo!	Search.com
	WebCrawler

Specialized Search Engines	
General	*News*
AllSearchEngines (UK directory of subject-specific search engines)	Google News Moreover (updates news hourly)
Search Engine Colossus (an international directory of search engines)	Newspapers.com (directory of international, national, college, and business papers)
	World News Network
	Yahoo! News
Media	
ClipArt Searcher	
GMP3	*Business*
Google Images	AllBusiness
Google Videos	WoW Yellow Pages Business Directory
Yahoo! Image Search	
Yahoo! Video Search	

account the popularity of websites, their contents, where the keywords appear on the page, and other sites that link to that page, presenting you with the most relevant matches for your keywords.

Because there is no single index to or abstract of material on the Web, you should expect to use a wide range of search engines. Given the vastness of the Internet, it only makes sense that multiple search engines are available, as Table 8.3 shows. These include general and metasearch engines (which combine multiple search engines) as well as specialized, subject-specific search engines. Like the Web itself, search engines are constantly expanding as they index more pages in cyberspace each second.

Of course, search engines will vary in coverage and relevance for your project. Some are faster than others, while many incorporate results from multiple engines simultaneously. To some extent, all search engines overlap. Moreover, no single search engine is comprehensive. Researchers estimate that almost 40 percent of the Web is not even indexed by the most popular search engines. But by using a variety of search engines to conduct keyword searches, you will access a broader scope of sources and thereby increase your chances of finding the information you need.

Following are descriptions of four of the most frequently used search engines listed in Table 8.3.

1. **Google** (www.google.com) is currently one of the world's largest and easiest-to-use search engines, indexing over eight billion pages. Google offers specialized engines such as Google Government, Google University, Google Microsoft (which searches only sites related to Microsoft), Google Apple (which searches Apple Macintosh–related sites), and Google Local (to search for businesses in your area).
2. **Yahoo!** (www.yahoo.com), one of the earliest search engines, also includes a human-edited, subject-based directory of the Web. Yahoo! suggests related keywords and allows you to open links in new windows.
3. **Ask.com** (www.ask.com) provides a variety of personalized search options, including MyStuff, to save and organize your results in folders; view your search history; and share, print, and repeat searches.
4. **Bing** (www.bing.com) provides one of the most comprehensive searches, including image, audio, and video links, and divides findings into categories.

Keyword Search Shortcuts

To use search engines effectively, you need to know how to conduct a keyword search, which is crucial to your success. When you enter a keyword into a search engine, it searches through its indexed websites and presents you with those that feature your keyword in the title, heading, meta tags, or text of the page. Your choice of keywords is vital since the results you get depend on the keywords you plug into the search engine.

Follow the guidelines below to conduct successful keyword searches:

1. **Be specific.**

   - Narrow your subject. Although it might be tempting to start off with a broad subject like *business travel*, you may find yourself confronted with millions of hits that range from earning travel miles to news about corporate balance sheets. Take a look at Figure 8.11, which shows the results from a search that is too general. A search on the words *business travel* yielded over 302 million results. Before you start your search, narrow your focus and thereby restrict your search.
   - Use multiple keywords. Choose at least two or three significant keywords to specify your search. For example, if you key in just the word *virus*, you might receive vast amounts of information on everything from avian flu and the West Nile virus to computer problems. Specify *computer virus* to narrow your search.
   - Don't use prepositions. Most search engines automatically exclude *of, to, in, from*, and *on*.

2. **Modify your keywords.**

   - Refine your search. Unlike the vague keywords *business travel*, specify the type of information you want to receive by refining your topic. Search for *business travel affordable* or *business travel convenient* to net more precise, pertinent information. Figure 8.12 shows how advanced search options can help you narrow your keyword search. Here the search was narrowed by using multiple and more specific search words: *business travel affordable convenient*.

FIGURE 8.11 Example of a General Search That Yields Too Many Results

Keyword is too broad

Too many hits

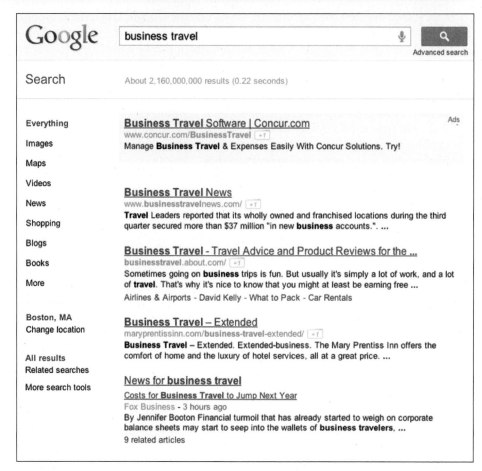

Reprinted courtesy of Google, Inc.

- Find relevant synonyms. Try to locate possible synonyms or alternative keywords that might give you additional pathways to gain information on your topic. Many search engines even suggest ways to refine your search using other keywords. Take advantage of these suggestions.

3. **Use Boolean connectors.** Boolean connectors, such as AND, OR, and NOT, are essential to limit and guide your search by reducing unrelated search results.

- Inserting *AND* between keywords will limit your results to those containing all your keywords. For example, *business AND travel AND affordable* will produce only records that contain all three terms. Note that some search engines, such as Google, automatically search all keywords and therefore don't require the AND command.

FIGURE 8.12 Example of a More Refined Google Search

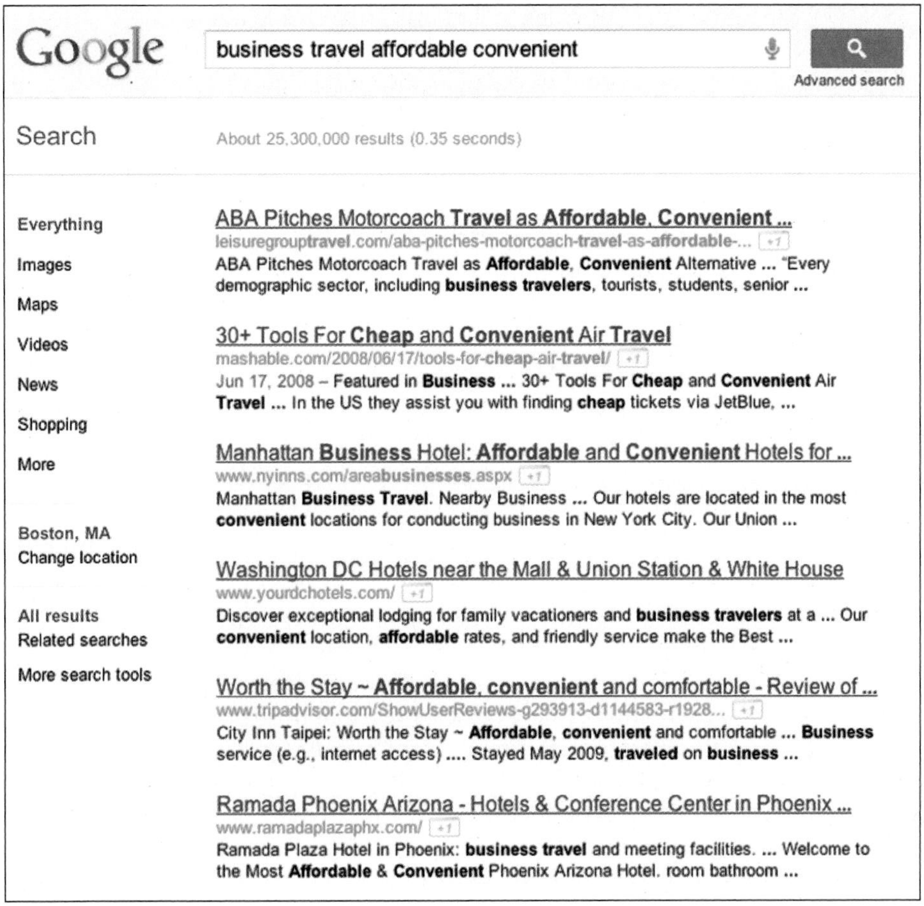

Use multiple keywords to restrict search

Click on search results

Reprinted courtesy of Google, Inc.

- *OR* will produce records containing either or both keywords. *Business OR travel* will list pages containing either *business* or *travel* or both words.
- *NOT* as part of your keyword search excludes unwanted pages. *Telecommuting benefits NOT environmental* indicates that you are not concerned with how telecommuting saves energy or reduces pollution. Instead, you are looking for benefits to employers and employees.

4. **Use delimiters.** To refine your search even further, use the following *delimiters*, sometimes called *wild-card characters*:

- Quotation marks around a string of keywords limit your term to that particular phrase. For example, *Wells Fargo* without quotation marks would pull up sites for Wells Engineering; Fargo, North Dakota; and Fargo Wells.

- A plus sign (+) between terms can be used to replace the Boolean AND command, locating results that include all keywords. Similarly, a minus sign (−) can replace the OR command in many search engines.
- An asterisk (*) at the end of a term broadens your search beyond that afforded by the keyword. For example, the term *employ** will return all terms that use *employ* as the stem of a keyword, such as *employment*, *employer*, and *employability*.
- Some search engines, such as Google, allow you to use an asterisk within quotation marks to indicate a missing word. Searching for *"Australian * technology"* will find results including keyword phrases ranging from *Australian medical technology* to *Australian information technology*.

Internet Search Shortcuts

In addition to the keyword search strategies above, there are a number of shortcuts that allow users to access frequently requested information quickly and easily on the Internet. Many of these techniques take advantage of advanced search options available on major search engines. Of course, depending on the type of Internet resources available to you, and the constantly changing Web terrain, you will doubtless discover many more shortcuts. But the following ones give you additional ways to access information faster and more efficiently.

- Use Google Alerts by entering your email address and a keyword term into the Alerts system to be notified when your search term appears in the top ten or twenty search results for a Google News or Web search. This service is particularly useful for ongoing business research because it provides constant updates on the latest information relevant to your topic.
- Select the images, audio, or video options from your search engine to obtain thumbnails of all indexed images, MP3 files, or video files that relate to your keywords.
- To find sites in a particular language, file format (Word, PDF, PowerPoint), or date, use advanced search options available on most search engines.
- To obtain frequently sought-after information on weather, time zones, and flight times, just type in such keywords as "Chicago weather," "time Tokyo," or "Delta 346." Or you can key in the name of the city with its ZIP code to obtain a current weather report or one projected over the next thirty days.
- To locate a map, type in the street address of the area you want and a street map will automatically appear.
- To receive a selection of Web definitions taken from a variety of respected sources, type in "define" and then your keyword into Google.
- Choose country-specific search engines, such as those available at Yahoo! or Google, to access information about companies, laws, news, and international relations essential for communicating with international readers.
- To check stock quotes, key in the stock exchange symbol of the company you are interested in, e.g., TWX for Time Warner.
- To locate information on well-known people, Ask.com provides an instant biography with a photo, along with links to encyclopedia articles and official websites.

- Enter advanced delimiters to specify the exact type of domain you want to search. Searching Google for *"internet security" site:gov* will retrieve all .gov (government) sites with information on Internet security. Similarly, searching for *"international training" site:edu* will pull up all .edu (education) sites with information on international training.

Evaluating Websites

The criteria used to evaluate websites are similar to those you follow to assess print documents (see page 329). Just as not everything you read in print is accurate and unbiased, not everything on the Internet will be correct, up-to-date, objective, or useful. The fact that something is posted on the Web does not make it correct—after all, anyone can post practically anything. Remember that much of the information on the Internet is placed there without a peer-review process—that is, facts may not have been checked, sources may not be authoritative, and ideas and opinions may be biased and not be backed up by solid evidence. Overall, the quality of information you find on the Internet can range from unsupported, slanted, unethical, or just plain wrong, to authoritative, well supported, and objective.

Keep in mind, too, that as with print sources, websites may be created by special interest groups or individual businesses. The information (e.g., statistical data) you find at a special interest site may be clearly skewed toward or against a particular point of view, and a website created by a company is, of course, going to be prejudiced in the firm's own behalf, because it wants to sell its products and services. You will have to read, assess, and differentiate carefully the Web material you want to include in your report.

To do effective workplace research, you have to make critical inquiries and judgments. Don't be afraid to seek out and test different viewpoints, and never be afraid to verify what you find on the Web by asking for the opinion of a librarian or a specialist in your company. Note how in Figure 8.13 (page 376) the Food and Drug Administration cautions pet owners to evaluate online veterinary pharmacies according to several important criteria.

Here are some questions to determine whether websites (and print sources, too) are credible, accurate, up-to-date, relevant, and objective.

Credible

- What credentials does the author or the organization post? Read the "About Us" or "About Me" page of any website critically to learn more about the author's qualifications and awards or the organization's history, mission statement, and track record.
- Does the website refer to the most respected and current research being done in the field?
- Is the author discussed in other sources, including reference works and highly respected websites? Are there links to or from the author's or company's website to those sites?
- Is it possible to contact the author or organization with questions or requests for further help? If the website provides no contact information, contains many dead links, or features undocumented information, be suspicious.

FIGURE 8.13 Advice on Evaluating Online Veterinary Pharmacies

Advice When Buying Pet Medicines Online

When buying from online veterinary pharmacies, keep an eye out for red flags. Be careful if the . . .

- **Site does not require veterinary prescriptions for prescription drug orders.**

Websites that sell prescription veterinary medicines without valid veterinary prescriptions for them are breaking the law. Online questionnaires or consults don't take the place of valid veterinary prescriptions. Sites that sell drugs without requiring valid veterinary prescriptions rob both you and your pet of the protection provided by a veterinary physical exam.

- **Site has no licensed pharmacist available to answer questions.**

Does a professional answer your questions about your pet's medicines?

- **Site does not list physical business address, phone number, or other contact information.**

If something goes wrong with your order, can you get in contact with the site? How?

- **Site is not licensed by the State Board of Pharmacy where the business is based.**

If the business is based in the U.S., check to see if it is properly licensed in the state where it is based by contacting that state's Board of Pharmacy. Contact information is available on the National Association of Boards of Pharmacy (NABP) website: http://www.nabp.net/boards-of-pharmacy.

- **Site's prices are dramatically lower than your veterinarian's or other websites' prices.**

If it seems too good to be true, then it probably is.

- **Site ships you medicine that you didn't order or that looks very different from what your pet normally takes.**

Don't give these medicines to your pet! Contact the site immediately!

Always Check for Site Accreditation

Be sure the site is accredited. The National Association of Boards of Pharmacy (NABP) created a voluntary accreditation program called Vet-VIPPS (Veterinary-Verified Internet Pharmacy Practice Sites). Vet-VIPPS accredited pharmacies must meet strict criteria, including protecting patient confidentiality, quality assurance, and validity of prescription orders.

Source: Adapted from www.fda.gov/animalVeterinary/ResourcesforYou/AnimalHealthLiteracy/ucm203000.htm.

Accurate

- How have the data been gathered, recorded, and interpreted? Are they complete, valid, logical, and consistent with the procedures and protocols of your profession?
- Do the facts, dates, and statistical information match those found in other sources, such as on other websites or in traditional research sources such as almanacs and encyclopedias (see pages 363 and 364)? Again, do not rely on only one website for your facts and figures.
- Are all visuals clear, presented without distortion, and consistent with the information presented (see pages 465–467)?
- Is the website full of typographical, grammatical, or spelling errors? If so, it may be unreliable in other matters, such as statistics, conclusions, and recommendations.

Up-to-Date

- When was the content written, placed on the website, and last updated? If the website does not provide such information, its content is likely to be out-of-date.
- Does the website content refer to current events, studies, and experiments, or do all the references go back too many months or years? In some business and technical fields, information needs to be updated weekly, or even daily.
- Is the information given on the website contradicted, supplemented, or labeled out-of-date on other more recent websites and in print reference works?

Relevant

- Is the website's content directly related to the subject of your report, or is it only remotely close to your topic and thus not central to your inquiry? Remember, your boss wants bottom-line information to answer specific questions.
- Does the website offer enough of the right type of data to be useful, or is it sketchy?
- Will others accept or trust the numbers and interpretations you have gathered from the website?
- How does the website help or hurt the case you are trying to build? Even a respected competitor's figures can give you valid and relevant information you can convincingly apply to your report.

Objective

- Does the website openly and clearly state its purpose? Does it claim to give objective information, or is it really just trying to sell something or present a biased point of view? Again, read the "About Me" or "About Us" page carefully.
- Is the website legitimate? Verify whether a company or organization is legitimate by checking with the National Consumers League, www.nclnet.org.

- Does the website acknowledge opposing viewpoints, and does it freely and clearly admit its own limitations and agendas?
- Does the author use a professional, objective tone, or is there unwarranted humor or sarcasm?
- Is the site free from sexist, racist, and other demeaning stereotypical language or visuals?
- Is this website part of a reputable online discussion community (a listserv or chat room), or is it a platform for one individual's or organization's viewpoints?

Note Taking

Once you have consulted appropriate print and online sources, you need some systematic way to record relevant information from them to use in preparing the outline and drafts of your business report. But before you can outline or draft, you have to organize and classify the information from your research efficiently. And even after you start to draft your work, you will probably need to continue your research and take additional notes as you investigate new and relevant sources.

The Importance of Note Taking

Note taking is the crucial link between finding sources, reading and responding to them, and writing your business report. At this stage, you are gathering crucial background information to build and support your report. Never trust your memory to keep all of your research facts straight. Taking notes is time well spent.

Take your time. Do not be too quick in getting it done or too eager to begin writing your report. Careless note taking could lead to serious omissions, inconsistencies, or even contradictions between what different parts of your business report say.

Notice how the business research report at the end of this chapter and the one in Chapter 15 on international employees in the workforce (pages 701–719) incorporate a variety of important facts, interpretations, and even visuals from the sources that the writers consulted, including blogs, surveys, and interviews. Before these research details could be included in the report, the writer had to take careful notes.

How to Take Effective Notes

Effective note taking requires you to (a) identify only the most relevant points, (b) exclude irrelevant or inessential ones, (c) summarize key information concisely and accurately, and (d) document the results. Follow these guidelines to make note taking easier and more efficient:

1. Photocopy or scan hard copy or download print or online articles, sections of books, reports, and other sources that you consult and mark relevant quotations, statistics, and other information you may incorporate into your work.

Use different colored pens, highlighters, or computer colors to highlight or underline different types of information or to indicate where such information may go in your work (e.g., background material, methods, visuals, etc.).

2. Record all quotations verbatim.
3. Bookmark any sites on the Internet that you know will be important for your work and that you may be likely to return to. (See the Tech Note below.)
4. Cut and paste information from online sources directly into your word-processing program for reference, being careful to include exact source information.

What to Record

To take effective notes, you have to know how to summarize a great deal of information accurately. Study the guidelines on writing summaries in Chapter 9 (pages 425–426). Here are the types of information you need to record in your notes.

1. **Include full bibliographic information for each source.** For *books*, list author, title, city of publication, publisher, date of publication, edition (if not the first), and page numbers. For *journal articles,* include author, title, publisher, volume number, date, and page numbers. For *websites*, record author, title, name of the site, URL, and the date you accessed the site.
2. **Copy quotations, names, facts, dates, and statistics accurately from the source.** Be sure, too, that you record the author's words correctly. You may want to quote them verbatim in your paper or report. Always compare what you have written down with the original.
3. **Distinguish quotations from paraphrases.** Place quotation marks around any words, sentences, or extended quotations you record directly to avoid the risk of plagiarizing (see page 384).

Tech Note

Using Bookmarks

Bookmarks, also called *favorites*, are direct links to preselected websites that are stored in a Web browser. Just as you slip a sheet of paper into a book or turn down the corner of a page of a print magazine to mark your place, you can use your Web browser's bookmark feature to keep track of useful sites so that you can quickly return to them as you continue your research. To create a bookmark, simply click on the bookmark or favorites option on your Web browser's pull-down menu while you have the webpage open, and then name the bookmarked page. Most browsers will automatically fill in a name based on information supplied by the website, but you may want to supply a more specific name so that you can remember later exactly why you bookmarked the site. To return to the website later, select the bookmark, and your Web browser will automatically take you to that page.

4. **Indicate in your notes why the material you quoted or paraphrased is significant.** You may not remember later why the material is significant, so help your memory by leaving yourself a reminder—"use in introduction."
5. **Clearly mark separate works by the same author.** If you are using two or more works by the same author in your research, be sure to indicate which quotations or paraphrases apply to which works.
6. **Identify and record only the most relevant and useful points.** You cannot copy down everything in your notes. Include only what you plan to actually use in your report.
7. **Indicate where graphics might be effective.** If you would like to add a graphic or visual to your report to support your report's findings or to clarify points for your audience, make a note to yourself so that you remember later.

To Quote or Not to Quote

Before recording information from sources, ask yourself three questions:

1. How much do I need to quote directly?
2. Where can I shorten a quote by using ellipses?
3. When should I paraphrase instead of quote?

Tech Note

Electronic Note-Taking Software

Electronic note-taking software offers a fast and convenient way to organize your notes, saving you from worrying about losing note cards or scraps of paper. The four most popular electronic note-taking programs are Microsoft OneNote, Evernote, Google Notebook, and GoBinder. Although each is slightly different (some allow users to download or upload audio and video), most have the following features in common to help researchers input, organize, search, save, print, and share their notes:

- Copy and paste text and images from Internet sources. Many programs automatically include the URL.
- Keyboard your own notes or, if you have neat handwriting, write directly on the tablet PC with a stylus (a pen-like tool). First check that your PC tablet has handwriting recognition software.
- Flag or color-code your notes to highlight and differentiate them.
- Organize your notes into categories, folders, or outlines.
- Reorganize your notes using a drag-and-drop feature.
- Search your notes for keywords or flags to locate relevant information.
- Back up your notes to make sure you don't lose them.
- Print your notes, export them to your office program, or email them to others.
- Collaborate with others who have the same program via a "user list."

Incorporating Quotes

A safe rule to follow is this: Quote sparingly. Do not be a human scanner. If you incorporate too many quotations in your report, you will simply be transferring the author's words from the book, article, or website to your paper. That will show that you have read the work but not that you have successfully evaluated it. Do not use direct quotations simply as filler. Save them for when they count most:

- When an author has summarized a great deal of significant information concisely into a few well-chosen sentences
- When a writer has clarified a difficult concept exceedingly well
- When an author has made his or her chief statement or thesis

For example, the note in Figure 8.14 contains a brief, well-worded, and significant statement by an author. It is not necessary to quote verbatim all the evidence leading to that statement. The note stands well on its own.

If you are uncertain about exactly how much to quote verbatim, keep in mind that no more than 10 to 15 percent of your report should be made up of direct quotations. Remember that when you quote someone directly, you are telling your readers that these words are the most important part of the author's work as far as you are concerned. Be a selective filter, not a large funnel.

Using Ellipses

Sometimes a sentence or passage is particularly useful, but you may not want to quote it fully. You may want to delete some words that are not really necessary for your purpose. These omissions are indicated by using an *ellipsis* (three spaced dots within the sentence to indicate where words have been omitted). Here are some examples.

Full Quotation:	"Diet and nutrition, which researchers have studied extensively, significantly affect oral health."
Quotation with Ellipsis:	"Diet and nutrition . . . significantly affect oral health."

FIGURE 8.14 Note Containing a Direct Quotation

Iona Crandell, "Importance of Internet Medicine," *Journal of Emergency Medicine* 15 (Sept. 2012): 32.

"The Internet is a primary source of medical information for consumers. CybMed, an Internet marketing firm, estimates that more than 80 million people in 2011 consulted the Web for a variety of health-related information. Most users searched popular sites such as Medscape to look up the signs and symptoms of their medical problems. Consumers also flocked to PharmInfo.net, an FDA website, to find information on new drugs and their possible side effects. These websites are the closest thing to a doctor who makes house calls."

When the omission occurs at the end of the sentence, you must include the end-of-sentence punctuation after the ellipsis. In the following example, note how the shortened sentence ends with four spaced dots: the closing period and the three dots for the ellipsis.

Full Quotation: "Decisions on how to operate the company should be based on the most accurate and relevant information available from both within the company and from the specific community that the establishment serves."

Quotation with Ellipsis: "Decisions on how to operate the company should be based on the most accurate and relevant information available. . . ."

At times you may have to insert your own information within a quotation. This addition, known as an *interpolation*, is made by enclosing your clarifying identification or remark in brackets inside the quotation; for example, "It [the new transportation network] has been thoroughly tested and approved." Anything in brackets is not part of the original quotation.

Paraphrasing

Most of your note taking will be devoted to paraphrasing rather than writing down direct quotations. A *paraphrase* is a restatement in your own words of the author's ideas. Even though you are using your own words to translate or restate, you still must document the paraphrase because you are using the author's facts and interpretations. You do not use quotation marks, though. When you include a paraphrase in your paper, you should be careful to do four things:

1. Be faithful to the author's meaning. Do not alter facts or introduce new ideas.
2. Follow the order in which the author presents the information.
3. Include in your paraphrase only what is relevant for your paper. Delete any details not essential for your work.
4. Use paraphrases in your report selectively. You do not want your report to be merely a restatement of someone else's ideas.

Paraphrased material can be introduced in your paper with an appropriate identifying phrase, such as "According to Dampier's study," "To paraphrase Dampier," or "As Dampier observes." The note shown in Figure 8.15 paraphrases the following quotation:

While the effects of acid rain are felt first in lakes, which act as natural collection points, some scientists fear there may be extensive damage to forests as well. In the process described by one researcher as "premature senescence," trees exposed to acid sprays lose their leaves, wilt, and finally die. New trees may not grow to replace them. Deprived of natural cover, wildlife may flee or die. The extent of the damage to forest lands is extremely difficult to determine, but scientists find the trend worrisome. In Sweden, for example, one estimate calculates that the yield in forest products decreased by about one percent each year.[1]

---

[1]Bill Dampier, "Now Even the Rain Is Dangerous," *International Wildlife* 10, pp. 18–19.

FIGURE 8.15 Note Containing a Paraphrase

Acid rain (body of report)
damage to forests

Dampier, www.rainenviron.com

Acid rain is as dangerous to the forests as to the lakes. Victims of "premature senescence," the trees become defoliated and die with no new trees taking their place. Without the trees' protection, wildlife vanishes. Although the exact damage is hard to measure, Swedish scientists have observed that in their country forest products decreased by 1 percent yearly.

© Cengage Learning 2013

Documenting Sources

Documentation is at the heart of all the research you will do on the job. To document means to furnish readers with information about the print and electronic sources you have used for the factual support of your statements, including books, journals, newspapers, surveys, reports, websites, and other resources such as listservs and email. Without proper documentation, you will not be able to persuade a customer to buy your product or service, and you will not convince your boss that you are doing your best work.

Documentation is an essential part of any research you do for at least four reasons:

1. It demonstrates that you have done your homework by consulting experts on the subject and relying on the most authoritative sources to build your case persuasively.
2. It shows that you are aware of the latest research in your field, thus lending credibility and authority to your conclusions and recommendations.
3. It gives proper credit to those sources and avoids plagiarism (see page 384). Citing works by name is not a simple act of courtesy; it is an ethical requirement and, because so much material is protected by copyright, a point of law.
4. It informs readers about specific books, articles, or websites you used so they can locate your source and verify your facts or quotations.

The Ethics of Documentation: Determining What to Cite

As a researcher, you have to be sure about what information you must cite and what information you do not need to cite. Before you start consulting sources, you have to be very clear about the ethical standards involved in documentation. The following sections will give you a useful overview to make the documentation process more understandable and easier to follow.

What Must Be Cited

To ensure that your business report avoids any type of plagiarism and maintains high ethical and professional standards, follow these guidelines:

- If you use a source and take something from it, document it. Document any direct quotations, even a single phrase or keyword.
- Stay away from *patchworking*—using bits and pieces of information and passing them off as your own—which is also an act of plagiarism. Always put quotation marks around anything you take verbatim, and document it.
- If any opinions, interpretations, and conclusions expressed verbally or in writing are not your own (e.g., you could not have reached them without the help of another source), you must document them.
- Even if you do not use an author's exact words but still get an idea, concept, or point of view from a source, document that work in your report.
- Never alter any original material to have it suit your argument. Changing any information—names, dates, times, test results—is a serious offense.
- If you use statistical data you have not compiled yourself, document them.
- Always document any visuals—photographs, graphs, tables, charts, images downloaded from the Internet (and if you construct a visual based on someone else's data, you must acknowledge that source, too).
- Never submit the same research paper for one course that you wrote for another course without first obtaining permission from the second instructor.
- Do not delete an author's name when you are citing or forwarding an Internet document. You are obligated to give the Internet author full credit.

What Does Not Need to Be Cited

Be careful not to distract readers with unnecessary citations that only demonstrate your lack of understanding of the documentation process and can undercut the professionalism of your report. There is no need to cite the following:

- Common-knowledge scientific facts and formulas, such as "The normal human body temperature is 98.6 degrees Fahrenheit" or "H_2O is the chemical formula for water."
- Readily available geographical data, such as elevation of mountains; depths of lakes, rivers, etc.; population; mileage between two places; and so on.
- Well-known dates, such as the date of the first moon landing in 1969.
- Factual historical information, such as "George W. Bush was the 43rd president of the United States."
- Proverbs from folklore, such as "The hand is quicker than the eye."
- Well-known quotations, such as "We hold these truths to be self-evident . . . ," although it may be helpful to the reader if you mention the name of the person being quoted.
- The Bible, Koran, or other religious texts, but provide a reference to the text and to the portion of the text quoted in parentheses (for instance, *New Jerusalem Bible*, Exod. 2.3).

■ Classic literary works, but again reference the original author and the name of the work parenthetically—for instance, Twain, *The Adventures of Huckleberry Finn* (Chapter 4), or Shakespeare, *The Merchant of Venice* (5.3.15). Indicate, though, from which edition you took the quotation.

Documentation Styles

Most professional associations have created their own guidelines for documenting sources. For example, the Council of Biology Editors has outlined a style for biologists in *The CBE Manual for Authors, Editors, and Publishers*; the *Harvard Law Review* provides a style manual for legal professionals in *The Bluebook: A Uniform Style of Citation*; and the National Association of Social Workers uses *Writing for the NASW Press: Information for Authors*. But the two most widely used documentation styles are MLA (Modern Language Association) and APA (American Psychological Association), as detailed in their publications:

■ *MLA Handbook for Writers of Research Papers*, 7th ed. (New York: Modern Language Association, 2009), www.mla.org/style
■ *Publication Manual of the American Psychological Association*, 6th ed. (Washington, DC: American Psychological Association, 2010), www.apastyle.org
■ *APA Style Guide to Electronic References* (Washington, DC: American Psychological Association, 2007)

MLA is used primarily in the humanities, while APA is used in psychology, nursing, social sciences, and several technological/scientific fields. In business, however, whether you use MLA or APA will be determined by your employer. Because MLA and APA are the most well-known and accessible documentation styles, many businesses prefer to rely on one or the other, or they adapt or modify these methods to suit the company's needs and those of its clients. The long business report at the end of this chapter follows MLA style; the long report on multinational employees in Chapter 15 (pages 701–719) follows APA style.

What MLA and APA Have in Common

While there are differences between MLA and APA, they share two basic features:

1. Both require in-text citations within your researched report. In-text citations, also known as *parenthetical citations*, are brief citations placed directly in your report, in or at the end of the sentence requiring a reference, to indicate the source of the information. These citations include only the most essential information (to avoid interrupting the reader's train of thought), but they let readers know precisely where you borrowed information. An in-text citation also tells readers that they can expect a full bibliographic reference to appear at the end of your report.

2. Both include a complete list of sources cited at the end of the report. MLA uses a *Works Cited list* at the end of the report, while APA calls for a *References list*. These lists provide detailed bibliographic information about all of the sources you cited in your research. Every source you cite in the text must also be included in these lists, with some exceptions in APA style (interviews, surveys, personal conversations, emails, and IMs).

Beyond these two shared points, however, MLA and APA follow different citation formats. The next two sections of this chapter will show you how to cite sources (both in the text and at the end of a report) using MLA and APA styles.

Using MLA In-Text Citations

This section will help you prepare in-text citations according to MLA guidelines. It will also describe the specific information you have to give readers and where to put it.

When you insert in-text citations in your report, keep in mind that every citation in your report must correspond to a work you cite in full at the end of your report in the Works Cited list. And, correspondingly, every work you cite in the Works Cited list must be included as an in-text citation in your report. Note that when you cite a page number parenthetically, MLA does not use *page*, *pages*, *p.*, or *pp*. You just list the author's last name and page number, without a comma between them (e.g., Jones 32). Here are some guidelines and examples:

- If you do not mention the author by name in your sentence, then you must include the author and the page number in parentheses at the end of the sentence:

    About 5 percent of the world's population has diabetes mellitus, and 25 percent of the world's population carries the gene for the disease (Walton 56–57).

- But if you do refer to the author by name in your sentence, supply only the page number consulted in parentheses at the end of the sentence. It is redundant to cite the author's name both in the sentence and parenthetically.

    According to Tamara deTorres, inflation in New Zealand is expected to increase rapidly by 2012 (27).

- When you cite a work written by two authors, list both authors, connected by *and*:

    While fax machines are still used to transmit copies of important documents such as contracts and signed letters, email attachments frequently take the place of faxes (Perez and Klein 25–27).

- If the work was written by three authors, list all of the authors, separated by commas, the last two connected by a comma followed by *and*:

    Tourism has increased by 9 percent this last quarter, thanks to individuals passing through our state on their way to the National Association of Black Business Leaders Conference (Muscovi, Klein, and Philips 57–58).

- When the work you consulted has more than three authors, list just the first author's name followed by *et al.* (Latin for "and others"):

  The principles of ergonomics have revolutionized the design of office furniture (Brodsky et al. 345–47).

- For a work by two authors with the same last name, provide both authors' first and last names, connected by *and*:

  New York City has the highest percentage of residential turnover per year (Joyce Yates and Peter Yates 66).

- Sometimes you will have to cite two or more works by the same author. To distinguish one work from the other, provide the author's last name, followed by a comma and a shortened version of the particular work's title (in the example below, the complete title of the work has been shortened from *Compensations and Benefits*):

  Many firms offer employees 401K stock purchase plans (Howe, *Compensations* 16).

- Occasionally, you may have to cite two works in the same sentence. When you do, list both sources, in alphabetical order, separated by a semicolon. As a rule, though, do not overload readers with citations.

  The use of salt domes to store radioactive wastes has come under severe attack (Jelinek 56–57; McPherson and Chin 23–26).

- When the work you consulted was written by a company, organization, or government agency, use a shortened version or abbreviation of the name of the organization. In the example below, *Commission* replaces *Commission on Wage and Price Control*. You would replace *American Medical Association* with *AMA* if it is listed this way in the Works Cited list.

  Salaries for local electricians were at or above the national average (Commission 145).

- If the work you've referenced has no author—not even the name of a company, organization, or government agency—use a shortened version of the title and the page number:

  Customers surveyed tended to focus on the product's problems rather than on its benefits ("Unicom Survey" 22–23).

- Since websites do not have page numbers, all that you have to list parenthetically is the name of the author, or if you use the author's name in your sentence, then simply list a shortened title of the website:

  D'Arcy claims that by 2015 transportation in Boston will be even more streamlined to accommodate a projected increase in riders ("Boston Trans.").

- For a quotation consisting of four or fewer typed lines, include the quotation within your sentence (using quotation marks), and place the in-text citation at the end of the sentence, followed by a period:

Pilmer has observed that "renovation of downtown areas is one of the most pressing issues facing city government" (16).

- For a quotation longer than four typed lines, set the quoted material apart from the text by indenting it one inch on the left side and eliminating the quotation marks. Place the in-text citation after the quotation and the period, as in the following example:

L. J. Ronsivalli offers this graphic analogy of how radiation can penetrate solid objects:

> One might wonder how an x-ray, a gamma ray, or a cosmic ray can penetrate something as solid as a brick wall or a piece of wood. We can't see that within the atomic structures of the brick wall and the wood there are spaces for the radiation to enter. If we look at a cloud, we can see its shape, but because distance has made them too small, we can't see the droplets of moisture out of which the cloud is made. Much too small for the eye to see, even with the help of a microscope, the atomic structure of solid materials is made of very small particles with a lot of space between them. In fact, solids are mostly empty spaces. (20–21)

Using APA In-Text Citations

Like MLA, APA uses both in-text citations and a References list at the end of the report that supplies full publication information. As in the earlier discussion of MLA style, this section will show you what to include and where, according to the APA guidelines.

As in MLA style, any APA in-text citations in your report must also appear in the References list at the end of your report and vice versa. However, unlike in MLA, personal communications (see page 390) appear only in the text in APA style. In addition, APA requires the publication year rather than the page number in in-text citations (with a comma between the author's name and publication year). When using APA documentation, follow the specific guidelines below.

- When you do not mention the author by name in your sentence, then you must include the author's name and the unabbreviated year of publication in parentheses at the end of the sentence:

About 5 percent of the world's population has diabetes mellitus, and 25 percent of the world's population carries the gene for the disease (Walton, 2007).

- But if you do refer to the author by name in your sentence, supply only the year of publication of the work in parentheses:

According to Tamara deTorres (2012), inflation in New Zealand is expected to increase rapidly by 2015.

- When you cite a work written by two authors, list both authors, connected by an ampersand (&):

While fax machines are still used to transmit copies of important documents such as contracts and signed letters, email attachments frequently take the place of faxes (Perez & Klein, 2008).

- If the work was written by three to five authors, list all of the authors, separated by commas, the last two connected by a comma and an ampersand:

Tourism has increased by 9 percent this last quarter, thanks to individuals passing through our state on their way to the National Association of Black Business Leaders Conference (Muscovi, Klein, & Philips, 2010).

- When the work you consulted has six or more authors, list just the first author's name followed by *et al.* (Latin for "and others"):

The principles of ergonomics have revolutionized the design of office furniture (Brodsky et al., 2007).

- For a work by two authors with the same last name, provide each author's first initials followed by a period and connected by an ampersand:

New York City has the highest percentage of residential turnover per year (J. Yates & P. Yates, 2012).

- Sometimes you will have to cite two or more works by the same author. To distinguish one work from the other, simply provide the copyright year that matches the appropriate work:

Many firms offer employees 401K stock purchase plans (Howe, 2008). However, corporate contributions to employee tax-sheltered accounts have decreased over the last three years (Howe, 2011).

- Occasionally, you will have to cite two works in the same sentence. When you do, list both sources, in alphabetical order, separated by a semicolon. As a rule, though, do not overload readers with more than two or three citations.

The use of salt domes to store radioactive wastes has come under severe attack (Jelinek, 2010; McPherson & Chin, 2009).

- When the work you consulted was written by a company, an organization, or a government agency, use the full name of the organization in your first citation, with the abbreviated name in square brackets. Use the abbreviated name without the square brackets in subsequent citations.

First Citation

Salaries for local electricians were at or above the national average (Commission on Wage and Price Control [Commission], 2011).

Subsequent Citations

Salaries for custodians, however, remained steady (Commission, 2011).

- If the work you've referenced has no author—not even the name of a company, organization, or government agency—use a shortened version of the title and the copyright year:

Customers surveyed tended to focus on the product's problems rather than on its benefits ("Unicom Survey," 2006).

- When you cite an online work, include the author and date but not the name of the website:

D'Arcy claims that by 2015 transportation in Boston will be even more streamlined to accommodate a projected increase in riders (2011).

- APA regards unpublished interviews, surveys, focus groups, emails, telephone conversations, text messages, and IMs as "personal communications." Document personal communications in the text, but do not include them in your References list. Give the initials and surname of the person with whom you communicated either in your sentence or in the citation, specify the type of communication, and include the month, day, and year:

Dr. T. K. Patavi indicated that this information came from the preliminary report submitted last month (interview, April 25, 2011).

- For a quotation shorter than forty words, include the quotation within your sentence (using quotation marks), and place the in-text citation at the end of the sentence, followed by a period:

Pilmer has observed that "renovation of downtown areas is one of the most pressing issues facing city government" (2011).

- For quotations longer than forty words, set the quotation apart from the text by indenting it one-half inch on the left side and eliminating the quotation marks. Place the year after the author's name and the page number after the period at the end of the quotation:

L. J. Ronsivalli (2012) offers this graphic analogy of how radiation can penetrate solid objects:

> One might wonder how an x-ray, a gamma ray, or a cosmic ray can penetrate something as solid as a brick wall or a piece of wood. We can't see that within the atomic structures of the brick wall and the wood there are spaces for the radiation to enter. If we look at a cloud, we can see its shape, but because distance has made them too small, we can't see the droplets of moisture out of which the cloud is made. Much too small for the eye to see, even with the help of a microscope, the atomic structure of solid materials is made of very small particles with a lot of space between them. In fact, solids are mostly empty spaces. (pp. 20–21)

Preparing MLA Works Cited and APA Reference Lists

Whether you follow MLA or APA, you will need to list your sources at the end of your report, on a new page, under the title of Works Cited (MLA) or References (APA) at the top and then arrange the list alphabetically by the authors' last names (except when no author is listed). To see a complete Works Cited list, look at the example at the end of the business report in this chapter (pages 412–413). A complete APA References list can be found at the end of the long report on multicultural workers in Chapter 15 (pages 716–719).

Both MLA and APA advise that you begin each citation flush to the left margin (but indent subsequent lines one-half inch) and separate each item within the citation with a period and one space. Always double-space within and between the entries.

MLA and APA differ in the way you include two or more works by the same author. MLA does not repeat the author's name in a subsequent entry but replaces the author's name with three hyphens (- - -) and a period. When your APA References list includes two or more works by the same author, give the author's name for each entry, and if the publication year is the same, use lower case letters (*a*, *b*, etc.) after the copyright year to distinguish them.

There are many more differences between MLA and APA styles. Table 8.4 points out major differences between the two methods of documentation in terms of where you place information, capitalization, punctuation, and use of italics and quotation marks.

Sample Entries in MLA Works Cited and APA Reference Lists

Book by One Author

MLA Spraggins, Marianne. *Getting Ahead: A Survival Guide for Black Women in Business.* Indianapolis: Wiley, 2009. Print.

APA Spraggins, M. (2009). *Getting ahead: A survival guide for black women in business.* Indianapolis, IN: Wiley.

TABLE 8.4 Basic Differences Between Preparing an MLA Works Cited List and APA References List

	MLA	APA
Author	• List author's last name first, followed by a comma, and then first name and (if applicable) middle name or initial.	• List author's last name first, followed by a comma, and then cite only the first initial and middle initial (if known).
	• For two or three authors, invert only the first author's name (e.g., Smith, John, and Jose Alvarez), and connect the last two authors' names with *and*.	• For multiple authors, invert all authors' names, and separate the last two names with an ampersand (&).
	• For more than three authors, cite just the first author listed on the work (Smith, John) and add *et al.* ("and others"), or you can provide all names in full in the order in which they appear on the title page or byline.	• For more than seven authors, invert the first six authors' names, insert an ellipsis (. . .) and then list the name of the last author (also inverted).
Title	• Italicize the full title of the book, newspaper, journal, or magazine, including any subtitles.	• For a book, italicize the full title, and capitalize only the first word of the title and any proper names. If there is a subtitle, place it after the main title, followed by a colon, and capitalize only the first word of the subtitle.
	• Capitalize all words in the title except for prepositions and articles unless the book or journal begins with one of these.	• For newspapers, journals, or magazines, italicize the full title; capitalize all words in the title except for prepositions and articles.
	• Enclose titles of journal, newspaper, and magazine articles in double quotation marks.	• Do not enclose titles of newspaper, journal, or magazine articles in quotation marks.
	• Capitalize all words in the journal, newspaper, or magazine article title except for prepositions and articles.	• Capitalize only the first word of the article title (even if it is a preposition) and any proper nouns.
Volume and Page Numbers	• For articles, cite the volume and the issue number (separated by a period), followed by the year in parentheses: 52.1 (2012). For newspapers and magazines, use only the date—12 Aug. 2010. Then include page numbers without a "p." or "pp.": 91–100.	• Put the volume number of the journal or magazine in italics, with the issue number (not in italics) in parentheses immediately following, without a space. Then insert a comma and include page numbers: *12*(3), 87–102.

TABLE 8.4 (Continued)

	MLA	APA
Publication	• For books, give the city of publication. Then, after a colon, supply the publisher's name, followed by a comma, the year of publication, and a period. • Include the publication medium for all entries (e.g., Print, Web, PDF, etc.) at the end of the publication information.	• Place the date of publication in parentheses immediately after the author's name. Then add a period after the closing parenthesis. • For books, provide the city and two-letter abbreviation for the state, and then include the publisher's name after a colon—for example, Detroit, MI: MegaPress.
Web Sources (Websites, Blogs, etc.)	• You do not need to include URLs, but indicate the website name, sponsor or publisher, date of publication, and medium of publication, followed by the date of access.	• Insert URLs in place of page numbers with the following designation: Retrieved from [and then list the URL].
Personal Interview	• Provide the name of the person interviewed (last name first), followed by the type of interview that was conducted (email; in person) and the date.	• Interviews, conversations, and presentations are not included in APA reference lists, but you must still cite them within your paper.

Book by Two Authors

MLA   Wu, Melody, and Trent Tucker. *China's Role in the Global Economy.* Denver: Tradevision P, 2011. Print.

APA   Wu, M., & Tucker, T. (2011). *China's role in the global economy.* Denver, CO: Tradevision Press.

Book by Three Authors

MLA   Mallahi, Kamel, Kevin Morrell, and Geoffrey Wood. *The Ethical Business: Challenges and Controversies.* New York: Macmillan, 2010. Print.

APA   Mallahi, K., Morrell, K., & Wood, G. (2010): *The ethical business: Challenges and controversies.* New York, NY: Macmillan.

Book by Four or More Authors (MLA)

MLA   Berkowitz, Harry A., et al. *Collaborating Effectively and Efficiently: A Case Study.* Los Angeles: Ridgeway, 2010. Print.

Book by Eight or More Authors (APA)

APA   Berkowitz, H. A., Barner, P. L., Choi, D. G., Osler, T. O., Ruiz, J., Rowell, C. F., . . . Emmons, W. D. (2010). *Collaborating effectively and efficiently: A case study*. Los Angeles, CA: Ridgeway.

Book Published Online

MLA   Marcus, Bonnie. *Advancing Women's Leadership.* Head over Heels: Women's Business Radio, 2009. Web. 10 Mar. 2011.

APA   Marcus, B. (2009). *Advancing women's leadership* [Adobe Digital Editions version]. Retrieved from http://womenssuccesscoaching.com/wp-content/uploads/2010/01/HeadOver_eBook1-Mar10F.pdf

Edited Collection of Essays

MLA   Chavez, Lisle, and Ted Nowaki, eds. *Urban Planning and People Oriented Space.* Boston: Academic P, 2011. Print.

APA   Chavez, L., & Nowaki, T. (Eds.). (2011). *Urban planning and people oriented space.* Boston, MA: Academic Press.

Work Included in a Collection of Essays

MLA   Papademos, Lucas. "The Effects of Globalization on Inflation, Liquidity, and Monetary Policy." *International Dimensions of Monetary Policy.* Ed. Jordi Gali and Mark Gertler. Chicago: U of Chicago P, 2010. 593–608. Print

APA   Papademos, L. (2010). The effects of globalization on inflation, liquidity, and monetary policy. In J. Gali & M. Gertler (Eds.), *International dimensions of monetary policy* (pp. 593–608). Chicago, IL: University of Chicago Press.

Book by a Corporate Author

MLA   Computer Literacy Foundation. *PCs in the Classroom.* 3rd ed. New York: Technology P, 2011. Print.

APA   Computer Literacy Foundation. (2011). *PCs in the classroom* (3rd ed.). New York, NY: Technology Press.

Article in a Professional Journal

MLA   Fieseler, Christian, Matthes Fleck, and Miriam Meckel. "Corporate Social Responsibility in the Blogosphere." *Journal of Business Ethics* 91.4 (2010): 599–614. Print.

APA   Fieseler, C., Fleck, M., & Meckel, M. (2010). Corporate social responsibility in the blogosphere. *Journal of Business Ethics, 91*(4), 599–614.

Article in a Print Magazine

MLA   Kurowska, Teresa. "Is the Boss Watching Every Keystroke You Make?" *Today's Workplace* Oct. 2011: 47+. Print.

APA   Kurowska, T. (2011, October). Is the boss watching every keystroke you make? *Today's Workplace, 47*, 72–73.

Article in an Online Professional Journal

MLA   Mayer, Gloria, and Michael Vallaire. "Enhancing Written Communication to Address Health Literacy." *Online Journal of Issues in Nursing* 14.3 (2009). n. pag. Web. 30 Sept. 2009.

APA   Mayer, G., & Vallaire, M. (2009, September 30). Enhancing written communication to address health literacy. *Online Journal of Issues in Nursing, 14*(3). doi:10.3912/OJIN.Vol14No02ManOS

Article in a Print Newspaper

MLA   Korkki, Phyllis. "Finding a Job by Starting a Business." *New York Times* 31 Jan. 2010: BU2. Print.

APA   Korkki, P. (2010, January 31). Finding a job by starting a business. *New York Times*, p. BU2.

Online Encyclopedia Article

MLA   Kling, Arnold. "International Trade." *Concise Encyclopedia of Economics.* 2nd ed. Library of Economics and Liberty, 2008. Web. 27 Mar. 2010.

APA   Kling, A. (2008). International trade. In *Concise encyclopedia of economics* (2nd ed.). Retrieved March 27, 2010, from http://www.econlib.org/library/CEE/html

Unsigned Online Encyclopedia Article

MLA   "Inflation." *Encyclopaedia Britannica Online.* Encyclopaedia Britannica, 2010. Web. 27 Mar. 2011.

APA   Inflation. (2010). In *Encyclopaedia Britannica online.* Retrieved March 27, 2011, from http://www.britannica.com.

Unsigned Article in a Print Magazine or Newspaper

MLA   "The Green Machine." *Economist* 13 Mar. 2010: 7–8. Print.

APA   The green machine. (2010, March 13). *Economist,* 7–8.

Article in an Online Newspaper or Magazine

MLA   Zakaria, Fareed. "'Swing for the Fences': Energy Secretary Steven Chu on Boosting Technology." *The Washington Post.* The Washington Post Company, 29 Mar. 2010. Web. 31 Mar. 2010.

APA   Zakaria, F. (2010, March 29). "Swing for the fences": Energy secretary Steven Chu on boosting technology. *The Washington Post.* Retrieved from http://www.washingtonpost.com/wp-dyn/content/article/2010/03/29/AR2010032901892.html

Government Document

MLA   United States Middle Class Task Force. *Green Jobs: A Pathway to a Strong Middle Class.* Washington: GPO, 2009. Print.

APA   United States Middle Class Task Force. (2009). *Green jobs: A pathway to a strong middle class* (Publication No. 0851-K-02). Washington, DC: U.S. Government Printing Office.

Website

MLA Home page. Natl. Council of La Raza, 2010. Web. 28 Mar. 2010.

APA When referencing an *entire* website, APA style is to provide the URL in the body of the text and not list it in the References section.

Radio

MLA *Serious Money.* Hosts Renee Hanson and James Ayala. Financial News Radio. KFNN, Phoenix, 7 Apr. 2010. Radio.

APA Hanson, R. (Host), & Ayala, J. (Host). (2010, April 7). *Serious money.* Phoenix, AZ: KFNN.

Television

MLA "No Frills Business Travel." Prod. Jeff Nathenson. *Business Traveler*. CNN. 14 Apr. 2010. Television.

APA Nathenson, J. (Producer). (2010, April 14). No frills business travel [Television series episode]. In *Business Traveler*. Atlanta, GA: CNN.

Podcast

MLA Gunther, Marc, prod. "Coca-Cola's New PlantBottle Sows Path to Greener Packaging." *Greenbiz Radio.* Greener World Media, 1 Dec. 2009. Web. 28 Mar. 2010.

APA Gunther, M. (Producer). (2009, December 1). *Coca-Cola's new PlantBottle sows path to greener packaging* [Audio podcast]. Retrieved from http://www.greenbiz.com/podcast/2009/12/01/coca-cola-new-plantbottle-sows-path-greener-packaging

Email

MLA Frazer, Tim. "Site Inspection Report for Landsdowne Corners." Message to the author. 12 July 2011. Email.

APA Emails are not included in the References list. They are cited in the text as a personal communication.

Brochure

MLA Gao, Hubert. Coping with Carpal Tunnel Syndrome. New York: Beth Israel Hospital, 2012. Print.

APA Gao, H. (2012). Coping with carpal tunnel syndrome [Brochure]. New York: NY: Beth Israel Hospital.

Survey

MLA Guttierez, Joseph. "Market Survey for Duron, Inc." Survey. n.p. 10 Apr. 2011. Print.

APA Surveys are unpublished personal communications not included in the References list.

Lecture or Speech

MLA Phillips-Ricks, Jonathan. National Association of Black Business Leaders Conference. New York, 15 Aug. 2012. Lecture.

APA Phillips-Ricks, J. (2012, August 15). Lecture presented at the National Association of Black Business Leaders Conference, New York, NY.

Press Release

MLA American Council of Organic Farmers. *New Ways to Eliminate Chemicals from Home Gardens.* Omaha: ACOF, 31 Mar. 2011. Print.

APA American Council of Organic Farmers. (2011, March 31). *New ways to eliminate chemicals from home gardens* [Press release]. Retrieved from aaof.org

Map or Graphic

MLA Fineberg, Donald. *Sonoma, California.* Map. Sonoma: Professional Maps, 2011. Print.

APA Fineberg, D. (2011). *Sonoma, California* [Map]. Sonoma: Professional Maps.

Motion Picture

MLA *Understanding Diabetes: From Diagnosis to Cure.* Dir. Jayne T. Cahill. Healthcare Videos. 2010. DVD.

APA Cahill, J. T. (Director). (2010). *Understanding diabetes: From diagnosis to cure* [DVD]. Allentown, PA: Healthcare Videos.

A Business Research Report

The rest of this chapter consists of a business report written by members of the marketing team at New Horizons Development, Inc., a real estate group with holdings across the country. The marketing team was asked to create a plan to attract tenants to Sawmill Ridge, a new apartment complex that New Horizons was constructing in the Dallas/Fort Worth area. Study the report on pages 399–413 to see how the team members successfully combined primary and secondary sources to research and write their business report. Also note how the report documents these sources in the text and references them on the Works Cited list using the MLA format. Compare the documentation style in this report to the long report in Chapter 15 (pages 701–719), which uses APA in-text documentation and includes a References list. While these two reports illustrate MLA and APA guidelines, keep in mind your boss may ask you to adapt these styles to conform to your company's policies on documentation. See Chapter 9 (pages 442–444) for more on abstracts and Chapter 15 (pages 692–695) for more on preparing a table of contents.

A Letter of Transmittal

Distinctive, relevant letterhead

New Horizons

Building Apartment Homes Since 1992

2800 Taylor Blvd.
Fort Worth, TX 76003
817-555-3300
www.newhorizons.com

April 12, 2011

Talia Martinez-Ryals, Director
New Horizons, Southeast
170 Waters Drive
Tucson, AZ 85749-3001

Dear Director Martinez-Ryals:

Indicates that plan was requested by the reader

We are happy to include the attached marketing plan for Sawmill Ridge that you commissioned. The development is scheduled to open in three months in this rapidly growing community of Arlington, Texas.

Specifies purpose, scope, and importance of the plan

Our plan projects launching a campaign to market Sawmill Ridge for its community atmosphere and benefits as well as its proximity to expanding major employers and retail centers. We have determined that our most appropriate target audience will be young professionals.

Emphasizes that the plan is based upon careful research

Our report is based on extensive demographic and market research. For the last two months, we have thoroughly investigated the South Arlington area, interviewed numerous business and community leaders, conducted focus groups, and explored current and projected demographics.

Invites reader's questions

Thank you for asking us to prepare this report for New Horizons and you. We would be happy to answer any questions you have about this report or supply you with further information. We look forward to hearing from you.

Respectfully,

Adrienne Hong

Adrienne Hong

Tyrell Carpenter

Tyrell Carpenter

Margarita Gonzales

Margarita Gonzales

Encl. 1

A Marketing Plan for Sawmill Ridge, 2011

Adrienne Hong, Tyrell Carpenter,
and Margarita Gonzales

Prepared for
Talia Martinez-Ryals, Project Director, New Horizons
Southeast, Inc.

April 12, 2011

Abstract

This report contains a marketing strategy for our new Sawmill Ridge development in South Arlington, scheduled to open in three months. Based upon extensive primary and secondary research, our strategy for long-term success is to promote this new apartment complex by fostering a sense of community and emphasizing the proximity of the complex to employers, schools, and shopping. These benefits will attract our target audience of young professionals (ages 24–34) whose jobs, income, family demographics, and lifestyle will help guide us in determining how and where we advertise Sawmill Ridge. We need to take advantage of both Internet and print media, including developing a customized Sawmill Ridge website that will provide links to social networking sites and offer news about weekly and monthly activities. Essential to our strategy as well is developing a follow-up program to retain renters, and thus reduce turnover expenses, by providing the services and amenities that our target audience appreciates and expects. Whether attracting potential renters or minimizing vacancies, we have to put into practice what current research identifies as the key element in promoting an apartment complex—responsive follow-up calls. If accepted, our plan promises maximum occupancy at Sawmill Ridge with a minimum loss of tenants when leases expire.

Table of Contents

Logically divides report into sections

Gives readers a quick glance at the organization of the report

iii

1

Introduction: Strategic Plan

Two years ago, New Horizons Development, Inc., purchased a large tract of land (21.2 acres) in the Dallas/Fort Worth area to develop Sawmill Ridge, a Class A apartment community with a 54-building complex. The opening of Sawmill Ridge is scheduled for completion in three months. As a result of our overall objectives for this project, we have created a marketing plan for the next phase of Sawmill Ridge to attract tenants to the development.

Our plan, described in this report, is based on a key marketing strategy linking Sawmill Ridge to community development. We provide background information about market conditions and location, describe the target audience, and outline plans for advertising the development. Our report also covers how we need to follow up with inquiries and how we can best retain residents when their lease expires.

Building Community

To attract and retain tenants to Sawmill Ridge, our plan will be based on the marketing approach "Building Community by Offering Neighborhood Convenience." Selling the community is as important to our marketing plan as selling the development itself. Having access to neighborhood conveniences is especially attractive to our target audience, analyzed below. Consequently, the theme of community involvement is woven throughout our plan.

Background

The rental market conditions in the greater Dallas/Fort Worth area, as well as Sawmill Ridge's proximity to three major employers and many new retail establishments, have shaped the focus and goals of our marketing plan.

Rental Market Conditions: Supply and Demand

Sawmill Ridge is opening at a very favorable time. The Dallas/Fort Worth rental market has continued to thrive for over two decades (D'Argento, *Contemporary Assessment* 211–13), and apartment net leasing activity in this area has hit a five-year high in the past twelve months, with absorption reaching 24,390 units (Evinson 211). As Figure 1 on page 2 shows, the annual demand included a strong fourth quarter performance as apartment occupancy in the Dallas Metro area increased by approximately 5,620 units. During the fourth quarter of 2010, occupancy in the Dallas/Fort Worth area tightened significantly, standing at 92.7%. These figures point to a steady impressive increase since the beginning of 2010. O'Connor and Associates (pocconor.com), which provide monthly data on apartment trends, record an impressive 3.1 percent growth in apartment occupancy in Arlington throughout 2010.

2

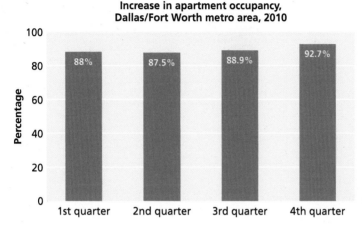

Figure 1 Quarterly Dallas/Fort Worth Occupancy Rates (2010)

Further contributing to this strong demand, apartment communities have lost fewer residents to first-time home purchases over the last year (North Central Texas Council).

Most important, though, in fueling this sizable demand for apartments, Dallas/Fort Worth has added a large block of new jobs. The U.S. Bureau of Labor Statistics reported that from October 2009 to October 2010, more than 24,000 jobs were generated in the area (*Employment and Earnings Report* 111). According to Barbara Corcoran, a real estate expert frequently appearing on *The Today Show*, "The single most important factor in selecting housing is the location of jobs in the area." Corroborating this view, a recent article in *Time* magazine declared that Texas was "the job leader … adding more jobs faster than any other state. The reason may be … housing, drawing people to the state and creating jobs" ("The Economy" 20). We can take advantage of this increase in jobs.

Location: Selling the Community

As the map in Figure 2 on page 3 illustrates, Sawmill Ridge is located in South Arlington, near the intersection of Pike Blvd. and Interstate 20, a major east-west corridor in the Metro area (South Arlington). Dallas is 9 miles to the east, while Fort Worth is only 11 miles to the west. Moreover, Sawmill Ridge is close to major retail centers and several large corporations and businesses.

Proximity to Retail Centers

Essential to our marketing campaign, Sawmill Ridge is located in a highly desirable retail corridor in Arlington. The Parks Mall, which houses more

3

Figure 2 Map of Sawmill Ridge and the South Arlington Area

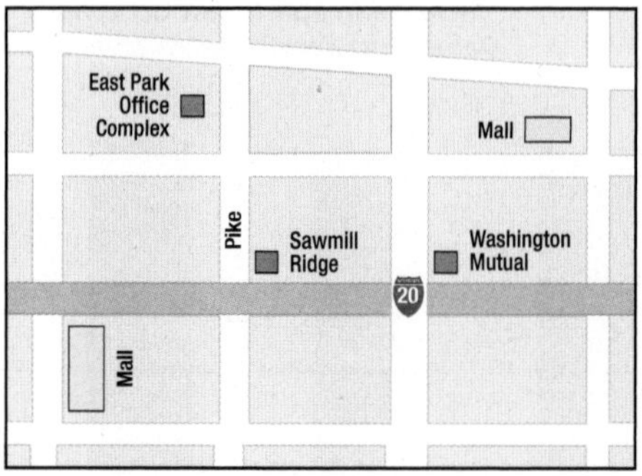

Map provides readers with a clear visual of the site location

than two million square feet of retail shopping, is only 1.2 miles north of Sawmill Ridge. A new upscale retail center, the Highlands, with an additional 900,000 square feet of retail space, is currently being built only 1.8 miles from Sawmill Ridge. Based on our interview with Highland's Associate Director, companies like Ethan Allen, Golf Galaxy, Talbot's, and Studio Movie Grill have already purchased retail space in the new center (Weir). Other retailers—possibly the Body Shop, Gemstone Jewelers, Filene's Basement, and Best Buy—have been negotiating retail space and are leasing units quickly (Stratton). Having so many diverse and popular retailers nearby substantially increases the attractiveness of Sawmill Ridge and further emphasizes the neighborhood feeling for our subdivision.

Successful blend of primary research (interview) and secondary research (articles)

Our Arlington Area Quality of Life survey (available on our intranet) of over 700 area residents, conducted between November 15 of last year and January 31 of this year, highlights this fact. Response rate was high—over 30% of those surveyed responded. Ninety-five percent of the area residents indicated that they visit the Parks Mall weekly, and an additional 86% plan to visit the Highlands just as frequently after it opens. Only 12% were concerned that the Highlands would not cater to their needs (Hong et al., survey).

Primary research (survey) provides useful statistics not available elsewhere

Proximity to Employers

Sawmill Ridge's proximity to major employers is also a tremendous advantage in terms of rental appeal. Our campaign will make maximum use of this benefit. Among the area's largest employers is Washington Mutual, whose central office with 1,500 employees is within easy walking distance of Sawmill Ridge. Another major group of employers can be found at East Park Office Complex. A recent study projects that over twenty-five tenants will employ more than 350 individuals there, a very convenient location since

Report supplies essential data to support the plan

4

the complex is less than a mile from Sawmill Ridge (D'Argento, "Booming Market").

Second work by previously cited author requires title

We can also target our campaign to market our development to current employees at these firms. Our semiannual focus group, which surveyed eleven employees from Washington Mutual and eight employees at the three largest firms at the East Park Office Complex, concluded that more than half of the employees at these companies are dissatisfied with their long commute to work and would strongly consider relocating to Sawmill Ridge. Only 20% of focus group participants claimed they would be "somewhat likely" to relocate (Hong et al., focus group).

Focus group provides otherwise unavailable information

Targeting the Right Demographics

Based on the large numbers of varied retailers and businesses found in the South Arlington area, we will tailor our marketing strategy to attract young professionals (ages 24–34) with per capita incomes between $27,600 and $38,500 per year, individuals already working at these businesses or most likely to be employed there. Because many individuals in this population range have families with small children or are young single professionals, Sawmill Ridge will emphasize the security and satisfaction with the community that our development offers. (Three top-rated elementary schools and one junior high school are within a 2.4-mile radius of Sawmill Ridge.)

Identifies target audience

Primary research cited

Our plan is firmly based on the demographics of the area. The U.S. Census Bureau report (*American FactFinder*) and local real estate studies (Kearney-Schwartz; Evinson) provide the following statistics, relevant to the 5-mile area surrounding Sawmill Ridge, reinforcing why and how our plan needs to reach this target audience:

Goverment agency listed like a corporate author

- The total population within this radius is 380,085, and the median age is 29 years old.
- 37.4% of the population is composed of single-earner or dual-earner family households with teenage children, while 19.8% of these households have children under 12 years old.
- The approximate size of renter-occupied housing units for the average household is 2.56 people.
- The average household income is $42,847, higher than the national average by 12 percent. Of this population, 33% are in management-level positions, 25.7% work in non-management but well-paying office jobs, and 17% hold entry-level positions.

Target demographic solidly backed up by federal and local statistics; website does not need page number

Based on these South Arlington demographics, we should easily be able to attract this target group to Sawmill Ridge, offering them the comfort, con-

5

© Cengage Learning 2013

Work with no author listed

venience, and security they demand from a large home apartment developer such as New Horizons. ("New Complex" D11+).

Advertising Strategy

We plan to advertise both on the Web and through print media to reach the maximum number of potential tenants for Sawmill Ridge, with a special emphasis on Web advertising.

Web Marketing

Cites only bottom-line numbers

Citation for Punji includes article title because author has two works listed in Works Cited

As the major medium in attracting clients, the Web offers a wide range of options and benefits for our plan. In fact, our marketing plan depends heavily on New Horizons using Web marketing. About 72% of Americans use Web-based sources on a regular basis (Punji, "E-commerce" 48). Of those, an estimated 60% rely on the Internet almost exclusively to research basic information, make everyday purchases, and research matters of personal and financial importance, including home ownership and apartment rentals (Poulos 43–67).

All statistics and key points solidly backed by research

Most relevant for our plan, research further shows that 85% of professionals looking for rentals rely on Web-based media as their resource versus 13% who use print resources. Yet Alvarez and Pick find that 21% use Web and print resources with many moving toward using apps on their mobile phones (4–5). According to Terry Slattery, in his posting at **ForrentPress.com**, "Mobile marketing is quickly spreading as a leading method of communication to engage consumers." Similarly, Sue McCallister ("Mobile Apps Can Assist Homebuyers, Renters") claims that listing apartment rentals with embedded links in mobile apps will direct traffic to developers' sites. We need to explore the possibility of contracting with a Web-based mobile provider. Right now, potential renters find notifications via all three media: print, online, and mobile. Regardless of the route, if a Web inquiry, for instance, is followed up within the first eight hours of being received, a leasing professional has a 75% chance of renting the space to that client, a much higher success rate than with prospective tenants responding to print advertisements (Bruckner 45).

Costs

In terms of bottom-line costs, Internet advertising is highly advantageous. The Internet ads will cost us only a fraction of what we would have to spend on a print source. Table 1 on page 6 shows the prices of print versus Internet advertisements for a quarter page community-targeted newspaper ad or comparable Internet ad over a one-month period.

6

Table 1. Print versus Internet Advertising Rates (February 2011)—
One-Quarter Page/Internet Comparable Rates

	Print	**Internet**
Dallas Monitor (Sunday)	$950 (4 consecutive Sunday ads)	$252 (one month—continuous)
Dallas Tribune (Sunday)	$825 (4 consecutive Sunday ads)	$450 (one month—continuous)
Dallas Daily News (Sunday)	$825 (4 consecutive Sunday ads)	$450 (one month—continuous)
Dallas Business News (business weekly)	$725 (4 weeks)	$375 (one month—continuous)
El Mundo (Hispanic weekly)	$675 (4 weeks)	$300 (one month—continuous)
Dallas Defender (African American weekly)	$675 (4 weeks)	$300 (one month—continuous)
Dallas Alternative (arts and alternative weekly)	$650 (4 weeks)	$275 (one month—continuous)

Table clearly spells out differences between print and Internet ad rates

Table appears at right place in report and is easy to find and follow

No source line listed for table because it is original work

The Most Effective Internet Rental Sites

To determine the most attractive Internet sites, we met with consultants at E-pointe, a firm New Horizons used successfully on previous projects. Our team also attended a seminar regarding the top Internet rental sites in the Dallas/Fort Worth area to determine which were most frequently accessed by renters (Je-mun). Based on the information we gathered from these sources, we believe that the following four real estate-specific sites are the most effective venues for advertising Sawmill Ridge:

Attending seminar represents an important part of secondary research

Speech cited by speaker's name

- **www.rent.com**—The number one national rental source with a popular Dallas/Fort Worth area section.
- **www.apartments.com**—The number two rental source in America, again with a Dallas/Fort Worth area section.
- **www.rentclicks.com**—Another national site, which, although new, is impressive due to its high placement results on major search engines.
- **www.DFWapartments.com**—A rental source accessed more often than any other site by Dallas/Fort Worth residents.

Bulleted list breaks up text and improves readability

Creating a Custom Website

In addition to advertising on sites such as those identified above, we need to follow through with our "Creating Community" theme by designing our own

7

website targeted to reach the demographic audience most likely to rent at Sawmill Ridge. Creating our own website is essential to target our audience and enhance our professional image (Punji, "Image"). Developing our website will help us in the following ways:

1. A well-designed site can help New Horizons surpass other developers whose Internet presence is weak in the Arlington area.
2. It will provide our on-site personnel with a powerful marketing tool (Gilbert). Naturally, all our print media, signage, and stationery will include our URL to give Sawmill Ridge maximum exposure.

3. Reflecting New Horizons' green philosophy, our website will feature images of Sawmill Ridge's eco-friendly landscapes and energy-saving features.
4. Our site will include an application form that prospective tenants can fill out and submit online, considerably simplifying the leasing process for both metro and out-of-town renters. When potential customers complete the form online, we intend to waive the application fee of $50.00, thus "encouraging the Web shopper to action," as Alvarez and Picknote term it (224).

5. A Sawmill Ridge website will further allow us to give prospective tenants the convenient option of paying their security deposit and monthly rent online.
6. The new site will also feature a Twitter feed that enables Sawmill Ridge residents to communicate with a large audience of socially conscious people who will spread the word about our development. By connecting with local environmental groups, we can build a solid network of people interested in New Horizons' commitment to the Arlington area.

7. Sawmill Ridge will have its own Facebook page as well with a link to our website, further advancing our marketing objective of "building community."

In designing our website, we hope to profit from and be consistent with similar sites that other divisions in our company have created. Figure 3 shows the homepage, located at **www.newhorizonstucson.com,** for a comparable development in Tucson, Arizona, which is targeted to a similar audience. This website receives over 3,100 hits a week and incorporates the following features we think should be included on the website for Sawmill Ridge:

- Advertisements for discounts and specials at local businesses
- Information and links to area social, recreational, and educational resources
- A "Residents' Page" with uploaded pictures from parties, comments from residents, and news about upcoming events at Sawmill Ridge
- Design and feature updates on a regular basis

8

Figure 3 Home Page for New Horizons Tucson Complex

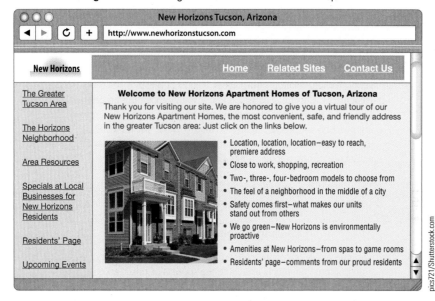

© Cengage Learning 2013

Easy-to-navigate, user-friendly site

Provides relevant links

Writing is clean and concise

Background color makes writing easy to read

Print Advertising

Even though we will use Web marketing as our major medium to attract new tenants, we should not neglect print media entirely. Our young professional audience in the Dallas/Fort Worth area still regularly reads and subscribes to several glossy print publications. Studying this print market, we believe that the following are the chief publications in which ads for Sawmill Ridge should be run:

Assures reader that print media will not be neglected

- *The Arlington Advocate*. As Table 1 indicates, the cost for ads in newspapers is very high. Yet *The Arlington Advocate* will combine the cost of print ads and those on its website. After interviewing Darius Tant, manager of *The Arlington Advocate*'s advertising department, we negotiated favorable terms for a three-month contract. We can run two ads a month for $750.00 and receive coverage in print and online. We should also consider advertising in the major Dallas/Fort Worth newspapers to reach individuals seeking to relocate to the metroplex area.
- *Apartment Guide Magazine*. One of the most popular rental publications for prospects in the Dallas/Fort Worth area, with a circulation of 125,000, the AGM is readily available at many locations in the Sawmill Ridge area. A free advertisement on their website, located at **www.apartmentguide.com**, is included with the cost of a print ad in

Lists in easy-to-read bulleted segments the top five print advertising options

Research on price and circulation backs up argument and convinces reader to accept plan

9

Gives reader clear idea about quantities, distribution, and importance of publications

Apartment Guide. A one-page ad in *Apartment Guide* costs $850 per month. Sawmill Ridge is also negotiating with *The College Apartment Guide*, a subsidiary of *AGM*, to target Dallas/Fort Worth students seeking to obtain employment in the Arlington area after graduation.

- **Direct Mailers**. Professionally made direct mailers will be created and distributed to comparable communities to attract residents from these areas. We anticipate printing about 10,000 of these.
- **Brochures**. Carefully designed and visually appealing brochures, such as those we use at our other sites, will be given to potential tenants during property tours and can also be distributed to local businesses and major employer relocation departments for our outreach marketing. We anticipate printing about 5,000–6,000 of these.

These last three marketing tools can be prepared relatively inexpensively in-house thanks to our new high-speed printers and Photoshop software.

Creating a Follow-Up Program

Authors plan ahead for second phase of marketing strategy

Following up every telephone call, email inquiry, or mobile text communication from potential tenants is crucial to the success of this marketing plan. When we receive a query, an expression of interest, or a referral, we need to gather all relevant contact information from potential clients, including cell, home, and work phone numbers; email address; and street address. In addition, we should try to set up appointments immediately unless a prospective resident is unable to commit, in which case a Sawmill Ridge representative should call to arrange an appointment.

Uses secondary research (published article) to document consequences of not following plan

Even more important, follow-up contact is a must for prospective tenants who have visited Sawmill Ridge. Accordingly, each visitor to our development will receive both a follow-up phone call (or email) and a thank-you card. As Kelly Talos emphasizes in an article in *Rental Professionals Today*, 47% of rental communities do not follow up with prospective residents effectively, resulting in numerous missed opportunities (90).

Retaining Renters

Research goes beyond just renting units

Our marketing plan also takes into account ways to retain residents at Sawmill Ridge beyond their initial leasing agreements. We need to make sure we retain our renters, maintain the overall occupancy of the development, and reduce turnover expenses. As the remarks below from Roger Cho, in *Resident Developer's Monthly*, indicate, developers need to link keeping residents with initial lease signing:

Long quotation indented 1 inch with no quotation marks

Developers and renters often neglect the importance of retaining residents, perhaps the most basic ingredient in the leasing recipe. . . . Retention is particularly important in apartment

10

complexes, where, depending on monthly rental rates, the owners can expect to lose as much as 15% of their annual revenue as a result of turnover . . . and as much as 8% of annual revenue due to residents withholding rent when their satisfaction is low and repairs are neglected. Those numbers may seem standard, but add them together and you stand to lose nearly 25% annually—a percentage you can drastically reduce with a smart retention plan. (34)

To minimize turnover and to maximize occupancy, our plan addresses the expectations of our audience and is consistent with our marketing strategy of "Creating Community":

- To accommodate our residents' schedules, the office will open early (6:00 a.m.) and close late (7:00 p.m.) five days a week. This policy should help residents unable to visit the office during typical nine-to-five business hours.
- Sawmill Ridge will host at least four functions each month, including a monthly resident brunch, all announced on our Facebook and Twitter pages. In months with greater chances for decreased occupancy, the number of activities should increase. Resident functions should appeal to all residents. A resident activity team will survey renters for areas of interest and will set up events.
- Thanks to an agreement we made with the Pet Motel in South Arlington, Sawmill Ridge residents will receive a 15% discount on kennel care.
- Emergency maintenance service will be provided 24/7. Maintenance technicians will be instructed to leave follow-up cards in each apartment home after they complete the service request.
- All residents will receive greetings on their birthday, including candy on their doorstep and a birthday card in their mailbox.
- A monthly newsletter (also available online), such as the one for New Horizons Tucson, can supplement our website, showing our appreciation to tenants, providing information about upcoming events and policy changes, and creating a strong community atmosphere.

Conclusion

We believe that our twofold approach—targeting the interests of our ideal demographic audience and building a sense of community—will be effective in renting the apartments at the new Sawmill Ridge development. As part of this approach, we will market Sawmill Ridge online, in print media, and through mobile apps to reach and to appeal to this large audience of potential renters. We will emphasize customer follow-up during the decision-making process, and we will continue to foster a sense of community and professionalism long after residents have signed their leases to ensure maximum occupancy of our development.

11

Works Cited
list begins on
a new page
with the title
centered at the
top

<h2 align="center">Works Cited</h2>

Alvarez, Jesus, and Mary Ellen Pick. *eMarketing Tools for Interactive Communication*. Columbus, OH: Capital City P, 2011. Print.

Bruckner, Hans. "Don't Forget to Follow Up." *Real Estate Today* Mar. 2011: 45. Print.

List ordered
alphabetically
by author

Cho, Roger. "Retaining Renters." *Resident Developer's Monthly* Oct. 2010: 34+. Print.

Corcoran, Barbara. "The Real Estate Market." Prod. Victor Talbott. *The Today Show* NBC. KVOA, Tucson, 28 Feb. 2010. Television.

D'Argento, Camille. *A Contemporary Assessment of Dallas/Fort Worth Real Estate*. Dallas: Dallas UP, 2010. Print.

Second and
subsequent
lines indented
five spaces

---. "South Arlington's Booming Market." *Dallas Real Estate Journal Online* 5.3 (22 Nov. 2007). Web. 1 Feb. 2009. n. pag.

"The Economy." *Time*, "Briefing." 4 Apr. 2011: 20. Print, Web.

Evinson, Lakita. "Leasing at a Five-Year High in Our Region." *Dallas/Fort Worth Today* Jan. 2010: 211. Print.

Gilbert, Ray. Personal interview. 4 Oct. 2010.

Hong, Adrienne, Tyrell Carpenter, and Margarita Gonzales. "Arlington Area Quality of Life Survey." Survey. Arlington, TX. 31 Jan. 2011. Print.

---. "Notes." Focus group. Arlington, TX. 12 Dec. 2010. Print.

Je-mun, Yun. "Selling Dallas on the Web." Dallas Association of Real Estate Brokers conference, Dallas. 14 Sept. 2009. Lecture.

Kearney-Schwartz, Marianne. "2007 Local Real Estate Statistics." *Dallas Real Estate Journal* 6.1 (23 Jan. 2011): 15–19. Print.

McCallister, Sue. "Mobile Apps Can Assist Homebuyers, Renters." *The M-Commerce Guide*. 4 May 2010. Web. 8 Nov. 2010.

"New Complex State of the Art." *Dallas Times* 20 Jan. 2011: D11+. Print.

North Central Texas Council of Governments. *2010 Population Estimates*. Arlington, TX: Author, 2010. Print, Web.

O'Connor and Associates. "Dallas/Fort Worth Market Update." Nov. 2010. 1 Dec. 2010. Print.

Poulos, Paul. "Internet Real Estate Trends." *Real Estate Today*. Ed. Michael R. Rosenfeld. New York: Jossey-Bass, 2007. 43–67. Print.

12

Punji, Kamlesh, "E-Commerce Profits from Soaring Rise in Internet Use." *Contemporary Marketing Times* 16 (Jan. 2011): 47–51. Print.

---. "Your Internet Image." *Punji's Picks: Blog*. 30 Oct. 2011. Web. 16 Dec. 2010.

Slattery, Terry. "ForRent Launches New GPhone Application." *Forrent.com Press Blog*. 29 June 2009. Web. 14 Dec. 2010.

South Arlington, Texas. Map. Dallas: MapMakers, 2011. Print.

Stratton, Daniella-Kay. "Highlands Center on the Move." *Dallas Life* Nov. 2010: 341. Print.

Talos, Kelly. The Importance of Brokerage Follow-Through." *Rental Professionals Today* 16.5 (2010): 90. Print.

United States Bureau of Labor Statistics. *Employment and Earnings* 57.12 (Dec. 2010). Washington, DC: GPO, 2010. Print.

United States Census Bureau. *American FactFinder*. 15 June 2010. Web. 1 Nov. 2010.

Weir, Heidi. Personal interview. 25 Mar. 2011.

Two works by same author

Unsigned map listed by map title

Organizations listed alphabetically by name

Conclusion

This chapter has introduced you to some very basic yet essential strategies and tools for doing primary and secondary research on the job. Clearly, you need to rely on a host of resources—print and online reference works, databases, various search engines, websites, blogs, interviews, and surveys. Relying on these research tools and strategies, you will have the most up-to-date, most thorough, and most relevant answers to the questions you are asked to investigate and the problems you need to solve on the job. Exploring all these resources will prepare you to write the types of documents—websites, instructions, short and long reports, proposals—discussed in later chapters. Studying the business report included in this chapter will show you the types of research your employer will expect you to do.

✓ Revision Checklist

Process of Research
- [ ] Identified ways to research a significant, timely, and restricted topic.
- [ ] Formulated a mission statement to develop a clear sense of the purpose of research for employer.
- [ ] Consulted a wide variety of research materials.
- [ ] Conducted database, online catalog, and Internet searches.
- [ ] Used both primary and secondary research methods as needed.
- [ ] Evaluated the relevance and validity of my sources.
- [ ] Met with employers, reference librarians, or experts in my field.
- [ ] Networked with colleagues and boss to meet all deadlines.
- [ ] Documented sources accurately and completely.

Primary Research
- [ ] Conducted, or observed directly, an experiment or visited a site to describe and evaluate events, places needed for research.
- [ ] Set up an appointment with an authority on my topic. Prepared a list of appropriate questions beforehand.
- [ ] Interviewed experts, customers, and relevant government officials, used prepared questions, and stayed focused on the subject. Asked important follow-up questions.
- [ ] Identified a target audience for surveys.
- [ ] Created and distributed questionnaire surveys using only valid, accurate questions and then interpreted my respondents' answers.

Secondary Research
- [ ] Searched online catalogs for necessary research materials.
- [ ] Used databases to find relevant articles and other research materials.

☐ Obtained full text of appropriate articles and other materials in hard or electronic copy.

☐ Checked relevant reference materials, such as almanacs, abstracts, encyclopedias, maps.

☐ Located relevant government sources using USA.gov, the Library of Congress, or the U.S. Government Printing Office.

☐ Read and evaluated periodicals, books, government documents, blogs, and websites.

Internet Searches

☐ Searched for pertinent information using several search engines.

☐ Used multiple specific, concrete keywords to conduct Internet searches.

☐ Used search strategies such as Boolean connectors or delimiters to narrow and focus my keyword searches.

☐ Conducted a directory search or used e-library sources.

☐ Downloaded electronic sources available via library's online catalog or databases.

☐ Located webpages relating to my topic.

☐ Continued to refine my search using synonyms, alternative keywords, or more specific terms to access only the most useful material for my project.

☐ Joined a newsgroup or surveyed corporate blogs to extend my research.

Taking Notes

☐ Took careful notes and identified precisely the source from which my information came.

☐ Recorded all quotations accurately.

☐ Paraphrased fairly, ethically representing original material.

☐ Distinguished my comments and responses clearly from those of my sources.

☐ Incorporated information from notes into appropriate places in my document.

☐ Used correct punctuation with direct quotations, especially ellipses and brackets for interpolations.

Documentation

☐ Gave full and proper credit to sources consulted and cited in my work.

☐ Avoided plagiarism by supplying complete and accurate documentation of all sources quoted, paraphrased, or consulted for the paper or report.

☐ Recorded all direct quotations accurately; included page references where applicable.

☐ Paraphrased information correctly and acknowledged sources fully and accurately.

☐ Double-checked spelling of authors' and publishers' names and accuracy of all pertinent publication information.

☐ Followed MLA or APA documentation method consistently in preparing Works Cited or Reference lists.

☐ Included all necessary in-text (parenthetical) references; cited each parenthetical reference fully and in correct alphabetical order in Works Cited or Reference lists.

☐ Made sure all works referred to in my report were included in Works Cited or Reference lists.

☐ Alphabetized Works Cited and Reference entries correctly.

☐ Documented all Internet sources properly, giving access dates.

Exercises

1. Choose, define, and restrict a topic based on a problem or issue you might deal with in one of the following divisions of a company:
   a. IT
   b. human resources/diversity
   c. security
   d. marketing
   e. accounting
   f. health care/health risks
   g. energy/utilities
   h. transportation
   i. environment

   Discuss the steps you took to narrow the topic, the audience you would be writing for, and the types of questions that audience may have.

2. Based on the problem you identified in Exercise 1, select an expert relevant to that field to interview for primary research. Confer with classmates or co-workers to decide on whom to interview. Prepare a list of ten questions for the interview, remembering to stay focused on your topic.

3. Visit an appropriate office, plant, agency, environmental site, or other location relevant to the topic you chose in Exercise 1.

4. Plan a direct observation experiment in a laboratory or other appropriate location to further investigate the topic you selected in Exercise 1. Construct an outline for this experiment, considering the types of information you expect to record and how it will enhance your research.

5. Construct a questionnaire to gather more information about the topic you selected in Exercise 1 or one of the following:
   a. Internet security
   b. online purchasing
   c. business etiquette

    d. safety in the workplace

    e. global business trade

    f. corporate image/mission

    Come up with a list of ten questions and vary the format to include multiple-choice, yes/no, ranking, and open-ended questions. Think about how you will select your participants, the type of information you want to gather, and how it will enhance your research.

6. Working in a group, assume you are employed by a company that is marketing a new product or service. Your group is conducting market research into possible competitors. Choose three or four different brands of a product or service that is comparable to yours. Prepare a short report in which your group observes, tests, and analyzes these competing products or services, including packaging, contents, pricing, and endorsements. Make recommendations about how the product or service offered by your company can excel in the marketplace.

7. Gather a focus group to plan an advertising campaign for the product or service you selected in Exercise 6. Write a short report to your instructor on outcomes.

8. Prepare a list (providing full bibliographic information) of fifteen articles for the restricted topic you selected in Exercise 1. Use at least one of the online databases discussed in this chapter (pages 358–362).

9. Using the secondary research strategies explained in this chapter, investigate recent developments in online credit card fraud. Narrow your topic to a specific aspect of credit card fraud, such as prevention, detection, prosecution, and so on. Find the following:

    a. two recent and reliable Web sources using one of the search engines mentioned on pages 370–371

    b. two government sources (e.g., articles, reports, statistics, handbooks) using USA.gov

    c. two online articles using an e-library listed on page 357

    d. two online newspaper articles

    e. one corporate website using a directory such as those listed on pages 364–365

10. Using the search tools discussed in this chapter, locate the following items related to your major or job. Select titles that are most closely related to your career and explain how and why they would be useful to you. Prepare a separate bibliographic citation for each title.

    a. titles of three important journals that are available in print and online

    b. an abstract of an article appearing in one of the journals

    c. a term in a specialized dictionary

    d. a description or an illustration in a specialized encyclopedia

    e. an article in an international newspaper available online

    f. a training film or recording made after 2009

    g. three U.S. government documents released after 2009

11. Assume you have to write a brochure about one of the following topics, introducing it to an audience of consumers. Using the resources and databases discussed in this chapter, prepare a working bibliography that contains at least ten relevant sources. After gathering and reading those sources, prepare the text and a visual for the brochure and submit them with your bibliography to your instructor. This assignment may be done as a collaborative writing exercise.
   a. virtual reality
   b. fiber optics
   c. Internet auctions (e.g., eBay)
   d. interactive television
   e. robotics in medicine
   f. the greenhouse effect
   g. computer dating
   h. laser surgery
   i. globalization
   j. diversity in the workplace
   k. DNA testing
   l. ethics of business blogging
   m. airport security
   n. tablet PCs

12. Using appropriate references discussed in this chapter, answer any five of the following questions. After your answer, list the specific works you consulted. Supply complete bibliographic information. For books, indicate author or editor, title, edition, place of publication and publisher, date, and volume and page numbers. For journals and magazines, include volume and page numbers; for newspapers, precise date and page numbers. For Internet sites, provide complete URL.
   a. What is nanotechnology?
   b. How many calories are there in an orange?
   c. List three interviews that Hillary Clinton granted in 2010 or 2011.
   d. What is the boiling point of coal tar?
   e. What was the headline in the *New York Times* the day you were born?
   f. List three publications on outdoor recreation issued by the U.S. Department of the Interior from 2010 to the present.
   g. What was the population of Spokane, Washington, in 2010?
   h. List three articles, published between 2010 and 2011, on the advantages of electronic signatures.
   i. Who discovered the neutrino?
   j. What is the first recorded (printed) use of the word *ozone*?
   k. Who edited the second edition of the *Encyclopedia of Psychology*, published in 1994?
   l. Approximately how much rainfall does Sierre Leone receive each year?
   m. How many factories does Toyota have in the United States?
   n. Who is the head of public relations for the Red Cross?
   o. Name five plants that have the word *fly* as part of their common name.

   p. What are the names and addresses of all the four-year colleges in the state of South Dakota?

   q. Who is India's current head of state?

   r. What is the current membership of the American Dental Association?

   s. What are the names of the justices who currently serve on the U.S. Supreme Court?

13. Write a paraphrase of two of the following paragraphs:

   a. Deep-fat frying is a mainstay of any successful fast-food operation and is one of the most commonly used procedures for the preparation and production of foods in the world. During the deep-frying process, oxidation and hydrolysis take place in the shortening and eventually change its functional, sensory, and nutritional quality. Current fat tests available to food operation managers for determining when used shortening should be discarded typically require identification of a change in some physical attribute of the shortening, such as color, smoke, foam development, etc. However, by the time these changes become evident, a considerable amount of degradation has usually taken place.[2]

   b. E-mail marketing is one of the most effective ways to keep in touch with customers. It is generally cost-effective, and if done properly, can help build brand awareness and loyalty. At a typical cost of only a few cents per message, it's a bargain compared to traditional direct mail at $1 or more per piece. In addition, response rates on e-mail marketing are strong, ranging from 5% to 35% depending on the industry and format. Response rates for traditional mail averages in the 1% to 3% range.

      One of the benefits of e-mail marketing is the demographic information that customers provide when signing up for your e-mail newsletter. Discovering who your customers really are—based on age, gender, income, and special interests, for example—can help you target your products and services to their needs. Points to consider when creating your e-mail newsletter:

- HTML vs. plain text: Response rates for HTML newsletters are generally far higher than plain text, and graphics and colors tend to make the publications look far more professional. The downside is that HTML email is slower to download, and some e-mail providers may screen out HTML email.
- Provide incentive to subscribe: Advertise the benefits of receiving your newsletter to get customers to sign up for your newsletter, such as helpful tips, informative content, or early notification of special offers or campaigns.
- Don't just sell: Many studies suggest that email newsletters are read far more carefully when they offer information that is useful to the customers' lives rather than merely selling products and services. Helpful tips, engaging content, and humor are often expected to accompany email newsletters.

---

[2]Vincent J. Graziano, "Portable Instrument Rapidly Measures Quality of Frying Fat in Food Service Operations," *Food Technology* 33, page 50. Copyright © by Institute of Food Technologists. Reprinted by permission.

- Limit questions: As each demographic question you ask may reduce the number of customers signing up, it's best to limit the amount of information you solicit.[3]

c. Phishing is a form of identity theft in which victims are tricked into turning over their personal information to criminals through bogus email and websites. Phishing schemes, which rely on spam email, emerged in 2004 as a method of capturing personal information to use in identity theft. In phishing schemes, consumers receive email that purports to convey some urgent message about their financial accounts. Recipients are encouraged to respond promptly by clicking a link in the message to what are imitations of legitimate, trusted websites. As a result, the online consumer, believing he or she is connected to a legitimate enterprise, divulges personal financial information, which is diverted to the location of the criminal perpetrator. Essentially, phishers have hijacked the trusted brands of well-known banks, credit card issuers, and online retailers, to obtain valuable personal financial information that can be misused or sold to others for the same purpose. Although later phishing attacks have been more generic, they are based on a similar pretext.

Spam email easily reaches thousands if not millions of unsuspecting consumers at a time. With perhaps 5 percent of recipients estimated to divulge personal financial information, the spam scams are lucrative. In November 2004, the Anti-Phishing Working Group (APWG) reported over 1,500 new phishing attacks that month and a 28 percent average monthly growth rate in phishing sites from July through November. The APWG also identified a new form of fraud-based websites that pose as generic e-commerce sites, rather than brand-name sites, and perpetrate loan scams, mortgage frauds, online pharmacy frauds, and other banking frauds. While the United States hosts the largest number of phishing sites, South Korea, China, Russia, Nigeria, Mexico, and Taiwan have also been identified as hosts. Federal law enforcement and others work with foreign counterparts to take down offending sites.

Thus, phishing attacks and "spoofed" email afford criminals an easy and cheap means of obtaining sensitive personal information from consumers, which can be very lucrative even if the false website is shut down within 48 to 72 hours.[4]

14. Ask a professor in your major what he or she regards as the most widely respected periodical in your field. Find a copy of the periodical, and explain its method of documentation (providing examples). How does it differ from the MLA method?

---

[3]Small Business Administration, "Starting Your Own Business." *E-Marketing*. 2005. http://www.sba .gov/starting_business/marketing/emarketing.html.

[4]Anti-Phishing Working Group (APWG), "Phishing Attack Trends Report" (July 2004). See http:// www.antiphishing.org. APWG is an industry association focused on eliminating the identity theft and fraud resulting from phishing. It reports regularly on the form and volume of phishing scams.

15. Put the following pieces of bibliographic information in proper form according to the MLA method of documentation for Works Cited. Correct errors in formatting, punctuation, and so on.

a. *New York Times.* "Shareholders Suffer at Cisco." page 2, April 6, 2011. Rob Cox and Robert Cyron.

b. Enterprise & Society. 11(4)2010. "The Good Consumer: Credit Reporting and the Intervention of Financial Identity in the United States, 1840–1940." Josh Lauer. Pages 686–694.

c. *Electronic News.* Access date: 4 April, 2011. Colleen Taylor. "Chip Market Sales Show Growth in March." http://www.edn.com/article/CA6437823 .html?text=march

d. "Managing Your Time." Blog post. Aug. 10, 2012. Being Smart at Work Blog. Debra Horowitz. www.bsaw@yahoo.com.

e. "Bill Gates College Tour: Billionaire Encourages Students to Teach." Post Date: April 21, 2010. http://www.huffingtonpost.com/2010/04/21/bill-gates-college-tour-b_n_546938.html. Huffington Post. Access Date: 11 April 2010.

f. "Court Won't Lift Stay on Arizona Law." http://www.forbes.com/feeds/ap/2011/04/11/general-us-arizona-immigration-appeal_8402691.html. *Forbes .com.* Bob Christie. Access Date: 11 April 2011. Post Date: April 11, 2011.

g. "401(k) Plans for Small Businesses." http://www.dol.gov/ebsa/publications/401kplans.html. U.S. Department of Labor. Access Date: 11 April 2011. Revision Date: October 2010.

h. "A Guide to Middleware Architecture and Technologies." Pages 41–89. Ian Gorton. *Essential Software Architecture.* Published in November 2010 by Springer Press in New York.

i. "Managing Channel Profits." *Marketing Science.* Abel P. Jeuland. Pages 29–48. Volume 27, Issue 1. Jan/Feb 2011.

j. Volume 31, Issue 5. "The Trade Policy Jungle: A Survival Guide for Academic Economists." *The World Economy.* Simon J. Evenett. Pages 498–516. May 2011.

k. "What Scares Google." Page 28. Kevin Maney. *The Atlantic Monthly.* September 2009, volume 304.

l. *Inc.com.* Access Date: 11 April 2011. Post Date: 11 April 2011. "5 Tips for Running a Part Time Business." http://www.inc.com/guides/201101/5-tips-for-running-a-part-time-business.html. Eric Markowitz.

m. Personal interview. 13 Sept. 2011. Marsha Keys, CEO, Biltmore Polymers, Chicago, IL.

n. *Nation.* September 2010. "Doing Green Jobs Right." Pages 20–22. Volume 291. Amy B. Dean.

o. *The International Trade Journal.* "The Effect of Import Penetration on Labor Market Outcomes: The Case of Austrian Manufacturing Industry." 2011. Volume 25, Issue 2. Özlem Onaran. Pages 163–204.

p. April 2011, Volume 40, Issue 3. *Money.* "Why Buy Trouble." Beth Braverman and Veronica Crews. Pages 78–83.

q. "How Loyalty to Customers Led to Storied Bank's Fall." Pages B1–B3. Robin Sidel. November 27, 2010. *Wall Street Journal.*

    r. Mark Steyn. March 2011. Pages 26–29. "Canada's Economic Folly." *USA Today*.

    s. *EH.Net Encyclopedia*. Haupert, Michael J. http://eh.net/encyclopedia/article/haupert.mlb. "The Economic History of Major League Baseball." Edited by Robert Whaples. Revision Date: February 1, 2010. Access Date: April 11, 2011.

    t. Emily Brandon and Kerry Hennon. *U.S. News and World Report*. "Where to Launch a Second Career." Volume 7, Issue 49. Pages 36–43. October 2010.

    u. "Giving Time to Give Thanks for Benefits." Kelly Butler. Page 7. *Employee Benefit News*. November 2010.

    v. "Profits on an Overseas Holiday." *Bloomberg Businessweek*. March 21, 2011. Peter Coy, Jesse Drucker, Karen Weise, Diane Brady, Richard Rubin. Pages 64–69.

16. Put the bibliographic references you listed in MLA format in Exercise 15 into APA format for a References list.

17. Select one article from the periodical you chose for Exercise 14. Convert the bibliographic information for that article into the MLA parenthetical style.

18. The following passage contains mistakes in the MLA method of documentation. Find the mistakes and explain how to correct them.

> More and more companies are allowing employees to "telecommute" (see Smith; Dawson; Brown; Gura and Keith; and Allen). One expert defines telecommuting as "home-based work" (13). Having terminals in their homes "allows employees to work at a variety of jobs" ("New Employment Opportunities"). It has been estimated that currently 900,000 employees work out of their homes (Pennington, p. 56). That number is sure to increase as computer-based businesses multiply in the late 1990s (Brown). In one of her recent articles on telecommuting, Holcomb (167) found that "in the last year alone 43 companies in the metropolitan Phoenix area made this option available to their employees."
>
> Employees who telecommute cite a variety of benefits for such an arrangement (see in particular articles by Gura, Smith, and Kaplan). One employee of a mail order company whose opinion was quoted observed that "I can save about 15–17 hours a week in driving time" (from Allen). Working at home allows the telecommuting employee to work at his or her optimum times ("The Day Does Not Have to Start at 9:00 a.m."). Also, in articles by Kaplan and Keith the benefits of not having to leave home are emphasized: "A telecommuting parent does not have to worry about child care" (39). Telecommuting may "be here to stay" (quoted in a number of different website sources).

19. Submit your preliminary list of references (your tentative Works Cited page) for a long report to your instructor.

CHAPTER 9

Summarizing Information at Work

A summary is a brief restatement of the main points of a book, report, website, article, laboratory test, PowerPoint presentation, meeting, or convention. A summary saves readers hours of time because they do not have to study the original work or attend a conference. A summary can reduce a report or an article by 85 to 95 percent (or even more) and can capture the essential points of a three-day convention in a one-page memo. Moreover, a summary can tell readers whether they should even be concerned about the original; it may be irrelevant for their purposes. Finally, since only the most important points of a work are included in a summary, readers will know they have been given the crucial information they need.

Access chapter-specific interactive learning tools, including quizzes and more in your English CourseMate, accessed through www.cengagebrain .com.

Summaries in the Information Age

Thanks to the Web and other communication technologies, we have an abundance of information. It would be impossible to locate, classify, understand, and assess all this information without the help of summaries. They can be found all around you.

- Google, one of the most frequently used search engines, retrieves positive "hits" by looking for keywords that summarize a source and help users determine whether the material is relevant for their purpose.
- A home page on the Web is in essence a summary of the various pages to which it is connected.
- Television and radio stations regularly air one- or two-minute "news breaks" that summarize in a few paragraphs the major stories of the day, such as CNN (www.CNN.com).
- Blogs, such as those shown in Figures 4.10 and 4.11 (pages 156–157 and 161–162) frequently summarize the results of many weeks of research and decision making.
- Today's e-communication technologies, such as IMs and text messages, require you to summarize information concisely for readers who have small screens and tight schedules.

The Importance of Summaries in Business

Summaries are vital in the world of work. They get to the main points—the bottom line—right away for busy readers, giving them the big picture.

On the job, writing summaries for employers, co-workers, and customers is a regular and important responsibility. Chapter 14 discusses a variety of reports—periodic, sales, progress, trip, test, and incident—whose effectiveness depends on a faithful summary of events. You may be asked to summarize a week-long business trip in one or two pages or a two-hour teleconference in just a few paragraphs. You may have to condense a proposal to fit a one-page format for an organization. A busy manager may ask you to read and condense a ninety-page report so that she or he will have a knowledgeable overview of its contents. Or you may be asked to summarize the main features of a competitor's product or a new model you saw at a trade show. You may be asked to write a news release—another type of summary vital for your organization's image (pages 444–451)—for your employer's website.

Figure 9.1 is a summary of a long report evaluating child-care facilities. Note how it concisely identifies the main purpose and conclusions of the report.

FIGURE 9.1 A Summary of a Long Report on Child-Care Facilities

HIGH MARKS FOR ON-SITE CARE

The best place to find high-quality child care may be your workplace, according to a new study by Burud & Associates, a California-based work/life benefits consulting firm. The study of 205 work-site child-care programs found that such centers are eight times more likely than other facilities to meet the high standards set by the National Association for the Education of Young Children (NAEYC).

Forty-one percent of work-site centers open at least two years are NAEYC-accredited—compared to only 5 percent of all child-care centers. Other key findings include:

• Ninety-two percent of work-site centers offer infant care.
• Workplace centers provide "substantially better" employee benefits—which help the centers recruit and retain high-quality caregivers.
• One in four centers is open past seven p.m. and one in seven is open before six a.m. and/or offers weekend care.

© Cengage Learning 2013

Source: First appeared in *Working Mother*, May 1999. Reprinted by permission.

Contents of a Summary

The chief problem in writing a summary is deciding what to include and what to omit. Determine what is most relevant for your audience and its purpose. As we have seen, a summary is a streamlined review of *only* the most significant points. You will not save your readers time if you simply rephrase large sections of the original. That will simply supply readers with another report, not a summary.

Make your summary lean and useful by briefly telling readers the main points: purpose, scope, conclusions, and recommendations. A summary should straightforwardly and accurately answer readers' two most important questions:

1. What are the findings of the report or meeting?
2. How do the findings apply to my business, research, or job?

Note how the summary in Figure 9.1 successfully answered those two questions by indicating to working mothers where the best child care is likely to be found and why.

How long should a summary be? While it is hard to set down precise limits about length, effective summaries are generally no more than 5 to 10 percent of the length of the original. The complexity of the material being summarized and your audience's exact needs can help you to determine an appropriate length. The following suggestions will guide you on what to include and what to omit in your summary.

What to Include in a Summary

1. **Purpose.** A summary should indicate why the article or report was written or why a hearing meeting was held. Your summary should give readers a brief introduction (even one sentence will do) indicating the main purpose of the report or event.
2. **Essential specifics.** Include only the names, costs, titles, places, or dates essential to understanding the original.
3. **Conclusions or results.** Emphasize what the final vote was, the result of the tests, or the proposed solution to the problem.
4. **Recommendations or implications.** Readers will be especially interested in important recommendations—what they are, when they can be carried out, and why they are necessary, or why a plan will not work. They may look to your summary to determine what action needs to be taken.

What to Omit from a Summary

1. **Opinion.** Avoid injecting opinions—your own, the author's, or the speaker's. You distract readers from grasping main points by saying that the report was too long or missed the main point, that a salesperson from Detroit monopolized the meetings, or that the author digressed to blame the Land Commission for failing to act properly. A later section of this chapter (pages 438–442) will deal with evaluative summaries.

2. **New data.** Stick to the original article, report, book, or meeting. Avoid introducing comparisons with other works or conferences; readers will expect an unbiased digest of only the material being summarized.

3. **Irrelevant specifics.** Do not include biographical details about the author of an article that might be included in "Notes on Contributors." This information plays no role in the reader's understanding of your summary.

4. **Examples.** Illustrations, explanations, and descriptions are unnecessary in a summary. Readers will want to know outcomes, results, and recommendations, not the illustrative details supporting or elaborating on those results.

5. **Background.** Material in introductions to articles, reports, and conferences can usually be excluded from a summary. Such "lead-ins" prepare the reader for a discussion of the subject by presenting background information, not the big picture readers expect to see in a summary.

6. **Jargon.** Technical definitions or jargon in the original document may confuse rather than clarify the essential information for general readers.

7. **Reference data.** Exclude information found in footnotes, bibliographies, appendixes, tables, or graphs.

Preparing a Summary

To write an effective summary, you need to proceed through a series of steps. Basically, you will have to read the material carefully, making sure that you understand it thoroughly, to identify the major points, and, finally, to put the essence of the material into your own words. Follow these steps to prepare a concise, useful summary.

1. **Read the material once in its entirety to get an overall impression of what it is about.** Become familiar with large issues, such as the purpose and organization of the work and the audience for whom it was written. See if there are visual cues— headings, subheadings, words in italic or boldface type, sidebars—that will help you to classify main ideas. Also look for a conclusion and any mini-summaries within the article or report.

Tech Note

Using Software to Summarize Documents

Your word-processing software can help you prepare your summary. First, download and open an existing digital file of the material (article, report, technical paper) you need to summarize. Then, "save as" a new file with a new filename. If all you have is hard copy, do an OCR scan so you can create a digital file to edit. As you read through it on your screen, cut nonessential material. The first time through, it is easier to cut what you don't want than it is to select exactly what you do want. Using this method of editing, you can delete single sentences or several paragraphs at a time. Also, highlight key points as you read through the original material. Highlighting during the first pass can guide you on your second reading as you attempt to include only relevant material for your summary.

2. Reread the material. Read it a second time or more if necessary. Locate all of the main points, and underline them. (If the work is a book or in a journal that belongs to your library, work with a photocopy so you can underline.) To spot the main points, pay attention to the key transitional words, which often fall into predictable categories:

- Words that enumerate: *first, second, third, initially, subsequently, finally, next, another*
- Words that express causation: *accordingly, as a result, because, consequently, subsequently, therefore, thus*
- Words that express contrasts and comparisons: *although, by the same token, despite, different from, furthermore, however, in comparison, in contrast, in addition, less than, likewise, more readily, more than, not only . . . but also, on the other hand, the same is true for, similar, unlike*
- Words that signal essentials: *basically, best, central, crucial, foremost, fundamental, important, indispensable, in general, leading, major, obviously, principal, significant*

Pay special attention to the first and last sentences of each paragraph. Often the first sentence of a paragraph contains the topic sentence, and the last sentence summarizes the paragraph or provides a transition to the next paragraph.

Also be alert for words signaling information you do *not* want to include in your summary, such as the following:

- Words announcing opinion or inconclusive findings: *from my personal experience, I feel, I admit, in my opinion, might possibly show, perhaps, personally, may sometimes result in, has little idea about, questionable, presumably, subject to change, open to interpretation*
- Words pointing out examples or explanations: *as noted in, as shown by, circumstances include, explained by, for example, for instance, illustrated by, in terms of, learned through, represented by, such as, specifically in, stated in*

3. Collect your underlined material or notes and organize the information into a draft summary. At this stage do not be concerned about how your sentences read. Use the language of the original, together with any necessary connective words or phrases of your own. Key the draft into your computer. *Expect to have more material here than will appear in the final version.* Do not worry; you are engaged in a process of selection and exclusion. Your purpose at this stage is to extract the principal ideas from the examples, explanations, and opinions surrounding them.

4. Read through and revise your draft(s) and delete whatever information you can. As you revise, see how many of your underlined points can be condensed, combined, or eliminated. You may find that you have repeated a point. Check your draft against the original for accuracy and importance. Be sure to be faithful to the original by preserving its emphases and sequence. Put quotation marks around any direct quotations. But try to avoid direct quotation wherever possible at this stage.

5. Now put the revised version into your own words. Again, make sure that your reworded summary has eliminated nonessential words. Connect your sentences with words that show relationships between ideas in the original (*also, although, because, consequently, however, nevertheless, since*). Compare this version of your summary with the original material to double-check your facts.

6. Do not include remarks that repeatedly call attention to the fact that you are writing a summary. You may want to indicate initially that you are providing a summary, but avoid such remarks as "The author of this article states that water pollution is a major problem in Baytown" and "On page 13 of the article three examples, not discussed here, are found."

7. Edit your summary to make sure it is fair, clear, and concise. Compare your summary with the original to make sure you have captured the key points, especially those in the conclusion and recommendation sections. Check as well that your summary is coherent. Tell the reader how one point flows into another. Proofread your summary carefully.

8. Identify the source you have just summarized. Include pertinent bibliographic information in the title of your summary or in a footnote or an endnote. This gives proper credit to the original source and informs your readers where they can find the complete text if they want more details.

Case Study

Summarizing an Original Article

Figure 9.2, a 2,500-word article entitled "Virtual Reality: The Future of Law Enforcement Training," appeared in the *FBI Law Enforcement Bulletin* and hence is of primary interest to individuals in law enforcement administration. Assume you have been asked to write a summary of the FBI article for your boss, a police chief in a medium-sized city who would be interested in incorporating virtual reality segments into the police academy training program.

By following the steps outlined above, you would first read the article carefully two or three times, underscoring or highlighting the most important points, signaled by key words. Note what has been underscored in the article. Also study the comments in the margins to see why certain information is to be included or excluded from the summary.

After you have identified the main points, extract them from the article and, still using the language of the article, join them into a coherent working draft summary, as in Figure 9.3. Then shorten and rewrite the working draft in your own words to produce the compact final version of your summary, as shown in Figure 9.4. Only 154 words long, the final summary is 6 percent of the length of the original article and records only major conclusions relevant to the audience for the article.

FIGURE 9.2 An Original Article with Important Points Underscored for Use in a Summary

Virtual Reality: The Future of Law Enforcement Training

Jeffrey S. Hormann

A late night police pursuit of a suspected drunk driver winds through abandoned city streets. The short vehicle chase ends in a warehouse district where the suspect abandons his vehicle and continues his flight on foot. Before backup arrives, the rookie patrol officer exits his vehicle and gives chase. A quick run along a loading dock ends at the open door to an apparently unoccupied building. The suspect stops, brandishes a revolver, and fires in the direction of the pursuing officer before disappearing into the building. The officer, shaken but uninjured, radios in his location and follows the suspect into the building.

    Did the officer make a good decision? Probably not by most departments' standards. Whether the officer's decision proves right or wrong, the training gained from this experience is immeasurable, that is, provided the officer lives through it. Fortunately for this officer, <u>the scenario occurred in a realistic, high-tech world called virtual reality</u>, where <u>training</u> can have <u>a real-life impact without the accompanying risk</u>.

Traditional Training Limitations

Experience may be the best teacher, but in real life, police officers may not get a chance to learn from their mistakes. To survive, they must receive training that prepares them for most situations they might encounter on the street. <u>However</u>, because <u>many training programs</u> emphasize repetition to produce desired behaviors, they <u>may not achieve the intended results</u>, especially after students leave the training environment. Thus, the more realistic the training, the <u>greater the lessons</u> learned.

    <u>Additionally</u>, even some in law enforcement may fall prey to the effects of what has come to be termed "The MTV Generation." As products of this generation, today's young officers purportedly have short attention spans requiring new, nontraditional training methods. The key to teaching this new breed is to provide fast-paced, <u>attention-getting instruction</u> that is <u>clear, concise, and relevant</u>.

Training with Virtual Reality

<u>Virtual reality</u> can <u>provide</u> the type of <u>training that today's law enforcement officers need</u>. By completely immersing the senses in a computer-generated environment, the <u>artificial world becomes reality to users</u> and greatly enhances their training experiences.

    <u>Although</u> considerable research and development have been conducted in this field, <u>only a limited amount has applied directly to law enforcement</u>. The apparent reason simply is that, for the most part, <u>law enforcement has not asked for it</u>.

    <u>Because virtual reality</u> technology is <u>relatively new</u>, most <u>law enforcement administrators know little about it</u>. They know even less about what it can do for

Margin notes:
- Omit scenario—example of background; an opener
- Include important observation
- Major distinction
- Omit explanation and example
- Include significant qualification
- Emphasize author's main point
- Important reason for its neglect by law enforcement
- Restatement of main point above

(Continued)

FIGURE 9.2 (Continued)

Note parallel items with key words signaling important applications

Include definition

Important explanation

Significant phrase

Omit specific pieces of equipment

Omit example

Omit further examples

Major conclusion

Major value to audience of administrators

Emphasize significant advantages in training

their agencies. By <u>understanding what virtual reality is, how it works</u>, and <u>how it can benefit them</u>, law enforcement administrators can become significantly involved in the development of this important new technology.

What Is Virtual Reality?

<u>Simply stated, virtual reality</u> is <u>high-tech illusion</u>. It is a computer-generated, three-dimensional environment that engulfs the senses of sight, sound, and touch. Once entered, it becomes reality to the user.

Within this virtual world, users travel among, and interact with, objects that are wholly the products of a computer or representations of other participants in the same environment. Thus, the limits of this virtual environment depend on the sophistication and capabilities of the computer and the software that drives the system.

How Does Virtual Reality Work?

Based on data entered by programmers, computers create virtual environments by generating <u>three-dimensional images</u>. Users usually view these images through a head-mounted device, which, for instance, can be a helmet, goggles, or other apparatus that restricts their vision to two small video monitors, one in front of each eye. Each monitor displays a slightly different view of the environment, which gives users a sense of depth.

<u>Another device</u>, called a position tracker, monitors users' physical positions and provides input to the computer. This information instructs the computer to change the environment based upon users' actions. <u>For example</u>, when users look over their shoulders, they see what lies behind them.

<u>Because</u> virtual reality users remain stationary, they use a <u>joy stick</u> or trackball to move through the virtual environment. Users <u>also</u> may wear a special glove or use other devices to manipulate objects within the virtual environment. <u>Similarly</u>, they can employ virtual weapons to confront virtual aggressors.

To enhance the sense of reality, some researchers are experimenting with tactile feedback devices (TFDs). TFDs transmit pressure, force, or vibration, providing users with a simulated sense of touch. <u>For example</u>, a user might want to open a door or move an object, which in reality, would require the sense of touch. A TFD would simulate this sensation. At present, <u>however</u>, it is important to remember that these devices are <u>crude</u> and somewhat <u>cumbersome to use</u>.

Uses for Virtual Reality

In <u>today's competitive business environment</u>, organizations continuously strive to accomplish tasks faster, better, and inexpensively. This especially holds true in training.

Virtual reality is <u>emerging rapidly</u> as a <u>potentially unlimited</u> method for providing <u>realistic</u>, <u>safe</u>, and <u>cost-effective training</u>. <u>For example</u>, a firefighter can battle the flames of a virtual burning building. A police officer can struggle with virtual shoot/don't shoot dilemmas.

Within a virtual environment, <u>students</u> can <u>make decisions</u> and act upon them <u>without risk</u> to themselves or others. By the same token, <u>instructors</u> can critique

FIGURE 9.2 (Continued)

students' actions, enabling students to review and learn from their mistakes. This ability gives virtual reality a great <u>advantage over most conventional training methods</u>.

The Department of Defense <u>(DOD) leads public and private industry</u> in <u>developing virtual reality training</u>. <u>Since</u> the early 1980s, DOD has actively researched, developed, and implemented virtual reality to <u>train members</u> of the <u>armed forces</u> to fight effectively in combat.

<u>DOD's current approach</u> to virtual reality training <u>emphasizes team tactics</u>. Groups of military personnel from around the world engage in combat safely on a virtual battlefield. Combatants never come together physically; <u>rather</u>, simulators located at various sites throughout the world transmit data to a central location, where the virtual battle is controlled. Basically, it costs less to move information than people. Consequently this form of training has proven quite cost-effective.

An <u>additional benefit</u> to this <u>type of training</u> is that <u>battles</u> can be <u>fought under varying conditions</u>.

Virtual battlefields <u>re-create real-world locations</u> with <u>interchangeable characteristics</u>. To explore "what if" scenarios, participants can modify enemy capabilities, terrain, weather, and weapon systems.

Virtual reality <u>also can re-create actual battles</u>. Based on information from participants, the Institute for Defense Analyses re-created the 2nd Armored Cavalry Regiment Offensive conducted in Iraq during Operation Desert Storm. The success of the virtual re-creation became apparent when, upon viewing the simulations, soldiers who had fought in the actual battle reported the extreme accuracy of the event's depiction and the feeling of reliving the battle. <u>Clearly, virtual reality holds great potential</u> for accurate review and analysis of <u>real-world situations</u>, which would be <u>difficult to accomplish</u> by <u>any other method</u>.

Preliminary studies, for instance, show that military units perform better following virtual reality training. <u>Even though</u> virtual environments are only simulations, the complete immersion of the senses literally overwhelms users, totally engrossing them in the action. <u>This realism</u> presumably plays a <u>major role</u> in the <u>program's success</u> and <u>likely</u> will prove positive in future endeavors. <u>In fact</u>, due to its success in training multiple participants in group combat situations, DOD plans to train infantry personnel individually with virtual reality fighting skill simulators.

Law Enforcement Training

While virtual reality has proven its value as a training and planning tool for the military, <u>applications for this technology reach far beyond DOD</u>. In varying but key ways, many military uses can <u>transfer to law enforcement</u>, including training in firearms, stealth tactics, and assault skills.

Unfortunately, few organizations have dedicated resources to developing virtual reality for law enforcement. <u>According</u> to a recently published resource guide, more than <u>100 companies</u> currently are <u>developing and/or selling virtual reality hardware or software</u>. However, <u>none</u> of these firms <u>mentioned law enforcement uses</u>.

<u>Further</u>, a review of relevant literature revealed numerous articles on virtual reality technology, but only a few addressed law enforcement applications. <u>Yet</u>,

Use only main points relevant to target audience of administrators

Note main military advantage

Omit examples

Major conclusion signaled by key word "clearly"

Omit example

Key word "major" signals relevant idea for audience

Omit military application

Significant parallel points

Omit statistics

Subordinate idea

(Continued)

FIGURE 9.2 (Continued)

Restatement of major point

virtual reality <u>clearly offers law enforcement benefits</u> in a number of areas, including pursuit driving, firearms training, high-risk incident management, incident recreation, and crime scene processing.

Pursuit Driving

Include application but omit examples

<u>Pursuit driving</u> represents one area in which <u>virtual reality application</u> has become <u>reality for law enforcement</u>. Law enforcement personnel identified a need and provided input to a well-known private corporation that developed a driving simulator equipped with realistic controls.

Omit specific mechanism and explanation of operation of screen mechanism

The simulator provides users with realistic steering wheel feedback, road feel, and other vehicle motions. The screen possesses a 225-degree field of view standard, with 360-degree coverage optional. <u>As noted in demonstrations</u>, simulations can involve one or more drivers, and environments can alternate between city streets, rural back roads, and oval tracks. The vehicle itself can change from a police car to a truck, ambulance, or a number of others.

Note cost efficiency again

Virtual reality driving simulators provide police departments invaluable training at a <u>fraction of the long-term cost of using actual vehicles</u>. In fact, the simulator is being used by a number of police departments around the country.

Include major advantage but exclude specific example

During the past year, for example, the Los Angeles County Sheriff's Office Emergency Vehicle Operations Center (EVOC) has used a four-station version of the driving simulator to train its officers. The simulators help students develop judgment and decision-making skills, while providing an environment free from risk of injury to students or damage to vehicles. Still, as the EVOC supervisor cautions, <u>virtual reality training</u> should <u>complement, not replace</u>, actual behind-the-wheel instruction.

Note major distinction for training purpose

Firearms Training

New subtopic; include advantage but omit examples

In another way, virtual reality could <u>greatly enhance</u> shoot/don't shoot <u>training simulators</u> currently in use, such as the Firearms Training System, a primarily two-dimensional approach that possesses limited interactive capabilities. A <u>virtual reality system</u> would <u>allow officers</u> to enter any <u>three-dimensional environment</u> alone or as a member of a team and confront computer-generated aggressors or other virtual reality users.

Include significant points on advantages

Evaluators could specifically observe the <u>training from any perspective</u>, including that of the officers, or the criminal. The <u>training scenarios</u> could involve actual building floor plans or local city streets, and criteria <u>such as</u> weather, number of participants, or types of weapons could be altered easily.

Next three reasons to use virtual reality signaled by keywords: in addition, also, and likewise

High-Risk Incident Management

<u>In addition</u> to weapons training, virtual reality <u>could prove invaluable for SWAT team members</u> before <u>high-risk tactical assaults</u>. Floor plans and other known facts about a structure or area could be entered into a computer to create a virtual environment for commanders and team members to analyze prior to action.

Incident Re-creation

Law enforcement agencies could <u>also collect data</u> from victims, witnesses, suspects, and crime scenes <u>to re-create traffic accidents</u>, shootings, and other crimes.

FIGURE 9.2 (Continued)

The virtual environment created from the data could be used to refresh the memories of victims and witnesses, to solve crimes, and ultimately, to prosecute offenders.

Crime Scene Processing
Virtual reality crime scenes could <u>likewise</u> be used to <u>train both detectives and patrol officers</u>. First, students could search the site and retrieve and analyze evidence <u>without ever leaving the station</u>. Then, actual crime scenes could be re-created to add realism to training or to evaluate prior police actions.

Omit examples

Is Virtual Reality Virtually Perfect?
<u>Though</u> virtual reality may appear to be the <u>ideal law enforcement tool</u>, as with any new technology, <u>some drawbacks exist</u>. <u>Currently</u>, areas of concern range from cumbersome equipment to negative physical and psychological effects experienced by some users. Fortunately, <u>however</u>, the <u>field is evolving and improving constantly</u>, and as <u>virtual reality gains widespread use</u>, most <u>major concerns should be dispelled</u>.

Crucial qualification and justification for using virtual reality in law enforcement training

Physical Limitations and Effects
<u>Because</u> computers currently are <u>not fast enough</u> to process large amounts of graphic information in real time, <u>some observers</u> describe virtual environments as "<u>slow-moving</u>." The human eye can process images at a much faster rate than a computer can generate them. In a <u>virtual environment</u>, frames are displayed at a rate of about 7 per second, an extremely slow speed when compared to television, which generates 60 frames per second. Users find the resulting choppy or slow graphics less than appealing.

Omit examples of limitations/effects

Source: Adapted from FBI Law Enforcement Bulletin, 64/no. 7: 7–12.

Make Sure Your Summary Is Ethical

Your supervisor will expect your summary to be honest, fair, and accurate, identifying the most crucial points of the article. Figure 9.5 on page 434 contains an unethical summary of Figure 9.2 that distorts the meaning and intention of the original by leading the reader to conclude that virtual reality is not valuable for law enforcement administrators—the very opposite point the author makes.

You can write an ethical summary by doing the following:

- Make sure your summary agrees with the original.
- Emphasize the main points the author makes.
- Do not omit key points.
- Be fair in expressing the author's conclusions/recommendations.
- Do not dwell on minor points to the exclusion of major ones.
- Do not let your own opinions distort or contradict the message of the original document.

FIGURE 9.3   A Working Draft Summary of the "Virtual Reality" Article in Figure 9.2

Law enforcement officers put their lives on the line every day, yet their training does not fully allow them to anticipate what they will find on the streets. Virtual reality will give them realistic, high-tech benefits of encountering criminals without any risks. Traditional training methods, which work through repetition, cannot equal the advantages of virtual reality when it comes to teaching officers the lessons they must learn to survive in the field. This new breed of officers is demanding the attention-getting, highly realistic training that virtual reality affords them. Virtual reality translates the artificial world of the computer into the real world. Yet even though much research has been done on virtual reality, it is new to law enforcement officials. Moreover, manufacturers have not marketed their technology to them. It is essential that these administrators know how virtual reality works and what it can do for them. Virtual reality has been defined as high-tech illusion through the computer user's perceived interaction with the real world. Working through sophisticated software, virtual reality gives users a three-dimensional (hearing, feeling, and seeing) view of the things and people around them. Virtual reality requires specific equipment including goggles/headsets, a tracker, a trackball, and special gloves. But these devices do have problems; at present, they are crude and can be cumbersome. Even so, virtual reality provides cost-effective and life-saving benefits for law enforcement administrators. Thanks to this technology, new officers will be able to make quicker and better decisions in the field. Virtual reality has already been tried by the Department of Defense; the armed forces have used it to re-create battlefield conditions, helping the troops better understand the enemy and its position. Yet virtual reality holds great appeal for other real-world applications, especially law enforcement. Unfortunately, the 100 companies that manufacture virtual reality equipment have neglected these law enforcement applications. Yet virtual reality easily accommodates law enforcement instruction. Driving simulators help officers prepare for high-speed chases. In Los Angeles County, such simulators complement more traditional training. Virtual reality can help officers in a variety of training missions—firearms, high-risk incidents, re-creating crimes, understanding the crime scene. Using virtual reality, officers never have to leave the station. Admittedly, virtual reality has drawbacks, but as this new technology improves, users should face fewer problems.

> **FIGURE 9.4** A Final, Effective Summary of the Article in Figure 9.2

Virtual reality offers benefits for law enforcement training that traditional methods cannot provide. This computer-generated technology simulates and re-creates real-life crime scenes without placing officers at risk. Thanks to virtual reality's three-dimensional world of sight, sound, and touch, officers enter the criminals' world to gain invaluable experience interacting with them. Because virtual reality has not been marketed for law enforcement use, administrators may not know about it. Yet it provides a cost-effective, realistic way to enhance training programs. The applications of virtual reality far exceed its military use of simulating battlefield conditions. Virtual reality allows administrators to give trainees hands-on experience in pursuit driving, firearms training, SWAT team assaults, incident re-creation, and crime scene processing. Officers can investigate a crime without ever leaving the station. Although virtual reality is an emerging technology with limitations, it is quickly improving and rapidly expanding. Administrators need to incorporate it into their curriculum to give officers field-translatable experiences.

© Cengage Learning 2013

Executive Summaries

An executive summary, found at the beginning of a proposal (Chapter 13) or a long report (Chapter 15), is usually one or two pages long (four to six concise paragraphs) and condenses the most important points from the proposal or report for a busy manager—the executive. Your employer wants bottom-line conclusions, not all the technical details. An executive summary is written to help the reader reach a major decision based on the report or proposal. Your job, in an executive summary, then, is to concisely tell your employer what findings the report includes, what those findings mean for the company or organization, and what action, if any, needs to be taken. Figure 9.6 (page 436), an executive summary of a report on software for a safety training program, directly advises a decision maker to purchase a safety software package. Managers use executive summaries so they will *not* have to wade through entire reports. An effective executive summary is like a report itself—self-contained and able to stand on its own.

What Managers Want to See in an Executive Summary

Executive readers are most concerned with managerial and organizational issues—the areas over which they have supervisory control. These readers will look for information on costs, profits, resources, personnel, timetables, and feasibility. Your

FIGURE 9.5 An Unethical, Misleading Summary of the Article in Figure 9.2

Nonessential introductory material

Distorts article

Dwells on specific virtual reality equipment at the expense of the main advantages

Reverses chronology of events; misrepresents the role of virtual reality

Deletes unit's name

Deletes specifications

One-sided; omits success of simulation

Focuses on limits rather than usefulness

Does not subordinate flaws

A rookie police officer makes many mistakes in pursuing subjects. Training can cover many realistic situations, but young officers in the MTV Generation have short attention spans. Given the research so far on virtual reality, it holds little promise for law enforcement use. Virtual reality has too many limitations, but it works interestingly through gloves, helmets, and goggles, and with a position tracker users can see over their shoulders. It even has a joy stick (like those in an amusement park) and a crude device—a TFD—that simulates touch (nice to have in a horror movie). Instructors can gain much from virtual reality because they can better criticize their trainees. In the early 1980s, the DOD used virtual reality to duplicate battlefield conditions. The 2nd Armored Cavalry Regiment Offensive won the Iraqi War because of virtual reality. But companies manufacturing virtual reality technology are not interested in law enforcement applications, another indication of its limitations. The Los Angeles Sheriff's EVOC used a driving simulator—offering a 225-degree field of view but it can be ordered with a 360-degree field—but expressed their caution about it. There have been limited interactions in the use of virtual reality for firearms training, though floor plans might have helped SWAT teams. Witnesses may need to refresh their memories with virtual reality. Again drawbacks exist. Computers are not as fast as the human eye in processing information.

© Cengage Learning 2013

summary must supply key information on the executive's four *E*'s: evaluation, economy, efficiency, and expediency. Executive readers will expect you to summarize large, complex subjects into easy-to-read, easy-to-understand information that they can act on confidently.

Organization of an Executive Summary

An executive summary must be faithful to the report while giving readers what they need (Figure 9.6). First, read the report carefully, plan what you want to include, and then draft and revise using valuable connective words (page 68). Clearly, you cannot write an executive summary of your report until after you have written the report itself.

FIGURE 9.6 An Executive Summary

A Report on Providing Better Training at Techtron Sites

Management has commissioned this report to investigate ways to prepare for the OSHA audits scheduled between February and June 2012, at our seven regional Techtron plants. Most directly, this report focuses on our ability to complete Phase One of ISO 14001 certification.

Starts with purpose of report

Currently, the Techtron safety training programs are inadequate; they are neither comprehensive nor up-to-date. We lack necessary software to instruct employees about the EPA and OSHA regulations and requirements that apply to hazardous materials or procedures used in our company. Consequently, safety violations have occurred with lockouts, confined spaces, fall protection, and the "Right to Know Law" concerning labeling of chemicals.

Identifies problem the report investigates and why it is important

Exploring better ways to conduct our training sessions, we purchased a copy of the software program **EPA/OSHA Trainer**, regarded as the best on the market (available from EDI @ $800 per copy). The **Trainer** offers effective guidelines on developing safety meetings and giving demonstrations. It also includes instructions, written in clear, nontechnical language, on how to identify, collect, and document hazardous materials. Additionally, the **Trainer** supplies the full text of EPA/OSHA regulations, with updates issued quarterly.

Explains the solution tested

Highlights benefits of trainer

To test the effectiveness of the **Trainer** software, we scheduled an internal audit at our Hendersonville site last month. After progressing through the **Trainer** module, a core group of employees interviewed by management successfully completed all required regulatory training. Subsequently, employees who had undergone such training were able to instruct and monitor the performance of other employees in the program.

Verifies effectiveness of solution

To ensure the safety of our employees and to compete in a global marketplace, Techtron must pass the OSHA 14001 certification. Purchasing seven additional copies of the **EPA/OSHA Trainer** software (7 @ $800 = $5,600) in the next month is a wise and necessary investment.

Ends by stressing action to be taken and by when

Follow this organizational plan when you write an executive summary:

1. **Begin with the purpose and the scope of the report.** For example, a report might be written to study new marketing strategies, to identify obsolete software, or to relocate a branch store.
2. **Relate your purpose to a key problem.** Identify the source (history) and seriousness of the problem.

3. **Identify in nontechnical language the criteria used to solve the problem.** Be careful not to include too much information or too many details.
4. **Condense the findings of your report.** Relate what tests or surveys revealed.
5. **Stress conclusions and possible solutions.** Be precise and clear.
6. **Provide recommendations.** For example, buy, sell, hire more personnel, relocate, or choose among alternative solutions. You may also indicate when a decision needs to be made.

The order of information in an executive summary does not have to follow strictly the order of the report itself. In fact, some executive summaries start with recommendations. Find out your boss's preference.

Evaluative Summaries

To write an evaluative summary, also called a *critique*, follow the guidelines below. As with executive summaries, you will be expected to provide a commentary on the material (that is, give your opinion).

Your employer may often ask you to summarize and assess what you have read. For example, you may have to condense and judge the merits of a report, paying special attention to whether its recommendations should be followed, modified, or ignored. Your company or agency may also ask you to write short evaluative summaries of job candidates, applications, sales proposals, or conferences.

Guidelines for Writing a Successful Evaluative Summary

To write a careful evaluative summary, follow these guidelines:

- Keep the summary short—5 to 10 percent of the length of the original.
- Blend your evaluations with your summary; do not save your evaluations for the end of the summary.
- Place each evaluation near the summarized points to which it applies so readers will see your remarks in context.
- Include a pertinent quotation from the original to emphasize your recommendation.
- Comment on both the content and the style of the original.

Evaluating the Content

Answer these questions on content for your readers:

1. **How carefully and completely is the subject researched?** Is the material accurate and up-to-date? Are important details missing? Exactly what has the writer left out? Where could the reader find the missing information? If the material is inaccurate or incomplete, is the whole work affected or just part of it?
2. **Is the writer or speaker objective?** Are conclusions supported by evidence? Is the writer or speaker following a particular theory, program, or school of

thought? Is that fact made clear in the source? Has the writer or speaker emphasized one point at the expense of others? What are the writer's qualifications and background?

3. **Does the work achieve its goal?** Is the topic too large to be adequately discussed in a single talk, article, or report? Is the work sketchy? Are there digressions, tangents, or irrelevant materials? Do the recommendations make sense?

4. **Is the material relevant to your audience?** How would the audience use it? Is the entire work relevant or just part of it? Why? Would the work be useful for all employees of your company or only for those working in certain areas? Why? What answers offered by the work would help to solve a specific problem you or others have encountered on the job?

You may want to review pages 375–378 on evaluating websites.

Evaluating the Style

Answer these questions on style for the readers of your evaluative summary:

1. **Is the material readable?** Is it well written and easy to follow? Does it contain helpful headings, careful summaries, and appropriate examples?

2. **What kind of vocabulary does the writer or speaker use?** Are there too many technical terms or too much jargon? Is it written for the layperson? Is the language precise or vague? Would readers have to skip certain sections that are too complicated?

3. **What visuals are included?** Charts? Graphs? Photographs? How are they used? Are they used effectively? Are there too many or too few?

Figures 9.7, 9.8, and 9.9 (pages 440–442) contain evaluative summaries. Note how the writers' assessments are woven into the condensed versions of the originals. Figure 9.7 is a student's opinion of an article summarized for a class in information management. Figure 9.8 is an evaluative summary in memo format collaboratively written by two employees who have just returned from a seminar. They have divided their labor, one writing the opening paragraph and the summary of "Techniques of Health Assessment" and the other doing the summaries of "Assessment of the Heart and Lungs" and "Assessment of the Abdomen." Together they drafted and revised the "Recommendations" and prepared the final copy of the memo.

Another kind of evaluative summary—a book review—is shown in Figure 9.9. Many journals and websites carry book reviews to inform their professional audience about the most recent studies in their field. Reviews condense and assess books, reports, government studies, websites, blogs, films, and other materials. The short review in Figure 9.9 comments on why the book is useful for the intended audience, analyzes the style, provides clarifying information, and explains how the book is developed. The Web contains numerous sites that run book reviews, including

FIGURE 9.7 An Evaluative Summary of an Article

Abbasi, Sami M., Kenneth W. Hollman, and Robert Hayes. "Bad Bosses and How Not to Be One." *Information Management Journal* 42.1 (2008): 52–56.

Identifies purpose of article

According to this practical and convincing article, the way employees are managed determines a company's success. The authors helpfully begin by describing the new twenty-first-century workplace where power has shifted from a top-down authoritative management style to one respecting employees as "knowledge workers" whose professional contributions are essential in a digital

Comments on style and organization

culture. The article then turns to a classification of six types of difficult bosses, ranging from incompetents, crooks, and bullies to dodgers, know-it-alls, and "walking policy manual[s]" who stick to a policy, however dated or contradictory. These bad bosses use intim-

Indicates why and how article is useful to diverse audiences

idation, manipulation, blame, conflict, and cover-ups to exert or protect their power. Effective bosses, on the other hand, remove fear from the workplace, build trust, encourage feedback, and act as advocates for their employees with upper management. Although aimed at information managers, the guidelines in this readable article apply to anyone who wants to be a good—or better—boss.

© Cengage Learning 2013

www.amazon.com	Editorial and reader reviews and publisher-supplied abstracts on over 100,000 business books
www.barnesandnoble.com	"Business Books" Web page offers direct links to over twenty-five subcategories, e.g., accounting, management and leadership, and women in business. Each individual book's page includes an overview, Barnes and Noble editorial review, and reader ratings (one to five stars)
www.businessbookreview.com	Detailed one-paragraph summaries on thousands of business books plus over 700 expertly written, eight-page summaries of top best business books
www.getabstract.com	Site dedicated to only business book summaries by industry-expert writers; abstracts focus on the main points of the book

A book review includes the most important and useful information—to a key audience—about a book or report. Reviews are also important for the kinds of information that they do *not* include: details and irrelevant (for the audience) information that would only clog a summary. For example, the review in Figure 9.9 indicates that the book is an excellent guide to team building, but it does not go into detailed descriptions of the author's recommended tips, strategies, and instruments.

FIGURE 9.8 A Collaboratively Written Evaluative Summary of a Seminar

SABINE MEMORIAL HOSPITAL

7200 Medical Blvd.
Sabine, TX 77231-0011
512-555-6734   www.sabinememorial.org

TO: Mohammed Lau, M.S.N.   SUBJECT: Evaluation of Physical
        Director of Nurses   Assessment Seminar

FROM: Elena Roja, R.N.   DATE: September 13, 2012
        Lee Schoppe, R.N.

On September 7, Doris Fujimoto, R.N., and Rick Poncé, R.N., both on the staff of Houston Presbyterian Hospital, conducted a practical and beneficial seminar on physical assessment. The one-day seminar was divided into three units: (1) **Techniques of Health Assessment**, (2) **Assessment of the Heart and Lungs**, and (3) **Assessment of the Abdomen**.

Gives overall structure of seminar

Techniques of Health Assessment
Four procedures used in physical assessment—inspection, percussion, palpation, and auscultation—were defined and demonstrated. Return demonstrations, used throughout the seminar, meant we did not have to wait until we went back to work to practice our skills. The instructors stressed the proper use of the stethoscope and the ways of taking a patient's medical history. We were also asked to take the medical history of the person next to us.

Describes and evaluates each part of seminar

Assessment of the Heart and Lungs
After we inspected the chest externally, we covered the proper placement of hands for percussion and palpation and the interpretation of various breath sounds. The instructors helped us find areas of the lung and identify heart sounds. However, the film, "Cardiopulmonary Feedback," on examining the heart and lungs was ineffective because it included too much information.

Explains why one part was unsuccessful

Assessment of the Abdomen
The instructors warned that the order of examination of the abdoman differs from that of the chest cavity. Auscultation, not percussion, follows inspection so that bowel sounds are not activated. The instructors clearly identified how to detect bowel sounds and how to locate the abdomen and palpate organs.

Continues to emphasize practical benefits of seminar

Recommendations
We strongly recommend a seminar like this for all nurses whose expanding role in the health care system requires more physical assessments. Although the seminar covered a wealth of information, the instructors admitted that they discussed only basics. In the future, however, it would be better to offer follow-up seminars on specific body systems (e.g., chest cavity, abdomen, central nervous system) instead of combining topics because of the amount of information involved and the time required for demonstrations.

Ends with endorsement by offering suggestions for improving seminar

FIGURE 9.9 A Book Review

Book Review

Team Troubleshooting: How to Find and Fix Team Problems
By ROBERT W. BARNER
Davies-Black Publishing, Palo Alto, Calif.
Representation: Anita Halton Associates, (949) 494-8564, 326 pages, $32.95

Overall Recommendation

DON'T BOTHER	BORROW	BUY

Engaging: 4   Innovative: 4   Usefulness: 5   Visual Aids: 4

Quantifies rating

Begins with purpose and scope of book

Discusses organization

Highlights main points of book

Praises book and gives specific reasons

Team building with a twist: While most books on the subject limit themselves to talk about getting a new team up and running or fixing a team after it breaks, Robert W. Barner deals with both and then goes a step farther. *Team Troubleshooting* promotes the concept of maintaining a healthy team through anticipating problems, performing regular team tune-ups and taking proactive measures to head off trouble before it begins. High points: strategies to foresee and develop plans for dealing with change, rather than becoming a victim of it; how-to tips to extend beyond the team itself to the problems of creating, maintaining and mending external relationships; ideas to cope with the fact that the team is not an isolated, independent entity, but is part of a bigger system and may be at the mercy of management and other forces.

The unusually user-friendly format makes the book truly serve as a guide.

Useful organization, clear chapter headings and cross-referencing help the reader find information quickly. Barner also does more than just name problems and offer quick fixes—he helps to assure accurate problem diagnosis by clearly describing both symptoms and underlying causes. Dozens of new instruments—not just the same tired old quizzes that appear in so many other teamwork books—lead both team leaders and members (and trainers or consultants involved in the "team" business) through exercises. They include scripting out scenario forecasts, identifying early warning signs of trouble, performing stakeholder analyses, and mapping relationships. It's been a long time since anything new has been said about teams— Barner's book deserves a space on the shelf of anyone truly interested in making a team work.

—*Jane Bozarth*

Abstracts

In addition to writing summaries, your employer may ask you to write abstracts. Abstracts are found in several key documents in the world of work and are a staple of the Internet.

Differences Between a Summary and an Abstract

The terms *summary* and *abstract* are often used interchangeably, resulting in some confusion. That problem arises because there are two distinct types of abstracts: *descriptive abstracts* and *informative abstracts*. An informative abstract is the same as a summary; it indicates what research was done, what conclusions were reached, and what recommendations were made. Look at the summary in Figure 9.4. It explains why virtual reality should be included in law enforcement training: because virtual reality gives officers field-translatable training. Informative abstracts are found at the beginning of long reports. Descriptive abstracts, however, do not give conclusions.

All abstracts share two characteristics: the writer never uses "I" and avoids footnotes.

Writing an Informative Abstract

An informative abstract is not as long as an executive summary, which gives more supporting details. As a part of your course work or your job, you will probably have to write informative abstracts for long reports (Chapter 15).

One way to approach writing the abstract of a report is to think of it as a table of contents in sentence form. A table of contents is, in effect, a final outline; it is easily fleshed out into an abstract, as Figure 9.10 (page 444) shows. On the left is a table of contents, and on the right is the abstract written from that outline.

This system works only if your table of contents is neither too detailed nor too skimpy. Starting off with a good outline of an article or a report provides the best beginning for your abstract. Make sure your sentences are complete and grammatical. Do not omit verbs, conjunctions, or articles (*the*, *a*, *an*). Proper subordination is essential. You should expect to condense a whole paragraph of the original to a sentence, an individual sentence to a phrase, and a phrase to a single word.

Writing a Descriptive Abstract

Unlike an informative abstract, a descriptive abstract is usually only a few sentences long; it does not go into any detail or give conclusions. As the name implies, a descriptive abstract provides information on what topics a work discusses but not how or why they are discussed. Busy readers rely on a descriptive abstract to decide whether they need to consult the work itself. Here is a descriptive abstract of the article summarized in Figure 9.4:

> Virtual reality can be used to teach law enforcement officers firearms training, SWAT team assaults, incident re-creation, and crime location processing. This training technology will be of interest to law enforcement administrators.

Figure 9.11 (page 445) reproduces two descriptive abstracts from the reference work *Information Science Abstracts* as well as two abstracts from the *Journal of Interactive Marketing*, a publication that includes abstracts as a way to help readers learn about research in this specialized discipline.

FIGURE 9.10 An Abstract Written from a Table of Contents

Table of Contents	Abstract

Need for Genetic Counseling
   Definition of Genetic Counseling
   Statistics on Genetic Counseling
	Genetic counseling is a service for people with a history of hereditary disease. One in 17 births contains some defect; one-fourth of the patients in hospitals are victims of genetic diseases (including diabetes, mental retardation, and anemia). One of every 200 children born has chromosome abnormalities.

Purpose of Genetic Counseling	Genetic counseling offers advice to parents who may give birth to children with genetic diseases and assistance to those with children already afflicted.

The Counseling Process
   Evaluating the Needs of the
    Counselees
   Taking a Family History
   Estimating the Risks
   Counseling the Family
	The first step in counseling is to evaluate the needs of the parents. A family history is prepared and risks of future children being afflicted are evaluated. The life expectancy and possible methods of treatment of any afflicted child also can be determined. Alternatives are presented.

Determination of a Genetic Disorder
   Amniocentesis
   Karyotyping
   Fluorescent Banding
   Staining
	Four prenatal tests are used to determine if a genetic disorder is present: amniocentesis, karyotyping, fluorescent banding, and staining.

Advantages of Genetic Screening
   Lower Cost
   Increased Availability
	The development of these four relatively simple methods has lowered the cost of genetic counseling and increased its availability.

Source: Reprinted by permission of Professor Mary Scotto.

Writing Successful News Releases

A *news release*, sometimes called a *press release* or *media release*, is another type of on-the-job document that requires you to summarize key information for a variety of readers. Basically, a news release is an announcement (usually one page or a single screen on the Web) about your company's or agency's specific product, services, or personnel. It should be crisp and highlight only the most important and relevant facts clearly and straightforwardly, as the other summaries you have studied in this chapter do.

Such releases are frequently posted on a company's or agency's website, as well as sent in the mail, as a fax, or automatically to your computer or smartphone through an RSS feed (if you have clicked on "Subscribe to RSS" at the company's website). They inform readers—the news media, potential and current customers, other agencies—of newsworthy events that promote the professional accomplishments of a company. Because your company's image can be enhanced or tarnished by the release you write, double-check all your facts, particularly names, dates, places, and any warranty or sales conditions. The best releases project a professional, customer-centered, and quality-focused corporate or agency image.

Figures 9.12 and 9.13 (pages 446–447) contain sample news releases. The release shown in Figure 9.12 was distributed over the Web, and the release in Figure 9.13

FIGURE 9.11 Descriptive Abstracts of Journal Articles

2002-01577

Mann, Charles C **Electronic paper turns the page.** *Technology Review* **104**(2): 42–48 (March 1, 2001) (ISSN: 1099-274X) In English.
Journal URL: http://www.techreview.com

Reports that the key to enabling electronic books to revolutionize the publishing industry is the development of electronic paper. Draws a parallel between the general availability of paper and the development of the first moveable-type printing press, noting that the technology to create the printing press existed for 100 years before the widespread use of ordinary paper provided the means to make the press practical. Notes that several companies are engaged in research and development of electronic paper, adding that at least one working prototype already exists. Indicates that the recent discovery of electrical conducting properties in plastic will enable researchers to overcome the issues of flexibility and cost that have slowed development in the past.

28680312

Murphy, Jeannette **Globalization: Implications for health information professionals.** *Health Information & Libraries Journal* **25**(1): 62–68 (March 2008) (ISSN: 1471-1834) (DOI:10.1111/j.1471–1842.2007.00761.x)

The article discusses how globalization processes may affect health information personnel. The author notes how globalization has led to a global economy and global culture through the transmission of services and information and the creation of international organizations and policies. She suggests that international organizations such as the World Bank and the World Health Organization will affect international health governance through the influence of health librarians. Information technology in fact may improve health education and communication of health information. Globalization has also led to outsourcing of health care services such as information processing. Medical tourism and physician mobility have increased.

Informs readers about which topics are discussed, but not how or why

Source: Reprinted with permission from *Information Science Abstracts*, Vol. 37, No. 4. Copyright © Information Science and Technology Abstracts.

Strategic and Ethical Considerations in Managing Digital Privacy

Ravi Sarathy and Christopher J. Robinson (August 2003), *Journal of Business Ethics,* 46(2), pp. 111–126.

Information about customers and prospects is readily available through a variety of digital sources. The questions a marketer must answer is how much of this available data should be used for commercial purposes and how much should remain privileged and off limits. In this paper the authors develop a model of the factors influencing privacy strategy. This model incorporates external, ethical, and firm-specific factors that impact customer privacy protection strategy formulation. The model is then applied to various scenarios to determine the firm's most likely customer privacy strategy. International implications of the model are also discussed.

Scovotti. (8, 13)

The Professional Service Encounter in the Age of the Internet: An Exploratory Study

Gillian Hogg, Angus Laing, and Dan Winkelman (2003), *The Journal of Services Marketing,* 17(5), pp. 476–495.

The Internet, by providing access to an unprecedented amount of healthcare-related information, is changing the balance of power in the relationship between healthcare consumers and professionals. Patients play a more active role in the relationship, interacting with healthcare professionals and other consumers to understand their illnesses. This situation changes the nature of the doctor/patient relationship—where the doctor becomes only one of the *advisors* in the service encounter. The implications of this research extend to other types of service encounters, where consumers may be engaging in virtual, parallel service encounters.

Short, 3-4 sentence paragraph

Uses objective language

Source: Journal of Interactive Marketing 18/1 (Winter 2004): 84, 87. Copyright © 2004. Reprinted with permission of John Wiley & Sons, Inc.

was distributed in hard-copy form. The various parts of a news release, which are discussed on pages 449–451, are labeled on both figures.

Subjects Appropriate for News Releases

News releases should be written only about newsworthy subjects, such as these:

1. **New products, services, or publications**
   - a new or improved model, line of equipment, or website
   - new or expanded technical or customer-friendly services
   - the entrance of your company into a different market
   - the application of the latest technology in creating a product

2. **New policies or procedures**
   - acquisition of new technologies to lower production costs
   - changes in production to improve safety, delivery, accuracy
   - cooperation and collaboration between two agencies or branches

3. **Personnel changes and awards**
   - appointments
   - promotions or recognition for winning an award
   - new hires
   - retirements

4. **New construction and developments**
   - plant or office openings
   - new satellite, branch, or overseas offices
   - expansion of an existing site

5. **Financial and business news**
   - company reorganization: merger, acquisition
   - stock reports
   - quarterly earnings and dividends
   - sales figures

6. **Ecofriendly (green) news**
   - energy conservation
   - water/air quality improvement
   - organic farming techniques
   - protection of endangered species
   - preservation of environmentally sensitive sites
   - policies or products for the office, home, etc.

7. **Special events**
   - training seminars
   - demonstrations of new equipment
   - community service programs
   - charity benefits
   - visits of national speakers
   - dedications

FIGURE 9.12 A News Release from the Web

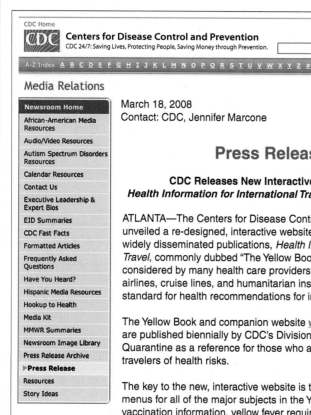

CDC Home

CDC **Centers for Disease Control and Prevention**
CDC 24/7: Saving Lives, Protecting People, Saving Money through Prevention.

[SEARCH]

A-Z Index A B C D E F G H I J K L M N O P Q R S T U V W X Y Z #

Media Relations

Newsroom Home
African-American Media Resources
Audio/Video Resources
Autism Spectrum Disorders Resources
Calendar Resources
Contact Us
Executive Leadership & Expert Bios
EID Summaries
CDC Fast Facts
Formatted Articles
Frequently Asked Questions
Have You Heard?
Hispanic Media Resources
Hookup to Health
Media Kit
MMWR Summaries
Newsroom Image Library
Press Release Archive
▶**Press Release**
Resources
Story Ideas

March 18, 2008   (404) 639-3286
Contact: CDC, Jennifer Marcone   http://bookstore.phf.org

Press Release

CDC Releases New Interactive Website for
Health Information for International Travel **(The Yellow Book)**

ATLANTA—The Centers for Disease Control and Prevention (CDC) unveiled a re-designed, interactive website of one of CDC's most widely disseminated publications, *Health Information for International Travel*, commonly dubbed "The Yellow Book." The Yellow Book is considered by many health care providers, travel professionals, airlines, cruise lines, and humanitarian institutions to be the gold standard for health recommendations for international travel.

The Yellow Book and companion website **www.cdc.gov/travel/yb** are published biennially by CDC's Division of Global Migration and Quarantine as a reference for those who advise international travelers of health risks.

The key to the new, interactive website is the use of drop-down menus for all of the major subjects in the Yellow Book, including vaccination information, yellow fever requirements, malaria information, geographic distribution, and health hints. Users can obtain customized reports for individual travel plans and locate particular subjects or destinations in the text without having to search through unrelated topics.

"International travelers will find the interactive Yellow Book website to be very helpful," said CDC Director Jeffrey P. Koplan, MD, MPH. "The new features make it easier to find information about preventive measures travelers can take to protect their health."

The handy, interactive design of the Yellow Book website incorporated many comments and suggestions from the public and health care providers seeking more customized information in a user-friendly format. CDC encourages users to submit comments about the new website through the "comments section" on the site's homepage.

30

http://bookstore.phf.org

Slug emphasizes importance of topic

Lead answers who, where, how, what, and why

Body emphasizes benefits for audience

Quotation from expert endorses new design and features

Conclusion invites reader interaction

Number 30 signals end of release

Centers for Disease Control and Prevention (CDC).

FIGURE 9.13 A News Release Circulated in Hard Copy

NEWS RELEASE

United States Department of Commerce • 1401 Constitution Avenue, NW •
Washington, DC 20230 • Web: http://www.commerce.gov

---

*Contact
information*

Release No. 0150.01

Cheryl Mendonsa
Cheryl.Mendonsa@technology.gov
202-482-8321

Marjorie Weisskohl
Marjorie.Weisskohl@technology.gov
202-482-0149

Slug

U.S. Department of Commerce Announces Website Focused on Resources for Tech-Based Economic Development (TBED)

Lead, or hook

Under Secretary of Commerce Phillip J. Bond today announced a new resource that
will help communities working to encourage the growth of technology companies
and build tech-based economies. A new website, the TBED Resource Center
(**http://www.tbedresourcecenter.org**), gives users the chance to learn from others'
experiences and to benefit from the latest research on creating a tech-based economy.

Body

The TBED Resource Center categorizes and provides links to more than 1,300 research
reports, strategic plans, best practices and impact analyses from state and federal
government, university researchers, and foundations. The website is a result of a
cooperative project of the U.S. Department of Commerce's (DOC) Office of Technology
Policy and State Science and Technology Institute, based in Westerville, Ohio.

*Quote from
authority*

"This website will be an invaluable tool to anyone involved in economic development,"
said Under Secretary Bond. "Access to information on what has worked in various
communities will help policymakers and business people make efficient use of valuable
time and resources, and provide productive ideas and contacts that others might find
useful." Bond made the announcement via live telecast hosted by Commerce's Economic
Development Administration and the National Association of Regional Councils.

*Emphasizes
benefits*

The user-friendly search options allow for easy navigation with abundant topics spanning
a wide field of interests. Users can search for reports based upon geography, topic, type,
or keyword with the option of selecting multiple fields. Numerous international reports
also are included, offering a globally diverse perspective in tech-based economic
development.

*Appropriate
positive tone*

Reports fall into one of the 35 topics, such as brain drain, education, entrepreneurship,
innovation, research and development, and work force.

To keep current with TBED trends, the website is continually updated as new reports are
released.

*Symbol #
signals end
of release*

#

Source: Material reprinted from the United States Department of Commerce.

News Releases About Bad News

While the topics above focus on a company's or organization's achievements—the positive contributions a firm or an agency makes to its customers and its community—not all the news you may be asked to announce is pleasant. At times you may have to report on events that are difficult: product recalls, work stoppages or strikes, layoffs, plant closures, limited availability or unavailability of products or parts, fires, computer viruses, alerts, higher prices, declining enrollment, or canceled events.

Even when you have to write releases about such unpleasant events, portray your company honestly and in the most professional and conscientious light. Regardless of the news, be accurate, ethical, honest and straightforward, and available.

Topics That Do Not Warrant a News Release

Not every event or change at your company or agency will warrant a news release. Releases about events that have already happened are old, tired news. Keep in mind, too, that it is unethical to write about some subjects (see pages 29–42 in Chapter 1). Following are topics that hold little or no interest or about which you should *not* write a release:

1. **Well-known products or services.** Don't repeat the obvious.
2. **Products or services still in the planning stages.** Not only would such news be premature, but releasing it ahead of time might meet with your boss's severe disapproval, not to mention your competition's delight.
3. **Controversial events or company problems.** Refrain from writing a news release on these subjects unless your boss instructs you to do so.
4. **A history of your department, division, or company.** Save remarks about your firm's history for the annual report to stakeholders.
5. **An obviously padded tribute to your boss, your company, or your customers.**

Organization of a News Release

The following sections contain guidelines for organizing and writing the different parts of your news release. Note how these various parts flow together in the news releases in Figures 9.12 and 9.13.

The cardinal rule in writing a news release is to put the most important piece of news first. Don't bury it in the middle or wait until the end. Everything in a news release should be arranged in descending order of importance so that your first paragraph contains only the most significant facts and ideas. Think of your release as an inverted pyramid with the top summarizing only the most crucial points.

The Three Parts of a News Release

The three components of a news release are the slug (headline), lead, and body.

The Slug, or Headline   One of the most crucial parts of your release is the headline, or *slug*. It should announce a specific subject for readers and draw them into it. Write a slug that entices or grabs your readers, but also informs them quickly and clearly. Note how the slugs in Figures 9.12 and 9.13 point readers to the main idea in the first sentence and use keywords to help readers track the subject.

The Lead   The first (and most significant) sentence of a news release is called the *lead*. It introduces and aptly summarizes your topic, sets the tone, and continues to keep readers' attention. For that reason it is also called the *hook*. See how the leads in Figures 9.12 and 9.13 grab readers' attention. The best leads easily answer the basic questions *Who? What? When? Where? Why?*—the five *W's*—and *How?* (or *How much?*). Not every lead will answer all these questions, but the more of them you do answer, the better your chances will be of capturing your audience's attention. Here are some effective leads that answer these crucial questions:

> *who*   *when*
> Massey Labs announced today that the FDA has approved the marketing
> *what*   *why*
> of its vaccine—Viobal—to retard recurrent lesions of herpes simplex.

> *who*   *what*
> Maryville Engineering, Inc., has been awarded a contract by Aerodynamics, Inc.,
> *why*
> to develop an acoustical system to measure and monitor stress levels at the
> *where*   *when*
> Knoxville aircraft plant, district manager Carmelita Stinn, P.E., announced today.

Misleading leads that deviate substantially from the guidelines just discussed annoy readers and risk losing their interest. Resist the temptation to start off with a question or with folksy humor.

Feeble humor:	Slap, slap, scratch, scratch, itch. This is how millions of people will handle their mosquito problems this summer.
Boastful start:	If you thought 2011 was great, you ain't seen nothin' yet!
Undirected question:	Does the thought of hypothermia send chills up your spine?
Exaggeration:	I bet you wonder how you ever did without the new Fourier scanner.

The Body   What kinds of information should you give readers in the second and subsequent paragraphs of your release? If the five *W's* are answered in your lead, the following paragraphs fill in only the most necessary supporting details. Regard your lead as a summary of a summary. The body of your release then amplifies the *Why?* and *How?* and may also get into the *So what?*

But avoid filling your news release with unnecessary technical details, such as scientific formulas or intricate speculations. Instead, relate your product or service to your targeted audience's needs by emphasizing the benefits to them without loading your news release with hype.

Use quotations selectively to clarify and highlight, not to apple-polish. Use a quotation only to report vital facts or to cite an authority. Do not turn your release into an interview with your employer or favorite customer.

Style and Tone of a News Release

The tone of your news release needs to be objective and take into account your audience's background. Keep your style simple and to the point. Here are some suggestions that will help you:

1. **Write concisely.** Use easy-to-read paragraphs of three to five sentences, and keep your sentences short. Under twenty words is ideal. Avoid long, convoluted sentences and fillers that start off with such phrases as "It has been noted that. . . ."

2. **Emphasize benefits to the reader.** Use graphic, understandable language that applies to the reader's life. Avoid unfamiliar jargon—it's deadly. A news release summarizing the benefits of a hospital counseling center will lose potential seminar participants by describing it in jargon: "Milford Hospital is pleased to offer an adventure-intensive counseling workshop that incorporates trust sequencing; high-element activities; and quantifiable decision modules."

3. **Keep your tone upbeat, easygoing, and direct; stress the human side.** But stay away from adjectives dripping with a pushy sell: *incomparable, fantastic, incredible*. Note how Figures 9.12 and 9.13 emphasize benefits to travelers and to communities interested in technology-based economic development, respectively.

As with other types of summaries described in this chapter, news releases give readers essential, useful information quickly and concisely. They provide the "big picture." And by doing so, they capture the reader's interest and emphasize the goals and contributions of the company or agency at the same time.

Conclusion

Summaries are a vital part of workplace writing, and knowing how to summarize a document—a report, proposal, or presentation—and select only the most important and relevant points for your audience is a prized job skill. Busy managers, clients, and even co-workers will depend on your summaries to give them the big picture, the bottom-line conclusions and recommendations they need to get the job done. Your summaries must be accurate, concise, relevant, and ethical. The process presented in this chapter for summarizing information will help you to write different kinds of summaries: As this chapter has also stressed, you need to be familiar with *executive summaries* that assist a manager in making a decision, *evaluative summaries* that summarize and assess what you have read, *informative abstracts* that let readers know what a report covers and the conclusions and recommendations it reaches, and *descriptive abstracts* that tell readers what topics a document covers but not how or why. A *news release* is another type of essential workplace summary that informs a large group of readers about important projects and accomplishments, as well as bad news subjects of crucial concern to your company's or organization's image and mission.

✓ Revision Checklist

- [ ] Read and reread the original thoroughly to gain a clear understanding of the purpose and scope of the work.
- [ ] Underlined key transitional words, main points, significant findings, applications, solutions, conclusions, and recommendations.
- [ ] Separated main points clearly from minor ones, background information, illustrations, and inconclusive findings.
- [ ] Omitted examples, explanations, and statistics from summary or abstract.
- [ ] Deleted information not useful to the audience because it is too technical or irrelevant.
- [ ] Changed language of original to my own words so I am not guilty of plagiarism.
- [ ] Made sure that emphasis of summary matches emphasis of original.
- [ ] Determined that sequence of information in the summary follows sequence of original.
- [ ] Added necessary connective words that accurately convey relationships between main points in original.
- [ ] Edited summary to eliminate wordiness and repetition.
- [ ] Cited source of original correctly and completely.
- [ ] Avoided phrases that draw attention to the fact that I am writing a summary or an abstract.
- [ ] Summarized material in informative summary objectively without adding commentary.
- [ ] Commented on both content and style in an evaluative summary.
- [ ] Interspersed evaluative commentary throughout the summary so that assessments appear near relevant points.
- [ ] Included a direct quotation in the evaluative summary or news release to illustrate or reinforce my recommendation.
- [ ] Ensured that descriptive abstract is short and to the point and does not offer a judgment.
- [ ] Prepared news release that projects a professional image of my employer.
- [ ] Summarized only essential information in news release.
- [ ] Arranged information from top down, with most important information first.
- [ ] Wrote clear, crisp slug and lead and concise body for news release.
- [ ] Saved word-processing file in document folder on computer.
- [ ] Sent news release on newsworthy topic to appropriate publications, organizations, and audiences.

Exercises

1. Summarize a chapter of a textbook you are now using for a course in your major field. Provide an accurate bibliographic reference for that chapter (author of the textbook, title of the chapter, title of the book, place of publication, publisher's name, date of publication, and page numbers of the chapter).

2. Summarize a lecture you heard recently. Limit your summary to one page. Identify in a bibliographic citation the speaker's name, the date, and the place of delivery.

3. Listen to a television network evening newscast and to a later news update on the same station. Select one major story covered on the evening news and indicate which details from it were omitted in the news update.

4. Write a summary of the marketing report in Chapter 8 (pages 398–413) or the business report in Chapter 15 on non-native speakers of English in the workforce (pages 701–719).

5. Bring to class an article from *Reader's Digest* and the original material it condensed, usually an article in a journal or magazine published six months to a year earlier. In a paragraph or two indicate what the *Digest* article omits from the original. Also point out how the condensed version is written so that the omitted material is not missed and how the condensation does not misrepresent the main points of the article.

6. Assume that you are applying for a job and that the human resources manager asks you to summarize your qualifications. In two or three paragraphs, indicate how your background and interests make you suited for the job. Mention the job by title at the beginning of your first paragraph.

7. Write a summary of one of the following articles.
    a. "Microwaves," in Chapter 1 (pages 47–49)
    b. "Protecting Personal Information," below

Protecting Personal Information: A Guide for Businesses

Small Business Administration

Companies keep sensitive personal information in their files—names, social security numbers, credit card or other account data—that identifies customers or employees. This information often is necessary to fill orders, meet payroll, or perform other necessary business functions. However, if sensitive data falls into the wrong hands, it can lead to fraud, identity theft, or similar troubles. Given the high cost of a security breach—losing your customers' trust and perhaps even defending yourself against a lawsuit—it is just plain good business to safeguard personal information.

Protect the Information That You Keep

What's the best way to protect the sensitive, personal information your company needs to keep? It depends on the kind of information and how it's stored. The most effective data security plans deal with four key elements: physical security, electronic security, password management, and employee training, all of which are discussed below.

Physical Security

Many data compromises happen the old-fashioned way—through lost or stolen paper documents. Often, the best defense is a locked door or an alert employee. Here are some helpful ways to protect information from falling into the wrong hands:

First of all, store paper documents or files, as well as CDs, floppy disks, Zip drives, tapes, and backups containing personally identifiable information, in a locked room or in a locked file cabinet. Limit access to employees with a legitimate business need. Control who has a key, and the number of keys. Then require that files containing personally identifiable information be kept in locked file cabinets except when an employee is working on the file. Remind employees not to leave sensitive papers out on their desks when they are away from their workstations, to put files away, log off their computers, and lock their file cabinets and office doors at the end of the day.

Next, implement appropriate access controls for your building. Tell employees what to do and whom to call if they see an unfamiliar person in the office or elsewhere on the premises. If you maintain off-site storage facilities, limit employee access to those with a legitimate business need. Know if and when someone accesses the storage site. For example, if you ship sensitive information using outside carriers or contractors, encrypt the information and keep an inventory of the information being shipped. Also, use an overnight shipping service that will allow you to track the delivery of your information.

Electronic Security

Computer security isn't just the realm of your IT staff. Make it your business to understand the vulnerabilities of your computer system, and follow the advice of experts in the field. To do this, identify the computers or servers where sensitive personal information is stored. Make sure, too, that you identify all connections to the computers where you store sensitive information. These may include the Internet, electronic cash registers, computers at your branch offices, computers used by service providers to support your network, and wireless devices like inventory scanners or cell phones.

Moreover, encrypt all sensitive information that you send to third parties over public networks (like the Internet), and consider encrypting sensitive information stored on your computer network or on disks or portable storage devices used by your employees. It is a smart idea to encrypt e-mail transmissions as well—within your company if they contain personally identifying information. In addition, regularly run up-to-date anti-virus and anti-spyware programs on individual computers and on servers on your network. Check expert websites (such as **www.sans.org**) and your software vendors' websites regularly for alerts about new threats and to learn about implementing new policies for installing vendor-approved patches to correct problems.

Most important of all, scan the computers on your network to identify and profile the operating system and open network services. If you find services that you don't need, disable them to prevent hacking or other potential security problems. For example, if e-mail service or an Internet connection is not necessary on a certain computer, close the ports to those services on that computer to prevent unauthorized access to that machine. When you receive or transmit credit card information or other sensitive financial data, use Secure Sockets Layer (SSL) or another secure connection that protects the information in transit.

Pay particular attention to the security of your Web applications—the software used to give information to visitors to your website and to retrieve information from them. Web applications may be particularly vulnerable to a variety of hacker attacks. In one variation called an "injection attack," a hacker inserts malicious commands into what looks like a legitimate request for information. Once in your system, hackers transfer sensitive information from your network to their computers. Be on guard against these attacks by more closely scrutinizing requests.

Password Management

Also essential to any business's security is password protection. You can control access to sensitive information by requiring your employees to use "strong" passwords. Technical security experts advise that the longer the password, the better. Because simple passwords—like common dictionary words—can be easily guessed, insist that your employees choose passwords with a mix of letters, numbers, and characters. Require an employee's user name and password to be different, and insist on frequent changes in passwords. Lock out users who don't enter the correct password within a designated number of log-on attempts. Password-activated screen savers can lock employee computers after a period of inactivity.

Another wise move is to warn employees about possible calls from identity thieves attempting to deceive them into giving out their passwords by impersonating members of your IT staff. Let employees know that calls like this are always fraudulent, and that no one should be asking them to reveal their passwords. Set up clear and effective procedures through which employees can communicate with IT staff.

When installing new software, make sure your company immediately switches vendor-supplied default passwords to a more secure strong password and cautions employees against transmitting sensitive personally identifying data—social security numbers, passwords, account information—via e-mail. Unencrypted e-mail is not a secure way to transmit any information.

Employee Training

Your data security plan may look great on paper, but it's only as strong as the employees who implement it. Take time to explain the rules to your staff, and train them to spot security vulnerabilities. Periodic training emphasizes the importance you place on data security practices. A well-trained work force is a company's best defense against identity theft and data breaches.

You can also create a "culture of security" by implementing a regular schedule of employee training. Update employees as you find out about new risks and vulnerabilities. Make sure training extends to employees at satellite offices and to temporary help and seasonal workers as well. If employees don't attend, consider blocking their access to the network. If you train employees to recognize security threats, you can cut down on the risks such thefts pose. Tell your employees how to report suspicious activity and publicly reward employees who alert you to vulnerabilities.

Finally, ask every new employee to sign an agreement to follow your company's confidentiality and security standards for handling sensitive data. Make sure these most recent members of your work force understand that abiding by your company's data security plan will be an essential part of their duties. And regularly remind all employees of your company's policy—and any legal requirement—to keep customer information secure and confidential.

Source: Federal Trade Commission, *Protecting Personal Information: A Guide for Businesses*. Washington, DC: GPO, 2008.

8. Write a descriptive abstract of the article you selected in Exercise 7.

9. Below are two sloppily written news releases that are poorly organized, unethical, and incorrectly formatted. Reorganize and rewrite them according to the guidelines presented in this chapter and those in Chapter 1 (pages 29–42) on making sure your writing is ethical.

a.   Friday   Alan Bowerstock

Metropolitan State University is a four-year urban institution of higher education offering majors in many fields. Located two miles west of Taylorsville Tech Park, MSU currently boasts more than 7,000 undergraduate students, more than our rival, Central Tech.

Among the many student services currently available at MSU is the Division of Career Placement; this division is located in the Student Services Building, Room 301, just across the hall from the Department for Greek Life.

In the last year, the Division has assisted more than 2,000 MSU students to find part-time jobs. Full-time jobs, too.

The goal of the Division is to help students earn money for their college expenses. The Division also wants to assist local businesses in contacting MSU qualified undergraduates.

The MSU family is well represented on the homepage. Every department from the University has been encouraged to report on only its most favorable activities. The Division of Career Placement is also on the homepage.

Counselors at the Division can help MSU students prepare a four- or five-line ad about their qualifications. The Division will run these ads in their homepage. The university will also run ads looking for job candidates from the state employment agency and the greater Taylorsville area.

b.   For General Information   Frank Day

J. T. Bushart, CEO of Bonnetti and Blount Construction for the last three years, asserted today that the company is devoted to progress and change. Bushart came to B & B Engineering after several years working for Capitol City Engineering, a less progressive firm.

Bonnetti and Blount is a leading firm of contractors and has worked for both national and international corporations.

The firm specializes in construction projects that require special expertise because of their challenges in difficult terrains.

B & B has just been awarded a contract to work on the 10 million dollar renovation of two major Fairfax dams. The firm anticipates hiring more than 200 new workers. These new employees will work on the dams that present dangerous conditions to residents.

When completed, the two new dams will further assist residents of Fairfax and Hamilton Counties receive all the necessary irrigation and hydroelectric energy they need.

Engineering sketches and blueprints are in the works.

10. Write an appropriate news release on one of the following newsworthy topics to be included on your company's website:
    a. hiring a new webmaster
    b. premiering a new product or service
    c. acquiring a smaller firm whose products and services are very different from those of your company
    d. providing an environmentally sensitive service that enhances life in the community in which your employer's headquarters is located
    e. promoting an employee who has been with your company for at least five years
    f. offering highly competitive warranties on a new line of products
    g. protecting a section of wetlands adjoining one of your company's construction sites

11. Write a news release on one of the following unpleasant topics, while still projecting a positive image of your company:
    a. inconveniences because of recent construction
    b. a temporary power outage in a neighborhood or city
    c. the temporary failure of one of your company's servers
    d. a change in the hours of operation of a store or plant
    e. an increase in insurance premiums
    f. a reduction in the number of times trash is picked up per week
    g. an order page is down on a frequently visited e-business site because of a computer virus
    h. a sports injury that has benched a star player on a local team for the next month
    i. a boil-water notice issued for a subdivision that depends on a local reservoir for its water supply

PART IV

Preparing Documents and Visuals

Designing Clear Visuals

Visuals are essential in the world of work. No matter where you work—in a health care facility, government agency, retail business, bank, and so on—your employer will expect you to be visually competent. That means you will have to interpret, use, and create visuals and graphics. Visuals often explain the job you have to do and help you report on how well you did it. Tables, graphs, charts, diagrams, maps, and photographs are an important, functional part of almost every workplace document, including employee handbooks, instructions, proposals, reports, blogs, newsletters, and websites. You may be called on to prepare a PowerPoint presentation once or twice a week, and the graphics you choose can significantly influence how well your ideas are received.

A company would not launch a new line of products or services without representing projected earnings in graphs and charts and offering customers relevant photos and drawings. Without visuals, workplace documents would be harder to understand, making it more difficult for you to persuade co-workers, managers, or customers to agree with your conclusions.

Visuals are crucial to the success of your written work. Note how many of the documents in this book include visuals. In fact, visuals work in conjunction with your writing to summarize, inform, illustrate, and persuade. The tables, graphs, and charts you use to simplify financial and other statistical information in a report, and the photos and other graphics you incorporate in them, have a major impact on how readers will see and judge your work. A poorly designed visual—one that is incorrect, incomplete, unclear, or unethical—signals to readers that your research and writing may also be flawed.

Visual Thinking in the Global Workplace

Our web-based, global culture highly prizes visual thinking, or seeing and presenting information in terms of spatial and organizational appearance. Experts estimate that as much as 80 percent of our learning comes through our sense of sight. The way we process information visually helps us to develop ideas, organize them, and even solve problems when they occur. As we saw in Chapter 2 (pages 52–54), visual devices such as clustering, mindmapping, and outlining help writers to discover, develop, and organize their ideas.

In our digital culture, whether we are navigating a website, using a media device, collaborating with colleagues, or communicating with clients, visuals play a major role in conveying information. Animation, color, icons, graphs, arrows, charts, and photographs all contribute to our understanding and evaluation. Visuals make documents informative, inviting, and easy to read.

Visuals also play a powerful role in symbolizing corporate identity and reinforcing brand loyalty in the minds of consumers across the globe; consider McDonald's golden arches, Nike's swoosh, Starbucks' mermaid, or Ralph Lauren's polo pony and rider. A company's own logo reveals a great deal visually about its image, values, and mission.

Chapter 10 surveys the kinds of visuals you will encounter most frequently in the world of work. It shows you how to read, locate, create, and write about them, and it also describes the types of visuals and visual configurations you can create and copy with graphics software packages. Finally, this chapter discusses the ethical ways to use visuals and explains how to select the most appropriate ones for global audiences. But as you have already seen, the discussion of visuals in this book is not confined just to this chapter. Visuals are discussed in other chapters as part of the process of preparing a variety of business documents—letters, instructions, proposals, reports, blogs, websites, and presentations.

The Purpose of Visuals

There are several reasons why you should use visuals. Each point is graphically reinforced in Figure 10.1.

1. Visuals arouse readers' immediate interest. They catch the reader's eye quickly by setting important information apart and by giving relief from sentences and paragraphs. Note the eye-catching quality of the visual in Figure 10.1.

2. Visuals increase readers' understanding by simplifying concepts. Visuals are especially important if you have to explain a technical process to a nonspecialist audience. They can make a complex set of numbers easier to comprehend, helping readers see percentages, trends, comparisons, and contrasts. Figure 10.1, for example, shows at a glance the growth of online purchases of computers over six years.

3. Visuals are especially important for non-native speakers of English and for multicultural audiences. Given the international audience for many business documents, carefully prepared visuals can make your communication with readers easier and clearer. As Terri Smith Ruckel's report shows, visuals can bridge cultures and continents with universally understood icons (see Figure 10.35 on page 505) and wordless instructions. Study the visuals in her report in Figure 15.3 (pages 701–719) to see further examples of easily understood visuals for an international audience.

4. Visuals emphasize key relationships. Through their arrangement and form, visuals quickly show contrasts, similarities, growth rates, and downward and upward movements, as well as fluctuations in time, money, and space.

FIGURE 10.1 A Line-and-Bar Chart Comparing Market Share of Online Computer Purchases with Those Purchased at Brick-and-Mortar Stores

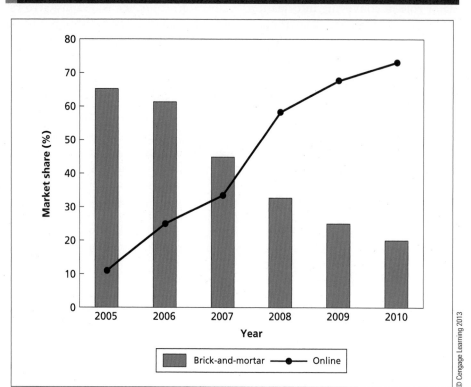

© Cengage Learning 2013

5. Visuals condense and summarize a large quantity of information into a relatively small space. A visual allows you to streamline your message by saving words. It can record a great deal of statistical data in far less space than it would take to describe those facts in words alone. Note how in Figure 10.1 the growth rates of two different types of businesses are concisely expressed and documented.

6. Visuals are highly persuasive. Visuals have sales appeal. They can convince readers to buy your product or service or to accept your point of view and reject your competitor's. A visual can graphically display, explain, and reinforce the benefits and opportunities of the plan you are advocating.

Types of Visuals and Their Functions

Table 10.1 shows some of the most common types of visuals found in workplace writing and explains how you can use them. Each of these visuals is discussed in this chapter.

TABLE 10.1   Types of Workplace Visuals

Table	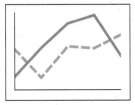	■ Presents large amounts of data in a compact space ■ Classifies data (figures, facts) into easy-to-read categories using rows and columns ■ Gives statistical data or verbal descriptions clearly and concisely ■ Allows readers to see comparisons and contrasts more quickly when numbers are embedded
Line graph	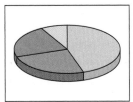	■ Transforms numbers into a picture (curves, shapes, patterns) that represents movement in time or space ■ Displays data that change often—costs, sales, rates, employment, production, temperature, etc. ■ Forecasts trends in terms of variables
Circle or pie chart	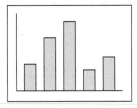	■ Uses wedges to show the parts/percentages (budgets, shares, time allotments, etc.) that make up the whole ■ Shows the proportion of each part to another; relates each part to the whole ■ Effective visual to show data from tables or charts ■ Easily understood by audiences worldwide
Bar chart	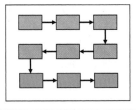	■ Uses vertical or horizontal bars to measure different data in space or time ■ Represents comparisons/changes through the comparative length of bars ■ Can be segmented (divided) to show multiple percentages within a single bar ■ Less technical than a graph
Flow chart	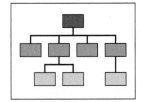	■ Reveals stages in an activity or process ■ Arrows provide accurate directions on the order in which steps are to be taken ■ Blueprints the actual steps to take, first to last, to complete a process/job ■ Identifies places where retracing or skipping steps is required
Organizational chart		■ Illustrates the relationship of one part of an organization to another ■ Shows structure of an organization from chief executive to divisions and departments to employees ■ Provides quick view of areas of authority and responsibility ■ Helps in routing material to appropriate readers

(Continued)

TABLE 10.1   (Continued)

Pictograph	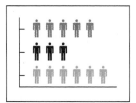	▪ Represents statistical data by means of pictures varying in size, numbers, or color ▪ Easy for a global audience to understand if the symbols/icons (pictograms) are chosen carefully ▪ Must supply a key stating how much each symbol represents
Map	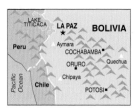	▪ Details specific geographic features (elevation, lakes/rivers, forests, etc.) ▪ Identifies roads, businesses (company locations, TV station, etc.) ▪ Displays census data (population density, number of households with children) ▪ Reveals location, distance, relationship of one place to another
Cutaway drawing	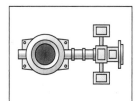	▪ Reveals the interior view of an object by removing the exterior view ▪ Explains how something works ▪ Uncovers an internal mechanism that a photo or simple drawing cannot
Exploded drawing	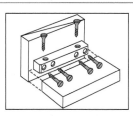	▪ Shows an object with parts separated to indicate the relationship of the parts to one another ▪ Illustrates how parts fit together ▪ Explains how a piece of equipment is constructed and should be assembled or disassembled
Photograph	 © Mutual Series Funds Trading Floor by Najlah Feanny/Corbis	▪ Demonstrates the actual appearance of something ▪ Accurately shows "before" and "after" scenes ▪ Pinpoints color accurately—no guesswork ▪ Cannot reveal interior parts
Clip art		▪ Provides ready-made icons, images, symbols, and pictures ▪ Available online or in print form ▪ Must be appropriate for audience, especially international readers ▪ Should not look unprofessional or cartoonish

Choosing Effective Visuals

Select your visuals carefully. You can find many of the visuals you need on the Internet, or you'll use a graphics program on your computer to create them. But you can also scan hard copy visuals and upload digital photographs (see the Tech Note on generating graphics on page 468). The following suggestions will help you to choose effective visuals:

1. **Include visuals only when they are relevant for your purpose and audience.** Never include a visual simply as a decoration or because it looks impressive. Each visual must have a definite function. A short report on fire drills, for example, does not need a picture of a fire station. Similarly, avoid any visual that is too technical for your readers or that includes more details than you need to show.

2. **Use visuals in conjunction with—not as a substitute for—written work.** Visuals do not take the place of words. In fact, you may need to explain information contained in a visual (see pages 468–469). A set of illustrations or a group of tables alone may not satisfy readers looking for a summary or a recommendation. See how the visual works in conjunction with the description of a magnetic resonance imager (MRI) in Figure 10.2 (page 466) to make the procedure easier to understand than if the writer had used only words or only the visual. This visual and verbal description are appropriately included in a brochure teaching patients about MRI procedures.

3. **Experiment with several visuals.** Select the visual that best suits your audience's needs. Evaluate a variety of options before you choose a particular visual. For instance, software such as PowerPoint (see pages 736–739) allows you to represent statistical data in several ways. Preview a few different versions of a visual or even different types of visuals to determine which one would be best.

4. **Always use easy-to-read visuals.** If readers have trouble understanding the function and arrangement of your visual, it probably is not appropriate and you need to change or revise it.

5. **Be prepared to revise and edit your visuals.** Just as you draft, revise, and edit your written work to meet your audience's needs, you may need to create several versions of your visual to get it right. Expect to change shapes or proportions; experiment with different colors, shadings, labels, and sizes. Replace any visuals that may distract the reader or contradict your message. Also replace or revise any visuals that are unethically distorted or exaggerated (see pages 496–503).

6. **Consider how your visuals will look on the page.** Visuals should add to the overall appearance of your work, not detract from it. Don't cram visuals onto a page or allow them to spill over your text or margins. Many word-processing programs help you move your visuals directly into your document so you can insert them properly. Go to pages 522–528 for advice on effective page layouts.

FIGURE 10.2 A Visual Used in Conjunction with Written Work

A PICTURE FROM THE INSIDE OUT

At the heart of the magnetic resonance imager is a large magnet that is big enough for you to lie inside. Look at the picture below. The **magnet** directs radio signals to surround sections of your body. When the signals pass through your body, they **resonate** (release a signal). Then your body's response is picked up by a receiver and sent to a computer. The computer analyzes the signal and converts it into a visual **image** of your tissues on a video screen.

1. The MR imager surrounds your body with a harmless **magnetic** field and radio signals that safely pass through your body.

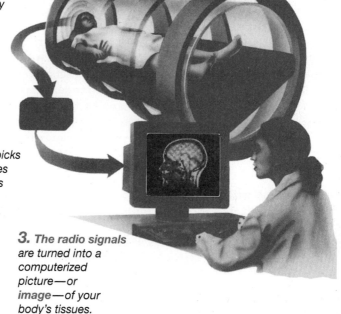

2. A receiver picks up and measures the radio signals that leave, or resonate from, your body.

3. The radio signals are turned into a computerized picture—or image—of your body's tissues.

Source: Reprinted by permission of Krames Communications.

Ineffective Visuals: What *Not* to Do

Here are some guidelines on what to avoid with visuals:

- Avoid visuals that include more details than your audience needs.
- Never use a visual that distracts from your work (for example, one that is too small, too large, does not use the right type of shading, etc.).
- Never use a visual that presents information that contradicts your work.
- Never distort a visual for emphasis or decoration by adding unnecessary lines, patterns, or bars.
- Be careful that you don't omit anything when you reproduce an existing visual.

FIGURE 10.3 An Ineffective Visual: Too Much Information Is Crowded into One Graphic

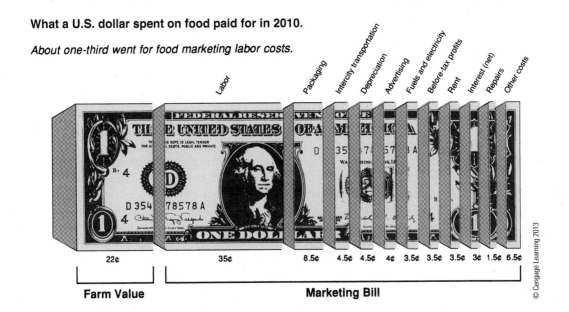

What a U.S. dollar spent on food paid for in 2010.

About one-third went for food marketing labor costs.

Labor — 22¢ — **Farm Value**

Labor, Packaging, Intercity transportation, Depreciation, Advertising, Fuels and electricity, Before-tax profits, Rent, Interest (net), Repairs, Other costs — 35¢ 8.5¢ 4.5¢ 4.5¢ 4¢ 3.5¢ 3.5¢ 3.5¢ 3¢ 1.5¢ 6.5¢ — **Marketing Bill**

© Cengage Learning 2013

- Never use visuals that stereotype (for example, avoid pictures of a workforce that excludes female employees).
- Don't use a visual that looks fuzzy, dotted, or streaked.

See how Figure 10.3 violates many of these rules. It divides an image of a U.S. dollar bill into too many slices. Confronted with so many different wedges the reader would have trouble identifying, separating, comparing, and understanding the costs.

Generating Your Own Visuals

You can easily generate charts, graphs, or tables by using the templates available in your word-processing software, such as Microsoft Word, and by selecting Insert Chart or Insert Table. In fact, some of the visuals in this book, especially those in the long report in Chapter 15, were created using such software. Graph and chart templates allow you to insert your raw numerical data into the appropriate blanks and then add titles, labels, and color-coded keys. Such software helps you to construct different visuals, including pie charts (discussed on pages 479–480) and flow charts (pages 484–485), and also allows you to show a visual from different perspectives and with different emphases, such as shading, 3-D, exploded view.

Spreadsheet programs, such as Microsoft Excel, also provide easy-to-use templates for creating graphs and charts using the numerical data you have entered into your spreadsheet. When using these software programs, though, avoid the temptation to make your visuals too showy. Although dazzling graphics may attract your

Scanners and Scanning

Scanners allow you to produce a digital copy of an image or a document and to store and use it on a computer or online. A scanner resembles a printer, and there are many printers that can both scan and print images and documents. Using technology similar to that of a photocopier, a scanner captures an image and creates a high-resolution digital reproduction that can then be altered or enhanced, printed, used in a document, or stored as a file. Scanners provide an efficient means of incorporating into your documents visuals that are unavailable in digital form, such as older photographs, sketched diagrams, or pictures from books. For example, if you need to include a map but have only a hard copy, you can scan the image and paste it within your document rather than attaching a photocopy.

Scanners can use different types of software. Using optical character recognition (OCR) software, scanners can convert hard-copy text documents into electronic files that can be edited with word-processing or spreadsheet programs. Today's OCR software ensures a virtually error-free conversion of text from hard copy to editable electronic file. With photo-editing software programs like Photoshop, scanned photos can be digitally retouched, color enhanced, or resized without loss of quality.

Scanners save a company time, money, and space by converting paper documents into digital files that can be searched by keywords or dates. Businesses regularly scan such hard-to-store documents as medical and tax records, contracts, blueprints, and so on. As discussed in Chapter 7, employers also scan thousands of résumés from job candidates to select those applicants they want to interview (see page 291). Scanning these documents, a company can quickly retrieve and easily reproduce them, while reducing the risk of losing paper copies.

reader's immediate attention, you don't want them to take over the report. Computer-generated graphs and charts should work with your words, not overshadow them.

Inserting and Writing About Visuals: Some Guidelines

Using a visual requires more of you as a writer than simply inserting it into your written work. You need to *use visuals in conjunction with what you write*. The following guidelines will help you to (1) identify, (2) cite, (3) insert, (4) introduce, and (5) interpret visuals for your readers.

Identify Your Visuals

Give each visual a number and a caption (title) that indicates the subject or explains what the visual illustrates. An unidentified visual is meaningless. A caption helps your audience to interpret your visual—to see it with your purpose in mind. Tell your readers what you want them to look for.

- Use a different typeface (bold) and size in your caption than what you use in the text or the visual itself.
- Include key words about the function and the subject of your visual in a caption.
- Make sure any terms you include in a caption are consistent with the units of measurement and the scope (years, months, seasons) of your visual.

Tables and figures should be numbered separately throughout the text—Table 1 or Figure 3.5, for example. (In the latter case, Figure 3.5 is the fifth figure to appear in Chapter 3.) Following are some examples of figure and table numbers and titles.

- Table 2. Paul Jordan's work schedule, January 15–23
- Figure 4.6. The proper way to apply for a small business loan
- Figure 12. Income estimation figures for North Point Technologies

Cite the Source for Your Visuals

If you use a visual that is not your own work, you have to acknowledge your source (a newspaper, magazine, textbook, blog, website, or individual). Always insert a URL when you use a visual from the Web. If your paper or report is intended for publication or for use on your individual website, you must also obtain permission to reproduce it from the copyright holder. Even when you use a visual from a noncopyrighted source, such as a U.S. government document or website, you still have to give credit to the source. Failure to do so is considered plagiarism. You can acknowledge your source in a caption (or title) below the visual or in a footnote or by an asterisk, as in Table 10.4 or in Figures 10.9 and 10.15. Finally you may find yourself creating a visual based on someone else's data. Again, you are obligated to acknowledge who collected the data and where the data came from. See pages 383–397 for further information on documentation.

Insert Your Visuals Appropriately

Since many of the images you use will come from outside sources, especially the Web, you have to incorporate them clearly and in appropriate places in your written work. (See the Tech Note on page 470.)

Here are some guidelines to help you incorporate the visuals in the most appropriate places for your readers:

- Never insert a visual *before* a discussion of it; readers will wonder why it is there. Use a sentence or two to introduce your visuals.
- Do not insert a visual without a reference to it in the text, and do not refer to a visual in your text and then fail to include it.
- Always mention in the text of your paper or report that you are including a visual. Tell readers where it is found: "below," "on the following page," "to the right," "at the bottom of page 3."
- Place visuals as close as possible to the first mention of them in the text. Try not to put a visual more than one page after the discussion of it. Never wait

two or three pages to present it. By inserting a visual near the beginning of your discussion, you help readers better understand your explanation.

- Use an appropriate size for your visual. Don't make it too large or too small. There is no need to enlarge a photograph if a smaller image is understandable. Gauge size by determining how much data you need to present and where the visual best fits into your discussion of it.
- If the visual is small enough, insert it directly in the text rather than on a separate page. If your visual occupies an entire page, place that page containing your visual on the facing page or immediately after the page on which the first reference to it appears.
- Center your visual and, if necessary, box it. Leave at least 1 inch of white space around it. Squeezing visuals toward the left or right margins looks unprofessional.
- Never collect all your visuals and put them in an appendix. Readers need to see them at the points in your discussion where they are most pertinent.

Tech Note

Importing Visuals from the Web

To incorporate visuals from the Internet in your report, you will need to save the image to your hard disk or copy it into your computer's memory. Saving an image to your hard disk should only be done if you wish to store the image for future use, since otherwise it will clutter your hard drive. If you wish to use an image only once, copy it to your computer's temporary memory and paste it into the document.

When importing images from online sources, keep the following guidelines in mind:

- Be careful not to resize your image so that it becomes distorted. Images are limited to a certain size by the number of pixels (or dots) per inch (dpi). If you use your word processor's resize tool to make the image too large, it will become fuzzy and look unprofessional.
- Your computer can store only one image at a time in its temporary memory. When you copy images from the Web, make sure you paste the image before you copy another, or it will be lost.
- If you save an image to your hard drive for use in a document, give it an easy-to-remember file name, since your computer will automatically take the code word (often a combination of numbers and letters, as in a serial number) as the file name for the image. Save the file as a name you can easily recall.
- Be sure to format your visual so that the text wraps around it and does not become obscured by the image.
- Many websites will not allow you to copy their images. When you right-click on an image, a dialogue box will appear informing you that the image is protected and cannot be copied or saved. To obtain this image, you will need to email the site's owner and request permission to use the visual.

Source: Lori Brister and Anna Gibson.

Introduce Your Visuals

Refer to each visual by its number, and if necessary, mention the title as well. In introducing the visual, though, do not just insert a reference to it, such as "See Figure 3.4" or "Look at Table 1." Relate the visual to the text it illustrates or helps explain. Here are two ways of writing a lead-in sentence for a visual.

> Poor: Our store saw a dramatic rise in the shipment of electric ranges over the five-year period as opposed to the less impressive increase in washing machines. (See Figure 3.)

This sentence does not tie the visual (Figure 3) into the sentence where it belongs. The visual just trails insignificantly behind.

> Better: As Figure 3 shows, our store saw a dramatic rise in the shipment of electric ranges over the five-year period as opposed to the less impressive increase in washing machines.

Mentioning the visual in this way alerts readers to its presence and function in your work and helps them to more easily understand your message.

Interpret Your Visuals

Help readers understand your visual by telling them what to look for and why. Let them know what is most significant about the visual. Mention any distinctive features, major parts, or crucial relationships. Do not expect the visual to explain itself. Inform readers what the numbers or images in your visual mean, how they make or prove a key point. What conclusions do you want readers to reach after seeing your visual?

In a report on the benefits of vanpooling, the writer supplied the following visual, a table:

TABLE 1   Travel Time (in minutes): Automobile versus Vanpool

Private Automobile	Vanpool
25	32.5
30	39.0
35	45.5
40	52.0
45	58.5
50	65.0
55	71.5
60	78.0

Source: U.S. Department of Transportation. *Increased Transportation Efficiency Through Ridesharing: The Brokerage Approach* (Washington, D.C., DOT-OS—40096): 45.

To interpret the table, the writer called attention to it in the context of a discussion on transportation efficiency.

> Although, as Table 1 above suggests, the travel time in a vanpool may be as much as 30 percent longer than in a private automobile (to allow for pickups), the total trip time for the vanpool user can be about the same as with a private automobile because vanpools eliminate the need to search for parking spaces and to walk to the employment site entrance.[1]

Two Categories of Visuals: Tables and Figures

Visuals can be divided into two categories—tables and figures. A *table* arranges information—numbers and/or words—in parallel columns and rows for easy comparison of data. Any visual that is not a table is considered a figure. *Figures* include graphs, circle charts, bar charts, organizational charts, flow charts, pictographs, maps, photographs, and drawings. Expect to use both tables and figures in your work.

Tables

Tables contain parallel columns and rows of information organized and arranged into categories to show, in a compact space, changes in time, distance, cost, employment, or some other distinguishable or quantifiable variable. Spreadsheet programs heavily rely on tables to convey information, as in Table 10.2. Tables also summarize material for easy recall—causes of wars; provisions of a law; or differences between a common cold, flu, and pneumonia, as in Table 10.3 (page 474). Observe how Table 10.3 condenses much information and arranges it in quickly identifiable categories.

Parts of a Table

To use a table properly, you need to know the parts that constitute it. Refer to Table 10.4 (page 475), which labels these parts, as you read the following:

- The main *column* is "Amount Needed to Satisfy Minimum Daily Requirement," and the *subcolumns* are the protein sources for which the table gives data.
- The *stub* is the first column on the left-hand side, below the column heading "Source." The stub lists the foods for which information is broken down in the subcolumns.
- A *rule* (or line) across the top of the table separates the title from the column headings and the column headings from the body of the table.

---

[1] James A. Devine, "Vanpooling: A New Economic Tool," *AIDC Journal.*

TABLE 10.2   A Spreadsheet Table

	A	B	C	D	F	G	H
1	Acct#	Account Description	YTD	April	March	February	Janua
2							
3	**Employee Costs**						
4	110	Payroll	$171,465	$29,750	$28,547	$27,540	
5	120	IRS/FICA/Wk comp/State/SDI	$46,295	$8,032	$7,707	$7,435	
6	130	Commissions	$56,436	$9,520	$8,875	$9,825	
7	140	Retirement Plan	$20,575	$3,570	$3,425	$3,304	
8	150	Insurance	$9,000	$1,500	$1,500	$1,500	
9							
11	**Subcontractors & Services**						
12	201	Telecommunication Services	$2,616	$436	$436	$436	
13	202	Design Consultants	$875	$500	$375	$0	
14	203	Photo/Video Services	$535	$45	$325	$45	
15	250	Graphic Services	$2,957	$765	$95	$375	
16	251	Photo/Stats	$1,612	$568	$755	$0	
17	252	Compositor	$1,453	$388	$325	$195	
18	253	Printing Services	$6,186	$951	$849	$325	
19	254	Legal & Accounting					
20							
21	**Supplies and Materials**						
22	301	Office Supplies	$875	$500	$732	$433	
23	302	Office Postage	$535	$45	$255	$325	
24	303	Office Equipment & Furniture	$2,957	$765	$78	$21	
25	304	Miscellaneous Supplies	$1,612	$568	$49	$36	
26							
27	**Facilities Overhead**						
28	405	Plant	$1,612	$568	$1,700	$1,700	

Sheet1 / Sheet2 / Sheet3 /

Guidelines for Using Tables

When you include a table in your work, follow these guidelines.

- Number the tables according to the order in which they are discussed (Table 1, Table 2, Table 3). Tables should be numbered separately from figures (charts, graphs, photos) in your text.
- Keep the table on the page where it is most appropriate. It is hard for readers to follow a table spread across different pages.
- Give each table a concise and descriptive title to show exactly what is being represented or compared.
- Use words in the stub (a list of items about which information is given), but put numbers under column headings.
- Supply footnotes, often indicated by small raised letters ([a], [b]), if something in the table needs to be qualified, for example, the number of cups of milk in Table 10.4 (page 475). Then put that information below the table.

TABLE 10.3   **Table Showing Differences Between a Common Cold, Influenza, and Pneumonia**

Symptoms	Cold	Influenza	Pneumonia
Fever	Rare	Characteristic high (100.4 – 104°F); sudden onset, lasts 3 to 4 days	May or may not be high
Headache	Occasional	Prominent	Occasional
General aches and pains	Slight	Usual; often quite severe	Occasionally quite severe
Fatigue and weakness	Quite mild	Extreme; can last up to a month	May occur depending on type
Exhaustion	Never	May occur early and prominent	May occur depending on type
Runny, stuffy nose	Common	Sometimes	Not characteristic
Sneezing	Usual	Sometimes	Not characteristic
Sore throat	Common	Sometimes	Not characteristic
Chest discomfort, cough	Mild to moderate; hacking cough	Can become severe	Frequent and may be severe
Complications	Sinus and ear infections	Bronchitis, pneumonia; can be life-threatening	Widespread infections of other organs; can be life-threatening, especially in elderly and debilitated persons

Source: Microbiology: Principles and Applications, 4th Edition, by Jacquelyn Black. Copyright © 1999 John Wiley & Sons, Inc. Reproduced with permission of John Wiley & Sons, Inc.

- List items in alphabetical, chronological, or other logical order.
- Arrange the data you want to compare vertically, not horizontally; it is easier to read down a column than across a series of rows.
- Place tables at the top (preferable) or bottom of the page, and center them on the page rather than placing them up against the right or left margin.
- Don't use more than five or six columns; tables wider than that are more difficult for readers to understand.
- When possible, round off numbers in your columns to the nearest whole number to assist readers in following and retaining information.
- Always give credit to the source (the supplier of the statistical information) on which your table is based.

TABLE 10.4   Parts of a Table

Table number →

TABLE 1 Efficiency of Some Protein Sources in Meeting an Adult's Minimum Daily Requirements				
Source	Percent of Protein	Percent of Amino Acids	Amount Needed to Satisfy Minimum Daily Requirement (grams)	(ounces)
Cheese[a]	27	70	227	7.2
Corn	10	50	860	30.0
Eggs	11	97	403	14.1
Fish[a]	22	80	244	8.5
Kidney beans	23	40	468	16.4
Meat[a]	25	68	253	8.8
Milk	4	82	1,311	45.9[b]
Soybeans	34	60	210	7.3

← *Title*
← *Rule*
← *Column heading*
← *Subheading*
Stub {

Source: From Starr/Taggart, Biology: *The Unity and Diversity of Life*, 4th ed. Copyright © 1987 Brooks/Cole, a part of ← *Origin of data* Cengage Learning, Inc. Reprinted with permission. www.cengage.com/permissions.

[a] = Average value
[b] = Equivalent of 6 cups } *Footnotes*

Figures

As we saw, any visual that is not a table is classified as a figure. The types of figures we will examine next are

- graphs
- circle, or pie, charts
- pictographs
- maps
- bar charts
- organizational charts
- flow charts
- photographs
- drawings
- clip art

Graphs

Graphs transform numbers into pictures with shapes, patterns, and shading. They take statistical data presented in tables and put them into rising and falling lines or steep or gentle curves. The three types of graphs are (1) simple line graphs, (2) multiple-line graphs, and (3) area graphs.

Functions of Graphs

Graphs vividly portray information that changes, such as

- sales
- costs
- trends
- distributions
- employment
- energy levels
- temperatures
- population

Graphs not only describe past and current situations but also forecast trends.

Graphs versus Tables

Because graphs actually show change, they are more dramatic than tables. You will make the reader's job easier by using a graph rather than a table. Many financial websites and print publications—the *Wall Street Journal* and *USA Today*, for example—open with a graph for the benefit of busy readers who want a great deal of financial information summarized quickly.

Simple Line Graphs

A simple graph consists of two sides—a *vertical axis (y-axis)* and a *horizontal axis (x-axis)*—that intersect to form a right angle, as in Figure 10.4. The vertical axis is read from bottom to top; the horizontal axis from left to right. The space between the two axes contains the picture made by the graph. Figure 10.4 shows the amount of snowfall in Springfield between November 2010 and April 2011. The vertical line represents the *dependent variable* (the snowfall in inches), the horizontal line, the *independent variable* (time in months). The dependent variable is influenced most directly by the independent variable, which almost always is expressed in terms of time or distance.

Multiple-Line Graphs

The graph in Figure 10.4 contains only one line of data. But a graph can have multiple lines to show how a number of dependent variables (different conditions, products, etc.) compare with one another. The six-month sales figures for three salespeople can be seen in the graph in Figure 10.5. The graph contains a separate line for each of the three salespeople. At a glance, readers can see how the three compare and how many dollars each salesperson generated per month. Note how the line representing each person is clearly differentiated from the others by symbols

FIGURE 10.4 A Simple Line Graph Showing the Amount of Snowfall in Springfield from November 2010 to April 2011

Graph is easy to read and follow

Variables clearly labeled

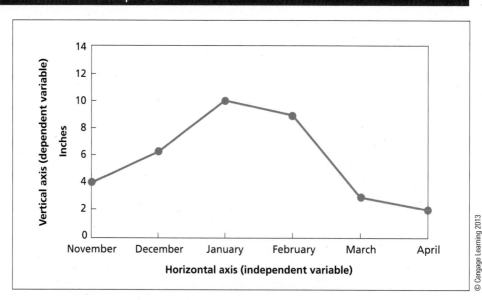

© Cengage Learning 2013

FIGURE 10.5 A Multiple-Line Graph Showing Sales Figures for the First Six Months of 2011 for Three Salespeople

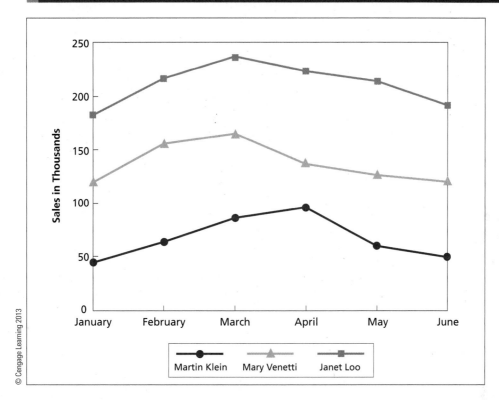

Uses different symbols and colors to represent each salesperson

Legend explains what different symbols and colors represent

© Cengage Learning 2013

and colors. Each line is clearly tied to a *legend* (an explanatory key below the graph) identifying the three salespeople.

Area Graphs

Figure 10.6 is another graphic representation of the information in Figure 10.5. The graph in Figure 10.6, known as an *area*, or *multiband*, *graph*, shows relationships among the three salespeople without providing the exact numerical documentation of Figure 10.5. In Figure 10.6 the spaces between the different curves are filled in, with colors keyed in the legend to identify the three salespeople.

Guidelines on Creating Graphs

1. Use no more than three lines in a multiple-line graph, so readers can interpret the graph more easily. If the lines run close together, use a legend to identify individual lines.
2. Label each line or color to identify what it represents for readers. Include a legend, as in Figure 10.5 and 10.6.
3. In a multiple-line graph, keep each line distinct by using different colors, dots or dashes, or symbols. Note the different symbols in Figure 10.5.

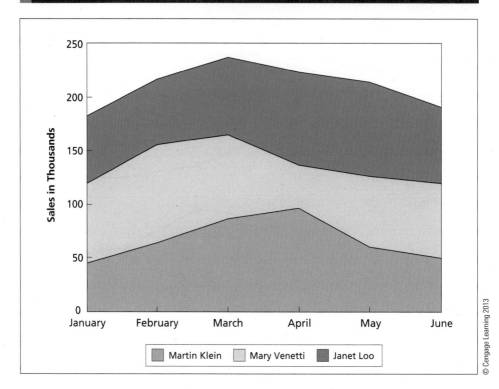

FIGURE 10.6 An Area, or Multiband, Graph of Figure 10.5, Showing the Performance of Three Salespeople

4. Make sure you plot enough points to show a reasonable and ethical range of the data. Using only three or four points may distort the evidence. (See pages 498–500.)
5. Keep the scale consistent and realistic. If you start with hours, do not switch to days or vice versa. If you are recording annual rates or accounts, do not skip a year or two in order to save time or be more concise.
6. For some graphs, there is no need to begin with a zero. A *suppressed zero graph* automatically begins with a larger number when it would be impossible to start with zero. For others, you may not have to include numbers beyond a certain point. Your subject and purpose will determine the range of data you need to show. If you use a suppressed zero graph, make sure you do so ethically. (See pages 497–503.)

Charts

Although charts and graphs may seem similar, there is a big difference between them. Graphs are usually more complex and plotted according to specific mathematical coordinates. Charts, however, do not display exact and complex mathematical data found in tables and graphs. Instead, they present an overall picture of how individual pieces of data (from a graph or table) fall into place to express relationships.

Among the most frequently used charts are (1) pie charts, (2) bar charts, (3) organizational charts, (4) flow charts, and (5) pictographs.

Pie Charts

Pie charts are also known as *circle charts*, a name that descriptively points to their construction and interpretation. Tables are more technical and detailed than pie charts. Figure 10.7 shows an example of a pie chart used in a government document. A table or graph with a more detailed breakdown of, say, a city's budget would be much more appropriate for a technical audience (auditors, budget and city planners).

The full circle, or pie, represents the whole amount (100 percent or 360 degrees) of something: the entire budget of a company or a family, a population group, an area of land, the resources of an organization or institution. Each slice or wedge represents a percentage or portion of the whole. A pie chart effectively allows readers to see two things at once: the relationship of the parts to one another and the relationship of the parts to the whole.

Preparing a Pie Chart   Follow these seven rules to create and present your pie chart:

1. **Keep your pie chart simple.** Don't try to illustrate technical statistical data in a pie chart. Pie charts are primarily used for general audiences.

2. **Do not divide the pie into too few or too many slices.** If you have only three wedges, use another visual to display them (a bar chart, for example, discussed below). If you have more than seven or eight wedges, you will divide the pie too narrowly, and overcrowding will destroy the dramatic effect. Instead, combine several slices of small percentages (2 percent, 3 percent, 4 percent) into one slice labeled "Other," "Miscellaneous," or "Related Items."

FIGURE 10.7 A Three-Dimensional Pie Chart Showing the Breakdown by Department of a Proposed City Budget for 2012

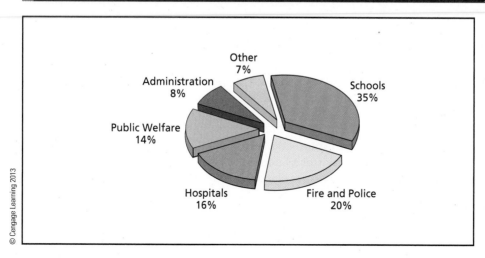

Other
7%

Administration
8%

Schools
35%

Public Welfare
14%

Hospitals
16%

Fire and Police
20%

Puts largest slice first

Size of slice determined by percentage; totals 100%

Uses a different color for each category

© Cengage Learning 2013

3. Make sure the individual slices total 100 percent, or 360 degrees. For example, if you are constructing a pie chart to represent a family's budget, the breakdown percentages might be as follows:

Category	Percentage	Angle of slice
Housing	25%	90.0°
Food	22%	79.2°
Energy	20%	72.0°
Clothes	13%	46.8°
Health care	12%	43.2°
Miscellaneous	8%	28.8°
Total	100%	360°

4. Put the largest slice first, at the 12 o'clock position. Then move clockwise with proportionately smaller slices. Schools occupy the largest slice in Figure 10.7 because they receive the biggest share of the budget.

5. Label each slice of the pie horizontally. Key in the identifying term or quantity inside unless it would be too hard to read. In that case, label each slice outside the pie, as in Figure 10.7. Regardless, make sure the label is big enough to read. Do not put in a label upside down or slide it in vertically. If the individual slice of the pie is too small, draw a connecting line from the slice to a label positioned outside the pie.

6. Shade, color, or cross-hatch slices of the pie to further separate and distinguish the parts. Use distinctive colors or patterns to help readers differentiate one slice of the pie chart from another. Figure 10.7 effectively uses color. But be careful not to obscure labels and percentages; also make certain that adjacent slices can be distinguished readily from each other. Do not use the same color or similar colors for two slices.

7. Give percentages for each slice to further assist readers.

Bar Charts

A bar chart consists of a series of vertical or horizontal bars that indicate comparisons of statistical data. For instance, in Figure 10.8 vertical bars depict an increase in number of working mothers over time. Figure 10.9 (page 482) uses horizontal bars to depict the nation's top 20 metropolitan areas, based on building permits. The length of the bars is determined according to a scale that your computer software can easily compute.

Bar Charts, Graphs, and Tables—Which Should You Use? When should you use a bar chart rather than a table or a line graph? Your audience will help you decide. If you are asked to present statistics on costs for the company accountant, use a table. Because a bar chart is limited to a few columns, it cannot convey as much information as a table or graph. However, if you are presenting the same information to a group of stockholders or to a diverse group of employees, a bar chart may be more relevant and persuasive.

FIGURE 10.10 A Multiple-Bar Chart Showing Advertising Expenditures by Major Media, 2008–2011

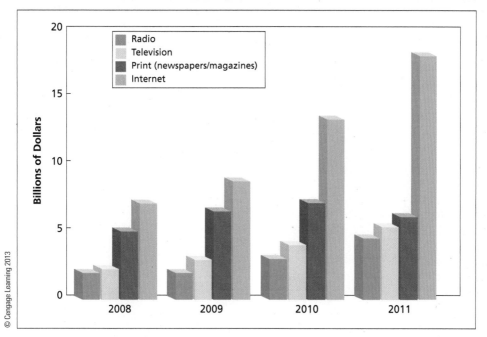

Legend explains what each bar represents

Avoids confusion by using only four bars in a group

© Cengage Learning 2013

FIGURE 10.11 A Segmented Bar Chart Representing Total Travel Expenditures for Weemco Communications, January 2012

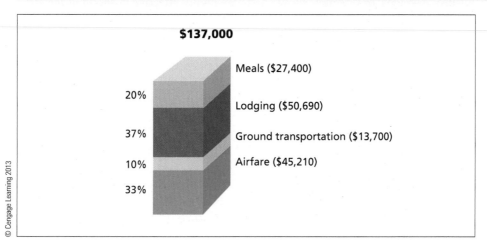

$137,000

Meals ($27,400)

20%

Lodging ($50,690)

37%

Ground transportation ($13,700)

10%

Airfare ($45,210)

33%

Entire bar equals the travel total

© Cengage Learning 2013

FIGURE 10.12 A Multiple-Bar, Segmented Bar Chart Showing Energy Consumption by Sector in Five States

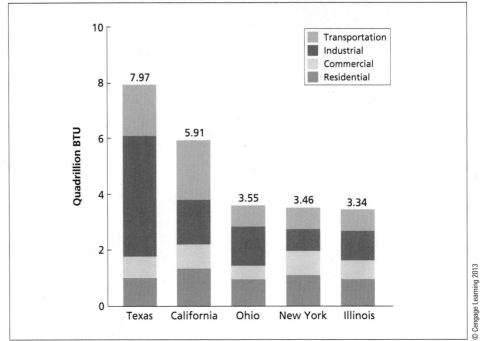

Shows multiple comparisons among many categories

Source: U.S. Energy Information Administration.

Flow Charts

A *flow chart* displays the stages in which something is manufactured, is accomplished, develops, or operates. Flow charts are highly effective in showing the steps in following a procedure. They can also be used to plan the day's or week's activities. A flow chart tells a story with arrows, boxes, and sometimes pictures. Boxes are connected by arrows to visualize the stages of a process. The direction of the arrows tells the reader the order and movement of events involved in the process.

Flow charts often proceed from left to right and back again, as in the flow chart below, showing the steps to be taken before graduation. Note that each step of the process is clearly labeled.

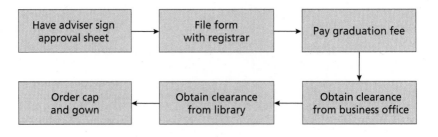

FIGURE 10.13 An Organizational Chart Representing Critical Care Nursing Services at Union General Hospital

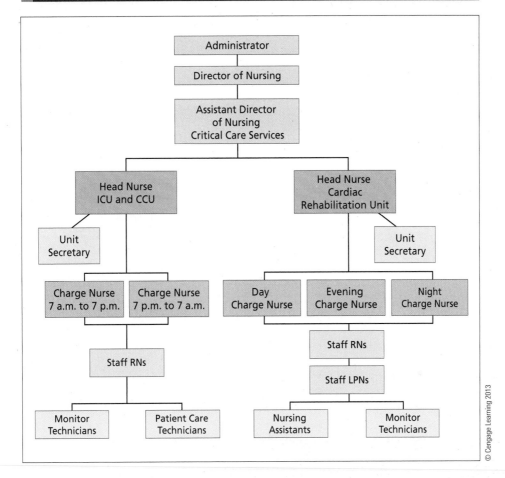

© Cengage Learning 2013

Flow charts can also be constructed to read from top to bottom. Programming instructions frequently are written that way. See, for example, Figure 10.14, which uses a programming flow chart to show the steps a student must follow in writing a research paper.

A flow chart should clarify, not complicate, a process. Do not omit any important stages, but at the same time do not introduce unnecessary or unduly detailed information. Show at least three or four stages, and make sure the various stages appear in the correct sequence.

Pictographs

Similar to a bar chart, a *pictograph* uses picture symbols (called *pictograms*) to represent differences in statistical data, as in Figure 10.15. A pictograph repeats the same symbol or icon to depict the quantity of items being measured. Each symbol

FIGURE 10.14 A Programming Flow Chart: Writing a Research Report

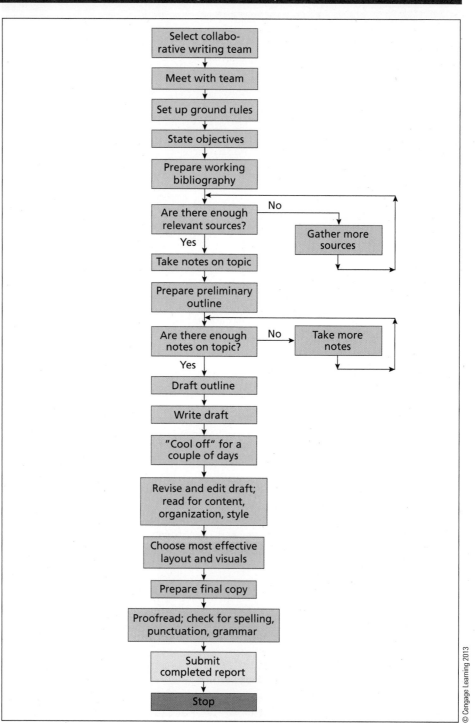

Process moves from start to finish

Each step concisely described

Arrows clearly tell readers what steps to take and when

Indicates when and why a step may have to be repeated

Each step included in a separate box

Process completed with last step

Select collaborative writing team

Meet with team

Set up ground rules

State objectives

Prepare working bibliography

Are there enough relevant sources? — No → Gather more sources

Yes

Take notes on topic

Prepare preliminary outline

Are there enough notes on topic? — No → Take more notes

Yes

Draft outline

Write draft

"Cool off" for a couple of days

Revise and edit draft; read for content, organization, style

Choose most effective layout and visuals

Prepare final copy

Proofread; check for spelling, punctuation, grammar

Submit completed report

Stop

© Cengage Learning 2013

FIGURE 10.15   A Pictograph Showing the Growth of One State's Retirement Assets

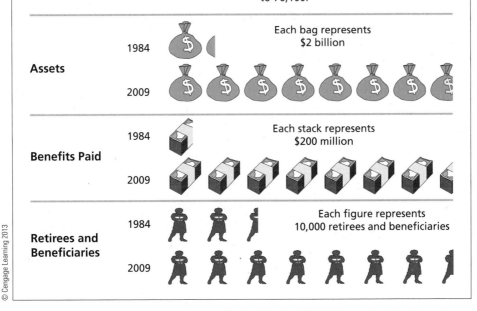

How does the Public Employees' Retirement System (PERS) compare today with 25 years ago?

During that time, PERS experienced spectacular growth, with both assets and benefits paid increasing more than sixfold.

- The market value of assets increased from $2.3 billion to $15.5 billion.
- PERS paid benefits of $94.3 million in FY '92. The total reached $1.525 billion during FY '09.
- Retirees and beneficiaries drawing benefits more than tripled, from 24,000 to 76,100.

Provides financial context/history to better understand visual

Assets

1984

2009

Each bag represents $2 billion

Specifies what each pictograph stands for

Benefits Paid

1984

2009

Each stack represents $200 million

Uses easily recognized pictographs

Retirees and Beneficiaries

1984

2009

Each figure represents 10,000 retirees and beneficiaries

Increases number, not size, of pictograph

© Cengage Learning 2013

Source: PERS Member Newsletter, February 2010. Official publication of the Public Employees' Retirement System of Mississippi. By permission of Public Employees' Retirement System.

stands for a specific number, quantity, or value. Pictographs are visually appealing and dramatic and are far more appropriate for a nontechnical than a technical audience.

Guidelines on Creating Pictographs   When you create a pictograph, follow these four guidelines:

1. Choose an appropriate, easily identifiable symbol for the topic.
2. Indicate the precise quantities each icon represents by placing numbers after or at the top of the visual.
3. Increase the number of symbols rather than their sizes because differences in sizes are often difficult to construct or interpret.
4. Avoid crowding too much information into a pictograph, such as the one in Figure 10.3, in which a U.S. dollar bill is divided into too many sections.

Maps

The maps you use on the job may range from highly sophisticated and detailed geographic tools to simple sketches such as the map in Figure 10.16, which shows the location of a town's water filter plants and pumping stations. This is a *large-scale map* that displays social, economic, or physical data for a small area. The map in Figure 10.17, however, gives a state-by-state account of the number of representatives in the U.S. House of Representatives in 2010.

You may have to construct your own map, like the one in Figure 10.16, scan one in a published source, or find one on the Internet. If you use a map that someone else created, be sure to obtain permission from the copyright holder.

Guidelines for Creating Maps

Follow these steps when you create a map:

1. Always acknowledge your source if you did not construct the map yourself.
2. Use dots, lines, colors, symbols, and shading to indicate features. Markings should be clear and distinct.
3. If necessary, include a legend, or map key, explaining dotted lines, colors, shading, and symbols, as in Figure 10.16.

FIGURE 10.16 A Map Showing the Location of Smithville Water Department's Water Filter Plants and Pumping Stations

Provides directional sign

Includes only relevant features

Supplies miles/ distance legends

Uses distinctive symbols

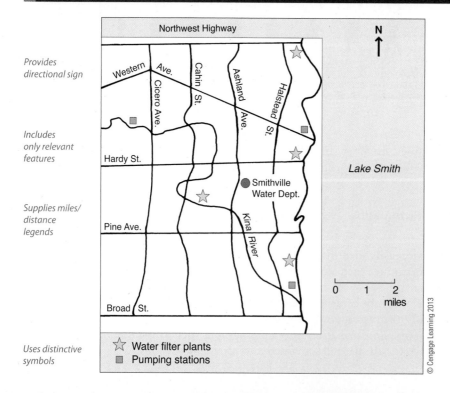

© Cengage Learning 2013

FIGURE 10.17 Number of U.S. Representatives by State, April 2010

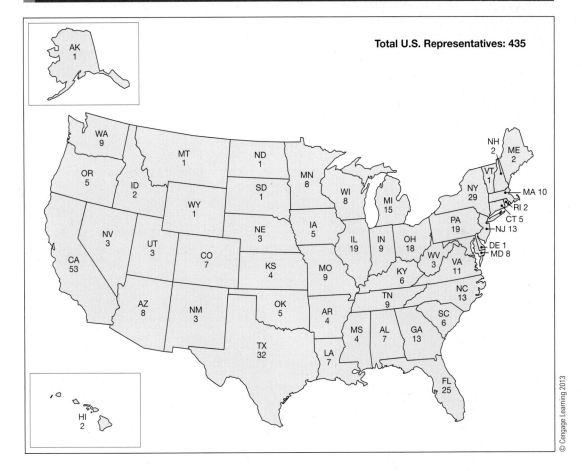

Total U.S. Representatives: 435

4. Exclude features (rivers, elevations, county seats) that do not directly relate to your topic. For example, a map showing the crops grown in two adjacent counties need not show all the roads and highways in those counties. A map showing the presence of strip mining or features such as hills or valleys needs to indicate elevation, but a map depicting population or religious affiliation need not include topographical (physical) detail.

5. Indicate direction. Conventionally, maps show north, often by including an arrow and the letter *N*, as in Figure 10.16.

Photographs

Correctly taken or scanned, photographs are an extremely helpful addition to job-related writing. A photograph's chief virtues are realism and clarity. Among its many advantages, a photo can

- show what an object looks like (Figure 10.18, page 490)
- demonstrate how to perform a certain procedure (Figure 10.19, page 492)

FIGURE 10.18 A Photo Showing What a Piece of Equipment Looks Like

© Corbis

- compare relative sizes and shapes of objects (Figure 10.19)
- compare and contrast scenes or procedures (Figure 10.20, page 493)

Guidelines for Taking Photographs

Whether you use a traditional or digital camera, observe the following guidelines when taking photographs:

1. **Take the photo from the most appropriate distance.** Decide how much foreground and background information your audience needs. For example, if you are photographing a three-story office building, your picture may misleadingly show only one or two stories if you are standing too close to the building.

2. **Select the correct angle.** Take the photo from an angle that makes sense, usually straight on so that the object or person photographed can be viewed in full.

3. **Include only the details that are necessary and relevant for your purpose.** Crop (edit out) any unnecessary details from the photograph.

4. **Provide a sense of scale.** So that readers understand the size of the object, include in the photograph a person (if the object is very large), a hand (if the object is small), or a ruler.

5. **Make sure you consider lighting and resolution.** Take your photograph in appropriate light so that the subject of your photograph is clear and crisp. If you are using a digital camera, take a high-resolution photograph for maximum clarity.

6. **Always ask for permission before you take a photograph of a person, place, or thing**, unless you are photographing a public park or building. Do not take photographs of copyrighted materials, for instance, a work of art, a page from a copyrighted book, or a movie or television screen. Also obtain permission from your boss if you are photographing your own company's equipment, sites, or designs.

To get a graphic sense of the effects of taking a photo the wrong and right way, study the photographs in Figures 10.21 and 10.22 (pages 494 and 495). In Figure 10.21 everything merges because the shot was taken from the wrong angle. The reader has

Digital Photography

Digital cameras and smartphones allow you to easily supply professional-looking, customized photos with your written work. Digital photography offers the following benefits in the global marketplace:

- Shows conditions or sites right away so you can include a photo in an incident or credit report—you do not need to wait for film to be developed
- Gives you more than one opportunity to take a picture—helps you select, highlight, and edit
- Saves expense and time of scanning a part of an image
- Allows you to send photographs easily and quickly over the Internet, store them on your computer or on a CD, and upload them to a website
- Allows you to edit photographs for color, sharpness, contrast, brightness, and size—you can also crop out unnecessary details and eliminate "red eye"

To achieve these goals, make sure that your digital camera is set to the correct photo-quality level. The poorer the quality, the more pictures your camera will hold. However, if you will be using the photographs for a newsletter or company magazine or if you wish to print photographic-quality images, make sure your camera is set on a higher quality level.

FIGURE 10.19 A Photo Showing How to Perform a Procedure and Comparing Relative Sizes and Shapes of Objects

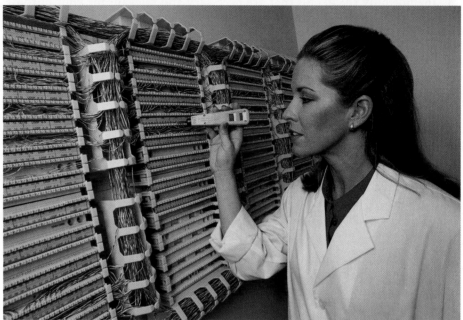

© Corbis

no sense of the parts of the truck, their size, or their function. A clear and useful picture of a hydraulic truck (often called a "cherry picker") used to cut high branches can be seen in Figure 10.22. The photographer rightly placed the truck in the foreground but included enough background information to indicate the truck's function. The worker in the bucket helps to show the relative size of the truck and the function of the equipment.

Drawings

Drawings can show where an object is located, how a tool or machine is put together, or what signals are given or steps taken in a particular situation. A drawing can be simple, like the one in Figure 10.23 (page 496), which shows readers exactly where to place smoke detectors in a house.

A more detailed drawing can reveal the interior of an object. Such sketches are called *cutaway drawings* because they show internal parts normally concealed from view. Figure 10.24 (page 497) is a cutaway drawing of an extended-range electric vehicle, the GM Volt. The drawing shows the passenger interior and the battery-powered engine, which allows the car to drive up to forty miles without the simultaneous use of gasoline.

Another kind of sketch is an *exploded drawing*, which blows the entire object up and apart, as in Figure 10.25 (page 498), to show how the individual parts are arranged. An exploded drawing is included in many equipment manuals and uses

FIGURE 10.20 Photos Showing Comparison/Contrast of Using Gasoline versus Electricity to Power an Automobile

FIGURE 10.21 A Poor Photograph—Taken from the Wrong Angle So That Everything Merges and Becomes Confusing

Jackieson/Shutterstock.com

callouts, or labels, to identify the components. The labels are often attached to the drawing with arrows or lines.

Guidelines for Using Drawings

1. Keep your drawing simple. Include only as much detail as your reader needs to understand what to do, be it to assemble a piece of furniture or to operate a mechanism.
2. Clearly label all parts so that your reader can identify and separate them.
3. Decide on the most appropriate view of the object to illustrate—aerial, frontal, lateral, reverse, exterior, interior—and indicate in the title which view it is.
4. Keep the parts of the drawing proportionate unless you are purposely enlarging one section.

Clip Art

Clip art refers to ready-to-use electronic images. These small, cartoon-style representations, such as the ones shown in Figure 10.26 (page 499), depict almost anything you can think of and pertain to almost every field—architecture, information

FIGURE 10.22 An Effective Photograph—Truck in Foreground, Enough Background Information, and a Worker to Show the Size and Function of the Truck

science, criminal justice, international marketing, sports, government, and health care. You can use clip art for a variety of projects. Figure 10.26 uses appropriate clip art in instructions to stockholders on how they can record their vote for a change in company policy.

Many graphics packages come with comprehensive clip art libraries grouped under such useful headings as energy, government, leisure, health, money, the outdoors, food, technology, and transportation. Microsoft Office includes a clip art library with links to a larger online database of free clip art. There are also free clip art and photo-illustration databases on the Internet, such as http://free-clipart.net.

Guidelines for Using Clip Art

When you use clip art, follow these guidelines:

   1. **Use simple, easy-to-understand images.** Select an image that conveys your idea quickly and directly. Avoid using an image of an unfamiliar object or of a drawing or silhouette that might confuse your audience.

   2. **Use clip art functionally.** Do not insert clip art as a decoration. Using too many pieces will make your work look unprofessional. Each piece of clip art should contribute significantly to, not compete with, your message.

FIGURE 10.23 A Simple Drawing Showing Where to Place Smoke Detectors in a House

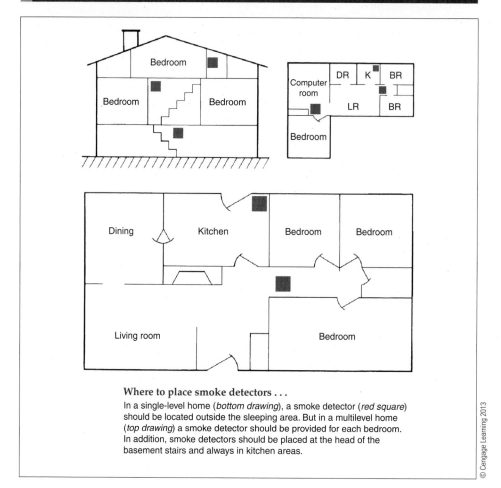

Where to place smoke detectors . . .
In a single-level home (*bottom drawing*), a smoke detector (*red square*) should be located outside the sleeping area. But in a multilevel home (*top drawing*) a smoke detector should be provided for each bedroom. In addition, smoke detectors should be placed at the head of the basement stairs and always in kitchen areas.

© Cengage Learning 2013

3. Make sure the clip art is relevant for your audience and your message. A clip art airplane does not belong in a technical report on fuel capacity or jet engine design.

4. Make sure your clip art is professional. Some clip art is humorous, even silly, which may not be appropriate for a professional business report or proposal.

Using Visuals Ethically

Make sure your visuals are ethical, whether you create or import them. Like your words, your visuals should represent you, your company, and the data truthfully. Ethical visuals convey and interpret statistical information and other types of data, products and equipment, locations, and even individuals without misinterpretation.

FIGURE 10.24 A Cutaway Drawing of a Hybrid Car

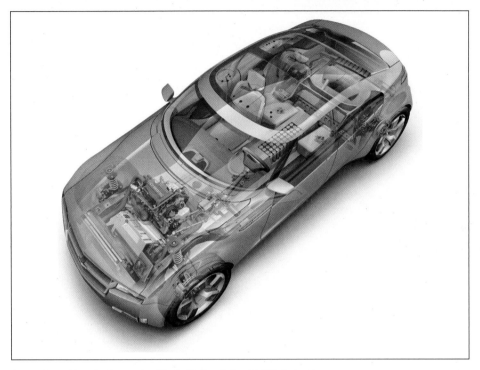

Source: General Motors Corporation. Used with Permission, GM Media Archives.

Your visuals should neither distort events and statistics nor mislead readers. Ethical visuals should be

- accurate
- honest, fair
- complete
- appropriate

- easy to read
- clearly labeled
- uncluttered
- consistent with conventions

Guidelines for Using Visuals Ethically

To ensure that your visuals are ethical, honest, accurate, and easy to read, follow these guidelines.

Photos

- Do not distort a photo by omitting key details; by misrepresenting dimensions, angles, sizes, or surroundings; or by superimposing one image over another.
- Do not take a photo of your most expensive, top-of-the-line product or model and pair it with the price of your cheapest product or model.

FIGURE 10.25 Drawing of a Disassembled Notebook Computer

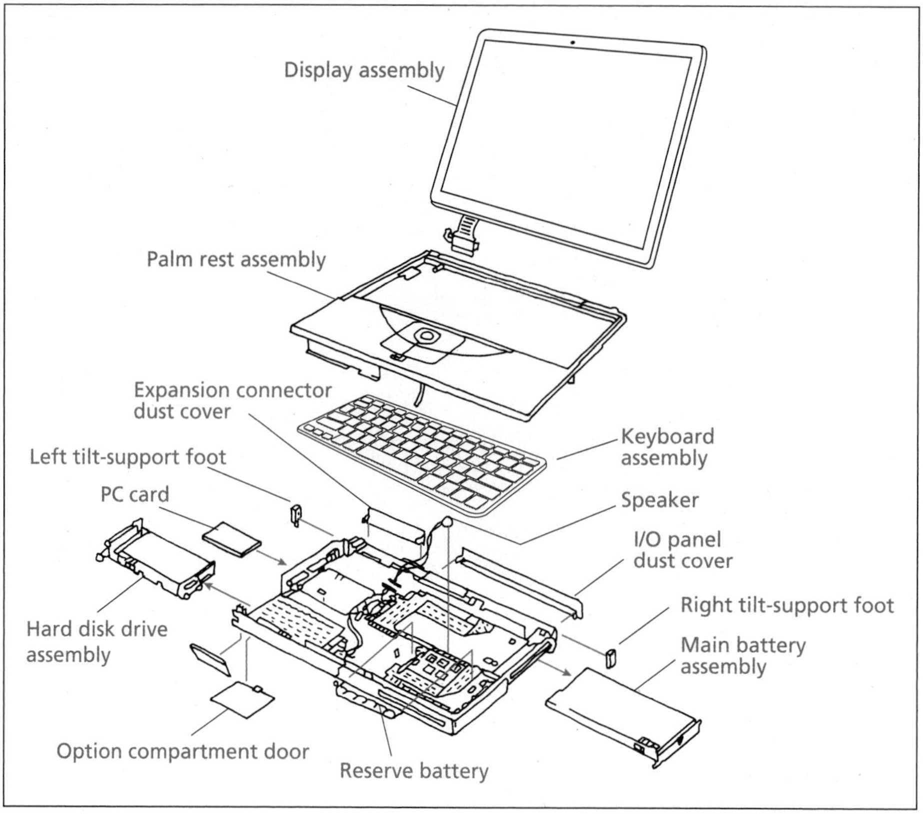

Display assembly

Palm rest assembly

Expansion connector dust cover

Left tilt-support foot

PC card

Keyboard assembly

Speaker

I/O panel dust cover

Right tilt-support foot

Main battery assembly

Hard disk drive assembly

Option compartment door

Reserve battery

- Do not misrepresent location—for example, taking a photo in a "doctored" or off-site location, studio, or lab and then identifying it as an "actual" location shot.
- Never take a photo of an individual for business purposes (e.g., for a newsletter, an ad, or the Web) without his or her permission.
- Don't counterfeit or subtly alter a company's logo and use it in a photo to sell, distribute, or promote an imitation as the real thing.

Graphs

- Don't distort a graph by plotting it in misleading or unequal intervals—for example, omitting certain years or dates to hide a decline in profits. Contrast the misleading interpretation in Figure 10.27 with the ethical revision in Figure 10.28 (page 500).
- Include information in correct chronological sequence along the horizontal axis.

FIGURE 10.26 Clip Art in a Business Document

How to Cast Your Vote on Corporate Issues

Vote in Person

The stockholders' meeting: 1007 South Broadway Blvd., Rye, NY, May 21, 2012, 10:00 am EST. Come and present your paper copy of the ballot and proper identification.

Vote by Internet

To vote *now* by Internet, go to www.voteonline.com.

Vote by Cell phone

To vote *now* by telephone, call 1-800-555-7171.

Vote by Mail

Mark, sign, and date your voting form and return it in the postage-paid envelope we have provided.

- Don't switch the type of information usually given along the vertical axis with that of the horizontal axis.
- Don't project growth or increases based on unreliable or invalid data.
- Don't misrepresent data or trends by making increments along the vertical axis too limited, leaving a much smaller (and incomplete) area to represent. When data are plotted this way, readers are unethically led to misinterpret the numbers—to read that there was little loss in revenue, or no change in sales, for example. For instance, if the horizontal axis begins at $5 and advances to $6 a share, you leave only an intentionally small and misleading area to measure. If stocks fell below $5 a share, your graph would unethically not represent those declines.

Bar Charts

- Don't use color or shading to mislead or distort—for example, shading one bar to make it more prominent than the others.
- Make sure the height and width of each bar truthfully represent the data. That is, don't make one of the bars larger to maximize the profits, products, or sales in any one year.
- Show bars for every year (or other appropriate period) covered. Note how the unethical bar chart (and accompanying text) in Figure 10.29 violates this rule, but the chart in Figure 10.30 ethically represents the data (see page 501).

FIGURE 10.27 An Unethical Graph and Misleading Interpretation

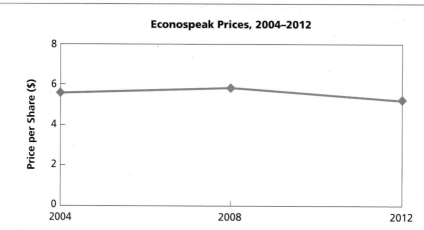

Econospeak Prices, 2004–2012

Econospeak's stock prices during 2004–2012 have been stable, resting securely at about $5.60. The graph above illustrates the stability of Econospeak's stock. Given our steady market, we believe shareholders will be confident in our recent decision to proceed with Econospeak's further expansion into global markets.

© Cengage Learning 2013

FIGURE 10.28 An Ethical Revision of Figure 10.27

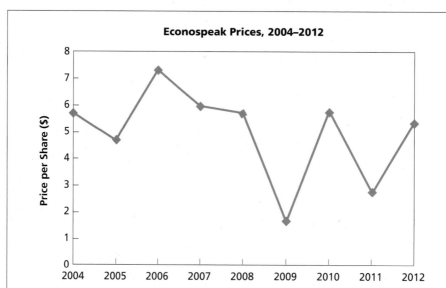

Econospeak Prices, 2004–2012

Econospeak's stock prices have not been as stable over 2004–2012 as we would have liked. The graph above illustrates the challenges the company has faced in the market in the past decade, resulting in fluctuation of prices. We believe, however, that Econospeak's further expansion into global markets will be possible by 2013.

© Cengage Learning 2013

FIGURE 10.29 An Unethical Bar Chart

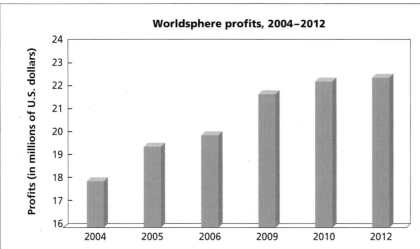

Profits at Worldsphere have seen a healthy increase over recent years. The above bar chart demonstrates the steady increase in profits, which have risen $4.5 million since 2004. Given the profit history of Worldsphere, our investors can be confident of future profits and the security of their stock in our company.

© Cengage Learning 2013

FIGURE 10.30 An Ethical Revision of Figure 10.29

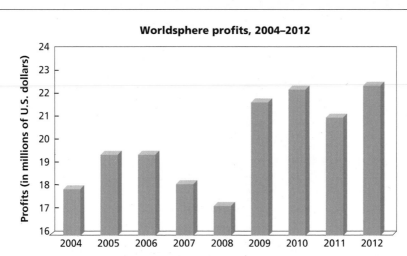

Although profits at Worldsphere have been volatile recently, we are now at our highest profit margin of the last eight years. The bar chart above shows the effects of market difficulties for the period 2004–2012, when the economy suffered major cutbacks. However, Worldsphere achieved a successful turnaround in 2009, with profits regaining strength due to advances in research and technology.

© Cengage Learning 2013

FIGURE 10.31 An Unethical Pie Chart

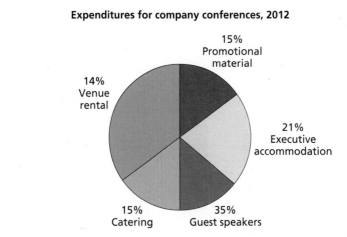

Expenditures for company conferences, 2012

In light of corporate reductions for company conferences this year, we have tried to keep costs down for catering and guest speakers.

FIGURE 10.32 An Ethical Revision of Figure 10.31

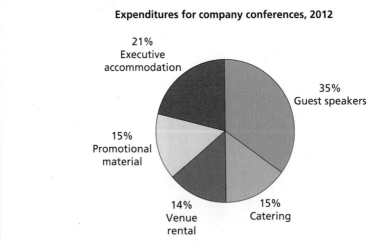

Expenditures for company conferences, 2012

Though they account for the largest share of expenditures for company conferences this year, the fees we paid for guest speakers may actually net our greatest advantages. These speakers have shown us ways to maximize our profits in expanding markets.

Pie Charts

- Don't use 3-D to distort the thickness of one slice of the pie and thereby misleadingly deemphasize other slices.
- Avoid concealing negative information (losses, expenses, etc.) by including the information in another category, or slice, or lumping it into a category marked "Other" or "Miscellaneous."
- Make sure percentages match the size of the slices of the pie chart. Study Figures 10.31 and 10.32 (on page 502). Note how a larger expense for guest speakers (35% of budget) is unethically misrepresented in Figure 10.31 by using a smaller-sized wedge, while the expenses for venue rental (14%) are actually less than for guest speaker expenses but are drawn larger to misrepresent costs.

Drawings

- Avoid any clutter that hides features.
- Label all parts correctly.
- Do not omit or shadow any necessary parts.
- Tell readers what to look for, as in Figure 10.23.
- Draw an object accurately. Indicate if your drawing is the actual size of the object or equipment or if it is drawn to scale. Provide a scale.

Pictographs

- Choose icons that are culturally and ethnically appropriate, tasteful, and free from stereotyping.
- Make sure the size or shape of your pictograph accurately represents the financial or other statistical information it is meant to represent.
- Do not distort the size of a pictograph, increasing or decreasing its height or width, to suggest a larger or smaller quantity, whether in sales, population, services, etc. Note how in Figure 10.33, each dollar-sign pictograph is twice as big as the preceding one, leading readers to conclude that sales have more than doubled in succeeding quarters. But the size of the pictograph in Figure 10.33 does not truthfully represent the actual increase in sales. The revision of the visual in Figure 10.34 shows a far less impressive but far more honest sales record.

Using Appropriate Visuals for International Audiences

Whether you are writing for an expanding international business community in South America or China, or for multicultural readers in the United States, you will have to prepare numerous documents that require visuals. These can range from instructions containing warning and caution statements to tables, graphs, charts, and photos for proposals, reports, and online presentations.

FIGURE 10.33 An Unethical Pictograph

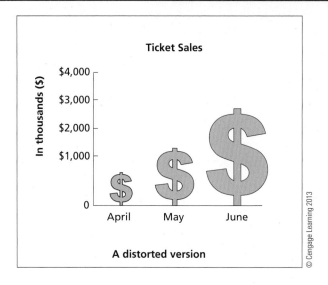

A distorted version

© Cengage Learning 2013

FIGURE 10.34 An Ethical Revision of Figure 10.33

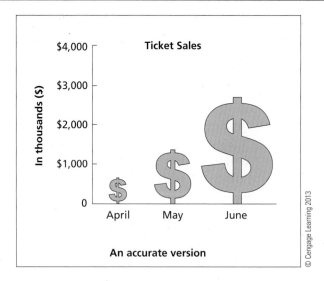

An accurate version

© Cengage Learning 2013

Choose your visuals for international readers just as carefully as you do your words. Visuals should be designed to help international audiences understand your message clearly and without bias. Visuals and other graphics must be consistent with your message, not detract from or distort it.

Visuals Do Not Always Translate from One Culture to Another

While there are some internationally recognized icons, such as those in Figure 10.35, keep in mind that visuals do not automatically transfer from one culture to another. They do not always have the same meaning for all readers across the globe. Like words, visuals and other graphic devices may have one meaning or use in the United States and a radically different one in other countries around the world.

To avoid confusing or offending an international audience, investigate the meaning and cultural significance of a visual before you use it. Consult a native speaker from your audience's country to see if your visuals are culturally acceptable; then follow up with your collaborative team and any graphic artists assigned to the project to ensure that every visual is appropriate and accurate.

Guidelines for Using Visuals for International Audiences

To communicate appropriately and respectfully with international readers through visuals and other graphics, follow these guidelines:

1. **Do not use any images that ethnically or racially stereotype your readers.** A large clothing manufacturer was criticized by the Asian American community when it released T-shirts depicting two men with slanted eyes and rice paddy hats as Laundromat owners. Similarly, depicting Native Americans through clip art images of red-faced chiefs is insulting. Rather than using an ethnic or racial pictograph, use neutral stick figures or nonbiased clip art.

2. **Be respectful of religious symbols and images.** To a U.S. audience, a cross represents a church, first aid, or a hospital, and when the cross is a large red one on a white background, the Red Cross. But the symbol for that humanitarian agency in Turkey and in other Muslim countries is the crescent. Portraying a smiling Buddha to sell products is considered disrespectful to residents in Southeast Asia.

3. **Avoid using culturally insensitive or objectionable photographs.** What is acceptable in photographs in the United States may be taboo elsewhere in the world. A photograph portraying men and women eating together at a business conference is unacceptable in Saudi Arabia. Moreover, sitting with one leg crossed over the other is regarded as a disrespectful gesture in many countries in the world. It may be safest not to show photographs of people to an international audience.

FIGURE 10.35 Internationally Recognized Icons

4. Avoid any icons or clip art that international readers would misunderstand. In the United States, an owl can stand for wisdom, thrift, and memory, while in Japan, Romania, and some African countries it is a symbol for death. A software program showing an icon of a mailbox (below) to represent email confused readers in other countries who thought the icon represented a birdhouse. A better alternative would be an icon of an envelope.

Similarly, the Mac icon for "trash" (a garbage can) confused some international readers since not all garbage cans look alike.

5. Be cautious about using images or photos with hand gestures, especially in manuals or other instructional materials. Many gestures are culture-specific; they do not necessarily mean the same thing in other countries that they do in the United States. Table 10.5 lists cultural differences around the globe for some common gestures.

6. Don't offend international readers by using colors that are culturally inappropriate. Emphasizing the color red in a document aimed at Native Americans or the color yellow in a document for Asians or Asian Americans is likely to be seen as offensive. Yet red in China symbolizes happiness while yellow in Saudi Arabia signifies strength. Green is regarded as a sacred color in Saudi Arabia.

Always research (on the Internet or by consulting a representative from the audience you want to reach) if the color scheme you have chosen is appropriate for your target audience. For example, purple is the color of death and mourning in Thailand, as is white in China. Although orange is the symbolic color of Northern Ireland, you should avoid it when writing to readers whose culture does not value that color.

7. Be careful when using directional signs and shapes. While road signs tend to be fairly recognizable throughout the world—for example, the octagon is generally understood as the shape for the stop sign—there are country-specific signs and code books. For instance, a pennant-shaped sign, signaling a no-passing zone on American highways, may not have the same meaning in Nigeria or India. The symbol below points to a railroad crossing for American readers but would baffle an audience in the Czech Republic.

TABLE 10.5   Different Cultural Meanings of Various Gestures

Gesture	Meaning in the United States	Meaning in Other Countries
OK sign (index finger joined to thumb in a circle)	All right; agreement	Sexual insult in Brazil, Germany, Russia; sign for zero, worthlessness in France
Thumbs-up	A winning gesture; good job; approval	Offensive gesture in Muslim countries
Waving or holding out open palm	Stop	Obscene in Greece—equivalent to throwing garbage at someone
Pointing with the index finger	This way; pay attention; turn the page	Rude, insulting in Japan, Sri Lanka, and Venezuela; in Saudi Arabia used only for animals, not for people
Nodding head up and down	Agreement, saying yes	Greek version of saying "no"; in China means "I understand," not "I agree"
Motioning with index finger	Signal to "come over here"	Insulting in China—instead, extend arm and wave to ask "come over"

8. Avoid confusing an international audience with punctuation and other writing symbols used in the United States. Not all cultures use a question mark (?) to end a sentence asking a question, to represent the Help function in a computer program, or for an FAQ link on a website. Similarly, ellipses (. . .) and slashes (and/or) may not be a part of the language your international audience reads and writes. Be careful about using the following graphic symbols, familiar to writers and speakers of U.S. English but not necessarily to an international audience:

#	pound	&	ampersand (and)
©	copyright	*	asterisk

Include a glossary or a key to these symbols or any other graphics your reader may not understand, or revise your sentences to avoid these symbols.

Conclusion

This chapter has introduced you to the types of visuals you can expect to use frequently in the world of work and has given you guidelines on how to construct or import, label, insert, introduce, and cite them in your writing. Following the guidelines in this chapter will make your documents more businesslike, more persuasive, and easier for your readers to follow. Equally important, whenever you use a visual, always keep in mind the importance of the same ethical standards that you follow in your written work. You also need to be respectful of your audience's culture and the context in which your visual will be included.

✓ Revision Checklist

☐ Selected most effective type of visual (table, chart, graph, drawing, photograph, clip art) to represent information the audience needs.
☐ Drafted and edited visual until it met readers' needs.
☐ Selected right amount of detail to include in visual.
☐ Made sure every visual is attractive, clear, complete, and relevant.
☐ Gave each visual a number, a caption (title), and, where necessary, a legend and callouts.
☐ Inserted visual as close as possible to the relevant discussion that goes with it—that is, on the same or facing page.
☐ Surrounded each visual with adequate white space.
☐ Inserted page number where visual can be found.
☐ Introduced and interpreted each visual in appropriate place in report or proposal.
☐ Explained what visual shows and why it is important.
☐ Made sure every photograph is clear and was taken from the proper angle.
☐ Acknowledged sources for visuals and gave credit to individuals on whose statistical data the visual was based.
☐ Secured written permission to use visual in published work.
☐ Used and interpreted visuals ethically and appropriately, free from any bias, stereotype, or anything culturally offensive.
☐ Did not distort or skew any visual to misrepresent data.
☐ Selected visuals and colors that respect the cultural traditions of international readers.

Exercises

1. Bring to class printouts of three or four home pages that use especially effective visuals. In a short memo or email (three or four paragraphs) to your instructor, indicate why and how each visual is appropriate for and convincing to a particular audience. What would each home page look like without its visual?

2. Find a website you think includes poorly designed or inappropriate visuals. Explain to your instructor why these visuals are ineffective, and redesign two of them.

3. Locate a print document that is visual-poor, and select an appropriate visual to accompany it. Justify your choice of visual and its inclusion in an email to your instructor.

4. Record the highest temperature reached in your town for the next five days. Then collect data on the highest temperature reached in three of the following cities over the same period: Boston, Chicago, Dallas, Denver, Los Angeles, Miami, New Orleans, New York, Philadelphia, Phoenix, Salt Lake City, San Francisco, Seattle. (You can get this information from a printed newspaper or on the Internet.) Prepare a table showing the differences for the five-day period.

5. Go to a supermarket and get the prices of four different brands of the same product (a candy bar, a soft drink, a box of cereal). Present your findings in the form of a table.

6. A government agency supplied the following statistics on the world production of oranges (including tangerines) in thousands of metric tons for the following countries during the years 2008–2011: Brazil, 2,005, 2,132, 2,760, 2,872; Israel, 909, 1,076, 1,148, 1,221; Italy, 1,669, 1,599, 1,766, 1,604; Japan, 2,424, 2,994, 2,885, 4,070; Mexico, 937, 1,405, 1,114, 1,270; Spain, 2,135, 2,005, 2,179, 2,642; and the United States, 7,658, 7,875, 7,889, 9,245. Prepare a table with that information, and then write a paragraph in which you introduce and refer to the table and draw conclusions from it.

7. Keep a record for one week of the number of miles you walk, ride, or drive each day. Then prepare a line graph depicting that information.

8. Prepare a table to show the following statistical data: According to the 2000 census, the town of Ardmore had a population of 34,567. By the 2010 census, the town's population had decreased by 4,500. In the 2000 census, the town of Morrison had a population of 23,809, but by the 2010 census, the population had increased by 3,689. The 2010 census figure for the town of Berkesville was 25,675, which was an increase of 2,768 from the 2000 census.

9. Prepare a line graph for the information in Exercise 8.

10. Prepare a bar graph for the information in Exercise 8.

11. Write a paragraph introducing and interpreting the following table.

Year	Soft Drink Companies	Bottling Plants	Per Capita Consumption (gallons)
1940	750	750	10.3
1945	578	611	12.5
1950	457	466	18.6
1955	380	407	17.2
1960	231	292	15.9
1965	199	262	15.7
1970	171	229	15.4
1975	118	197	16.0
1980	92	154	18.7
1985	54	102	21.1
1990	43	88	23.1
1995	45	82	25.3
2000	37	78	27.6
2005	34	72	30.1
2010	31	70	32.3

12. According to a municipal study in 2012, the distribution of all companies classified in each enterprise industry in that city is as follows: minerals, 0.4%; selected services, 33.3%; retail trade, 36.7%; wholesale trade, 6.5%; manufacturing, 5.3%; and construction, 17.8%. Make a circle chart to represent the distribution, and write a one- or two-paragraph interpretation to accompany and explain the significance of your visual.

13. Construct a segmented bar chart to represent the kinds and numbers of courses you took in a two-semester period or during your last year in high school.

14. Prepare a bar chart for the different brands of the product you selected in Exercise 5. Write a paragraph introducing your chart.

15. Find a pictograph in a math or business textbook, in a magazine (try *Businessweek*, *Newsweek*, or *U.S. News & World Report*), or on a website from the Department of Labor Statistics. Make a bar graph from the information contained in the pictograph, and then write a paragraph introducing the bar graph and drawing conclusions from it.

16. Construct an organizational chart for a business or an agency you worked for recently. Include part-time and full-time employees, but indicate their titles or functions with different kinds of shapes or lines. Then attach the chart to a brief email to your employer explaining why this kind of organizational chart should be distributed to all employees. Focus on the types of problems that could be avoided if employees had access to such a chart.

17. Prepare a flow chart for one of the following activities:
   a. jumping a "dead" car battery
   b. giving an injection
   c. making a reservation online
   d. using an iPhone to check the status of a flight
   e. checking your credit online
   f. putting out an electrical fire
   g. joining a chat group on the Internet
   h. preparing a visual using a graphics software package
   i. changing your email account password
   j. starting a personal blog
   k. filing for an extension to pay state income tax
   l. any job you do

18. Draw an interior view of a piece of equipment you use in your major, and then identify the relevant parts using callouts.

19. Prepare a drawing of one of the following tools, and include appropriate callouts with your visual.
   a. printer
   b. iPod
   c. pliers
   d. stethoscope
   e. swivel chair
   f. DVD player
   g. ballpoint pen
   h. flashlight
   i. high-definition TV
   j. pair of eyeglasses

20. Prepare appropriate visuals to illustrate the data listed below. In a paragraph immediately after the visual, explain why the type of visual you selected is appropriate for this information.

   a. Life expectancy is increasing in the United States. This growth can be dramatically measured by comparing the number of teenagers with the number of older adults (over age 65) in the United States during the past few decades. In 1970 there were approximately 28 million teenagers and 20 million older adults. By 1980 the number of teenagers climbed to 30 million, and the number of older adults increased to 25 million. In 1990 there were 27 million teenagers and 31 million older adults. By 2010 the number of teenagers had leveled off to 23 million, but the number of older adults soared to more than 36 million.

   b. Researchers estimate that for every adult in the United States, 3,985 cigarettes were purchased in 1985; 4,100 in 1990; 3,875 in 1995; 3,490 in 2000; and 2,910 in 2005.

21. Find a photograph that contains some irrelevant clutter. The marketing department of your company wants to use the photograph. Write a letter to the department head explaining what to delete and why.

22. Below are five examples of poorly prepared visuals with brief explanations of how they were intended to be used. Redo one of the visuals to make it easier to read and to better organize the information. Write a paragraph to accompany your new visual.

   a. To illustrate a report on problems that pilots have encountered with a particular model of jet engine.

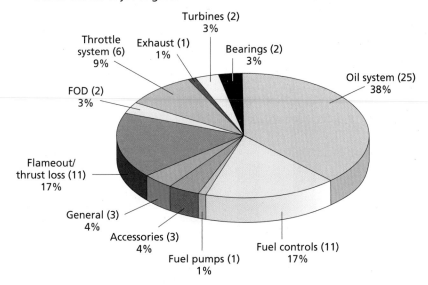

b. To show that a hiking trail is compatible with wheelchair access laws.

c. To show the percentage of white oak acorns in each quality category, by year, that were collected from traps from 2006 to 2010 in national forests.

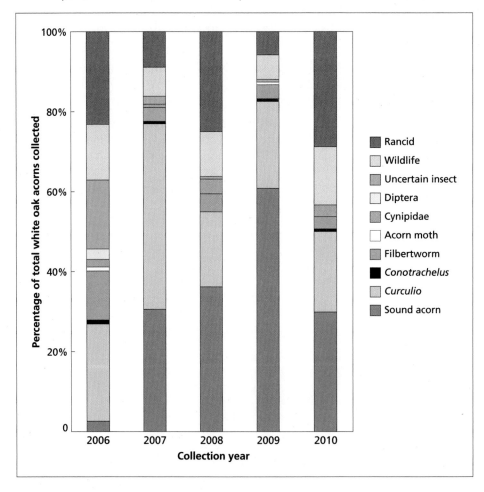

d. To accompany a report on where a business invested its funds.

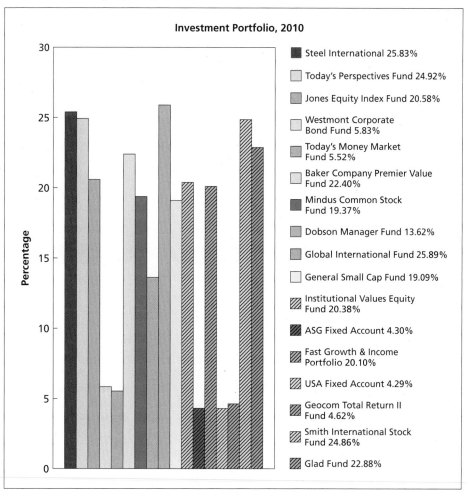

Investment Portfolio, 2010

Steel International 25.83%

Today's Perspectives Fund 24.92%

Jones Equity Index Fund 20.58%

Westmont Corporate Bond Fund 5.83%

Today's Money Market Fund 5.52%

Baker Company Premier Value Fund 22.40%

Mindus Common Stock Fund 19.37%

Dobson Manager Fund 13.62%

Global International Fund 25.89%

General Small Cap Fund 19.09%

Institutional Values Equity Fund 20.38%

ASG Fixed Account 4.30%

Fast Growth & Income Portfolio 20.10%

USA Fixed Account 4.29%

Geocom Total Return II Fund 4.62%

Smith International Stock Fund 24.86%

Glad Fund 22.88%

e. to encourage programmers to write applications for the BlackBerry platform.

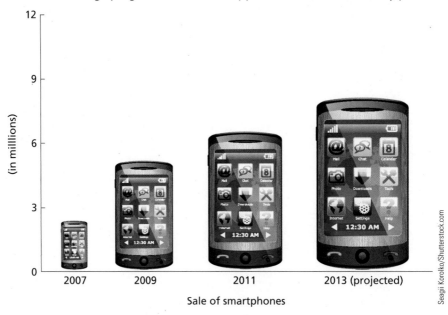

Sale of smartphones

23. You work for a large manufacturer of industrial heat pumps and have been asked to help write a section of a report on the increased business your firm has been doing overseas. Based on the sales figures below for the years 2010, 2011, and 2012 (listed in that order) for each of the following countries, prepare two different yet complementary visuals. Also, supply a one-page description and interpretation of the statistics represented in your visuals. You may work collaboratively with one or more students in your class to prepare the visuals and to write the section of the report on international sales.

Argentina, 45, 53, 34; Australia, 78, 90, 115; Bolivia, 23, 43, 52; Brazil, 29, 34, 35; Canada, 116, 234, 256; China, 7, 100, 296; Denmark, 65, 54, 87; England, 256, 345, 476; France, 198, 167, 345; Germany, 234, 398, 429; Holland, 65, 80, 89; Italy, 49, 52, 97; Japan, 67, 43, 29; Korea, 55, 43, 28; New Zealand, 12, 69, 114; Norway, 33, 92, 104; Switzerland, 164, 266, 306; Sweden, 145, 217, 266.

In your written report, take into account trends, shifts in sales, and possible consequences for further marketing, and conclude with a specific recommendation to your employer.

24. Supply an appropriate piece of clip art to accompany each of the boat safety tips in the following list. Explain your choices in a short report to your instructor.

Top 10 Boating Safety Tips

1. **Wear your lifejacket or PFD.** An approved Personal Flotation Device (PFD) is required by law for each person on board. Remember: It won't work if you don't wear it.

2. **Boating and booze don't mix.** Alcohol impairs your ability to make good, quick decisions. This is critical when operating a fast and easily maneuverable personal watercraft.

3. **Know your craft.** Study the manufacturer's manual and practice handling your craft under experienced supervision and in open water well away from other boaters.

4. **Take a boating safety course.** Learn common boating rules, regulations, and safe practices.

5. **Look out.** Ride defensively. Collisions with other boats or stationary objects such as rafts or docks are the number one cause of personal watercraft injuries.

6. **Watch the weather.** Check the weather forecast before starting out. Be alert for the wave, wind, and cloud changes that signal the approach of bad weather.

7. **Be prepared for cold water.** Cold water robs your body heat 25 times faster than cold air of the same temperature does. If you fall off your craft into cold water, get back on board immediately.

8. **Know the area.** Do not assume the water is clear of obstructions. Rocks, shoals, sandbars, and submerged pilings can seriously damage the craft or those on board. Check marine charts and stay in marked channels.

9. **Carry safety equipment.** Besides approved PFDs and a sound-signaling device (like a whistle), carry a tow rope and, when operating on a large body of water, some small type-B flares in a watertight container.

10. **Don't ride at night.** Most personal watercraft do not have the lights that the law requires for night riding; therefore, it is unsafe to be on one after dark.

Source U.S. Coast Guard.

---

25. You have been asked to create a logo, including a visual, for one of the following new businesses opening in your town. Explain your design and why you think it is effective.
   a. landscaping firm
   b. home health care company
   c. studio offering musical instrument lessons
   d. outdoor apparel manufacturer
   e. toxic waste removal company
   f. math and science tutor

26. Explain why the following visuals and graphic symbols would be inappropriate in communicating with an international audience and how you would revise a document containing them:
   a. clip art showing a string tied around an index finger
   b. a picture of a man with a sombrero on a website for Pronto Check Cashing Company
   c. clip art of a lightbulb and the logo "Smart Ideas" for a CPA firm

d. clip art showing someone crossing the middle finger over the index finger (wishing sign)
e. a drawing of a cupid figure for a caterer
f. a sales brochure showing a white and blue flag for a French audience
g. a poster showing a women's track and field team used to advertise a brand of footware to an Arabic-speaking audience
h. a satisfied customer making the "okay" gesture in an ad aimed at a Japanese audience
i. an image of a rabbit on a website for an automobile manufacturer to stress how fast its cars are
j. a photograph of a roll of Scotch tape to show international readers that your company can solve problems quickly
k. a photograph of a baseball umpire holding up his hands to ask readers to repeat a step in a set of instructions
l. a drawing of a white glove to sell home and carpet cleaning supplies
m. a piece of clip art showing a rooster for a business that opens early in the day
n. the letters *a, b, c* in a box to indicate that a directory is alphabetically arranged
o. a pair of scissors moving along a rectangle of dotted lines to indicate a merchant offers prospective customers a money-saving coupon

CHAPTER **11**

Designing Successful Documents and Websites

The success of your document (letter, proposal, report) or website depends as much on how it looks as what it says. In designing print documents and websites, you need to project a positive, professional image of yourself, your company, and your product. A report filled with nothing but thick paragraphs of type, with no visual clues to break them up or to make information stand out, is sure to intimidate readers and turn them away. They will conclude that your work is too complex and not worth their effort or time. And if there is one thing that the world of work especially dislikes, it is someone who cannot get to the main point—the bottom line—quickly.

Websites and documents need to look user-friendly, signaling to your audience that your message is

- easy to read
- easy to follow and understand
- easy to recall

Don't bog your audience down with long, unbroken paragraphs or crowded screens of information. Instead, separate material into smaller units that are visually appealing. You can help readers find key points at a glance by *chunking* (writing smaller paragraphs) and using lists, boldface type, and bullets. That way, they can find your ideas easily the first time through or on a second reading if they have to double back to check or verify a point.

This chapter will show you how to design professional-looking, reader-friendly documents and websites that are a credit to you and your company.

Access chapter-specific interactive learning tools, including quizzes and more in your English CourseMate accessed through www.cengagebrain.com.

Organizing Information Visually

Today's Web-based culture prizes visual thinking in the design of all kinds of documents. The way you organize and visualize information can help readers move quickly and clearly through your document. Take a quick look at Figures 11.1 and 11.2 on pages 518 and 519–521. The same information is contained in each figure. Which visually appeals to you more? Which do you think would be easier to read? Which is better designed? As the two figures show, design and layout play a crucial role in an audience's overall acceptance of your work.

FIGURE 11.1 A Poorly Designed Document

No title

Single-spacing makes document difficult to read

Lack of headings in color or boldface makes it hard for readers to organize material

Arbitrary font changes confuse readers

Uneven presentation of numbers

Lack of adequate margins makes document look dense and complex

Unnecessary italics are confusing

Inconsistent use of italics and boldfacing

The results for the recent cholesterol screening at *our company's Health Fair* were distributed to each employee last week. Many employees wanted to know more about cholesterol in general, the different types of cholesterol, what the results mean, and the foods that are high or low in cholesterol.

We hope the information provided below will help employees better answer their questions concerning cholesterol and our cholesterol screening program.

High cholesterol, along with high blood pressure and obesity, is one of the primary risk factors that may contribute to the development of coronary heart disease and may eventually lead to a heart attack or stroke. Cholesterol is a fatty, sticky substance found in the bloodstream. Excessive amounts of the bad type of cholesterol can deposit on the walls of the heart arteries. *This deposit is called plaque, and over a long period of time plaque can narrow or even block the blood flow through the arteries.*

Total cholesterol is divided into three parts—LDL (low-density lipoprotein), or bad cholesterol; HDL (high-density lipoprotein), or good cholesterol; and VLDL (very low-density lipoprotein), a much smaller component of cholesterol you don't have to worry about. Bad (LDL) cholesterol forms on the walls of your arteries and can cause a lot of damage. Good cholesterol, on the other hand, functions like a sponge, mopping up cholesterol and carrying it out of the bloodstream.

You should have received *three cholesterol numbers.* One is for your HDL (or good cholesterol) and the other is for your LDL, or bad cholesterol, reading. These two numbers are added to give you the third, or composite, level of your total cholesterol. You are doing fine.

As you can see, a total cholesterol reading of below 200 is considered safe. Continue what you have been doing. If your reading falls in the moderate risk range of 200–239, you need to modify your diet, get more exercise, and have your cholesterol checked again in six months. *If your reading is above 240, see your doctor.* You may need to take cholesterol-lowering medication, if your doctor prescribes it. Reducing your total cholesterol by even as little as 25% can decrease your risk of a heart attack by 50%.

The Surgeon General recommends that your LDL, or bad cholesterol, should be below 130; and your HDL, or good cholesterol, needs to be at least above 36. Ideally, the ratio between the two numbers should not be greater than 5 to 1. That is, your HDL should be at least 20% of your LDL. The higher your HDL is, the better, of course. So even if you have a high LDL reading, if your HDL is correspondingly high you will be at less risk.

One of the easiest ways to decrease your cholesterol is to modify your diet. Cholesterol is found in foods that are high in saturated fat. *Saturated fat comes from animal sources and also from certain vegetable sources.* Foods high in bad cholesterol that you should restrict or avoid, include whole milk, red meat, eggs, cheese, butter, shrimp, oils such as palm and coconut, and avocados. Generally, food groups low in cholesterol include fruits, vegetables, and whole grains (assorted wheat breads, oatmeal, and certain cereals), *lean meats (fish, chicken)*, and beans.

The goal of our cholesterol screening is to help each employee lower his or her cholesterol level and eventually reduce the risk of heart disease. **Besides the advice given above,** you can do the following: get regular aerobic exercise—bicycling, brisk walking, swimming, rowing—for at least 30 minutes 3–4 times a week. But get your doctor's approval first. *Eat foods low in cholesterol* but high in dietary fiber (beans, oatmeal, brown rice). Maintain a healthy weight for your frame to lower your body fat. Minimize stress, which can increase cholesterol. Learn relaxation techniques.

FIGURE 11.2 An Effectively Designed Document with the Same Text as Figure 11.1

Cholesterol Screening

The results for the recent cholesterol screening at our company's Health Fair were distributed to each employee last week. Many employees wanted to know more about cholesterol in general, the different types of cholesterol, what the results mean, and the foods that are high or low in cholesterol. We hope the information provided below will help employees better answer their questions concerning cholesterol and our cholesterol screening program.

Determining Risk Factors

High cholesterol, along with high blood pressure and obesity, is one of the primary risk factors that may contribute to the development of coronary heart disease and may eventually lead to a heart attack or stroke. Cholesterol is a fatty, sticky substance found in the bloodstream. Excessive amounts of the bad type of cholesterol can deposit on the walls of the heart arteries. This deposit is called **plaque** and over a long period of time plaque can narrow or even block the blood flow through the arteries.

Separating Types of Cholesterol

Total cholesterol is divided into three parts: (1) **LDL** (low-density lipoprotein), or bad cholesterol; (2) **HDL** (high-density lipoprotein), or good cholesterol; and (3) **VLDL** (very low-density lipoprotein), a much smaller component of cholesterol you don't have to worry about. Bad (LDL) cholesterol forms on the walls of your arteries and can cause a lot of damage. Good cholesterol, on the other hand, functions like a sponge, mopping up cholesterol and carrying it out of the bloodstream.

Title clearly set apart from text with capitalization, larger font

Text is double-spaced with more ample margins, making it more readable

Page does not look cluttered

Headings in color and larger font divide material into easy-to-follow units for readers

Consistent use of one font

Only key words being defined are boldfaced

Types of cholesterol are helpfully labeled with numbers

(Continued)

FIGURE 11.2 (Continued)

Understanding Your Cholesterol Results

Concise paragraph provides clear opening for new section

You should have received three cholesterol numbers. One is for your **HDL** (or good cholesterol), and the other is for your **LDL** (or bad cholesterol) reading. These two numbers are added to give you the third, or composite, level of your total cholesterol.

Cholesterol levels can be classified as follows:

Emphasizes range of risk categories by setting them apart in a shaded box

Minimal Risk	Moderate Risk	High Risk
below 200	200–239	above 240

Paragraph clearly defines and distinguishes numbers

As you can see, a total cholesterol reading of below 200 is considered safe. You are doing fine. Continue what you have been doing. If your reading falls in the moderate risk range of 200–239, you need to modify your diet, get more exercise, and have your cholesterol checked again in six months. If your reading is above 240, see your doctor. You may need to take cholesterol-lowering medication, if your doctor prescribes it. Reducing your total cholesterol by even as little as 25% can decrease your risk of a heart attack by 50%.

Includes additional space between sections

Knowing the Relationship Between Bad and Good Cholesterol

The Surgeon General recommends that your **LDL**, or bad cholesterol, be below 130. And your **HDL**, or good cholesterol, needs to be at least above 36. Ideally, the ratio between the two numbers should not be greater than 5 to 1. That is, your **HDL** should be at least 20% of your **LDL**. The higher your **HDL** is, the better, of course. So even if you have a high **LDL** reading, if your **HDL** is correspondingly high you will be at less risk.

Paragraphs are neither too long nor too short

Recognizing Food Sources of Cholesterol

Headings consistently formatted

One of the easiest ways to decrease your cholesterol is to modify your diet. Cholesterol is found in foods that are high in saturated fat. Saturated fat

Page is numbered in footer

FIGURE 11.2 (Continued)

comes from animal sources and also from certain vegetable sources. Foods high in bad cholesterol that you should restrict include

1. whole milk
2. red meat
3. eggs
4. cheese
5. butter
6. shrimp
7. oils such as palm and coconut
8. avocados

Generally, food groups low in cholesterol include fruits, vegetables, and whole grains (wheat breads, oatmeal, and certain cereals), lean meats (fish, chicken), and beans.

Realizing It Is Up to You

The goal of our cholesterol screening program is to help each employee lower his or her cholesterol level and eventually reduce the risk of heart disease. Besides the advice given above, you can do the following:

- Get regular aerobic exercise—bicycling, brisk walking, swimming, rowing—for at least 30 minutes 3–4 times a week. But get your doctor's approval first.
- Eat foods low in cholesterol but high in dietary fiber (beans, oatmeal, brown rice).
- Maintain a healthy weight for your frame to lower your body fat.
- Minimize stress, which can increase cholesterol. Learn relaxation techniques.

A double-spaced, numbered list helps readers easily identify foods with bad cholesterol

Easy-to-follow examples of foods low in cholesterol

Heading signals conclusion

Bulleted list serves as both conclusion and plan for future action

3

Source: Thanks to Sgt. Mannie E. Hall of the U.S. Army for creating this document.

Your company stands to win or lose points with customers as a result of your design choices. A visually appealing document or website will enhance your company's reputation and improve its sales. A poorly designed one will not. Consumers will think your firm is old-fashioned, inflexible, hard to do business with, or just plain unprofessional. A poorly designed document suggests that your company does not care about specific problems customers may have over a policy or a set of instructions and that it does not respect the individuals it does business with.

Characteristics of Effective Design

As you read this chapter and apply its principles to your own work, make sure your documents and websites offer your audience the qualities of effective design on the left and avoid those on the right:

Effective Design

- visually appealing
- logically organized
- clear
- accessible
- varied
- relevant

Ineffective Design

- crowded
- disorganized
- hard to follow
- difficult to read
- boring, repetitious
- inconsistent

Figure 11.2 displays effective (positive) design characteristics, whereas Figure 11.1 displays ineffective (negative). Study the annotations to these figures to see how the errors in Figure 11.1 were corrected in Figure 11.2. By modeling your documents on the layout and design of Figure 11.2, you can guarantee that your message will be clearly understood and well received.

Desktop Publishing

Desktop publishing programs, sometimes referred to as *page layout software*, provide an inexpensive alternative to a professional print shop. Because desktop publishing software permits users to design page layouts, include visuals, and produce

Tech Note

Desktop Publishing Programs

Word-processing programs such as Microsoft Word and Corel WordPerfect have evolved into more and more powerful desktop publishing programs. In addition, files created in word-processing programs can be imported into sophisticated desktop publishing programs such as Adobe InDesign, Apple Pages, Microsoft Publisher, and QuarkXPress, as can graphics created or edited in Adobe Photoshop, Corel PaintShop Photo Pro, Adobe Illustrator, CorelDRAW, and Microsoft PhotoDraw.

high-quality final copies, you can create printed documents right in your own home or office.

With desktop publishing you can

- select from templates specifically designed for reports, newsletters, brochures, and other formats
- design your own templates if you create similar documents often
- take advantage of numerous font choices
- delete, insert, and move entire blocks of text
- integrate various changes in typeface—such as bold, italics, and underlining
- vary type sizes
- justify margins
- change line spacing
- center words, titles, or lines of text
- break and number pages
- arrange text in multiple columns
- insert sidebars, tables, footnotes, headers, footers, and sidebar quotes
- import graphics such as drawings, photographs, icons, and logos from your hard drive, a CD, or the Web
- insert graphics available within your software program, such as clip art (usually located in a separate clip art file—see page 527), special symbols and shapes (typically located in a Symbols menu), and illustrations, charts, and graphs you have created using the drawing tools of your software program.

Study the advice given in Figure 11.3 by a publications manager who designs his company's newsletter.

Type

There are many different styles of type (see pages 530–532). All computers come equipped with software that contains a large number of typefaces. Additionally, more than 1,000 high-quality fonts are available over the Internet at such sites as http://1001freefonts.com and www.fonts.com.

Templates

Desktop publishing software sometimes includes predesigned templates. Templates, like those shown in Figure 11.4, are page layout patterns for reports, newsletters, brochures, and other marketing and communications documents. A template for a report, for example, would contain all the headings and divisions you need. You can also create and save your own template for an original format you use frequently.

Graphics

Like other visuals, graphics work in conjunction with your words. A document without visuals or graphics may look boring, confusing, or unattractive. You will have to judge how and when a graphic can improve your message and help you convince your reader.

FIGURE 11.3 Designing a Company Newsletter: Advice from a Publications Manager

Designing a Newsletter: Advice from a Pro

My name is Drew Harris, and I am the publications manager at Mellon IT, a firm that has a strong commitment to designing professional-looking documents. I oversee the publication of the company's monthly newsletter, available in a variety of formats—electronically, posted on our website, and available in a limited number of hard copies to save paper and reduce costs—for management, clients, and vendors. I do both editorial and graphic design work on the newsletter, ensuring that the text is readable and the visuals are crisp, clear, and relevant.

Doing the Layout for Each Issue

The first thing I do is to plan the way each issue will look—that is, the layout of both text and graphics. I use a template with a two-column vertical grid on each page. Into the grid I fit three or four stories as well as our regular features, including

- a question-and-answer section
- a staff profile
- a calendar
- a boxed insert with a safety tip

I have to calculate how much space to give each of these items so that the newsletter look balanced, uncrowded, and consistent. But before inserting stories and features, I include all masthead (publication) information—the title of the newsletter, the Mellon logo, the volume number and date, and the company's physical and URL addresses.

Sequencing Each Issue

Each issue leads with the most important story inserted in the left column, where readers often start looking at the issue. Every story has to be in the same typeface (font) and type size; otherwise, the page will look unprofessional and sloppy. So the title will stand out, I include it in boldface color type, followed by a byline in small caps. If a story is continued on another page, a "jumpline" in italics must tell readers where, and of course, I must save room on the continuing page. I also include a sidebar, always shaded and boxed.

FIGURE 11.3 (Continued)

Using Space

But I do more than fit text into the available space. I also have to fit space in and around each story or feature. Double-spacing between paragraphs gives our newsletter an uncrowded look. I leave 3 picas of space between columns on a page so text in one column does not crash text or graphics in the other column. I also leave room for a header, which carries the page number and the title of the newsletter, and for a generous footer, or bottom margin, to frame the contents.

Using Visuals and Color

Visuals must relate to the right story or feature, be clear, and be properly sized. This always involves some experimentation. A visual that is too big dwarfs a story, and one that is too small is hard to read. Embedded images can play havoc with the design. If a reader's browser does not have the right plug-in installed, he or she will see "Image cannot be displayed." To be safe, I translate text and graphics into HTML and plain text.

In photo captions, I always include the names of the people in the photo to identify them for our readers. I leave plenty of space around each photo and other visuals to set them apart from the text. We use clip art, particularly to identify regular features such as the calendar and Q&A section, but it has to be relevant and professional looking.

Like visuals, color can make or break a newsletter. Color has to be tasteful, improve readability, be functional, not used just for decoration. The Q&A section, for instance, carries a blue banner, while the safety tip is aptly framed in red.

Editing for Accuracy

Design and content influence each other. To verify that information is correct and complete, I track down missing names and titles, dates for company events and employee presentations, and captions for figures. Spelling a person's name two different ways in the same story or issue embarrasses Mellon. An especially awkward error occurred once when a story omitted a hyperlink in the middle of a paragraph, leaving a gaping hole. You can never proofread too much.

FIGURE 11.4 Examples of Templates

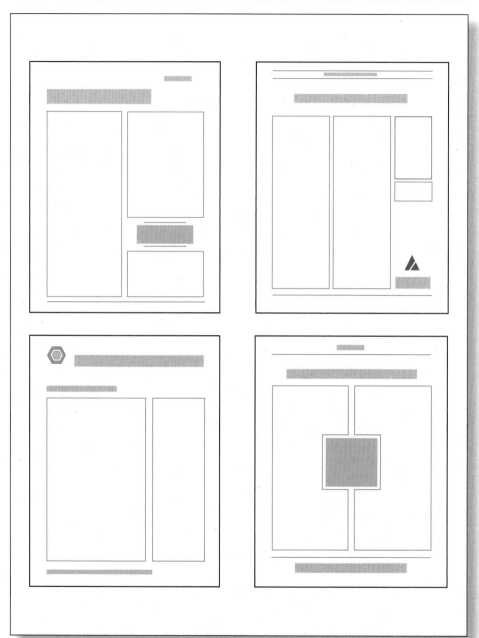

A graphics program allows you to draw shapes and lines that can be manipulated (skewed, enlarged) and offers options for sophisticated use of color and shading. With a drawing program, you can create diagrams, charts, and illustrations that can be saved as electronic files and imported into word-processing or desktop publishing documents.

You can move and place graphics anywhere in a document. Graphic design programs provide customized graphs, charts, tables, and expanded font (design) sizes and shapes.

Here are some types of graphic tools available to you:

1. Drawing tools. Desktop publishing programs contain tools that allow you to create a variety of shapes, rules, borders, and arrows. Having drawn a box, circle, triangle, or whatever, you can use other tools to fill in or alter the appearance of the shape or to add words to it; then you can rearrange all the elements into a graphic that communicates quickly and effectively.

2. Icons. Icons are symbols or visual representations of concepts or actions. The skull and crossbones on a container of poison is an icon that warns of danger. Many highway signs are icons that tell us quickly what to expect ahead: an S-curve, merging traffic, a hospital. *Graphic icons* are simply pictures that communicate directly. No matter what language we speak—and even if we cannot read—icons tell us at a glance which restroom to use or how to fasten the seat belt in an airplane. A nation's flag is an icon, and so are many religious symbols and most company logos.

Computer software often relies on icons to help us perform common actions without having to remember keyboard commands. Arrows in the scroll bars move us easily around a window; a file folder helps us group related documents; and a trash can holds files to be deleted later. Icons in the menu bar make it easy to print a file, open an address book, search for specific text, cut copy, or send an instant message. Examples of public service information icons are shown below:

  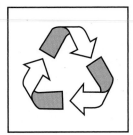

3. Clip art. Clip art is a type of simple drawing, often classified by themes, that can be imported (copied) from a CD or the Internet to your document. (See pages 494–496 in Chapter 10.) Some clip art packages offer as many as 500,000 images, arranged into such diverse categories as animals, computers, holidays, famous people, food, and various businesses and technologies. Clip art is widely used to make business documents attractive and appealing.

4. Stock photos and art. Stock photos and art on a CD or on the Web can be imported—sometimes for a permissions fee, sometimes for free—for use in your documents or on your website.

Before Choosing a Design

Before you begin designing a document, you need to know exactly what you are designing, for whom, and how. Note that planning is the first thing the publications manager does in Figure 11.3. Think carefully about how your work will look on the page, what you want to achieve, and what your resources are. Always get your plans approved by your boss or the print publications committee. Here are questions to ask:

1. Who is your target audience? Will your readers be local residents or global consumers? What do they have in common—gender, location, educational level? You must consider how to reach and appeal to this group before you decide on a "look and feel" for your document. Print size, style, and so forth should be selected based on the audience's characteristics and needs.

2. What is the purpose of your document? Will you have to motivate readers to look at it? Use an eye-catching cover page, attractive colors, and short paragraphs. Or is your document required reading for co-workers and therefore subject to "house style"? Are you designing a newsletter for a professional group? If so, must you comply with a set of guidelines about how, for example, your newsletter should be laid out?

3. How will your document be reproduced? Will it be set by a professional printer, or is it an internal document that can be printed out and distributed? This will affect the number and type of colors you use and the type and size of paper you select.

4. How will it be distributed? Will it fit in a standard size envelope? Will it need to be folded more than once? Do you need larger envelopes or a smaller page size? Will the document be read by more than one reader or more than once? If so, the paper will need to be sturdy.

5. How much will it cost? Financial decisions are crucial in making design choices. Color printing can be much more expensive than standard black-and-white printing. If you use color, try to stick to one or two colors throughout to keep the costs down. Talk to your boss about the costs of reproduction before you decide on anything.

The ABCs of Print Document Design

The basic features of print document design are

- page layout
- typography, or type design
- heads and subheads
- graphics
- color

Every page of your document should integrate those five elements. The proper arrangement and balance of page layout elements, type, heads, graphics, and color involve the same level of preparation that you would spend on your research, drafting, revising, and editing. Your document design should complement—not detract from—your message. Just as you research your information, you have to research and experiment in order to adopt the most effective design for your document.

Page Layout

Text and white space must be arranged pleasingly on each page of your document. Too much or too little of one or the other can jeopardize the reader's acceptance of your message. To design an effective and attractive page layout, pay attention to these four elements.

   1. White space. White space, which refers to the open areas on the page that are free of text and visuals, can help you increase the impact of your message. If you pack too much information or too many design elements into a document, you will distract the reader from the message you want to convey. Skimping on white space can frustrate a reader eager to locate parts of a document quickly and to process them easily. White space can entice, comfort, and appeal to the reader by

   - attracting the reader's attention
   - assuring the reader that information is presented logically
   - announcing that information is easy to follow
   - assisting the reader to organize information visually
   - allowing the reader to forecast and highlight important information

Compare Figures 11.1 and 11.2 again (see pages 518 and 519–521). Which document was designed by someone who understands the importance of white space?

   2. Margins. Use wide margins, usually 1 to 1½ inches, to "frame" your document's text and visuals with white space. Margins prevent your document from looking cluttered or overcrowded. If your document requires binding, you may have to leave a wider inside margin (2 inches).

   3. Line length. Most readers find a text line of 10 to 14 words, or 50 to 70 characters (depending on the type size you choose), comfortable and enjoyable to read. Excessively long lines that bump into the margins signal that your work is difficult to read. In the following example, note how the extra-long lines unsettle your reading and tax your eye movement; they signal rough going.

To succeed in the world of business, workers must brush up on their networking skills. The network process has many benefits that you need to be aware of. These benefits range from finding a better job to accomplishing your job more easily and efficiently. Through networking you are able to expand the number of contacts who can help you. Networking means sharing news and opportunities. The Internet is the key to successful networking.

Conversely, do not print a document with overly short or extremely uneven lines.

> To succeed in the world of
> business, workers must
> brush up on their networking skills. The
> network process has many
> benefits you need to be
> aware of.

Readers will suspect your ideas are incomplete, superficial, or even simple-minded.

4. Columns. Document text can be organized in either single-column or multicolumn formats. Memos, letters, and reports are usually formatted without columns, whereas documents that intersperse text and visuals (such as newsletters and magazines) work better in multicolumn formats.

Typography

Typography consists of font (also called *typeface*), font size, font styles, justification, lists, and captions.

Font

The readability of your text is crucial. Select a font, therefore, that ensures your text is

- legible
- attractive
- functional
- appropriate for your message
- complementary with accompanying graphics

Fonts are also characterized as *serif* (the short cross-lines at the ends of some letters) or *sans serif* (without the serifs). The most familiar fonts are the following four. Times New Roman and Palatino are serif fonts; Helvetica and Arial are sans serif fonts.

Times New Roman	Arial
Helvetica	Palatino

The font you use can make your document look businesslike or too casual. Avoid using a font that looks like script, and don't mix and switch fonts. The result makes your work look amateurish and disorganized, as in Figure 11.1.

Font Size

Font size options are almost unlimited, depending on your software package and printer capabilities. Font size is measured in units called *points*. There are 72 points to the inch. The bigger the point size, the larger the type. Never print your letter or report in 6- or 8-point type, like a newspaper ad, or in a size larger than 12-point type. Here are some suggestions.

Times New Roman 8 point = footnotes or endnotes

Helvetica 12 point = letters, reports, and emails

Palatino 14 point = headings

Arial 22 point = title of your report

Font Styles

Font styles include roman, boldface, italics, underlining, and small caps.

Avoid overusing boldface and italics. Use them only when necessary and not just for decoration. Do *not* underline the text unless absolutely necessary. Not

---

Roman

Boldface

Italics

<u>Underlining</u>

SMALL CAPS

---

only will too many special visual effects make your work harder to read, but you will lose the dramatic impact these features have to distinguish and emphasize key points that rightfully deserve to be set in boldface or italic type.

Justification

Sometimes referred to as *alignment*, justification consists of left, right, full, and centered options. Left-justified (also called *unjustified* or *ragged right*) text is preferred because it allows the space between words to remain constant, making the text easier to read. In full-justified text (both left and right margins are aligned) the word spacing varies from line to line. Left justification gives a document a less formal look than full justification. Narrow columns of text should be set left-justified to avoid awkward gaps between words and excessive hyphenation.

Our website offers consumers a mall on the Internet. It gives shoppers access to our products and services and makes buying easy and fun. Our website offers consumers a mall on the Internet. It gives shoppers access to our

Left-justified text

Our website offers consumers a mall on the Internet. It gives shoppers access to our products and services and makes buying easy and fun. Our website offers consumers a mall on the Internet. It gives shoppers access to our

Right-justified text

Our website offers consumers a mall on the Internet. It gives shoppers access to our products and services and makes buying easy and fun. Our website offers consumers a mall on the Internet. It gives shoppers access to our products

Full-justified text

Our website offers consumers a mall on the Internet. It gives shoppers access to our products and services and makes buying easy and fun. Our website offers consumers a mall on the Internet. It gives shoppers access to our

Centered text

Lists

Placing items in a list helps readers by dividing, organizing, and ranking information. Lists emphasize important points and contribute to page design that is easy to read. Lists can be numbered, lettered, or bulleted. Take a look at the memos, proposals, and reports in Chapters 4, 13, 14, and 15 that effectively use lists.

Captions

Captions are titles that help a reader identify a visual and quickly explain the nature of the picture or other graphic. (Chapter 10 discusses using captions with visuals in a document.)

Heads and Subheads

Heads and subheads are brief descriptive phrases that signal starting points or major divisions in your document. They provide helpful road signs for readers charting their course through a document, as in Figure 11.2. Heads divide, or chunk, your document into its major parts, sections, or segments. Note how many of the figures in this book include heads and subheads. They immediately attract attention and quickly inform readers about the function, scope, purpose, or contents of your document and its individual sections. Moreover, they help readers prioritize information by emphasizing the main points they need to look for and remember. Without heads and subheads your work will look unorganized and cluttered.

How to Write Heads and Subheads

It takes time to write appropriate heads and subheads and determine where to place them in your document. Following the writing process described in Chapter 2, first map out what you want to say. By carefully outlining your work and then revising it, you can determine how many sections you will need and what kinds of information each should contain. In your final copy, every major section will require a head; and each subdivision will use a subhead. By looking at your table of contents (see pages 692–695), which is a blueprint of your document, readers will know at a glance how you have logically divided your work.

How to Format Heads and Subheads

To design a document with logical heads and subheads, follow the guidelines below. Refer to Figure 11.2 as you study them.

1. To signal that you have formatted your document into parts, you have to separate one section from another by inserting white space between the sections to make room for the head. Leave at least two additional spaces above and below a head to set it off from a previous section.
2. Be consistent in the way you print each type of head—that is, center each head or align it flush with the left margin.
3. Use larger type size for heads and subheads than for text; major heads should be larger than subheads. If your text is in 10-point type, your heads may be in 16-point type and your subheads in 12- or 14-point type.

Fourteen-Point Head
Subhead in 12-Point
Use larger type for heads than you do for text; major headings should be larger than subheads. If your text is in 10-point type, your heads may be in 14-point type and your subheads in 12.

---

4. To further differentiate heads from subheads, use all capital letters, initial capital letters (capitalize the first letter of each important word), boldface, or italics.
5. If you are using a color printer, consider using a second color for major heads.
6. Never end a page with a head. Always allow for at least two lines of text on the bottom of the page after a head.

How to Keep Heads and Subheads Grammatically Parallel

Heads and subheads should parallel each other grammatically. Note how the following headings, from a poorly organized proposal from the Acme Company, are not parallel:

- What Is the Problem?
- Describing What Acme Can Do to Solve the Problem
- It's a Matter of Time. . . .
- Fees Acme Will Charge

- When You Need to Pay
- Finding Out Who's Who

Revised, the heads are parallel and easier for readers to understand and follow:

- A Brief History of the Problem
- A Description of Acme's Solutions
- A Timetable Acme Can Follow
- A Breakdown of Acme's Fees
- A Payment Plan
- A Listing of Acme's Staff

Grammatically parallel heads are easier to follow and further demonstrate that you have logically structured your document.

How to Make Heads Functional

Heads not only alert readers to major sections of your work, but they also summarize what readers will find in each section. To make sure your heads and subheads are functional, follow these three guidelines:

1. Keep your heads and subheads concise, but avoid using vague, one-word titles such as "Conditions." Readers will not know if you are referring to the weather, financial markets, or the terms of an agreement.
2. Don't overuse heads; having too many heads is as bad as having too few. Note how even in a short document such as Figure 11.2, several heads help readers better follow the discussion about watching their cholesterol.
3. Make sure each of your heads and subheads matches the divisions and subdivisions, as well as page references, found in your table of contents (pages 692–695).

Graphics

The following list identifies some common graphics included in desktop publishing software packages.

1. **Boxes.** Boxes isolate or highlight text or visuals. Table 10.4 (page 475) is enclosed within a box.
2. **Rules.** Rules are classified as either vertical or horizontal. Vertical rules might be used to separate columns of text, while horizontal rules might separate sections introduced by subheads.
3. **Letterhead and logo.** A company's corporate image is represented and symbolized by its letterhead, usually consisting of a graphic, or icon, integrated with the company name. Notice the variety of logos used in letters in Chapters 5 and 6, especially Figures 6.3 and 6.8 (pages 214 and 226). Often the company's street address, email address, and website URL are included in the letterhead. Company letterhead conveys the firm's message and expresses its character. Typically, logos and sometimes the entire letterhead are imported as graphics files. Letterhead and logos should creatively set one company apart from others.

Using Color

Nothing attracts attention and communicates more quickly and more powerfully than color. It may be one of the most important tools you can use to increase sales, gain faster project approval, or provide a major advantage over competitors. Color helps to sell ideas 85 percent more effectively than black-and-white communications.

Color can help you organize written information and therefore enhance readability. Color visually breaks up long segments of text and can tie important ideas together. For example, colored boxes help readers locate important information quickly. Use color for borders and graphic accents, headings, titles, key words, Internet addresses, sidebars, rules, and boxes that link related facts, figures, or information. See Drew Harris's advice about color in Figure 11.3 (pages 524–525).

Guidelines on Using Color Effectively

Here are some guidelines to follow when you use color in your documents:

- Estimate how the color will look on the page—colors look different on the screen than they do on a sheet of paper. Print a sample page to get a clear view.
- Make sure the text color contrasts sharply with the background color.
- Use no more than two or three colors on a page unless there are photographs, illustrations, or graphics.
- Too many bright colors overwhelm the eye, so use them sparingly—only to call attention to important elements.
- Select "cool" colors, such as blue, turquoise, purple, and magenta, for backgrounds. Avoid light blue text, which is hard to read against a dark background, or yellow text on a light blue background.
- Use colors that respect an international reader's cultural heritage. See page 506.

Poor Document Design: What *Not* to Do

Up to this point, we have introduced the various elements of effective document design. Now, in contrast, we'll look at what you need to avoid. By knowing what looks bad from a reader's point of view, you will be much better able to design a document that works well visually for your readers and reflects your company's best image.

Figure 11.1, a poorly designed document, illustrates many of the following common mistakes in document design. But note how Figure 11.2, which contains the same information as Figure 11.1, incorporates many of the effective techniques just described. Avoid the following errors when you design your document:

1. **Insufficient white space.** Narrow margins and limited space between headings and text are classic mistakes. A lack of white space between paragraphs or heads or around the borders of a document tells a reader, "Roll up your sleeves. This will be tough, unenjoyable reading."

2. **Inappropriate line length.** Excessively long lines are hard to read and signal to your audience that your work is highly complex and unrewarding. Overly short lines, however, make your work seem incomplete and immature.

3. Overuse of visuals. Too many visuals (boxes, rules, and clip art) can crowd your pages. Establish a balance between text and visuals. Never use visuals just for decoration or to fill space. Review the guidelines on using visuals on pages 466–467 and 468–472.

4. Mixed typefaces. The large number of fonts available in desktop publishing packages makes it easy to use too many of them. Be aware of this temptation, as the result often looks amateurish and disorganized. Don't use more than two type families on a page.

5. Hard-to-read fonts for body copy. Some fonts, such as Avant Garde, may look great in headlines but are difficult to read in the text of your document.

6. Small type and tight leading. Most typefaces work best for body copy when sized between 9 and 11 points. Leave plenty of line space (leading) between each line, too. A good rule of thumb is to set leading two or three points larger than the body text size (e.g., 12-point leading with 10-point type).

7. Too few or no heads and subheads. Without heads, Figure 11.1 is very difficult to follow. But as we saw in Figure 11.2, heads and subheads signal starting points and major divisions; they thus provide helpful landmarks for readers.

8. Excessive spacing. Too much space distorts the document and its message. Leaving three or four spaces between consecutive lines of text signals to readers that your ideas may be lightweight and not well researched. Also, including too much space around a visual or around the borders of your text can call into question the overall professionalism of your work.
To avoid such errors, follow these tips:

- Use just one space after a period.
- Don't indent the first line of text after a head or subhead.
- Avoid full justification with narrow columns of text.
- Eliminate excessive spacing in lists between the bullets, numbers, or symbols and the actual text entries.

9. Misused capitals, boldface, or italics. Printing an entire document in all capital letters makes it hard to read. (But printing heads or subheads in capitals will differentiate them from your text.) Similarly, avoid overusing boldface or italics. Not only will too many special effects make your work difficult to follow, but you will also lose the dramatic impact those attributes have.

Writing for and Designing Websites

In today's global marketplace, you can apply the skills you have just learned in designing documents and visuals to an online environment (websites, blogs, etc.).* In fact, companies often ask their employees to write for and prepare visuals for a corporate website, and while you won't be expected to construct such a website on

*Some of the material in this section comes from Michael Tracey, of Bay St. Louis, Mississippi, who designs websites and builds and upgrades computer hardware.

your own, you will likely be part of a team of specialists in IT, graphics, marketing, security, and engineering that is responsible for your firm's Web presence.

To be a helpful member of this team, you must keep up with the latest features of website design and understand how your company can incorporate them. You may also be asked to critique a competitor's website or to assist clients in preparing their own websites. This section of Chapter 11 stresses the principles and guidelines you need to follow to

- help readers navigate your website quickly, moving from one part to another easily
- ensure that your site is current
- make your site user-friendly for global readers as well as U.S. audiences
- create a visually attractive site
- keep your design and content ethical

The Organization of a Website: The Basics

The Web connects sets of information—pages—at millions of sites around the world. Some pages are quite long, while others take up just one computer screen. Websites can be created by individuals, corporations, government agencies, and so forth. Regardless of where it originates, a website has to arouse a reader's interest; show the appeal of a product, service, or topic; demonstrate the application of the product, service, or topic; and include a call to action.

A website can consist of one screen or many screens. In fact, a website for a government agency such as the Office of E-Government and Information Technology (www.whitehouse.gov/omb/e-gov) can translate into hundreds of pages of printed text. The home page is the thread that connects subsequent pages into a seamless presentation. To design and write an effective home page, you should always provide a menu that allows easy navigation to the rest of the site. Also, each page should include a menu button that links directly back to the home page, as illustrated below.

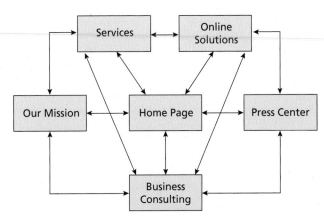

The home page is often but not always where visitors to a website will go first; if it catches their attention, they are likely to click on to your other pages. As you read the rest of this section, look at the sample home pages in Figures 11.5 through 11.8.

In the following sections of this chapter, we'll discuss the differences between a website and a printed document and help you to write for and contribute to the design of your own or your employer's website.

Case Study

Converting a Print Document for a Web Audience

Assume you work as an editorial assistant for the *FBI Law Enforcement Bulletin*, the journal that published the article "Virtual Reality: The Future of Law Enforcement Training" found in Figure 9.2 (pages 429–433). Because there has been great demand for copies of the article, your boss, the editor, has asked you to re-create this print document for a Web audience. It will have to meet the needs of a diverse group of readers—in law enforcement, community affairs, and education.

To do what your boss asks, you have to understand how a print document—the article—differs from a website. The article has thirteen sections presented in the following order:

1. Traditional Training Limitations
2. Training with Virtual Reality
3. What Is Virtual Reality?
4. How Does Virtual Reality Work?
5. Uses for Virtual Reality
6. Law Enforcement Training
7. Pursuit Driving
8. Firearms Training
9. High-Risk Incident Management
10. Incident Re-creation
11. Crime Scene Processing
12. Is Virtual Reality Virtually Perfect?
13. Physical Limitations and Effects

This is not how information looks on the Web or how it is processed by Web readers. A print document is usually read sequentially, page for page, whereas information on a website can be "navigated" using the search function or hyperlinks or both. This allows readers to customize their experience and focus on specific information. To convert the article into a website, you have to make it look like a Web document and organize it like one, as in Figure 11.5.

First, you will have to create a storyboard to decide how and what information is to be presented (see page 547), including a home page with a URL address, an eye-catching but professional image or photo, and navigational tools such as menus and hyperlinks. You might want to start the article on your home page and provide a "More" link to guide readers into your site.

You could convert each of the thirteen headings from the print article into hyperlinks, grouped under a navigation bar on the left side of the screen. Instead of turning pages, readers will click on the hyperlinks to jump to any section of the website. You could also transform the headings into tabs for a navigational bar along the top of the screen.

To allow visitors to comment on the site, you will need to provide a comment form or an email address for readers to send in their responses. You will also want to provide a link for contact information, so that interested readers can get in touch by mail, phone, or email.

FIGURE 11.5 The Home Page for a Website Based on the Virtual Reality Article in Chapter 9

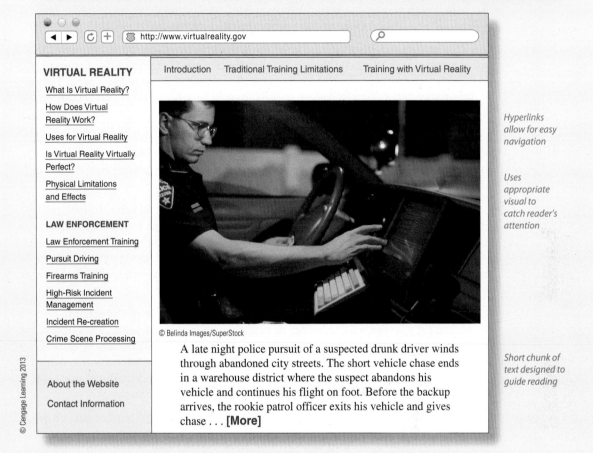

http://www.virtualreality.gov

VIRTUAL REALITY

What Is Virtual Reality?

How Does Virtual
Reality Work?

Uses for Virtual Reality

Is Virtual Reality Virtually
Perfect?

Physical Limitations
and Effects

LAW ENFORCEMENT

Law Enforcement Training

Pursuit Driving

Firearms Training

High-Risk Incident
Management

Incident Re-creation

Crime Scene Processing

About the Website

Contact Information

Introduction   Traditional Training Limitations   Training with Virtual Reality

© Belinda Images/SuperStock

A late night police pursuit of a suspected drunk driver winds through abandoned city streets. The short vehicle chase ends in a warehouse district where the suspect abandons his vehicle and continues his flight on foot. Before the backup arrives, the rookie patrol officer exits his vehicle and gives chase . . . **[More]**

*Hyperlinks
allow for easy
navigation*

*Uses
appropriate
visual to
catch reader's
attention*

*Short chunk of
text designed to
guide reading*

Web versus Paper Pages

Writing and designing a website requires the same emphasis on planning, drafting, formatting, and testing that preparing printed business documents does. Just as with these paper documents, writers must take into account spatial, rhetorical, and visual considerations when they prepare work for a website. Yet clearly, a website differs from a hard-copy memo, letter, or report in the way information is organized and visually illustrated.

Most important, websites are not fixed on the screen. They are designed so that readers can quickly scroll through them and click to related pages. Web pages can be changed, edited, and made available to a far greater audience far more quickly than any print source ever conceivably could be. Every website is really a work in progress. But in order to know how to write for a website, you need to understand how Web readers differ from print readers.

Web versus Print Readers

To help design and write for the Web, you need to recognize that websites are read differently from the way print documents are.

1. Readers do not generally go through a website word for word, hunting for information, carefully studying each sentence. They want to find information at a glance. On the average, Web readers spend 10 to 30 seconds scanning a page. They will not waste their time scrolling in the hopes of finally finding what they need. They'll just click to another site.

2. Web readers want articles, news stories, and features to be condensed and strategically arranged to give only essential information. Unlike in a print source, blocks of text are intimidating on the Web. Some experts advise using half the word count of a paper document for a website.

3. A Web audience will not necessarily read your entire website. In fact, they may not have even begun their search at your site. So you can't assume your Web audience will read from the start to the finish of your site. Don't write to build up a climax or postpone key facts. Web readers may not navigate through your pages in any predetermined order or even read all of its parts. They can start with the home page or a later page and then go to any other page on the site.

4. Text on the Web needs to be shorter because of the size of the screen. Text displayed on a screen will be harder to read than on a piece of paper. Lots of text, clip art, and images squeezed onto a webpage, coupled with the problem of glare on the computer screen, can lead to eyestrain and fatigue for Web readers more quickly than for print readers. Use shorter paragraphs with captions and headings, and ease readers' eye strain by inserting plenty of white space between chunks of text.

5. Web readers will expect navigational cues that readers of a print source do not have. A Web audience will be looking for visual markers, such as highlighted keywords to click on or bulleted lists, different colored text, commands (such as *search, click here, go to, go back, contact us*), arrows and crosses, or hyperlinks to other sites to visit. But do not simply insert "click here" without explaining why. Note how these cues are incorporated into the website in Figure 11.6.

Preparing a Successful Home Page

A home page has two main functions. First, it has to catch visitors' attention. If it fails to do that, everything else is a waste of time; your visitors will click their mouse and be gone. Second, it has to sell your company's product or service or promote your organization. That may seem backward, but unless you attract readers to your website, you cannot promote your business, share news, feature new products, and encourage shopping via your site.

Successful home pages accomplish these objectives clearly and effectively. For example, students applying for financial aid can find it a daunting process, but the design of the FinAid home page (Figure 11.6) keeps things upbeat and easy to follow. Even the image of the student "jumping for joy" contributes to the site's

FIGURE 11.6 Sample Home Page from FinAid

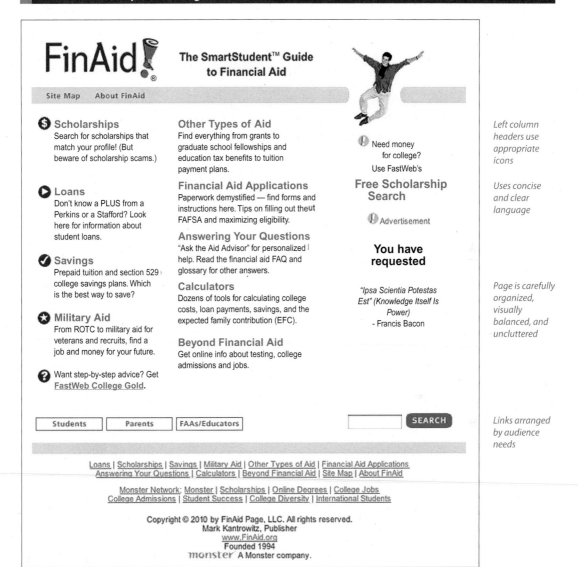

user-friendliness. The writing style and tone, for example, are appropriately conversational, friendly, and helpful. The FinAid home page also provides clear links to the various parts of the website. Similarly, the CDC home page in Figure 11.7 uses appropriate images for the "Health & Safety Topics" sections of the site—an airplane for "Travelers' Health" and a pictograph of a family for "Life Stages & Populations." The ICC home page in Figure 11.8 also includes relevant images for various divisions of the site.

FIGURE 11.7　The CDC Home Page with Examples of Navigational Links

Provides means to search site

Animated banner changes to highlight 5 different stories

Uses appropriate icons

Easy to navigate because of images and menus

White space ensures page does not look crowded

Writing is concise, clear, easy to understand

Courtesy of the CDC.

Designing and Writing for the Web: Eight Guidelines

Designing and writing for a website requires you to follow many of the rules you read about in Chapter 2 (pages 62–77) and earlier in this chapter. By adhering to the following eight guidelines, you can create an effectively designed and well-written website that will capture your audience's attention.

FIGURE 11.8 Home Page of the International Chamber of Commerce

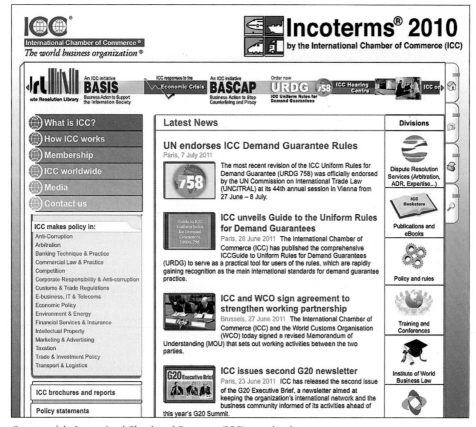

Courtesy of the International Chamber of Commerce (ICC), www.iccwbo.org.

Provides internal links and menus

Links provide access to additional information

Uses relevant, professional icons

Text uses brief chunks of prose and helpful images to break up page

Pleasing white space separates stories

Hyperlinks clearly marked

1. Make your site easy to find.

- Obtain a registered domain name for your site. Choose a professional domain name that clearly and quickly tells readers what you do. Avoid cute spellings and fanciful names, e.g., happihouse.com. Instead, use a clear, easy-to-remember name, e.g., toledohousepainters.com. A good place to start is the InterNIC website (www.internic.net), which provides updated information on Internet domain name registrations.

- Submit your website to the various search engines and directories to make sure the largest possible audience can find your site. See page 370 for some of the most useful search engines. Google, for instance, gives helpful tips.

- Make your site search-engine ready (see the Tech Note on page 546). To further optimize your chances of being listed by search engines, increase your crosslinking pages and update your site frequently.

2. **Make your site easy to navigate.**

- Help your visitors find their way easily through your site with logical and effective navigation tools such as

hyperlinks	search engines
navigation bars	button links
indexes and menus	rollover icons
site maps and tables of contents	previous, next, and back links

  Note how the CDC website in Figure 11.7 provides multiple navigation aids: clearly labeled sections, each illustrated with icons, and links to multimedia and tools.
- Test every link to make sure it is current, that it works, and that it connects to related sites. A website that does not identify its links or that contains a broken link will drive readers away.
- Don't overload your page with images, making it look crowded. See again how the FinAid website in Figure 11.6 uses a single image, six smaller icons, and boxes at the bottom left of the home page.

3. **Make your site informative.**

- Provide essential information on or through your home page, including your company's name, address, email, phone number, and corporate blog. The CDC website in Figure 11.7 offers audio and podcasts to visitors. The more helpful your site is, the more likely it will draw repeat visitors.
- Tell visitors what products or services you offer. Figure 11.6 offers tailor-made services to students, parents, and educators.
- Indicate what type of information can be obtained through your website, including links to your customer service and technical support. The FinAid website in Figure 11.6 easily locates information on scholarships, savings, military aid, and more.
- Offer readers different types of interaction—FAQs, bulletin boards, animated product demonstrations, and free email subscriptions. For example, the FinAid site offers dozens of useful calculators.

4. **Make your site easy to read for both native English speakers and international readers.**

- Put the most important point first in a seven- to eight-word headline (e.g., "The Smart Student Guide to Financial Aid," "World Malaria Day").
- Get to the point right away. Don't begin with a question or with background information.
- Write short descriptions of content—no more than three to four lines, as in the ICC home page in Figure 11.8 or the FinAid site in Figure 11.6.
- Provide headings with attention-grabbing keywords, bulleted lists, and numbered lists to help readers locate information quickly.
- Use plain, concise English—concrete nouns, action verbs. Keep sentences short, and use only the active voice. Write in a common sentence pattern: See the headlines and capsule summaries in Figure 11.8 or "Add a button or badge to your Web site" in Figure 11.7.

- Respect international readers by using, or converting to, units of measurement they are familiar with (see page 188).
- Include plenty of white space between sections.
- Insert scannable terms and hyperlinks; always highlight them to make them stand out.
- Select fonts that are easy to read (see page 530). Use larger fonts than you would for a print document.
- Select background colors that make your text easy to read. For example, don't use light yellow text against a white or light blue background or dark green lettering on a black background.

5. **Keep your site updated.**

- New information is vital for selling your product or service on a company website. Otherwise, readers have no reason to make return visits. Feature insider news updates about your business or preproduction information on products, services, community projects, and environmental efforts. Build in hyperlinks to product reviews, conferences, awards, and so on.
- Revise the design of your home page if your company offers a new product or service or a new promotion. Clearly, your site does not need a major design overhaul every week, but a new or revised home page alerts customers to the latest products and services. Note how the CDC home page in Figure 11.7 uses five different animated banners that change often to feature new events or programs that the CDC sponsors.
- Indicate when your site was last updated so readers will know your information is kept current.
- Periodically check your external links to make sure they work.

6. **Use images and icons effectively.**

- Arrange images and photos so they do not interfere with text or layout, as in Figure 11.7. Proportion is important for achieving a balance between different page elements.
- Choose appropriate icons or images to illustrate menus and page sections. See how Figure 11.6 uses easy-to-recognize icons for money and military aid, while each division of the ICC is identified in Figure 11.8 with an appropriate icon.
- Be conservative in using animations or anything that might be viewed as a gimmick because they may distract readers from other content on your page.
- Keep images proportional so that they are neither too big nor too small for the page.

7. **Encourage visitor interaction by soliciting feedback.**

- Ask readers to email you about your product, service, or website. Alert them to any relevant blog entries. And make sure a procedure for handling that feedback is developed within your organization.
- Include a feedback form or survey (see pages 346–353) with specifically targeted questions—including multiple-choice, pull-down menus, and comment boxes—about your website to encourage visitors to leave useful comments.

■ Conduct a usability test. Ask testers what they like about your website and what you could do better. Strive for interaction and connectivity.

8. **Make sure your website is ethical.**

■ Never post confidential or proprietary information.

■ Never post anything insulting or harassing, and never attack a competitor, a colleague, another department in your company, or a government agency.

■ Do not plagiarize from another Web (or print) source. If you include any information from another site—including quotations, visuals, or statistics—get permission, and acknowledge the source on your site.

■ Do not use sexist, racist, or other biased forms of language. Moreover, do not offend an international audience by using terms, names, or visuals that are insulting, stereotypical, or condescending. (See pages 505–507)

■ Never make false or exaggerated claims. Be honest and accurate. Earn your readers' trust.

■ Provide accurate and recent statistics and other data. Let readers know you have done your research.

Tech Note

Writing for Search Engines

Search engines do not "see" websites the way visitors do. They do not see designed pages or colorful graphics. Instead, they see only the HTML (hypertext markup language) source code, which is the "skeleton" of a website.

Because search engines see only the source code of a page, you need to include the most important keywords related to your website in the most likely places for a search engine to find them—near the top of your website in a section called the "head." Since search engines read websites from top to bottom, the keyword-containing text should appear in the "head" section and at the top of your page.

This holds true whether you are creating an electronic résumé (see pp. 290–295) or a company website. If your firm manufactures retractable awnings, you need to include the keywords *retractable* and *awnings* in the "head" tags and in the body text, thus proving definitively to the search engines that your page is indeed about "retractable awnings." Then, when a user visits a search engine to search for that keyword phrase, the likelihood of your site appearing toward the top of the rankings increases.

You will want to include your keyword phrases in the title, description, and keyword tags in the head section; the title and description tags are "selling" tools for your company. For instance, you might say "Discount Retractable Awnings" in your title tag to emphasize customer savings.

In the competitive online business world, the most important goal is to have customers, new and continuing, visit your company's website. Knowing how search engines work can ensure that customers can find your firm online.

Creating Storyboards for Websites and Other Documents

To help you plan the design of a website or a print document such as a brochure, newsletter, or even a handbook, you may find it useful to create a storyboard. The concept of *storyboarding* originated in the film industry as a way for directors to illustrate scenes in their films before actually shooting them. Storyboarding can be as simple as sketching what you want your webpage or other document to look like. Figure 11.9 illustrates successive versions of a storyboard for the home page of a website.

Acting as a map to your site or as a preliminary layout of a document, a storyboard gives you a clearer idea about the structure and navigation of each page of your website or document, and it allows you to plan the text and visuals. Like a draft of a report or proposal, your storyboard will become fuller the more you revise it—adding, moving, and linking content and graphics.

Here are some guidelines for effective storyboarding:

1. **Sketch the site or document pages on a piece of paper.** Label them according to various pages of the site or sections of the document, for example, Home Page, About Our Company, Contact Us, Place an Order.

2. **Plan the layout of each page.** Consider what content each page must have and how that information can be signaled through graphics. For instance, a home page should include your company's logo, a brief introduction to or history of the company, and information about the goods or services it offers.

3. **Determine basic design elements.** For instance, decide which fonts, backgrounds, and frames will look professional and appealing to customers. Then decide on size and height and where to place columns, images, sidebars, or highlighted areas.

4. **Build in navigational aids.** Make searching your website quick and effective. Always include hyperlinks to an email form so that readers can contact you. Ensure that your website is easy to navigate by drawing arrows on the storyboard between pages that should be linked together, as in Figure 11.9 (page 548). Every page should be linked to the home page. For brochures or other documents, make sure you insert clear and concise headings and subheadings, and provide contact information.

FIGURE 11.9 Successive Versions of a Storyboard for a Home Page

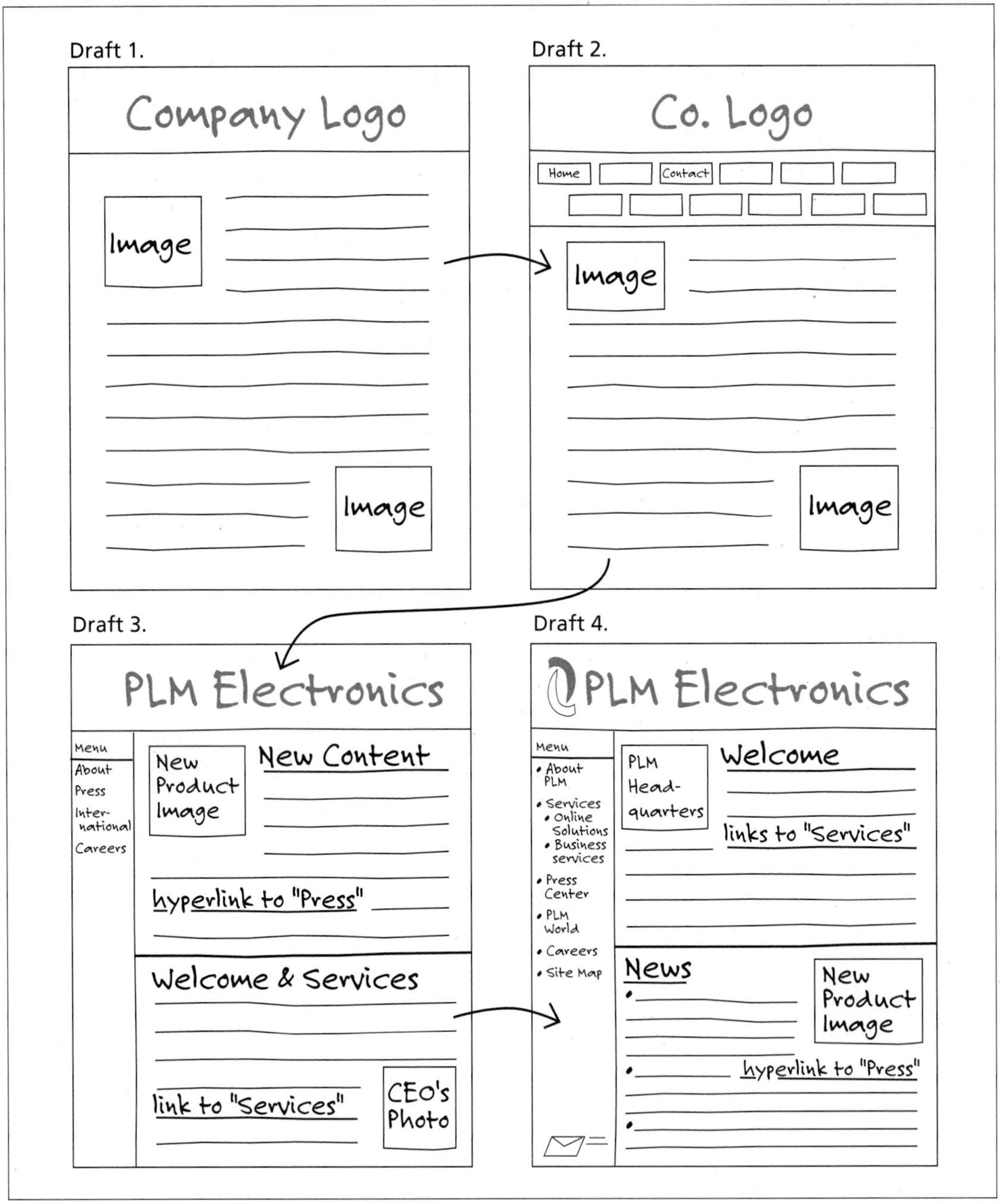

Draft 1.

Company Logo

Image

Image

Draft 2.

Co. Logo

Home   Contact

Image

Image

Draft 3.

PLM Electronics

Menu
About
Press
Inter-
national
Careers

New Product Image   New Content

hyperlink to "Press"

Welcome & Services

link to "Services"   CEO's Photo

Draft 4.

PLM Electronics

Menu
• About PLM
• Services
• Online Solutions
• Business services
• Press Center
• PLM World
• Careers
• Site Map

PLM Head-quarters   Welcome

links to "Services"

News
•
•
•
•
•

New Product Image

hyperlink to "Press"

PDFs and the Web

A Portable Document Format (PDF) file offers a variety of options to users. A PDF allows readers to see a snapshot of a document exactly as the original looks, complete with color, fonts, and visuals. As we saw, résumés are often sent as PDF files (see pages 290–291). Like a tracked document or a wiki, too, a PDF enables multiple readers to add comments to the document.

While navigating websites, you frequently encounter PDFs since they are so well suited for a Web environment. First, they allow documents originally written for print format to be easily posted on the Web without having to make complicated and difficult HTML conversions. Second, they can be saved as "read only" files and locked when posted online so that readers cannot make changes to the document. Having a fixed or set online form is especially helpful for companies or organizations who need a consistent and tested format. A website may have a link to a PDF for a set of instructions to download, or a link to a PDF with directions, or a page where you can download PDFs of various archival documents such as newsletters or company reports. Almost all restaurant websites post menus as PDFs. Third, PDFs can be modified to make them accessible to individuals with disabilities—that is, you can add audio, captions, and more, as the CDC does in Figure 11.7.

Converting documents into PDF is a fairly simple process and offers numerous options, including the ability to convert a batch of files into PDF simultaneously. You can create PDFs in Word, Excel, and most Web browsers via the "Print to PDF" function.

Four Rules of Effective Page Design: A Wrap-Up

The following four rules summarize the basic principles of effective document and Web design. If you adhere to them, your work, online or in print, will be professional looking.

1. Keep it simple. Do not overuse graphic effects and design styles in an effort to impress your audience; keep it simple so that you do not lose sight of the purpose of the document or website in an over-the-top design.

2. Be consistent. Use your layout and design elements consistently throughout your document and website. Repeat your design elements on each page, including fonts, text alignment, colors, and borders.

3. Make it clear. Make your message easy to read. Fancy designs may look good on paper, but they do not always add to the clarity of your document or website.

4. Remember that less is more. Limit the number of items on your page. An item can be a paragraph of text or a picture, as long as it provides a focal point on the page. Too much information on a Web or document page can make the document difficult to digest, and the information itself can be lost in the clutter.

Revision Checklist

Printed Documents

☐ Kept readers' busy schedules and reading time limits in mind when designing a document.

☐ Learned features of desktop publishing program.

☐ Arranged information in the most logical, easy-to-grasp order.

☐ Inserted heads and subheads to organize information for reader.

☐ Included only relevant visuals—clip art, icons, stock photos.

☐ Left adequate, eye-pleasing white space in text.

☐ Provided adequate margins to frame document.

☐ Justified margins, right, left, or center, depending on document and reader's needs.

☐ Maintained pleasing, easy-to-read line length.

☐ Kept line spacing consistent and easy on the readers' eyes.

☐ Chose appropriate typeface for message and document.

☐ Selected serif or sans serif font depending on message and readers' needs.

☐ Did not mix typefaces.

☐ Used effective type size, neither too small (under 10 point) nor too large (over 12 point), for body of text.

☐ Incorporated appropriate visual cues (for example, italics or boldface) for readers.

☐ Inserted heads and subheads to organize information for the reader.

☐ Made all heads and subheads parallel and grammatically consistent.

☐ Used lists, bullets, and numbers to divide information.

☐ Chose colors to make sure they look professional, contrast with background, and are appropriate for international readers.

Websites

☐ Planned location of text and visuals with a storyboard.

☐ Designed website so that it is easy to find on major search engines.

☐ Made sure navigation is clear and logical, not overly complex.

☐ Tested all links (internal and external) to make sure they are functioning so users can easily navigate the site.

☐ Identified all pages either with headings or with text that explains the purpose of each page.

☐ Ensured that the site is informative and relevant and that the content is current.

☐ Kept the site current by revising it frequently and including the most recent research.

☐ Used headings, subheadings, and white space to break information into readable chunks.

☐ Encouraged visitor interaction by soliciting feedback.

☐ Provided ways for reader to interact with the site, whether via email, a feedback page, or a blog where comments can be posted.
☐ Did not crowd images and text on the same page.
☐ Chose appropriate background colors so text is clear and easy to read and print.
☐ Made sure the site is ethical.

Exercises

1. Find an ineffectively designed print document—a form, a set of instructions, a brochure, a section of a manual, a catalog, a newsletter—and assume that you are a document design consultant. Write a sales letter (see pages 213–219) to the company or agency that prepared and distributed the document, offering to redesign it and any other documents they have. Stress your qualifications, and include a sample of your work. You will have to be convincing and diplomatic; precisely and professionally persuade your readers that they need your services to improve their corporate image, customer relations, and sales or services. Attach a copy of the document with your letter.

2. Bring two to three poorly designed documents to class. Working with a team of three or four students, determine which of the documents the group brought is the hardest to follow, the most unappealing, and the least logically arranged. After selecting that document, collaboratively write a memo to your instructor on what is wrong with the design and what you would do to improve its appearance and organization.

3. Then, as a group, redesign the document your group selected for Exercise 2. (Reformat it; add headings, spacing, and visual clues; include relevant clip art; and so on.) Submit both the original and the redesigned document to your instructor.

4. Redesign the document on page 552 to make it conform to the guidelines specified in this chapter. Reformat it; add headings, spacing, and visual clues; and provide a short introduction.

5. Redesign the document on page 553 to make it conform to the guidelines specified in this chapter.

A Poorly Designed Document for Exercise 4

WHY SHOULD YOU WASH YOUR HANDS?

Bacteria and viruses (germs) that cause illnesses are spread when you don't wash your hands.

If you don't wash your hands, you risk acquiring:

The common cold or flu

Gastrointestinal illnesses Shigella or hepatitis A

Respiratory illnesses

Should you wash your hands?

You need to wash your hands several times every day. Some important times to wash your hands are:

BEFORE

Preparing or eating food.

Treating a cut wound.

Tending to someone who is sick.

Inserting or removing contacts.

After

Using the bathroom.

Changing a diaper or helping a child use the bathroom (don't forget the child's hands)

Handling raw meats/poultry/eggs

Touching pets, especially reptiles

handling garbage

Sneezing or blowing your nose, or helping a child blow his/her nose

Touching any body fluids like blood or mucus

Being in contact with a sick person

Playing outside or with children and their toys

WHEN SHOULD YOU WASH YOUR HANDS?

There is a right way to wash your hands.

Follow these steps and you will help protect yourself and your family from illness.

Like any good habit, proper hand washing must be taught.

Take the time to teach it to your children and make sure they practice.

7

TTI's in the loop on effective detector placement

Ever sit in bumper-to-bumper traffic and wish they'd widen the roads so people could get through more quickly? Well, that costs a lot of money. Which is why transportation engineers who deal with traffic congestion and the problems it causes look for more cost-effective alternatives to get you where you're going—and faster.

TTI researchers recently completed a TxDOT/FHWA-sponsored study entitled *Effective Detector Placement for Computerized Traffic Management.* The research sought to expand and improve the use of inductance loop detectors (ILDs) to complement traffic signals, signal systems and other advanced traffic management systems. This is a cheaper congestion solution than building or widening a road.

An ILD is an electrical circuit containing a loop of copper wire embedded in the pavement. As a vehicle passes over the wire loop, it takes energy from the loop. If that change is large enough, a detection is recorded. Thus, we are able to collect data on the movement or presence of vehicles on the roadway. Advanced traffic management systems operate best with accurate information on how many vehicles are present and how fast they are traveling.

The primary goal of the recent project was to use loop detectors as an integral part of the congestion-reduction system. Traditional problems with ILDs were addressed—like crosstalk, or interference between two adjacent loops— and innovative new applications for ILDs in advanced traffic systems— like detecting wrong-way HOV-lane movements.

Other applications include using ILDs to move traffic more

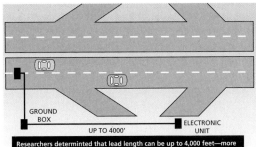

Researchers determined that lead length can be up to 4,000 feet—more than four times the accepted length on freeway entrance and exit ramps.

efficiently at diamond interchanges, at high-volume, high-speed approaches and on the freeway entrance ramps. The long-range contribution of the study is a set of guidelines for using ILDs in the situations listed above. As freeway management systems continue to evolve, the guidelines developed through the nine study reports will provide designers with practical information on the most effective placement of ILDs.

A major finding of the research deals with lead length, or the length of wire necessary to connect the loop to the detector electronic unit. The study showed that the loop can be placed more than 4,000 feet from the point of control—four times the currently accepted distance. This information will give traffic designers much more flexibility when integrating ILDs into their traffic system designs.

The researchers also made some important discoveries about using ILDs to measure speed. They found that the best speed trap is nine meters (two loops interconnected

with a timing device and spaced nine meters apart). They also determined that to get reasonably accurate and consistent speeds with an ILD, some things must be the same between a pair of loops: make, type or model of the detector units, sensitivity settings and loop configuration.

The findings from this research facilitate the use of loop detectors in managing traffic. And better management of driver frustration— just as important, even if less measurable than the congestion that causes it—is bound to follow.

Ultimately the three watchwords for this project were optimization, innovation, and implementation. Taking the tried-and-true and finding a better way to use it is, after all, the underlying building block for all engineering endeavors.

To order TTI Research Report 1392-9F, see the back page order form of this issue. For more information on loop detectors, contact Don Woods, 409/845-5792, FAX 409/845-6481 (E-mail: d-woods@tamu.edu).

Source: Texas Transportation Institute's Researcher; article author, Chris Pourteau. Reprinted with permission of the Texas Transportation Institute.

6. The home page of Stanley's Accounting Temps is ineffective. Write a one-page memo to your instructor specifying how Stanley's Accounting Temps might better address its online customers. Group your recommendations under the headings of content, design, and navigation. As a supporting document for your memo, design a new home page for the company by sketching it on a piece of paper or creating it on the computer.

A Poorly Designed Home Page for Exercise 6

7. Locate two websites that advertise a similar product or service. Analyze some of the elements each one uses, comparing the strengths and weaknesses of each site. Write a one-page memo to your instructor explaining which is the more effective site and why.

8. Find a website that you believe is ineffective. Using the four keys to effective writing (see pages 11–22), as well as your knowledge of Web design, write a one-page assessment of the site, discussing three or four changes you think would make it more effective. Attach a printed hard copy of the website with your assessment.

Writing Instructions and Procedures

Clear and accurate instructions are essential to the world of work. Instructions tell—and frequently show—how to do something. They indicate how to perform a specific task (draw blood; install new software); operate a machine (a pH meter; a digital camera); construct, install, maintain, monitor, adjust, or repair a piece of equipment (an incubator; a scanner). Everyone from the consumer to the specialist uses and relies on carefully written and designed instructions.

Access chapter-specific interactive learning tools, including quizzes and more in your English CourseMate accessed through www .cengagebrain.com.

Instructions, Procedures, and Your Job

As part of your job, you may be asked to write instructions, alone or with a group, for your co-workers or for the customers who use your company's services or products. When writing long, complex instructions, you certainly will be part of a team of engineers, programmers, document design experts, marketing specialists, and even attorneys. Your employer stands to gain or lose much from the quality and the accuracy of the instructions you prepare. The employees working at Shay Melka's company have to follow a host of instructions to collect and report business research (see Figure 8.1, pages 333–336).

While the purpose of writing *instructions* is to explain how to perform a task in a step-by-step manner, the purpose of writing procedures is slightly different. Often the two terms are incorrectly used interchangeably. *Procedures* refers to policies, duties, protocols, and guidelines that a business or organization expects its employees to follow. The email in Figure 1.1 (page 6), for example, outlines the procedures employees have to follow to receive tuition reimbursement for a course in business writing.

This chapter will first show you how to develop, write, illustrate, and design a variety of instructions. Then it will move into a discussion of writing procedures about job-related duties.

Why Instructions Are Important

Perhaps no other type of occupational writing demands more from the writer than do instructions because so much is at stake—for both you and your reader. The reader has to understand what you write and be able to perform the steps. You cannot afford to be unclear, inaccurate, or incomplete. Instructions are significant for many reasons, including safety, efficiency, and convenience.

Safety

Carefully written instructions get a job done without damage or injury. Poorly written instructions can cause an injury to the person trying to follow them and may result in costly damage claims or even lawsuits. Notice how the product labels in your medicine cabinet inform consumers how to take a medication safely. Without those instructions, consumers would be endangered by taking too much or too little medicine or by not administering it properly. To make sure your instructions are safe, they must be

- accurate
- consistent
- thorough
- clearly written
- effectively illustrated
- carefully organized

Your instructions also have to be legally proper. Companies have a legal and ethical obligation to prepare instructions that protect readers' safety. Instructions must

- specify what constitutes normal and proper use
- warn about misuse and identify potential risks and hazards
- describe under what conditions or settings a product can be used safely
- signal any cautions, risks, or dangers through prominently displayed symbols
- inform readers how to obtain further help or instruction if they have questions

Failure to provide such information in plain, clear language that readers can understand and easily follow is regarded by the courts to be as serious as manufacturing a defective product or not meeting code specifications. Several government agencies notify consumers about products that have been found to be unsafe. The U.S. Consumer Product Safety Commission (**www.cpsc.gov**) is a useful source for such information.

Instructions might also include boilerplate materials if available. Boilerplates are documents that your company has created and that can be revised, including warranties, guarantees, and specifications. Revise them periodically.

Efficiency

Well-written instructions help businesses run smoothly and efficiently. No work would be done if employees did not have clear instructions to follow. For example, without instructions on how to operate a piece of equipment, employees would not know how to get a job done. Imagine how inefficient it would be for a business if employees had to stop their work each time they did not have or could not understand a set of instructions. Equally alarming, what if employees made a number of serious mistakes because of confusing directions, costing a business sales, decreasing productivity, and increasing expenses? Giving readers helpful tips to make their work easier will increase their efficiency.

Convenience

Clear, easy-to-follow instructions make a customer's job easier and less frustrating. In the customer's view, instructions reflect a product's or service's quality and convenience. They can create goodwill or destroy it. How many times have you heard complaints about a company because its instructions were hard to follow? Poorly written and illustrated instructions will cost your customers time and you their business. Customers want instructions that are written in clear, plain language and that use photographs or simple drawings so they can assemble, install, or use a product right away. Instructions are also a vital part of "service after the sale." Owners' manuals, for example, help buyers avoid a product breakdown (and the headache and expense of starting over) and help them keep the product in good working order.

The Variety of Instructions: A Brief Overview

Instructions vary in length, complexity, and format. Some instructions are one word long: *stop, lift, rotate, print, erase.* Others are a few sentences long: "Insert blank disk in disk drive"; "Close tightly after using"; "Store in an upright position." Short instructions are appropriate for the numerous relatively nontechnical chores performed every day.

Detailed instructions may be as long as a page or a book. When your firm purchases a new server or a piece of earth-moving equipment, it will receive an instruction pamphlet or manual containing many steps, cautionary statements, and diagrams.

Instructions can be given in a variety of formats, as Figures 12.1–12.4 show. They can be paragraphs (Figure 12.1), employ visuals to illustrate each step (Figure 12.2), or use a numbered list (Figures 12.3 and 12.4). You will have to determine which format is most appropriate for the kinds of instructions you write. For writing that affects major policies or regulations, as in Figure 4.2 (page 135) or Figure 12.6 (page 564), you will most often send a hard-copy memo, which serves as a more permanent, legal notice than would an email.

Instructions Online

Many instructions are given online. If you purchase a new PC, for instance, online instructions will guide you in setting up and registering your computer. Websites for products and services often include hyperlinks to "help screens" that give consumers information on assembly and use. The popular Energy Star website, created by the U.S. Department of Energy and seen in Figure 12.5 (page 563), contains links to instructions for consumers on how to assess a home's energy needs and how

FIGURE 12.1 Instructions in Paragraph Form

Easy Temporary Join for Synthetic Ropes

Includes helpful visual

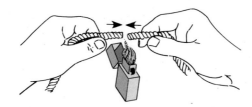

Directions written in clear, plain language

If you are faced with the problem of reeving a new halyard on a flagpole or mast, or through a block or pulley in an inaccessible location, the solution can be easy if both old and new lines are made of nylon or polyester (Dacron, Terylene, etc.). Simply join the ends of the old and new lines temporarily by melting end fibers together in a small flame (a little heat goes a long way). Rotate the two lines slowly as the fibers melt. Withdraw them from the flame before a ball of molten material forms, and if the stuff ignites, blow out the flame at once. Hold the joint together until it is cool and firm.

Source: R. I. Standish, *Parks*.

to insulate, heat, and cool it efficiently. The long set of instructions on installing a printer in Figure 12.12 combines print and online instructions.

Instructions are also provided online through videos on YouTube and other sites that actually show viewers how to assemble or install a product or perform another task. Someone talks you through each step, alerts you to potential problems, and gives helpful tips at various stages of the process. One smoke detector manufacturer, for example, hired a New Jersey fire chief to demonstrate how to properly assemble, locate and mount, maintain, and test its product. Viewers can watch videotaped instructions before attempting a task and can replay the video to double-check a step before going on. But remember that clear writing and careful planning are as essential in videotaped instructions as in printed ones.

Assessing and Meeting Your Audience's Needs

To assess your audience's needs, put yourself in your readers' position. In most instances, you will not be available for readers to ask you questions when they do not understand something. Consequently, they will have to rely only on your written instructions.

Do not assume that your readers have performed the process before or have operated the equipment as many times as you have. (If they had, there would be no

FIGURE 12.2 Instructions That Supply a Visual with Each Step

Proper Brushing

Proper brushing is essential for cleaning teeth and gums effectively. Use a toothbrush with soft, nylon, round-ended bristles that will not scratch and irritate teeth or damage gums.

1

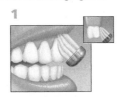

Place bristles along the gumline at a 45-degree angle. Bristles should contact both the tooth surface and the gumline.

Uses easy-to-follow steps

2

Gently brush the outer tooth surfaces of 2–3 teeth using a vibrating back and forth rolling motion. Move brush to the next group of 2–3 teeth and repeat.

Visuals help readers to follow directions

3

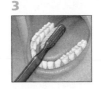

Maintain a 45-degree angle with bristles contacting the tooth surface and gumline. Gently brush, using back, forth, and rolling motion along all of the inner tooth surfaces.

Begins each step with strong, active verbs listed in color

4

Tilt brush vertically behind the front teeth. Make several up and down strokes using the front half of the brush.

Offers helpful hints

5

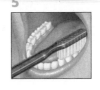

Place the brush against the biting surface of the teeth and use a gentle back and forth scrubbing motion. Brush the tongue from back to front to remove odor-producing bacteria.

Explains why a step is important

© Cengage Learning 2013

Source: Reprinted by permission of American Dental Hygienists' Association. Illustrations adapted and used courtesy of the John O. Butler Company, makers of *GUM* Healthcare products.

need for your instructions.) No one who has written a set of instructions ever disappointed readers by making directions too clear or too easy to follow. Remember, too, that your audience will often include non-native speakers of English, a worldwide audience of potential consumers.

How to Copy Files to a USB Flash Drive from Your PC or Notebook

Follow these instructions to copy your files to a flash drive from your PC or notebook. Refer to the photo of a USB flash drive below as you perform these instructions:

1. Insert the USB flash drive into a USB portal of your PC or notebook.

2. Find the folder or file to be copied to the USB flash drive, and right-click on it. NOTE: The folder or file will be highlighted, and a menu with "Open" at the top will appear.

3. Within the menu, move your cursor down to the "Send To" option. Here you will see a list of locations where you may send the selected folder or file.

4. Choose the USB flash drive location, and your folder or file will be automatically copied over. CAUTION: DO NOT REMOVE THE USB FLASH DRIVE AT THIS POINT, OR YOU WILL RISK DAMAGING IT.

5. Go to "My Computer" from the "Start" menu, and double-click on the USB flash drive. If the folder or files you selected in Step 2 are listed here, the copying was successful.

6. Eject the USB flash drive before removing it from the computer. To do so, go to "My Computer" again, right-click on the USB flash drive, and select the "Eject" option from the menu.

7. Remove the USB flash drive from the USB portal.

Uses numbered steps

Strong, active verbs give readers clear directions

Provides visual to assist readers

Inserts "Caution" at proper place

Tells reader how to determine if he/she did step accurately

Microsoft Clip Art

© Cengage Learning 2013

Key Questions to Ask About Your Audience

The more you know about your audience, the better your instructions can be. To determine your readers' needs, ask yourself the following questions:

- How and why will my readers use my instructions? (Co-workers, consumers, and experts in the field all have different expectations.)
- What language skills do they possess? Is English their first (native) language?
- How much do my readers already know about the product or service?
- How much background information will I have to supply?
- What steps will most likely cause readers trouble?
- What types of visuals do I need to supply (photographs for consumers; diagrams and spec sheets for technicians)?
- How often will readers refer back to my instructions—every day or just as a refresher?

- Where will my audience most likely be following my instructions—in the workplace, outdoors, in a workshop equipped with tools, or in their homes?
- What resources, such as special equipment or power sources, will my readers need to perform my instructions successfully?

Writing Instructions for International Audiences

Your instructions will often be aimed at a worldwide audience of potential customers, many of whom do not use English as their first language and may be following your instructions through a translation of them into their native language. Here are some useful guidelines when writing instructions for this diverse group of readers:

1. **Write in plain, simple language.** Use international English (see pages 183–199). Keep your sentences short. Use the active voice. Select words that have clear, unambiguous meaning. Do not use different words for the same thing.

2. **Use terms and units of measurement that your readers will understand.** Avoid abbreviations, acronyms, and jargon, and don't assume that international readers will use or understand the units of measurement common in the United States.

3. **Make sure all of your visuals are clear and culturally appropriate.** In some instances, your instructions may be given exclusively through visuals. (See pages 569–570 for some guidelines.)

4. **Be aware that colors can have different meanings.** Red, yellow, and green, for instance, may not convey the same meaning in other cultures that they do in the United States. (Refer to page 506.)

5. **Ask a non-native speaker to review your instructions.** This review will help ensure that your instructions are easy to understand because you used clear sentences and appropriate language.

Using Word-Processing Programs to Design Instructions

To make sure your instructions are easy to read, both for native as well as non-native speakers of English, take advantage of the following word-processing features, which will help you draft, revise, and format:

1. **Brainstorming and clustering to get ideas and steps down.** Find and compile the information you will need to include in the steps of your instructions, and think ahead about possible warnings, cautions, and any helpful hints that users may need to complete the steps (see pages 53–54).

2. **Use the Outline feature.** This feature of word-processing programs makes it easy to identify, order, and change the steps in a set of instructions, allowing you to try out different options quickly and easily.

FIGURE 12.4 Instructions in a Numbered List on How to Repair a Leaky Faucet

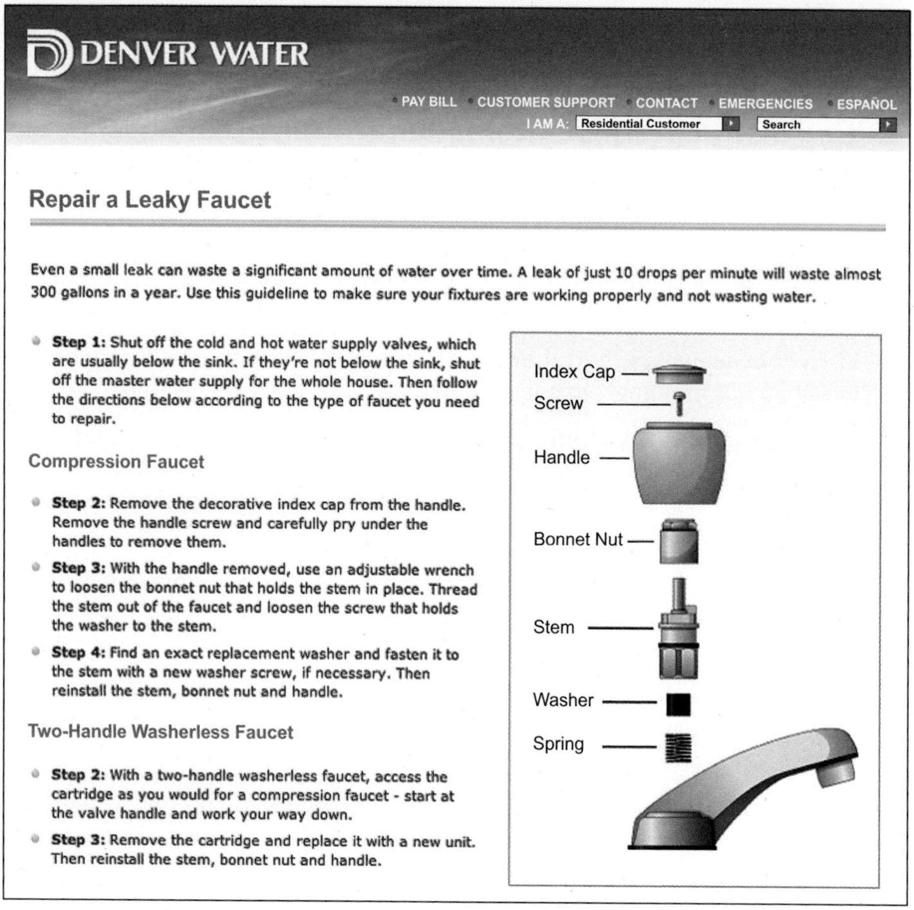

DENVER WATER

PAY BILL · CUSTOMER SUPPORT · CONTACT · EMERGENCIES · ESPAÑOL

I AM A: Residential Customer | Search

Repair a Leaky Faucet

Even a small leak can waste a significant amount of water over time. A leak of just 10 drops per minute will waste almost 300 gallons in a year. Use this guideline to make sure your fixtures are working properly and not wasting water.

- **Step 1:** Shut off the cold and hot water supply valves, which are usually below the sink. If they're not below the sink, shut off the master water supply for the whole house. Then follow the directions below according to the type of faucet you need to repair.

Compression Faucet

- **Step 2:** Remove the decorative index cap from the handle. Remove the handle screw and carefully pry under the handles to remove them.
- **Step 3:** With the handle removed, use an adjustable wrench to loosen the bonnet nut that holds the stem in place. Thread the stem out of the faucet and loosen the screw that holds the washer to the stem.
- **Step 4:** Find an exact replacement washer and fasten it to the stem with a new washer screw, if necessary. Then reinstall the stem, bonnet nut and handle.

Two-Handle Washerless Faucet

- **Step 2:** With a two-handle washerless faucet, access the cartridge as you would for a compression faucet - start at the valve handle and work your way down.
- **Step 3:** Remove the cartridge and replace it with a new unit. Then reinstall the stem, bonnet nut and handle.

Starts with helpful information about the importance of the instructions

Easy-to-follow steps

Exploded drawing shows relationship of parts to one another making disassembling easier

Recognizes differences in faucets

Courtesy of Denver Water.

3. **Choose a font that is easy to read.** Avoid fonts that may contain unusual lettering styles. Fonts like Arial or Helvetica provide clean and easy-to-read text for your instructions.

4. **Use numbered and bulleted lists.** Numbered steps help readers follow the sequence of your instructions, and bulleted lists break up text to make it easier to read.

5. **Provide adequate spacing.** Always double-space between steps in your instructions to set each one apart and to make reading and following the steps easier. Do not indent steps. Put any notes, cautions, or warnings right next to the step to which they pertain.

6. **Use boldface sparingly.** Use it to emphasize warnings, cautions, and notes (see page 531) so readers will not overlook the crucial messages they contain.

FIGURE 12.5 A Website with Links to Online Instructions on Making a Home More Energy Efficient

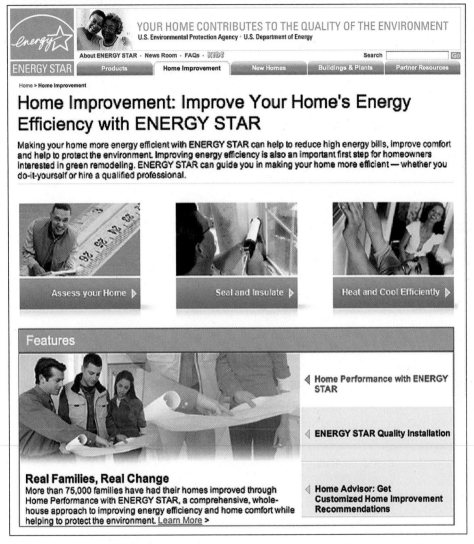

YOUR HOME CONTRIBUTES TO THE QUALITY OF THE ENVIRONMENT
U.S. Environmental Protection Agency · U.S. Department of Energy

About ENERGY STAR · News Room · FAQs · KIDS   Search [   ] [Go]

ENERGY STAR | Products | Home Improvement | New Homes | Buildings & Plants | Partner Resources

Home > Home Improvement

Home Improvement: Improve Your Home's Energy Efficiency with ENERGY STAR

Making your home more energy efficient with ENERGY STAR can help to reduce high energy bills, improve comfort and help to protect the environment. Improving energy efficiency is also an important first step for homeowners interested in green remodeling. ENERGY STAR can guide you in making your home more efficient — whether you do-it-yourself or hire a qualified professional.

Assess your Home ▷   Seal and Insulate ▷   Heat and Cool Efficiently ▷

Features

Home Performance with ENERGY STAR

ENERGY STAR Quality Installation

Real Families, Real Change
More than 75,000 families have had their homes improved through Home Performance with ENERGY STAR, a comprehensive, whole-house approach to improving energy efficiency and home comfort while helping to protect the environment. Learn More >

Home Advisor: Get Customized Home Improvement Recommendations

Hyperlinks to tips, checklists, and diagnostic tools

Page designed for easy access to online instructions

Courtesy of EnergyStar.gov/EPA.

7. **Be careful about the images you take from a clip art library.** Use only icons that will be immediately recognizable to a broad range of readers. Also, make sure the images are the right size—neither so big that they look frivolous nor so small that readers cannot see them clearly.

8. **Avoid the use of underlining for emphasis.** Underlined text may introduce confusion to your readers because it could be interpreted as a hyperlink.

9. **Use hyperlinks within instructions when necessary.**

Meeting Your Audience's Needs

The instructions contained in Figures 12.6 and 12.7 are addressed to two different audiences, each with separate needs. The Hercules memo in Figure 12.6 was sent to a technical audience— firefighters and supervisors—who needed instructions on a special process. Cliff Burgess's memo in Figure 12.7, however, went to all Wilcox employees, a more diverse group of readers. His helpful instructions do not require a list of equipment or materials or a complicated description of steps. Instead, his memo consists of an introduction, three bulleted instructions, a conclusion, and a nontechnical visual for his audience.

FIGURE 12.6   Instructions Alerting a Technical Audience to Special Circumstances

HERCULES

TO: All Shift Supervisors
              All Firefighters
FROM: Robert Ferguson   *R.F.*
SUBJECT: Operating Procedures During Energy Shortages
DATE: October 10, 2011

Explains why the instructions are important

The following policy has been formulated to help in maintaining required pressures during periods of low wood flow and severe natural gas curtailment.

Identifies conditions

All boilers are equipped with lances to burn residue or no. 6 fuel oil as auxiliary fuels. When wood is short, no. 6 oil should be burned in no. 2 and no. 3 boilers at highest possible rate consistent with smoke standards. To do this, take these steps:

Lists directions in numbered steps

1. Put no. 6 oil on no. 3 boiler lances.
2. Shut down overfire air.
3. Shut down forced draft.
4. Turn off vibrators.
5. Keep grates covered with ashes or wood until ash cover exists.

Alerts readers about what to expect

These steps will result in an output of 25,000 to 35,000 lbs./hr. steam from no. 3 boiler and will force wood on down to no. 4 boiler.

If necessary, the same procedure can be repeated on no. 2 boiler.

Source: Reprinted by permission of Hercules, Inc.

FIGURE 12.7 An Instructional Memo Listing Safety Precautions for a General Audience

WILCOX WORLDWIDE SYSTEMS

www.wilcox.com

TO: All Wilcox Employees
FROM: Cliff Burgess, Environmental Safety C.B.
RE: Precautions in Safely Using Video Display Terminal (VDT)
DATE: September 12, 2012

You may experience some possible health risks in using your computer video display terminal (VDT). These risks include sleep disorders, behavioral changes, danger to the reproductive system, and cancer.

The source of any risks comes from the electromagnetic fields (EMFs) that surround anything that carries an electric current—for example, copiers, circuit breakers, and especially VDTs. Your computer monitor is a major source of EMFs. Magnetic fields can go through walls as easily as light goes through glass.

Although EMFs may affect your health, you can considerably reduce your exposure to these fields by following these three simple steps:

- **Stay at least three feet (an arm's length) away from the front of your VDT.** (The magnetic field is significantly reduced with this amount of distance.)

- **Keep at least four feet away from the sides and back of someone else's VDT.** (The fields are weaker in the front of the VDT but much stronger everywhere else.) Refer to the following sketch, which you may want to post in your office.

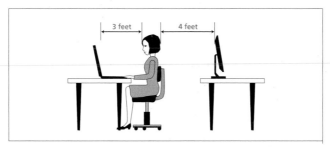

- **Switch your VDT/computer off when you are not using it.** (If the computer has to remain on, be sure to switch off the monitor; screen savers do not affect the exposure to **EMFs**.)

Our environmental safety team will continue to monitor and investigate any problems. Observing the guidelines above, however, will help you to take the necessary precautions in order to minimize your exposure to EMFs.

To view a demonstration of proper safety techniques, go to **www.wilcox .com/vdtsafety/**. Thank you.

**1215 Madison • St. Louis, Missouri 63174
314-555-4300 • FAX 314-555-4311**

Precise subject line

Introduction emphasizes reasons for instructions

Nontechnical explanation offers an analogy

Steps stand out through bullets, boldface, and spacing

Simple visual clarifies instructions

Conclusion reassures readers they are acting safely

Provides helpful URL

The Process of Writing Instructions

As we saw in Chapter 2, clear and concise writing evolves when you follow a process. To make sure your instructions are accurate and easy for your audience to perform, you must plan your steps, perform a trial run, write and test your draft, and revise and edit.

Plan Your Steps

Before writing, do some research to understand completely the process you are asking someone else to perform. Make sure you know

- the reason for doing something
- the parts or tools required
- the steps to follow to get the job done
- the results of the job
- the potential risks or dangers

If you are not absolutely sure about the process, ask an expert for a demonstration. Do some background reading and talk to or email colleagues who may have written or followed a similar instruction.

Perform a Trial Run

Actually perform the job (assembling, repairing, maintaining, dissecting) yourself or with your writing team. Go through a number of trial runs. Take notes as you go along, and be sure to divide the job into simple, distinct steps for readers to follow. Don't give readers too much to do in any one step. Each step should be *complete*, *sequential*, *reliable*, *straightforward*, and *easy* for your audience to understand and perform.

Write and Test Your Draft

Transform your notes into a draft (or drafts) of the instructions you want readers to follow. Then conduct a *usability test* by asking individuals from the intended audience (consumers, technicians) who have never performed the job to follow your instructions as you have written them. Ask participants to read your instructions aloud and ask questions. Observe where they run into difficulty—or get results different from yours.

Revise and Edit

Based on your observations and user feedback, revise your instructions to avoid

- missing steps
- too many activities in one step
- steps that are out of order
- unclear or incomplete steps

Edit the final copy of the instructions that you will give to readers. Consider whether your instructions would be easier to accomplish if you included visuals.

Analyzing the needs and the background of your audience will help you to choose appropriate words and details. A set of instructions accompanying a child's chemistry set would use different terminology, abbreviations, and level of detail than would a set of instructions a professor gives a class in organic chemistry.

General Audience:	Place 8 drops of vinegar in a test tube with a piece of limestone about the size of a pea.
Specialized Audience:	Place 8 gtts of CH_3COOH in a test tube, and add 1 mg of CO_3.

The instructions for the general audience avoid the technical abbreviations and symbols the specialized audience requires. If your readers are puzzled by your directions, you defeat your reasons for writing them.

Using the Right Style

To write instructions that readers can understand and turn into effective action, observe the following guidelines.

1. Make sure verbs are in the present tense and imperative mood. Imperatives are commands without the pronoun *you*. Note how the instructions in Figures 12.1 through 12.4 contain imperatives—"Reinstall the stem" instead of "You reinstall the stem." In instructions, deleting the *you* is not discourteous, as it would be in a business letter or report. The command tells readers, "These steps work, so do them exactly as stated." Choose imperative verbs such as those listed in Table 12.1.

2. Write clear, short sentences in the active voice. Keep sentences short and uncomplicated. Sentences under twenty words (preferably under fifteen) are easy to read. Note that the sentences in Figures 12.1 through 12.4 are, for the most part, under fifteen words. But do not omit articles (*a*, *an*, *the*) or any connective words (such as *and*, *but*, and *however*), which will make your instructions harder to follow.

3. Use precise terms for measurements, distances, and times. Indefinite, vague directions leave users wondering whether they are doing the right thing. Avoid vague words such as *frequently*, *occasionally*, *probably*, and *possibly*. The following vague direction is better expressed through precise revision.

Vague:	Turn the distributor cap a little. (*How much is a little?*)
Precise:	Turn the distributor cap one quarter of a rotation.

4. Include connective words as signposts. Connective words specify the exact order in which something is to be done (especially when your instructions are written in paragraphs). Words such as *first*, *then*, and *before* help readers stay on course, reinforcing the sequence of the procedures.

5. Number each step when you present your instructions in a list. You also can use bullets. Plenty of white space between steps also distinctly separates them for the reader.

TABLE 12.1 Some Helpful Imperative Verbs Used in Instructions

add	determine	hold	pass	rotate	tear
adjust	dig	include	paste	rub	thread
apply	display	increase	peel	run	tie
attach	double-click	insert	pick up	save	tighten
back up	download	inspect	plug	scan	tilt
blow	drag	install	point	scroll	trace
boldface	drain	lift	pour	scrub	transect
call up	drill	link on	press	select	transfer
change	drop	load	prevent	send	trim
check	ease	log on	print	set	turn
choose	eject	loosen	provide	shake	twist
clean	eliminate	lower	pry	shift	type
click	enter	lubricate	pull	shut off	unplug
clip	exit	maintain	push	slide	use
close	fasten	measure	raise	slip	ventilate
connect	find	mix	reboot	spread	verify
contact	flip	mount	release	squeeze	wash
copy	flush	move	remove	start	weigh
cover	follow	navigate	replace	stop	wind
create	forward	notify	reply	strain	wipe
cut	gather	oil	review	switch	wire
delete	group	open	roll	tab down	wrap

Using Visuals Effectively

Readers welcome visuals in almost any set of instructions. Visuals help readers get a job done more quickly and increase their confidence. A visual can help users

- locate parts, areas, and so on
- identify the size and placement of parts
- understand how to assemble parts effectively and easily
- illustrate the right and wrong way to do something
- identify possible sources of danger, injury, or malfunction
- determine whether a problem is serious or minor
- recognize normal and safe limits or ranges

The number and kinds of visuals you include will, of course, depend on the process or equipment you are explaining and on your audience's background and needs. Some instructions may require only one or two visuals. The instructions in

Figure 12.3 tell users how to copy files using a USB flash drive. The illustration clearly shows what a USB flash drive looks like. In Figure 12.2, each step is accompanied by a visual demonstrating a proper technique for brushing teeth.

One frequently used visual in instructions is an exploded drawing, like the one in Figure 12.4, which helps consumers see how various parts of a faucet fit together, or the one in Figure 12.8, which labels and shows the relationship of the parts of an industrial extension cord.

Take a look at the sample documents throughout this book to see further examples of how these and other visuals help readers perform a work-related task: photographs (pages 489–492), flow charts (pages 484–485), drawings (pages 492 and 494), clip art (pages 494–496), maps (pages 488–489), pictographs (pages 485–487), tables (pages 472–474), graphs (pages 475–478), and PowerPoints (pages 736–739).

Guidelines for Using Visuals in Instructions

Follow these guidelines to use visuals effectively in your instructions:

1. Set visuals off with white space so they are easy to find and read.
2. Place each visual next to the step it illustrates, not buried at the bottom of the page or on another page.

FIGURE 12.8 Exploded Drawing Showing How to Assemble an Industrial Extension Cord

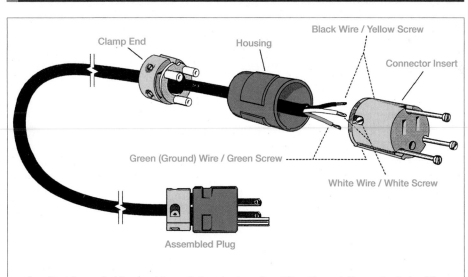

1. Run the end of the cord through the clamp end and then through the center hole of the housing.
2. Pull the cord through until it extends 2 inches beyond the housing.
3. Strip about $1\frac{1}{4}$ inches of outer insulation from the end of the cord.
4. Twist the exposed ends to prevent stray strands.

Source: Drawing courtesy of Sally Eddy.

3. Select a visual that is appropriate for your audience. For example, a photo of a person demonstrating a proper stretch technique is sufficient for a general audience, which does not need an elaborate medical illustration of the muscular system.

4. Assign each visual a number (Figure 1, Figure 2), and refer to visuals by figure number in your instructions.

5. Make sure the visual looks like the object the user is trying to assemble, maintain, run, or repair. Don't use a photo of a different model.

6. Always inform readers if a part is missing or is reduced in your visual.

7. Where necessary, label or number parts of the visual, as in Figure 12.4.

Refer to Chapter 10 for further guidelines on numbering visuals (pages 468–469) and inserting them in your document (pages 469–470).

The Six Parts of Instructions

Except for very short instructions, such as those illustrated in Figures 12.1 through 12.4, or for procedures related to policies or regulations (see Figure 12.13, pages 595–596), a set of instructions generally contains six main parts: (1) an introduction; (2) a list of equipment and materials; (3) the actual steps to perform the process; (4) warnings, cautions, and notes; (5) a conclusion (when necessary); and (6) a troubleshooting guide. The long set of instructions on installing a printer in Figure 12.12 contains most of these.

Introduction

The function of your introduction is to provide readers with enough *necessary* background information to understand why and how your instructions work. An introduction must make readers feel comfortable and well prepared before they turn to the actual steps.

What to Include in an Introduction

Not every introduction to a set of instructions will contain all five categories of information listed here. Some instructions will require less detail. You will have to judge how much background information to give readers for the specific instructions you write. Refer to Figure 12.9, which is an introduction to a guide for nurses who have to know how to use an infusion pump, as you read the following guidelines:

1. **State why the instructions are useful for a specific audience.** Many instructions begin with introductions that stress safety, educational, or occupational benefits. Here is an introduction from a set of safety instructions describing protective lockout of equipment.

> The purpose of these instructions is to provide plant electrical technicians with a uniform method of locking out machinery or equipment. This will prevent the possibility of setting moving parts in motion, energizing electrical lines; or opening valves while repair, setup, or cleaning work is in progress.

Note how Figure 12.9 highlights the safety and convenience of the equipment, helping nursing staff meet their patients' needs.

2. Indicate how a particular piece of equipment or process works. An introduction can briefly discuss the "theory of operation" to help readers understand why something works the way your instructions say it should. Such a discussion sometimes describes a scientific law or principle. An introduction to instructions on how to run an autoclave begins by explaining the function of the machine: "These instructions will teach you how to operate an autoclave, which is used to sterilize surgical instruments through the live additive-free stream."

The introduction in Figure 12.9 describes the different modes of the infusion pump and the pump itself.

3. Point out any safety measures or precautions a reader may need to be aware of. By alerting readers early in your instructions, you help them perform the process much more safely and efficiently. Again, the introduction in Figure 12.9 cautions the nursing staff about an audible alarm signal in the event of a malfunction.

4. Stress any advantages or benefits the reader will gain by performing the instructions. Make the reader feel good about buying or using the product by explaining how it will make a job easier to perform, save the reader time and money, or allow the reader to accomplish a job with fewer mistakes or false starts. Nurses learn from the introduction in Figure 12.9 that the infusion pump can be quickly programmed.

Note how the following introduction to a set of instructions for an all-in-one scanner encourages the reader to learn how to operate this system.

> Congratulations on choosing the MegaScan 5000 all-in-one laser copier, scanner, and fax to efficiently handle all of your electronic document needs. Wireless-ready and 2 terabytes of storage capacity. You can scan 40 documents per minute and perform document setup with easy to follow step-by-step instructions from a dialog box on your computer screen.

5. Provide hyperlinks. When readers will be following your instructions online, provide hyperlinks to any sites or materials they need to know about to follow your instructions. Note how the Energy Star website (Figure 12.5) includes hyperlinks for related resources, such as home improvement FAQs and insulation manufacturers. Similarly, provide relevant cross-references in printed instructions.

List of Equipment and Materials

Clearly, some instructions, as in Figure 12.1, do not need to inform readers of all equipment or materials they will need. But when you do, make your list complete and clear. For example, if a Phillips screwdriver is essential to complete one step, specify that type of screwdriver under the heading "Equipment and Materials"; do not list just "screwdriver." The "Tools Necessary" section of a manual in Figure 12.10 (page 573) shows an enlargement of a Phillips head screwdriver and clearly identifies, verbally and visually, other tools required to do the job.

Do not wait until readers are actually performing one of the steps to tell them that a certain type of drill or a specific kind of chemical is required. They may have to stop what they are doing to find the equipment or tool; moreover, the procedure may fail or present hazards if users do not have the right equipment at the right time.

FIGURE 12.9 Introduction to a Guide for Using an Infusion Pump

1Overview
Orientation

Gives function of equipment

Explains time-saving feature

Explains security option

Describes different modes or options

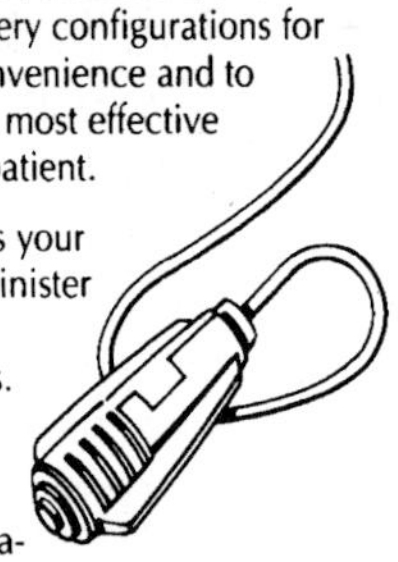

Explains convenience features

Emphasizes safety features

The LifeCare PROVIDER 5500 System is a portable infusion pump, specially designed to deliver analgesic drugs, antibiotics, and chemotherapeutics.

The pump can be programmed in either milligrams or cubic centimeters, and in four different delivery configurations for greater nursing convenience and to tailor precisely the most effective regimen for each patient.

Bolus Mode allows your patient to self-administer analgesia within programmed limits.

This is the traditional PCA delivery, "analgesia-on-demand," based on the patient's need.

Continuous Mode delivers a continuous "background" infusion with no additional PCA doses permitted.

Continuous-plus-Bolus Mode allows the patient to self-administer a Bolus dose in addition to receiving a simultaneous Continuous dose infusion.

Intermittent Mode delivers a specific dose (in cc or mg) at intermittent intervals over 24 hours.

You can also establish the "lockout" interval, the frequency with which a patient may receive a Bolus dose of analgesic drug.

The PROVIDER 5500 System records all settings in memory and can be quickly re-programmed to save nursing time when repeating established protocols or changing fluid reservoirs.

The portable system operates on battery power.

To minimize tampering and discourage theft, there is an optional locking security lockbox that also secures the system to an IV pole.

The audible alarm signals in the event of a malfunction, and the digital readout describes the malfunction.

● Compact and lightweight.

● Delivery rates between 0.1 cc and 250 cc per hour, in 0.1-cc increments.

Display Panel

● Individual display indicators appear only during programming and operation.

● Only on a *selective* basis.

● Tone sounds when activated.

● Runs on BATTERY POWER ONLY.

● Disposable Primary IV set with integral infusion cartridge.

Source: Reprinted by permission of Abbott Laboratories Hospital Products Division.

Steps for Your Instructions

The heart of your instructions will consist of clearly distinguished steps that readers must follow to achieve the desired results. Figure 12.12 contains a model set of instructions on how to set up a printer. Note how each step is precisely keyed to

FIGURE 12.10 List of Tools Needed for Instructions on Removing a Refrigerator Door

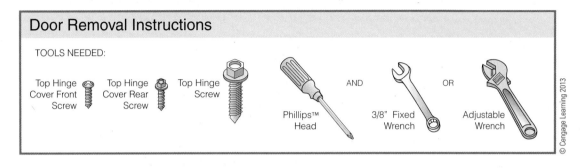

© Cengage Learning 2013

the visual, further helping readers perform the procedure. Refer to Figure 12.12 as you study this section.

Guidelines for Writing Steps

To help your readers understand your steps, observe the following rules.

1. Put the steps in their correct order, and number them. If a step is out of order or is missing, the entire set of instructions can be wrong or, worse yet, dangerous. Double-check every step, and number each one to indicate its correct place in the sequence of tasks you are describing.

2. Include the right amount of information in each step. Make each step short and simple. Giving readers too much information can be as risky as giving them too little. Keep in mind that each step should ask readers to perform a single task in the entire process. In the following example, note how the first version combines too many steps, while the revision corrects the problem:

Incorrect:
1. To access your voice mail, make sure you've listened to old messages and then press "1" to obtain your new messages.
2. When each new message is finished, press "7" to delete the message or "8" to store it in the archives. Press "2" to replay the message.
3. To review your saved messages, press "9." To end the call, press "#."

Correct:
1. To access your voice mail, press "1" to obtain your new messages.
2. When each new message is finished, press "7" to delete the message or "8" to store it in the archives. Press "2" to replay the message.
3. To review your saved messages, press "9."
4. To end the call, press "#."

3. Group closely related activities into one step. Sometimes closely related actions do belong in one step to help the reader coordinate activities and to emphasize their being done at the same time, in the same place, or with the same equipment.

Don't divide an action into two steps if it has to be done in one. For example, instructions showing how to light a furnace would not list as two steps actions that must be performed simultaneously to avoid a possible explosion:

Incorrect:	1.	Depress the lighting valve.
	2.	Hold a match to the pilot light.
Correct:	1.	Depress the lighting valve while holding a match to the pilot light.

Similarly, do not separate two steps of a computer command that must be performed simultaneously.

Incorrect:	1.	Press the CONTROL key.
	2.	Press the ALT key.
Correct:	1.	While holding down the CONTROL key, press the ALT key.

4. Give the reader hints on how best to accomplish the procedure. Obviously, you cannot do that for every step, but if there is a chance that the reader might run into difficulties, you should provide assistance. Particular techniques on how to operate or service equipment also help readers: "If there is blood on the transducer diaphragm, dip the transducer in blood solvent, such as hydrogen peroxide or Hemosol." You can also tell readers if they have a choice of materials or procedures, especially those that would give the best performance: "Several thin coats of paint will give a better finish than one heavy coat."

5. State whether one step directly influences (or jeopardizes) the outcome of another. Because all steps in a set of instructions are interrelated, you do not have to tell readers how every step affects every other. But stating specific relationships is particularly helpful when dangerous or highly intricate operations are involved. You will save the reader time, and you will stress the need for care. Forewarned is forearmed. Here is an example:

Step 2:   Tighten the fan belt. Failure to tighten the fan belt now will cause it to loosen and come off when the lever is turned on in Step 5.

Do not wait until Step 5 to tell readers that you hope they did a good job tightening the fan belt in Step 2. Information that comes after the fact is not helpful and could potentially be dangerous.

6. Where necessary, insert graphics to assist readers in carrying out the step. Almost every step in the set of long instructions in Figure 12.12 (pages 579–592) is illustrated with a drawing of the printer, an enlargement of a part, or a screenshot.

7. Your instructions might be translated into an international reader's language, as you can see in the warning statements in the next section.

Warnings, Cautions, and Notes

At appropriate places in the steps of your instructions, you may have to stop the reader to issue a warning, a caution, or a note. Warnings and cautions are mandatory text that you must provide to protect the user of the equipment from injury, or to protect hardware or software from damage. A note usually provides related information, such as an explanation, a tip, a comment, or other useful, but not imperative, information. Study the examples below as well as those in Figure 12.12, especially for Step 4, page 583, "Install ink cartridges."

Warnings

A warning ensures a reader's safety. It tells readers that a step, if not prepared for or performed properly, could seriously injure them, as the following warning does, or even endanger their lives.

WARNING: UNPLUG MACHINE BEFORE REMOVING PLATEN GLASS.

ADVERTENCIA: DESENCHUFE LA MAQUINA ANTES DE QUITAR EL VIDRIO.

Spanish translation

Cautions

A caution tells readers how to avoid a mistake that could damage equipment or cause the process to fail—for instance, "Do not force the plug."

Caution: Formatting erases all data on the disk

小心： 格式化会删除磁盘的所有资料

Chinese translation

Even diligent readers sometimes only skim or glance at a document. Some icons, like the ones below, universally convey "warning" or "caution" without requiring any text or explanation.

Notes

A note does not relate to the safety of the user or the equipment but does provide clarification, options, or a helpful hint on how to do the step more efficiently.

Turkish translation

At 20 degrees F, a battery uses about 68 percent of its power.

Eksi 7 derecede, bir pil enerjisinin yaklasik yuzde 68 ini kullanir.

Guidelines for Using Warnings, Cautions, and Notes

1. **Do not regard warnings and cautions as optional.** They are vital for legal and safety reasons to protect lives and property. In fact, you and your company can be sued if you fail to notify the users of your product or service of dangerous conditions that could result in injury or death.

2. **Put warnings and cautions as close as possible to the step to which they pertain.** (The exact placement may vary depending on the context and nature of the warning.) If you insert a warning or caution statement too early, readers may forget it by the time they come to the step to which it applies. Putting the notification too late exposes the reader, and possibly equipment as well, to risk.

3. **Put warnings and cautions in a distinctive format.** Warnings and cautions should be graphically set apart from the rest of the instructions. Use icons such as those shown above. Print such statements in capital letters, boldface, or different colors. Red is especially effective for warnings and cautions if your readers are native speakers of English. But remember that colors have different meanings in other countries (see page 506). If you expect your product to be used globally, you might consider rendering cautions using a different (but distinct) color, such as in Figure 12.12, where green is used to set cautions apart from the rest of the text. Also, make sure your audience understands an icon or a symbol, such as a skull and crossbones, an exclamation point inside a triangle, a circle with a line through it, or a traffic stoplight, often used to signal a warning, a hazard, or some other unsafe condition.

4. **Include relevant explanations to help readers know what to watch out for and what precautions to take.** Do not just insert the word *WARNING* or *CAUTION*. Explain what the dangerous condition is and how to avoid it. Look at the examples of cautions in Figure 12.12.

5. **Do not include a warning or a caution just to emphasize a point.** Putting too many warnings or cautions in your instructions will decrease their impact on readers. Use them sparingly—only when absolutely necessary—so readers will not be tempted to ignore them.

6. **Use notes only when the procedure calls for them and when they help readers.** See Figure 12.12.

Conclusion

Not every set of instructions requires a conclusion. For short instructions containing only a few simple steps, such as those in Figures 12.1 through 12.4, no conclusion is necessary. For longer, more involved jobs, a conclusion can provide a succinct wrap-up of what the reader has done, end with a single sentence of congratulations, or reassure readers, as the conclusion of Cliff Burgess's memo (Figure 12.7) does. A conclusion might also tell readers what to expect once a job is finished, describe the results of a test, or explain how a piece of equipment is supposed to look or operate. Figure 12.12 ends concisely but encouragingly with this sentence: "You're ready to print, copy, scan, and fax!" Always supply contact information and any hyperlinks, should a reader need further information.

Troubleshooting Guide

Instructions can also come with a section on troubleshooting to help readers when they encounter a problem. Often formatted as a table or chart, troubleshooting guides describe the problems that are most likely to occur and explain the easiest ways to correct them. Note how the troubleshooting guide in Figure 12.11 uses icons and boldfacing to help cable television users take appropriate action if they run into problems with their equipment (e.g., remote control) or reception (picture or sound). Troubleshooting tips can also be found within various steps of a set of instructions, as on page 578. Troubleshooting guides and tips help consumers avoid frustration and the expense of a service call.

Model of Full Set of Instructions

Study Figure 12.12, which is a full set of instructions for setting up an Epson printer. It includes most of the parts discussed in this chapter: an introduction; a list of materials; numbered steps; cautions and notes; and a conclusion.

Intended for a global audience, these instructions are a model of a user-friendly document. They are written and formatted to be easy to read and to perform. The language is clear and concise. Sentences are short and direct, yet the tone is free from any cultural bias.

Pay special attention to how the writer coordinates words, visuals, and colors to assist readers. Each step is clearly numbered and accompanied by an appropriate visual (e.g., enlarged drawing) or visual device (arrows, icons, directional symbols). Screen shots are inserted to help readers understand various steps they need to perform instructions online. Finally, caution statements and notes are inserted in the appropriate places to make sure readers install and set up their Epson printer safely.

FIGURE 12.11 Troubleshooting Instructions

TROUBLESHOOTING

My remote control isn't working

- Check to see if the batteries need to be replaced
- Make sure you have inserted the batteries correctly
- Remove any material that might get in the way of the signal (between the front of the remote and the cable box)

My cable box doesn't seem to be on, and the light isn't lit

- Make sure the power cord is plugged into both the cable box and the wall
- Plug in another working appliance to be sure the outlet is getting power

The light on the cable box is blinking and nothing is working

- The cable box must be activated over the phone or by a certified ABC Cable Technician
- If you have had the box activated and still have no picture, please call ABC Cable at 1-800-555-ABCT

Everything appears to be on, but my TV isn't showing a picture

- Check to be sure that your TV input is on "cable," or that your TV is on the correct channel to receive the input
- Make sure that all the cables (wall to cable box, cable box to outlet, cable box to TV, and TV to outlet) are firmly connected

The audio or menu is in the wrong language!

- Press and hold the "LANG" button on the remote until the desired language is listed on the screen.

If you need additional support, please go to
www.abccable.com/digitalcablehelp
or call ABC Cable at 1-800-555-ABCT

FIGURE 12.12 Complete Set of Instructions with Visuals Showing Parts Included

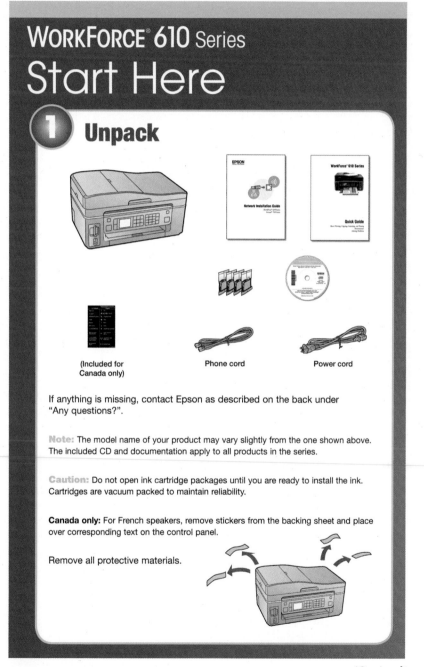

WORKFORCE® 610 Series
Start Here

1 Unpack

Visuals show contents

Unpacking serves as an introduction to the product

(Included for Canada only)   Phone cord   Power cord

If anything is missing, contact Epson as described on the back under "Any questions?".

Note reassures readers about use of different model names

Note: The model name of your product may vary slightly from the one shown above. The included CD and documentation apply to all products in the series.

Caution: Do not open ink cartridge packages until you are ready to install the ink. Cartridges are vacuum packed to maintain reliability.

Caution notice alerts readers to possible product damage

Canada only: For French speakers, remove stickers from the backing sheet and place over corresponding text on the control panel.

Remove all protective materials.

Colored arrows assist readers in unpacking contents

Courtesy of Seiko Epson Corp.

(Continued)

FIGURE 12.12 (Continued)

Uses two-level numbering— large circled numbers for the major sections and smaller boldfaced numbers for each step in those sections

2 Turn on and adjust

1 Connect the power cable.

Caution: Do not connect to your computer yet.

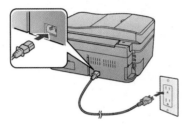

Enlarged drawing and contrasting colors assist readers to identify, connect, and move parts

2 Lift the control panel to raise it, then press the ⏻ **On** button.

Uses color to have note stand out

Note: To lower the control panel, squeeze the release lever underneath and push the control panel down. For more information on using and adjusting the control panel, see your *Quick Guide.*

Courtesy of Seiko Epson Corp.

FIGURE 12.12 (Continued)

3 **Make settings**

1   Select your language.

Press ▲ or ▼ to select the desired option.

Press **OK** when done.

Uses symbols, screens, and keypads to visualize and reinforce each step

2   Select your country/region, then press **OK**.

3   For the Daylight Saving Time setting, select **Summer** if your region uses Daylight Saving Time and it's currently in effect. (DST is effective from spring through summer.) Otherwise, select **Winter** to turn off the setting. Press **OK**.

Specifies and explains options

4   Press ▲ or ▼ to select the date format, then press ▶. Don't press **OK** yet.

Each step begins with imperative verb

Courtesy of Seiko Epson Corp.

(Continued)

FIGURE 12.12 (Continued)

Repeated use of red print alerts reader not to perform a step yet

5 Use the numeric keypad to set the date. Don't press **OK** yet.

 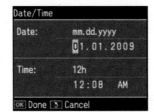

Each step includes written directions and often an appropriate visual

6 Press ▲ or ▼ to select the time format, then press ►. Don't press **OK** yet.

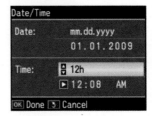

7 Use the numeric keypad to set the time.

 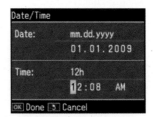

Uses boldface for emphasis

8 If you selected **12h** as the time format, press ▲ or ▼ to select **AM** or **PM**.

9 Press **OK** when done.

Provides helpful tip on changing date and time settings

Note: You can change the date and time settings by pressing the **Home** button, selecting **Setup**, selecting **Printer Setup**, then selecting **Date/Time**.

Courtesy of Seiko Epson Corp.

FIGURE 12.12 (Continued)

 Install ink cartridges

Note: Don't load paper before installing the ink cartridges.

1 Lift up the scanner.

2 Open the cartridge cover.

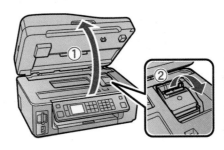

3 Shake the ink cartridges gently 4 or 5 times, then unpack them.

4 Remove only the yellow tape from each cartridge.

Caution: Don't remove any other seals or the cartridges may leak.

5 Insert the cartridges in the holder for each color.

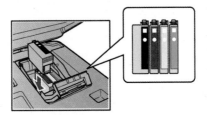

Courtesy of Seiko Epson Corp.

(Continued)

Gives important information before readers perform the steps

Directional arrows illustrate correct movement of parts

Specifies number of repetitions

Caution notice inserted in appropriate place with symbols showing wrong way to perform step

FIGURE 12.12 (Continued)

6 Press each cartridge down until it clicks.

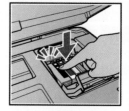

Visual clues reinforce the right way to perform step

7 Close the ink cartridge cover and press it down until it clicks.

Helps reader know when step is carried out successfully

8 Lower the scanner.

Uses concise language

9 Press the **OK** button to charge the ink. Charging takes about 3 minutes.

Note: Your product ships with full cartridges and part of the ink from the first cartridges is used for priming the product.

Note and caution notices help readers use product more economically and efficiently

Caution: Don't turn off the product while the ink system is charging or you'll waste ink.

Courtesy of Seiko Epson Corp.

FIGURE 12.12 (Continued)

⑤ Load paper

1   Open the paper support and pull up the extensions.

Generous white space makes steps easier to distinguish and to follow

2   Extend the output tray and raise the stopper.

Groups related activities in one step

Supplies helpful information on best way to use product

Note: If you are using legal-size paper, do not raise the stopper.

3   Hold the feeder guard forward, then squeeze the edge guide and slide it to the left.

Includes three visuals, two of them enlarged drawings, pinpointing exact places reader needs to recognize to perform the step

Courtesy of Seiko Epson Corp.

(Continued)

FIGURE 12.12 (Continued)

Visuals show right and wrong ways to load paper

Instructions written in concise and clear language

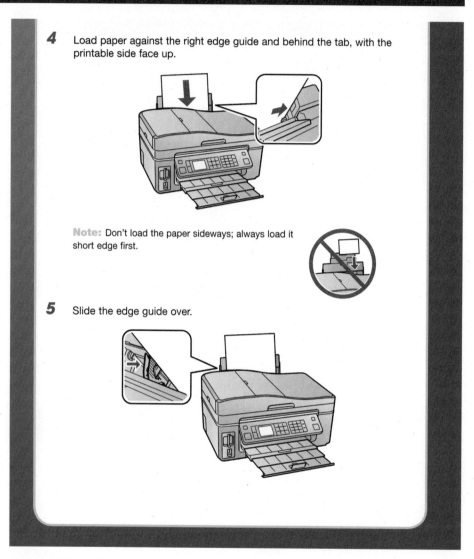

4 Load paper against the right edge guide and behind the tab, with the printable side face up.

Note: Don't load the paper sideways; always load it short edge first.

5 Slide the edge guide over.

Courtesy of Seiko Epson Corp.

FIGURE 12.12 (Continued)

6 Connect the phone cord

To use your product for faxing, connect the included phone cord from a telephone wall jack to the **LINE** port on the product. To connect a telephone or answering machine, see the *Quick Guide* for instructions.

Instructions clearly indicate product options

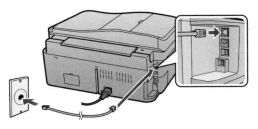

Note: If you're connecting to a DSL phone line, you must use a DSL filter or you won't be able to fax. Contact your DSL provider for the necessary filter.

7 Choose your connection

Section 7 alerts readers that they can use different types of connections

 Network (wireless or wired)

See your *Network Installation Guide*.
You can't install your software as described below.

Arrows point to where instructions for each type of connection can be found

OR

 USB connection

Make sure you have a USB cable (not included).
Then follow the steps below.

Courtesy of Seiko Epson Corp.   *(Continued)*

FIGURE 12.12 (Continued)

Provides brief introduction to ensure readers are using a USB connection

USB connection

Follow these steps to connect your product directly to your computer using a USB cable (not included).

Windows

Boldfacing and visuals reinforce key performance information

1   Make sure the product is **NOT CONNECTED** to your computer.

Shows readers what they will see on-screen if step is not performed properly

Note: If you see a Found New Hardware screen, click **Cancel** and disconnect the USB cable. You can't install your software that way.

Informs readers what to expect when carrying out specific step

2   Insert the WorkForce 610 Series software CD.

With Windows Vista®, if you see the AutoPlay window, click **Run SETUP.EXE**. When you see the User Account Control screen, click **Continue**.

3   Click **Install** and follow the on-screen instructions.

Courtesy of Seiko Epson Corp.

FIGURE 12.12 (Continued)

4 When you see this screen, select **Install driver for direct USB connection** and then click **Next**.

Steps 4 and 5 combine boldfacing, enlarged visuals, and on-screen instructions

5 When prompted, connect a USB cable. Use any open USB port on your computer.

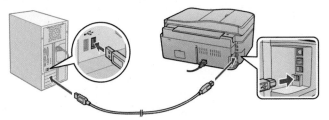

Note: If installation doesn't continue after a moment, make sure you securely connected and turned on the product.

Helpful hint if reader encounters a problem

6 Follow the on-screen instructions to install the rest of the software and register your product.

Be sure to register your product to receive these great benefits: 2-year limited warranty and 2-year toll-free customer support.*

7 When you're done, remove the CD.

You're ready to print, copy, scan, and fax! See your *Quick Guide* or online *Epson Information Center*.

Assures reader what to expect when instructions are finished

Courtesy of Seiko Epson Corp.

(Continued)

FIGURE 12.12 (Continued)

Macintosh®

First step for Macintosh users with prominent visual begins with most important task for this section

1 Make sure the product is **NOT CONNECTED** to your computer.

2 Insert the WorkForce 610 Series software CD.

3 Double-click the Epson icon.

4 Click **Install** and follow the on-screen instructions.

On-screen instructions assure readers that they are following directions properly

5 When you see this screen, select **Install driver for direct USB connection** and then click **Next**.

Courtesy of Seiko Epson Corp.

FIGURE 12.12 (Continued)

6 When prompted, connect a USB cable. Use any open USB port on your computer. Make sure the product is securely connected and turned on.

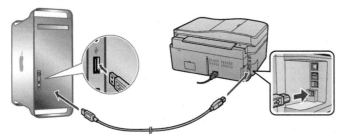

Mac OS® X 10.5 users: skip to step 9. Other users: continue with step 7.

7 When you see this screen, click **Add Printer.** Follow the steps on the right side of the screen to add the product.

8 Click **Next** again, if necessary.

Printed directions work in tandem with those on-screen

9 Follow the on-screen instructions to install the rest of the software and register your product.

Be sure to register your product to receive these great benefits: 2-year limited warranty and 2-year toll-free customer support.*

10 When you're done, eject the CD.

You're ready to print, copy, scan, and fax! See your *Quick Guide* or online *Epson Information Center*.

Functions as both a step— eject the CD— and a wrap-up

Courtesy of Seiko Epson Corp.

(Continued)

FIGURE 12.12 (Continued)

Any questions?

Quick Guide

Basic instructions for printing, copying, scanning, and faxing.

Provides information on how and where readers can receive further help

Online Epson Information Center

Click the desktop shortcut for easy access to the user's guide, FAQs, online troubleshooting advice, and software downloads. You can also purchase paper and ink.

Color helps to identify options for readers

Network Installation Guide and Video

Instructions on configuring the product for a network. For a video tutorial and other information about setting up a wireless network, go to: **epson.com/support/wireless**

On-screen help with your software

Select **Help** or **?** when you're using your software.

Epson Technical Support

Internet Support

Visit Epson's support website at **epson.com/support** and select your product for solutions to common problems. You can download drivers and documentation, get FAQs and troubleshooting advice, or e-mail Epson with your questions.

Speak to a Support Representative

Call (562) 276-4382 (U.S.) or (905) 709-3839 (Canada), 6 AM to 6 PM, Pacific Time, Monday through Friday. Days and hours of support are subject to change without notice. Toll or long distance charges may apply.

Boilerplate information

Courtesy of Seiko Epson Corp.

Writing Procedures for Policies and Regulations

Up to this point, we have concentrated primarily on instructions dealing with how to put things together; how to install, repair, or use equipment; and how to alert readers to mechanical or even personal danger.

But there is another similar type of writing that deals with guidelines for getting things done in the world of work: procedures. These concern policies and regulations found in employee handbooks and other internal corporate communications, such as on websites, in memos, in email messages, or on a company's intranet (a private computer network). Note, however, that some companies do not disseminate policy via email because it is perceived as less formal than hard copy and can easily be deleted. Figure 12.13 on pages 595–596 shows an example of a company's flextime procedures written in a memo format.

Some Examples of Procedures

Procedures deal with a wide range of "how-to" activities within an organization, including the following:

- accessing a company file or database
- preparing for an audit, a transition, a merger
- applying for family or medical leave
- dressing professionally at work or at a job site
- forwarding and routing information
- reserving a company vehicle or facility
- taking advantage of telecommuting options
- submitting a work-related grievance
- requesting travel expense reimbursement
- evacuating a building in case of a fire, flood, toxic spill, or other emergency
- fulfilling promotion requirements
- applying for a disability accommodation
- requesting a transfer within the company
- using company email

Policy procedures have a major impact on a company and its workers. They affect schedules, payrolls, acceptable and unacceptable behaviors at work, and a range of protocols governing the way the organization does business internally and externally. Procedures reinforce the company's or organization's mission and goals. See the letter from the CEO of IBM (Figure 1.11) stressing company protocols. Procedures also help an organization run smoothly and consistently. Adhering to them, all employees follow the same regulations and standards.

Meeting the Needs of Your Marketplace

As with instructions, you will have to plan carefully when you write a set of procedures. A mistake in business procedures can be as wide ranging and as costly as an error in a set of assembly instructions since poorly written procedures can land a company or its employees in significant financial and legal trouble.

To avoid such difficulties, spell out precisely what is expected of employees—how, when, where, and why they are to perform or adhere to a certain policy. Use the same strategies as for instructions discussed earlier in this chapter. Leave no chance for misunderstanding or ambiguity; be unqualifiedly straightforward and clear-cut. Determine what information employees need in order to comply with your company's regulations.

Many times procedures involve a change in the work environment. Help readers by including, whenever necessary, definitions, headings, some prefatory explanations, and an offer to assist employees with any questions they may have. Always present a copy of the procedures to management to approve or to revise before sending them to employees.

Case Study

Writing Procedures at Work

Scheduling employees' time is a major consideration in the world of work. Almost every company handbook devotes space to this issue. Figure 12.13 shows a memo from Tequina Bowers, a human resources director, notifying employees how they can take advantage of a new flextime schedule at work, including what they can and cannot do under the new policy. Note how Bowers divides her procedures into an introduction explaining when flextime will go into effect and what choices employees have, a section that clearly defines the concept of flextime, and finally the specific guidelines. These guidelines, while not sequential, function as a series of rules that employees have to follow, and ultimately these rules will affect the entire organization. Her job is to make sure the procedures are clear, do not contradict current company policy, and explicitly identify actions that a company will not tolerate; for example, switching hours with another employee.

The various regulations about what employees cannot do in flextime might be seen as the equivalents of the warning and caution statements discussed earlier (pages 575–576). Observe that Bowers's memo does not veer off to discuss benefits to the employer or to examine where flextime has been used elsewhere. Finally, this example of procedural writing protects NewTech Inc. legally by establishing the policies by which an employee's scheduled work time is clearly defined, delineated, and assessed. There's no room for guesswork in what Bowers has to write.

NewTech, Inc.
● ● ●

4300 Ames Boulevard, Gunderson, CO 81230-0999
303.555.9721   fax 303.555.9876
www.newtech.com

TO: All Employees
FROM: Tequina Bowers T.B.
                  Human Resources Schedule
DATE: March 21, 2012
SUBJECT: Opting for a Flextime Schedule

Effective sixty days from now, on May 30, 2012, employees will have the opportunity to go to a flextime schedule or to remain on their current 8-hour fixed schedule. This memo explains the new flextime option and establishes the procedures—guidelines and rules—you must follow if you choose this new schedule.

Flextime Defined

Flextime is based on a certain number of **core hours** and **flexible hours**. Our company will be open twelve hours, from 6:00 a.m. to 6:00 p.m. weekdays, to accommodate both fixed and flextime schedules. During this 12-hour period, all employees on flextime will be expected to work **eight and one-half consecutive hours**, which includes a half-hour for lunch.

Regardless of schedule options, all employees must work a **common core time from 10:00 a.m. to 4:00 p.m.**, but flextime employees will be free to choose their own starting and quitting times. For instance, they might elect to arrive at 8:00 a.m. and leave at 4:30 p.m., or they may want to start at 9:30 a.m. and leave at 6:00 p.m.

Flextime Guidelines and Rules

Employees are expected to understand their individual responsibilities and adjust their schedules accordingly. All flextime employees must adhere strictly to the following regulations and realize that the privileges of a flextime schedule will be revoked for violations.

Notifies readers of new policy and states purpose of the memo

Spells out precisely how company defines flextime—uses boldface for most important information

Provides helpful examples

Stresses employee responsibilities and consequences of violating rules

(Continued)

FIGURE 12.13 (Continued)

page 2

Carefully outlines what is acceptable according to new policy

What Flextime Employees Must Do

(1) Be present during core time, arriving and leaving the facility during their flexible work hours.

Numbered points make policy easier to understand, follow, and refer to in the future

(2) Observe a minimum unpaid half-hour lunch break each working day.
(3) Cooperate with their supervisors to make sure adequate coverage is provided for their department from 6:00 a.m. to 6:00 p.m.
(4) Notify supervisors at once of absences.
(5) Attend monthly corporate meetings even though such meetings may be outside their chosen flextime schedules.
(6) Adhere to company dress codes during any time they are at work, regardless of their flextime schedule.
(7) Agree to work on a flextime schedule for a minimum 6-month period.

Stipulates what new policy will not allow

What Flextime Employees Can't Do

(1) Be tardy during core time.
(2) Switch, bank, borrow, or trade flextime hours with other employees without the written approval of an immediate supervisor.
(3) File for overtime without a supervisor's approval.
(4) Self-schedule a vacation or leave by expanding flextime hours.
(5) Alternate between fixed time and flextime.

Explains steps to begin flextime

How Do You Sign Up for Flextime?

If you opt for a flextime schedule, first you need to obtain and complete a transfer of hours form from your supervisor. Next, you must bring the signed form to Human Resources (Admin. 201) to participate officially in this program.

Encourages feedback and questions

I will be happy to talk to you about this new work schedule and to answer any of your questions. Please call me at ext. 5121, email me at **tbowers@newtech.com**, or visit my office in Admin. 201. Thank you for your cooperation.

Some Final Advice

Perhaps the most important piece of advice to leave you with is this: Do not take *anything* for granted when you have to write a set of instructions or procedures. It is wrong and sometimes dangerous to assume that your readers have performed the procedures before, that they will automatically supply missing or "obvious" information, or that they will easily anticipate your next step or know what is expected of them at work without being informed. No one ever complained that a set of instructions was too clear or too easy to follow. Similarly, make sure that the procedures you may be called upon to write are easy to understand and to follow.

✓ Revision Checklist

- ☐ Analyzed my intended audience's background, especially why and how they will use my instructions.
- ☐ Tested my instructions to make sure they include all the necessary steps in their proper sequence.
- ☐ Ensured all measurements, distances, times, and relationships are precise and correct.
- ☐ Avoided technical terms if my audience is not a group of specialists in my field.
- ☐ Used the imperative mood for verbs and wrote clear, short sentences.
- ☐ Made my instructions easy to read and follow for an international audience.
- ☐ Chose effective visuals, labeled them, and placed them next to the step(s) to which they apply.
- ☐ Included relevant hyperlinks in online instructions to provide readers with help screens or further information.
- ☐ Made my introduction proportionate to the length and complexity of my instructions and suitable for my readers' needs.
- ☐ Included necessary background, safety, and operational information in the introduction.
- ☐ Provided a complete list of tools and materials my audience needs to carry out the instructions.
- ☐ Put the instructions in easy-to-follow steps and in the correct chronological/sequential order.
- ☐ Used numbers or bullets to label the steps and inserted connective words to reinforce order.
- ☐ Inserted warnings, cautions, and notes where necessary and included culturally appropriate icons and colors that make them easy to find and to understand.
- ☐ Supplied a conclusion that summarizes what readers should have done or reassures them that they have completed the job satisfactorily.

☐ Provided troubleshooting guide to help readers identify a problem and take appropriate steps to fix it.

☐ Clearly spelled out policies, protocols, responsibilities, restrictions, and consequences of procedures for readers.

☐ Defined any terms readers may be unfamiliar with in procedures and, where helpful, provided an example.

☐ Submitted a copy of procedures to administrators for their approval before distributing to employees.

Exercises

1. Find a set of instructions that does not contain any visuals but that you think should to make the directions clearer. Design those visuals yourself, and indicate where they should be inserted in the instructions.

2. In a technical manual in your field or in an owner's manual, locate a set of instructions that you think is poorly written and illustrated. In a memo to your instructor, explain why the instructions are unclear, confusing, or badly formatted. Then, revise the instructions to make them easier for the reader to carry out. Submit the original instructions with your revision.

3. Write a short set of instructions in numbered steps (or in paragraph format) on one of the following relatively simple activities.
   a. tying a shoe
   b. using an ATM to withdraw money
   c. unlocking a door with a key
   d. sending a text message from a cell phone
   e. planting a tree or a shrub
   f. sewing a button on a shirt
   g. removing a stain from clothing
   h. pumping gas into a car
   i. creating a blog
   j. logging onto your college library's server
   k. polishing a floor
   l. shifting gears in a car
   m. downloading a software upgrade
   n. posting a video to your blog

4. Write an appropriate introduction and conclusion for the set of instructions you wrote for Exercise 3.

5. Write a set of full instructions on one of the following more complex topics. Identify your audience. Include an appropriate introduction; a list of equipment and materials; numbered steps with necessary warnings, cautions, and notes; a

troubleshooting guide; and an effective conclusion. Also include whatever visuals you think will help your audience.

   a. scanning a document
   b. changing a flat tire
   c. testing chlorine in a swimming pool
   d. shaving a patient for surgery
   e. removing "red eye" from a digital photo
   f. surveying a parcel of land
   g. creating a slideshow of the digital photos you took on a job
   h. jumping a dead car battery
   i. using the Heimlich maneuver to help a choking individual
   j. filleting a fish
   k. creating a logo for a letterhead
   l. taking someone's blood pressure
   m. editing digital video
   n. painting a cabinet
   o. recording a podcast
   p. flossing a patient's teeth after cleaning
   q. backing up your computer files to an external hard drive

6. The following set of instructions is confusing, vague, and out of order. Rewrite the instructions to make them clear, easy to follow, and correct. Make sure that each step follows the guidelines outlined in this chapter.

Reupholstering a Piece of Furniture

   (1)   Although it might be difficult to match the worn material with the new material, you might as well try.
   (2)   If you cannot, remove the old material.
   (3)   Take out the padding.
   (4)   Take out all of the tacks before removing the old covering. You might want to save the old covering.
   (5)   Measure the new material with the old, if you are able to.
   (6)   Check the frame, springs, webbing, and padding.
   (7)   Put the new material over the old.
   (8)   Check to see if it matches.
   (9)   You must have the same size as before.
   (10) Look at the padding inside. If it is lumpy, smooth it out.
   (11) You will need to tack all the sides down. Space your tacks a good distance apart.
   (12) When you spot wrinkles, remove the tacks.
   (13) Caution: in step 11 directly above, do not drive your tacks all the way through. Leave some room.
   (14) Work from the center to the edge in step 11 above.
   (15) Put the new material over the old furniture.

P.S. Use strong cords whenever there are tacks. Put the cords under the nails so that they hold.

7. Write a set of procedures on "greening" a student union or an employee rest area. This exercise can be done collaboratively, with each member of the team taking a key area: lighting, heating/cooling, recycling, noise pollution, food services. Include a relevant visual with your procedure.

8. Write a set of procedures, similar to Figure 12.13, for one of the following policies or regulations:
    a. offering quality customer service over the phone or via the Web
    b. filing a claim for a personal injury on the job
    c. decorating an employee's personal space—what is and is not allowed?
    d. using the Internet at work
    e. ensuring confidentiality at work
    f. enrolling in mandatory courses to maintain a license or certificate
    g. playing music in the workplace
    h. going through an orientation procedure before beginning a new job
    i. following an acceptable company dress code
    j. allowing tattoos and body piercings in the workplace
    k. registering a domain name for a sponsored group at work or school
    l. receiving reimbursement for carpooling, taking public transportation, or riding a bicycle to work
    m. using a company vehicle
    n. going on a service call to a customer's home
    o. changing filters, parts, etc., on a periodic basis at work
    p. representing your company/organization at a professional meeting
    q. greening an office space
    r. submitting documentation for a promotion
    s. traveling with a pet on an airplane or a train
    t. requesting a refund for inadequate service

CHAPTER 13

Writing Winning Proposals

A proposal is a detailed plan of action that a writer submits to a reader or group of readers for approval to buy a product, try a service, upgrade technology, support a cause or project, and so forth. Proposal readers are usually in a position of authority—supervisors, managers, department heads, boards of private foundations, military or civic leaders—to endorse or reject the writer's plan. Proposals are among the most important types of job-related writing. Their acceptance can lead to improved working conditions, better use of technology, a more efficient and economical business, additional jobs and business for a company, or a safer environment.

Every proposal you write must exhibit a "can do" attitude, putting the reader and his or her company's needs at the center of your work. As you go through this chapter, keep in mind the slogan of Yates Engineering, which has won millions of dollars of business through its reader-centered proposals: "On time . . . within budget . . . to your satisfaction." Time, budget, and your readers' satisfaction and convenience are among the most important ingredients of a winning proposal. Advertisements such as the one in Figure 1.9 (page 28) often contain mini proposals appealing to a customer's need for a more economical and efficient way to do things. Notice how the advertisement in Figure 13.1 encourages potential clients to purchase a security package based on a variety of available options, from motion detection systems to video surveillance to well-trained officers.

Access chapter-specific interactive learning tools, including quizzes and more in your English CourseMate, accessed through www.cengagebrain.com.

Writing Successful Proposals: Some Examples

Proposals are written for many different purposes and many different audiences. Here are a few examples:

- to your boss, seeking authorization to hire staff, change a procedure, or purchase a new piece of equipment for the office (see Marcus Weekley's proposal in Figure 2.5, page 60, requesting that his firm upgrade their printers)
- to potential customers, offering a product or a service (such as offering to supply a fire chief with special firefighting gear, or offering an office manager a line of ergonomically designed furniture)
- to a government agency, such as the Department of the Interior, seeking funds to conduct a research project (for instance, requesting support to study the mating and feeding habits of a particular species, or asking for funds to discover ways to detect environmental hazards more quickly)
- to foundations to raise funds for a nonprofit organization (for example, a small art gallery requesting funding from an arts foundation)

601

FIGURE 13.1   An Example of a "Can Do" Attitude

CPS Security/www.cpssecurity.com.

Characteristics of Proposals

Proposals are characterized by three things:

1. They vary in size and scope.
2. They are persuasive plans.
3. They frequently are collaborative efforts.

Proposals Vary in Size and Scope

Depending on the job, proposals can vary greatly in size and scope. A proposal to your employer could easily be conveyed in a page or two. To propose doing a small job for a prospective client—for instance, redecorating a waiting room in an accountant's office—a letter with information on costs, materials, and a timetable might suffice. The sales letters in Figures 6.3 and 6.4 (pages 214 and 215) illustrate short proposals in letter format.

Longer, more formal proposals contain many sections similar to those found in reports (pages 702–718). An extremely large and costly project—for example, constructing a ten-story office building—requires a detailed proposal hundreds of pages long with appendixes on engineering specifications, detailed budgets, time-lines for various phases of construction, and even résumés of all key personnel who would be working on the project.

Grant proposals, where writers apply for funding—to staff a community center or run a rural clinic—must follow the detailed guidelines found in RFPs (pages 604–606) specified by the funding agency, which is often a government office or charitable foundation.

A discussion of long, elaborate business proposals and grant proposals is beyond the scope of this chapter. But the principles and techniques of audience analysis, organization, and drafting that this chapter does cover apply equally well to any longer project you may have to prepare by yourself or as part of a team at work or school.

Proposals Are Persuasive Plans

Proposals, whether large or small, must be highly persuasive to succeed. Without your audience's approval, your plan will never go into effect, however accurate and important you think it is. Enthusiasm is not enough to persuade readers; you have to supply hard evidence based on research. Your proposal must convince readers that your plan will help them improve their business, make their job easier, save them money, enhance their image, improve customer satisfaction, or all of these.

Competition is fierce in the world of work, and a persuasive proposal frequently determines which company receives a contract. Demonstrate why your plan and your company are better—more efficient, practical, economical—than a competitor's. In a sense, a proposal combines the persuasiveness of a sales letter (see Chapter 6), the documentation of a report (see Chapters 8 and 15), and the binding power of a contract, because if the reader accepts your proposal, he or she will expect you to live up to its terms to the letter.

You cannot write a successful proposal until you

1. fully understand your audience's needs or problems
2. prove why solving them is important to your audience's ongoing business
3. formulate a careful, detailed plan to solve these needs or problems
4. establish beyond doubt that you have the logic, time, technology, and personnel to solve the audience's precise problem
5. match your timetable and budget to your reader's

These five goals are not only vital to your persuasive plan but, when incorporated within a proposal that your reader accepts, become part of a legally binding agreement.

Proposals Frequently Are Collaborative Efforts

Like many other types of business and technical writing, proposals often are the product of teamwork and sharing. Even short in-house proposals (such as the one in Figure 13.6) are often researched and put together by more than one individual in the company or agency.

Often, individual employees will pull together information from their separate areas (such as IT, graphics and design, finance, marketing, sales, transportation, and even legal) to include it in a proposal that each team member then reads and revises until the team agrees that the document is ready to be released.

Types of Proposals

Proposals are classified according to (1) how they originate and (2) where they are sent after they are written. Distinctions are made between *solicited* and *unsolicited* proposals based on how they originate and between *internal* and *external* proposals based on where they are sent. Depending on your audience and your purpose, you may write an internal solicited or unsolicited proposal, or you may write an external solicited or unsolicited proposal. Solicited proposals often involve requests for proposals. Figure 13.2 provides a visual representation of the various types of proposals.

Solicited Proposals and Requests for Proposals

When a company has a particular problem to be solved or a job to be done, it will solicit, or invite, proposals. Accordingly, you do not have to spend time identifying the company's problem. The company will notify you and other competitors by preparing a *request for proposals* (RFP), which is a set of instructions that specifies the exact type of work to be done, along with guidelines on how and when the company wants the work completed.

Some RFPs are long and full of legal requirements and conditions. Others, like the examples in Figures 13.3 and 13.4, are more concise. RFPs are sent to firms with track records in the relevant field. RFPs are also printed in trade publications and put on the Web (see the Tech Note on page 611) to attract the highest number of qualified bidders for the job. The U.S. government publishes RFPs in the *Federal Register* or *Commerce Business Daily* (both are online), while private companies sometimes send their RFPs to *Business Daily*. No two RFPs are alike.

An RFP helps you understand what the customer wants. It is often extremely detailed and even tells you how the company wants the proposal prepared, for

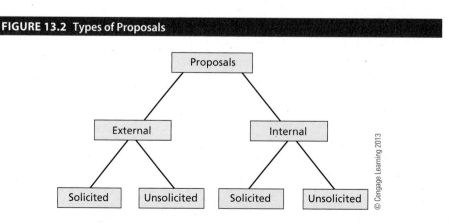

FIGURE 13.2 Types of Proposals

© Cengage Learning 2013

FIGURE 13.3 A Sample RFP for a Smaller Project

REQUEST FOR PROPOSALS

Mesa Community College is soliciting proposals to construct and to install 50 individual study carrels in its Holmes Memorial Library. These carrels must be highly serviceable and conform to all specification standards of the American Library Association (ALA) and the Americans with Disabilities Act. Proposals should include the precise measurements of the carrels to be installed, the specific acoustical and lighting benefits, Internet access, and the types and amount of storage space offered.

   Work on constructing and installing the carrels must be completed no later than the start of the Fall Semester, August 21, 2012. Proposals should include a schedule of when different phases of work will be completed and an itemized budget for labor, materials, equipment, and necessary tests to ensure high-quality acoustical performance. Contractors should detail their qualifications, including a description of similar recent work and a list of references. Proposals should be submitted in triplicate no later than March 1, 2012, to:

> **Dr. Barbara Feldstein**
> **Director of the Library**
> **Mesa Community College**
> **Mesa, CO 80932-0617**
> **BFeldstein@Mesa.edu**

© Cengage Learning 2013

example, what information is to be included (on backgrounds, personnel, equipment, budgets), where it needs to appear, and how many copies of the proposal you have to submit.

   Your proposal will be evaluated according to how well you fulfill the terms of the RFP. For that reason, follow the directions in the RFP exactly. Note that the solicited proposal in Figure 13.5 directly refers to the terms of the RFP. You should even use the language (specialized terms, specifically stated needs) of the RFP in your proposal to convince readers that you understand their requirements and to get them to accept your plan. If you have any questions, by all means call the agency or company so you do not waste your time or theirs by including irrelevant or unnecessary details in your proposal.

Unsolicited Proposals

With an unsolicited proposal, you—not the reader—make the first move. Unlike a solicited proposal, in which the company to which you are submitting the proposal knows about the problem, your unsolicited proposal has to convince readers that (1) there is a problem, (2) it is important enough to be corrected, and (3) you and your firm are the ones to solve it.

FIGURE 13.4 A Section of an RFP for a Larger Project

An RFP for the Fund for Rural America

The Fund for Rural America program supports competitively awarded research, extension, and education grants addressing key issues that contribute to the economic diversification and development of rural areas. The amount available for support of this program is approximately $9,500,000, from which approximately 15–20 awards will be made.

Program Description

Preservation of the economic viability of rural communities is the chief focus of the Fund for Rural America. The program focuses attention **on rural communities' twin challenges of rural community innovation** and **demographic change**. Challenges of an aging population, the arrival of new immigrant populations, . . . and workforce development all are critical issues that affect rural economies. **Rural communities may propose research, education, and extension/outreach projects that will increase our understanding of these demographic forces and develop the capacity to turn these challenges into economic promise.**

Rural Community Innovation

This program solicits research, education, and extension proposals that will **help rural Americans address existing and new problems in innovative ways. The goals of this program area are to generate relevant knowledge and transfer that knowledge to assist rural communities to diversify their economies, to develop and maintain profitable farms, firms, and businesses, to build community, to aid growth, to protect natural resources, and to increase family financial security.**

Projects such as the following combine multiple strategies for innovation that Fund for Rural America encourages in order to:
- help a community move to a more bio-based economy (carbon control credits, promoting locally based bio-based industries, linking bio-based materials to a diverse agriculture and community);
- include the use of land that offers development options (farmland preservation, farming on the urban fringe, rural-urban land use issues);
- institute policies that increase profits of small and minority farmers;
- use e-commerce applications for remote rural areas and minority populations, develop community information networks to support e-commerce and e-communities, and adapt e-commerce strategies for planning and social capital development;
- expand network capabilities among producers, businesses, entrepreneurs, families, individuals, nonprofit groups, community institutions, local government, and state and federal agencies;
- develop a new generation of computer-based planning tools (geospatial analysis, information stores, economic and land use blueprints) and the ability to apply and tailor them for place-sensitive development.

Source: U.S. Department of Agriculture.

FIGURE 13.5 A Sales Proposal in Response to a Request from a Company

Reynolds Interiors

250 Commence Avenue S.W. • Portland, OR 97204-2129
www.reynolds.com • 503-555-8733 • 503-555-1629

FOLLOW US ON FACEBOOK AND TWITTER

Letterhead advertises company's presence on social-networking sites

January 20, 2012

Mr. Floyd Tompkins, Manager
General Appliances
140 South Highway 11
Portland, OR 97222-1300

Dear Mr. Tompkins:

In response to your request for bids #GA01012012 posted on your website for an appropriate floor covering at your new showroom, Reynolds Interiors is pleased to submit the following proposal to meet your specific needs. We appreciated the opportunity to visit your showroom on January 14 in order to gather information to prepare this proposal.

Begins with a reference to company's request for bids

After carefully reviewing your specifications for a floor covering and inspecting your new facility, we believe that **Armstrong Classic Corlon 900** is the most suitable choice. We are enclosing a few samples of the Corlon 900 so you can see how carefully it is constructed.

Identifies best solution

Corlon's Advantages

Guaranteed against defects for a full three years, Corlon is one of the finest and most durable floor coverings manufactured by Armstrong. It is a heavy-duty commercial floor 0.085-inch thick for protection and durability. Twenty-five percent of the material consists of interface backing; the other 75 percent is an inlaid wear layer that offers exceptionally high resistance to the heavy, everyday traffic your showroom will see.

Describes product features that will benefit reader

Traffic tests conducted by the independent Contemporary Flooring Institute have repeatedly proved the superiority of Corlon's construction and resistance. You might want to visit the Institute's website (**www.cfi.org**) for a demonstration of how durable and versatile Classic Corlon flooring is.

Refers to an independent source to corroborate the benefits of the product

Another important feature of Corlon is the size of its rolls. Unlike other leading brands of similar commercial flooring—Remington or Treadmaster—Corlon comes in 12-foot-wide rather than 6-foot-wide

Distinguishes product from competitors'

(Continued)

FIGURE 13.5 (Continued)

Mr. Floyd Tompkins
Page 2

rolls. This extra width will significantly reduce the number of seams on your floor, thus increasing its attractiveness and eliminating the dangers of splitting or bulging.

Installation Procedures

Explains how job is done professionally

The Classic Corlon 900 requires that we use the inlaid seaming process, a technical procedure requiring the skill of a trained floor mechanic. Herman Goshen, our certified chief floor mechanic, has more than eighteen years of experience working with the inlaid seam process. His professional work and keen sense of layout and design have been consistently praised by our customers.

Installation Schedule

Gives realistic timetable

We can install the Classic Corlon 900 on your showroom floor during the first week of March, which fits the timetable specified in your request. The material will take three and one-half days to install but will be ready to walk on immediately. Be assured that your floor will be installed no later than March 7th. We recommend, though, that you do not move heavy equipment onto the floor for 24 hours after installation.

Costs

The following costs include the Classic Corlon floor covering, labor, and tax:

Itemizes all costs based on market conditions and reader's bid

750 sq. yards of Classic Corlon at $23.50/sq. yd.	$ 17,625.00
Labor (28 hrs @ $18.00/hr.)	$ 504.00
Sealing fluid (10 gals. @ $15.00/gal.)	$ 150.00
Subtotal	$ 18,279.00
Sales tax (5 percent)	$ 913.95
GRAND TOTAL	$ 19,192.95

Points out proposal comes in under budget—always a major consideration for buyers

Our costs are more than $300.00 below those you specified in your bid.

Reynolds's Qualifications

Reynolds Interiors has been in business for more than 28 years. In that time, we have installed more than 2,500 commercial floors in Portland and its suburbs. In the last year alone, we have served more than 60

FIGURE 13.5 (Continued)

Mr. Floyd Tompkins
Page 3

satisfied customers, including the new multipurpose Tech Mart facility in downtown Portland. Our designs have also been included in several commercial properties that have won awards from the Portland Architectural Review Board and have been showcased in such publications as *Portland Homes* and *Best Housing Plans, 2011–2012*. Reynolds has also consistently received high commendations from our many customers, and we would be happy to furnish you with a list of our references.

Establishes history of service and provides documented evidence of quality work

Thank you for the opportunity to submit this proposal to General Appliances. We are confident that you will be pleased with the appearance and durability of an Armstrong Classic Corlon 900 floor and our installation of it. If we can provide you with any further information about our service or Corlon flooring, or if you have any questions, please call us at 503-555-8733 or visit us at our website.

Thanks reader and encourages him to accept the proposal

Sincerely yours,

Neelow Singh

Neelow Singh
Sales Manager

Jack Rosen

Jack Rosen
Installation Supervisor

Doing all this is not as difficult as it sounds. See how the writers of the unsolicited proposal in Figure 13.6 identify a relevant problem for their readers. If your readers accept your identification of the problem, you have greatly increased the chances of their accepting your plan to solve it. Just remember that you will have to prove that solving the problem carries major benefits for your readers.

Internal and External Proposals

An internal proposal is written to one or several decision makers in your own organization who have to sign off or approve your plan. As you will see on pages 613–621, an internal proposal can deal with a variety of topics, including changing a policy or procedure, requesting additional personnel, or purchasing or updating equipment or software.

An external proposal is sent to a decision maker outside your company. It might go to a potential client you have never worked for or to a previous or current client. An external proposal can also be sent to a government funding agency, such as the Department of Agriculture, in response to its RFP (see Figure 13.4). External proposals tend to be more formal than internal ones.

Eight Guidelines for Writing a Successful Proposal

The following guidelines will help you persuade your audience to approve your plan. Refer to these guidelines and Figures 13.5 and 13.6 both before and while you formulate your plan.

1. **Approach writing a proposal as a problem-solving activity.** Your purpose should reflect your ability to identify and solve problems. Convince your audience that you know what their needs are and that you will meet them, as Neelow Singh and Jack Rosen do in Figure 13.5.

2. **Regard your audience as skeptical.** Even though you offer a plan that you think will benefit readers, do not be overconfident that they will automatically accept it as the best and only way to proceed. Brainstorm, alone or with your collaborative team, to anticipate and answer your readers' questions and objections. To determine whether your proposal is feasible, readers will study it carefully. If your proposal contains errors or inconsistencies, omits information, or deviates from what they are looking for, your readers will reject it.

3. **Research your proposal topic thoroughly.** A winning proposal is *not* based on a few well-meaning, general suggestions. To provide the detailed information necessary and to convince readers that you know your proposal topic inside and out, you will have to do your homework. Research your topic by studying the latest technology in the field, shopping for the best prices, comparing your prices and services with what the competition offers, verifying schedules, visiting customers, making site visits, and interviewing key individuals. Make sure that any technology or equipment you use or sell complies with all codes, specifications, and standards.

4. **Scout out what your competitors are doing.** Become familiar with your competitors' products or services, have a fair idea about their market costs, and be able to show how your company's work is better overall. Provide examples; offer a demonstration. Read competitors' websites and print publications (such as catalog and marketing brochures) very carefully. Let readers know you have done your homework on their behalf. See how the writing team for the business report in Chapter 8 (pages 399–413) researched its competition. Note, too, how the writers in Figure 13.5 prove that their product and service are superior to those of their competitors.

Online RFPs

RFPs are not only sent out to firms and published in trade magazines, but are often posted on numerous websites as well. The U.S. government, through its many departments and agencies, is an important source for RFPs. FedBizOpps (**www.fbo.gov**) claims that it offers "a comprehensive listing of government RFP's." In addition, the *Federal Register,* published daily by the U.S. government, issues hundreds of legal regulations as well as requests for proposals from various agencies. You can browse the *Federal Register* online via GPO Access (**www.gpoaccess.gov/fr**).

Another site to search for government RFPs on the Internet is *Commerce Business Daily* (*CBD*), which "lists notices of proposed government procurement actions, contract awards, sales of government property and other procurement information. A new edition of the *CBD* is issued every business day, and each edition contains approximately 500–1000 notices." The *CBD* website (**www.cbd-net.com**) lists the government's requests for, among other things, equipment, supplies, and a variety of services from assembling to maintaining equipment to dredging.

Here are a few of the agencies you can search in the *CBD* for RFPs:

- Department of Health and Human Services
- Department of Energy
- Department of Agriculture
- Defense Logistic Agency
- Library of Congress
- Department of Justice and other law enforcement agencies

5. **Prove that your proposal is workable.** The bottom-line question from your readers is "Will this plan work?" Your proposal should contain no statements that say, "Let's see what happens if we do *X* or *Y*." Analyze and test each part of your proposal to eliminate any quirks and to revise the proposal appropriately before readers evaluate it. What you propose should be consistent with the organization and capabilities of the company and should respect its corporate mission and culture. For instance, recommending that a small company of eighteen employees triple its workforce to implement your plan would be foolish and risky.

6. **Be sure your proposal is financially realistic.** "Is it worth the money?" is another bottom-line question you can expect from your readers. For example, recommending that your company spend $20,000 to solve a $2,000 problem is just not feasible. Note how Figure 13.5 stresses that the costs are under control. Above all, make readers believe that the benefits are worth the costs.

Document Design and Your Proposal

As we saw in Chapter 11, the overall design and layout of a document play a major role in its acceptance by an audience. This is especially true of a proposal—a key sales document for you and your company. Keep in mind that your proposal will be competing with many (perhaps 50 or 100) other proposals, and the first impression it makes should be attractive, logically organized, and reader friendly. If it is designed professionally and pleasingly, it will remain in the running. If not, your proposal may be rejected before your audience reads your first sentence.

Here are some guidelines to help you prepare an attractive and carefully designed proposal.

- Make sure you follow the RFP guidelines to the letter—in terms of spacing, title page, number of copies, appendixes, exhibits, and so on.
- Double-check to make sure your proposal looks professional. Use good-quality paper and a sturdy binding.
- Organize your proposal into sections that help readers identify and follow its various parts—for example, problem, solution, budget, timetable, personnel.
- Use plenty of clearly marked, logically ordered, and consistent headings (or, if necessary, subheadings) to separate sections of your proposal to help readers follow and understand your work easily, quickly, and clearly.
- Insert extra spacing between sections of your proposal so they stand out and show readers your work is organized.
- Use a professional-looking and easy-to-read font and type size. Stay away from script fonts and those with ornate designs. Do not try to cram more information in by resorting to an 8-point type.
- Include easy-to-follow indented lists, each item preceded by a bullet or an asterisk.
- Clearly label and insert all visuals in the most appropriate places in your proposal. See Figure 13.6, which does an especially effective job.
- Put budgets in easy-to-read tables; do not bury numbers in a paragraph of prose. Make sure each item in a budget is identified and highlighted or relegated to a footnote or an appendix.
- Keep your paragraphs to five to six sentences. Heavy blocks of prose slow readers down and make them think your work is dense and hard to follow. Consider your readers' comfort level.
- Do not fail to include and label any supporting documents or materials that are not a part of your proposal proper, for example, schedules, surveys, or samples, as in Figure 13.5.

7. **Be ethical.** Your proposal needs to follow all the guidelines for ethical conduct on pages 29–42. That means, essentially, that you must be trustworthy and truthful about all the claims you make about products, services, and contracts and that you will be professional and respond to any questions or problems your readers voice. Later sections of this chapter discuss the ethical standards you must adhere to when writing an internal proposal (pages 613–621) and a sales proposal (pages 621–629).

8. **Package your proposal attractively.** Make sure that your proposal is well presented (professional looking, inviting, and easy to read) and that all visuals are clear and appropriately placed. The visual appearance of your proposal can contribute greatly to whether it is accepted. Study the guidelines in the Tech Note on page 612.

Internal Proposals

The primary purpose of an internal proposal, such as the one shown in Figure 13.6, is to offer a realistic and constructive plan to help your company run its business more efficiently and economically.

On your job you may discover a better way of doing something or a more efficient way to correct a problem. You believe that your proposed change will save your employer time, money, or further trouble. (Note how Tina Escobar and Oliver Jabur identified and researched a more effective and less costly way for Community Federal Bank to conduct its business and to satisfy its customers in Figure 13.6.) Or your department head, manager, or supervisor may call your attention to a problem and ask you for specific ways to solve it.

Regardless of who identifies the problem, your proposal, generally speaking, will be an informal, in-house message. A brief (one- to three- or four-page) memo, as in Figure 13.6, or even a shorter email, should be appropriate.

Some Common Topics for Internal Proposals

An internal proposal can be written about a variety of topics, including the following:

- purchasing new or more advanced equipment to replace obsolete or inefficient computers, appliances, vehicles, and the like, or upgrading equipment technology
- obtaining new software and offering training sessions to show employees how to use it
- recruiting new employees or retraining current ones on a new technique or process
- eliminating a dangerous condition or reducing an environmental risk to prevent accidents—for employees, customers, or the community at large
- cutting costs—for services, supplies, transportation, advertising, etc.
- improving technology/communication within or between departments of a company or agency
- expanding work space or making it greener, more private, ergonomically comfortable and efficient for employees, or more inviting to customers

FIGURE 13.6 An Internal Unsolicited Proposal in Memo Format

EQUAL HOUSING
LENDER

POWELL
617-584-5200

MONROE
781-413-6000

LANGSTON
508-796-3009

COMMUNITY
FEDERAL BANK

www.comfedbank.com

TO: Michael L. Sappington, Executive Vice President
            Dorothy Woo, Langston Regional Manager
FROM: Tina Escobar, Oliver Jabur, ATM Services
DATE: June 10, 2011
RE: A proposal to install an additional ATM at the Mayfield
           Park branch within the next 60–90 days

PURPOSE

We propose a cost-effective solution to what is a growing problem at our Mayfield Park branch in Langston: inefficient servicing of customer needs and rising personnel costs. We recommend that you approve the purchase and installation, within the next two to three months, of another iSmart ATM at Mayfield. Such action is consistent with Community's goals of expanding branch banking services and promoting our image as a self-serve yet customer-oriented institution.

THE PROBLEM WITH CURRENT SERVICES
AT MAYFIELD PARK

Currently, we employ four tellers at Mayfield. But too much is being spent on personnel/salary for routine customer transactions. In fact, as determined by teller activity reports, nearly 25 percent of the four tellers' time each week is devoted to activities easily accommodated by the installation of another ATM. In the table below we break down teller activity for the month of May:

Teller #	Total Transactions	Routine Transactions
1	6,205	1,551
2	5,989	1,383
3	6,345	1,522
4	6,072	1,518
	24,611	5,974

Clearly, we are not taking advantage of our tellers' sales abilities when they are kept busy with routine activities. To compound the problem, we expect business to increase by at least 25 percent at Mayfield in the next few months, as projected by this year's market survey. If we do not

Clearly states why proposal is being sent

Acknowledges bank's corporate mission

Identifies problem by giving reader essential background information based on primary research

Provides easy-to-read table

FIGURE 13.6 (Continued)

Page 2

install an additional ATM, we will need to hire a fifth teller, at an annual cost of $29,900 ($23,000 base pay plus approximately 30 percent for fringes), for the additional 6,000 transactions we project.

Most important, though, customer needs are not being met efficiently at Mayfield. Recent surveys done for Community Federal by Watson-Perry, Inc., demonstrate that our customers are inconvenienced by not having an iSmart ATM at Mayfield. They are unhappy about long waits both at the ATMs and at the teller windows to do simple banking business, such as making deposits, withdrawals, and loan payments. Several discussions we had (on May 7, 24, and June 1) with manager Rachael Harris-Koyoto at Mayfield Park have confirmed customers' complaints.

Ultimately, the lack of another iSmart ATM at Mayfield hurts Community's image. With plentiful ATMs available to Mayfield residents at local stores and at other banks, our institution risks having customers and potential customers go elsewhere for their banking needs. We not only miss the opportunity of selling them our other services but also risk losing their business entirely.

A SOLUTION TO THE PROBLEM
Purchasing and installing an additional iSmart ATM at Mayfield Park will result in significant savings in personnel costs and time. Specifically, we can accomplish the following:

- Save money by not having to hire a fifth teller
- Use the additional iSmart ATM to further provide many of the services customers expect from our tellers, e.g., accept deposits, display image copies of checks on receipts, record total cash and check deposits
- Allocate teller duties more efficiently and productively by assisting customers with questions and transactions that cannot be handled through an ATM, such as opening a new account; purchasing bonds, CDs, and international currency; and setting up Internet banking services
- Increase time for tellers to cross-sell our services, including our line of banking products—annuities, mutual funds, debit cards, and global market accounts

Divides problem into parts—volume, financial, personnel, customer service

Emphasizes the expense if nothing is done

Cites important research

Verifies that problem is widespread

Emphasizes possible future problems

Relates solution to individual parts of the problem

Bulleted list makes benefits recommendation easy to follow

(Continued)

FIGURE 13.6 (Continued)

Page 3

- Service customer retirement options by having tellers track IRAs, 401(k)s, 403(b)s, and Roth IRAs
- Improve customers' satisfaction by giving them the option of meeting their banking needs electronically or through a teller
- Ease the stress on tellers at the high-traffic Mayfield Park branch

Shows problem can be solved and explains how

FEASIBILITY OF INSTALLING NEW iSMART ATMS

It is feasible to install another ATM at Mayfield. This location does not pose the difficulties we face at some of our older branches. Mayfield offers ample room to install a drive-up ATM in the stubbed-out fourth drive-up lane. In fact, it is away from the heavily congested area in front of the bank, yet it is easily accessible from the main driveway and the side drive facing Commonwealth Avenue, as the photograph below shows.

Photo shows location has room for additional ATM

Photo courtesy Taylor Wilson

Documents that work can be done on time and highlights advantage of doing it now

Judging experiences at the Powell branch, the iSmart ATM could be installed and operational within one to two months. Moreover, by authorizing the expenditure at Mayfield within the next month, you will ensure that additional ATM service is available before the busy Christmas season.

FIGURE 13.6 (Continued)

Page 4

COSTS

The costs of implementing our proposal are as follows:

Diebold drive-up iSmart ATM	$28,000.00
Installation fee	2,000.00
Maintenance (1 year)	1,500.00
	$31,500.00

Itemizes costs

This $31,500, however, does not truly reflect our annual expenditures. We would be able to amortize, for tax purposes, the cost of the installation of the ATM over five years. Our annual expenses would, therefore, actually be

Interprets costs for reader

$$\$30,000 \ (28,000 + 2,000)/5 \text{ years} = \$6,000 \text{ per year} + \$1,500 \text{ (maintenance)}$$

$7,500 per year

Compared with the $29,900 a year the bank would have to expend for a fifth teller position at Mayfield Park, the annual depreciated cost for the ATM ($7,500) in fact reduces by almost 75 percent ($22,400) the amount of money the bank will have to spend every year for much more efficient customer service.

Proves change is cost effective; provides specific financial evidence

CONCLUSION

Authorizing another ATM for the Mayfield Park branch is both feasible and cost effective. Endorsement of this proposal will save our bank more than $22,400 in teller services annually, decrease customer complaints, and increase customer satisfaction and approval. We will be happy to discuss this proposal with you anytime at your convenience, and answer any questions you may have.

Conclusion stresses benefits for the bank and its customers

Thank you for considering our plan. We look forward to hearing from you.

Thanks readers for studying the proposal

As this bulleted list shows, internal proposals cover almost every activity or policy that can affect the day-to-day operations of a company or an agency.

Following the Proper Chain of Command

Writing an internal proposal requires you to be sensitive to office politics. It may be wise first to meet with your boss to see if she or he has already identified the problem or has specific suggestions on how to solve it. If given the go-ahead, then you and your team need to provide your boss with a draft and ask for feedback.

But do not assume that your readers will automatically agree that there is a problem or that your plan is the only way to tackle it. To write a successful internal proposal, keep in mind the needs and preferences of your supervisor and others who may have to sign off on it. Remember that your employer will expect you to be very convincing about both the problem you say exists and the changes you are advocating in the workplace under his or her supervision. Don't rock the corporate boat by going over your supervisor's head, questioning his or her authority, or suggesting a plan that is too costly. Doing this, you are not showing respect for your company's mission, chain of command, or budget.

Ethically Resolving and Anticipating Corporate Readers' Problems

When you prepare an internal proposal, you need to be aware of the ethical obligations you have and the ways to meet them. Here are some important guidelines:

1. **Consider the implications of your plan company-wide.** The change you propose (transfers, new budgets or technology, new hires) may have sweeping and potentially disruptive implications for another office or division in your company.

2. **Keep in mind the possible impact of your change on co-workers from cultural traditions other than your own.** In addition to speaking to your employer, consult your human resources or cultural diversity director.

3. **Find out whether what you are proposing is within your company's budget.** How much can your department, branch, or office spend (e.g., on hardware, software, updates)? Check with your boss, and always monitor the price of your company's stock orders to make sure any expenditures are likely to be approved.

4. **Realize that your reader may feel threatened by your plan.** Be careful. Your plan may sound just right to you, but your boss may regard it as a piece of criticism leveled at him or her, your department, or the company as a whole. Think about the long-term effect your proposed change will have on your boss and co-workers. Don't override your boss or attempt to undermine existing authority.

5. **Take into account that your reader may have "pet projects" or predetermined ways of doing things.** Make every attempt to acknowledge them respectfully. You may even find a way to build on or complement such projects or procedures.

6. **Keep in mind that your boss may have to take your proposal farther up the organizational ladder for commentary and approval.** Again, refer to Joycelyn Woolfolk's description in Figure 3.6 (pages 99–100) of the chain of command at her organization.

7. **Never rely on someone else to supply the specific details on how your proposal will work.** For example, do not write an internal proposal that says the marketing, technical support, or human resources department could supply the necessary details for your proposal to work. That unfairly pushes the responsibility onto others.

Organization of an Internal Proposal

A short internal proposal follows a relatively straightforward plan of organization, from identifying the problem to solving it. Internal proposals usually contain four parts, as shown in Figure 13.6: *purpose*, *problem*, *solution*, and *conclusion*. Refer to the figure as you read the following discussion.

Purpose

Begin your proposal with a brief statement of why you are writing to your supervisor: "I propose that . . ." State right away why you think a specific change is necessary now. Then succinctly define the problem and emphasize that your plan, if approved by the reader, will solve that problem.

Problem

In this section, prove that a problem exists. Document its importance for your boss and your company; as a matter of fact, the more you show, with concrete evidence, how the problem affects your boss's work (and area of supervision), the more likely you are to persuade him or her to act.

Here are some guidelines for documenting a problem:

- Avoid vague (and unsupported) generalizations such as these: "We're losing money each day with this procedure." "Costs continue to escalate." "The trouble occurs frequently in a number of places." "Numerous complaints have come in." "If something isn't done soon, more problems will result." Figure 13.6 focuses on "rising personnel costs."
- Provide quantifiable details about the problem, such as the amount of money or time a company is actually losing per day, week, or month. Document the financial trouble so that you can show in the next section how your plan offers an efficient and workable solution. See the table in Figure 13.6.
- Indicate how many employees (or work-hours) are involved or how many customers are inconvenienced or endangered by a procedure or condition. The writers in Figure 13.6 researched teller activity and salary and used this information to document the importance of the problem.
- Verify how widespread a problem is or how frequently it occurs by citing specific occasions. Again, see how Escobar and Jabur cite evidence from the interview they conducted with branch manager Rachael Harris-Koyoto.

■ Relate the problem to an organization's image, corporate reputation, or in-fluence (where appropriate). Pinpoint exactly how and where the problem lessens your company's effectiveness or hurts its standing in the market. In-dicate who is affected and how the problem affects your company's busi-ness, as the writers do especially well in the second paragraph of the problem statement in Figure 13.6 on pages 614–615.

Solution or Plan

In this section, describe the change you propose and want approved. Tie your solu-tion (the change) directly to the problem you have just documented. Each part of your plan should help eliminate the problem or should help increase the productiv-ity, efficiency, or safety you think is possible.

Your reader will again expect to find factual evidence. Be specific. Do not give merely an outline of your plan or say that details can be worked out later. Supply details that answer the following questions: (1) Is the plan workable? and (2) Is it cost-effective?

To get the reader to say "Yes" to both questions, supply the facts you have gathered as a result of your research. For example, if you propose that your firm buy new equipment, do the necessary homework to find the most efficient and cost-effective model available, as the proposal writers in Figure 13.6 do about the bank purchasing and installing a new ATM.

■ Supply the vendors' names, the costs, major conditions of service and train-ing contracts, and warranties.
■ Describe how your firm could use the equipment or technology to obtain better or quicker results.
■ Document specific tasks the new equipment can perform more efficiently at a lower cost than the equipment now in use.

A proposal to change a procedure must address the following questions:

■ How does the new (or revised) procedure work?
■ How many employees or customers will be affected by it?
■ When can it go into operation?
■ How much will it cost the employer to change procedures?
■ What delays or losses in business might be expected while the company switches from one procedure to another?
■ What employees, equipment, technology, or locations are already available to accomplish the change?

As those questions indicate, your reader will be concerned about schedules, work-ing conditions, employees, methods, locations, equipment, and the costs involved in your plan for change. The costs, in fact, will be of utmost importance, and you may want to provide a separate and detailed "Costs" section in your proposal to demonstrate their importance, as the writers of the proposal in Figure 13.6 do on page 617. Make sure you supply a careful and accurate budget. Moreover, make those costs attractive by emphasizing how inexpensive they are compared to the cost of *not* making the change.

Conclusion

Your conclusion should be short—a paragraph or two at the most. Remind readers that (1) the problem is ongoing and serious, (2) the reason for change is justified and beneficial to your organization, and (3) action needs to be taken. Re-emphasize the most important benefits as Escobar and Jabur do in their proposal in Figure 13.6. Also indicate that you are willing to discuss your plan with the reader and want his or her feedback, a necessity in arguing for a corporate change at any level.

Sales Proposals

A sales proposal is the most common type of external proposal. Its purpose is to sell your company's products or services for a set fee. Whether short or long, a sales proposal is a marketing tool that includes a sales pitch as well as a detailed description of the work you propose to do. Figures 13.5 and 13.7 are examples of sales proposals.

The Audience and Its Needs

Your audience will usually be one or more executives who have the power to approve or reject a proposal. Your audience for a sales proposal may be even more skeptical than readers of an internal proposal, since they may not know you or your work. But you can increase your chances of success by trying to anticipate their questions, such as:

- Does the writer's firm understand our problem?
- Can the writer's firm deliver what it promises?
- Can the job be completed on time?
- Is the budget reasonable and realistic?
- Will the job be done exactly as proposed?
- Has the writer demonstrated his or her qualifications and trustworthiness?

Answer each of these questions by demonstrating how your product or service is tailored to the customer's needs.

Make sure, too, that your proposal has a competitive edge. Readers will compare your plan with those they receive from other proposal writers. Your proposal has to convince readers that the product or service your company offers is more reliable, economical, efficient, and timely than that of another company. Here is where your homework will pay off.

The key to success is incorporating the "you attitude" throughout your proposal. Relate your product, service, or personnel to the reader's exact needs as stated in the RFP for a solicited proposal or through your own investigations for an unsolicited proposal. *You cannot submit the same proposal for every job you want to win and expect to be awarded a contract.* Different firms have different needs.

FIGURE 13.7 An Unsolicited Sales Proposal

Professional, distinctive letterhead

OMEGA
Technologies, Inc.

203-555-6714   203-555-1732   techsupport@interserv.com

www.otech.com   Follow Us On Facebook and Twitter

August 3, 2011

Ms. Alexandra Tyrone-Phelps
Vice President, Operations
Gemini Design Solutions, Inc.
Hartford, CT 06631-7106

Dear Ms. Tyrone-Phelps:

Begins by thanking reader for using Omega's services and suggests ways Omega can help

Thank you for allowing Omega to service your Webmax desktops and assess your networking capabilities last week. While we were working at your facility, I saw several ways in which Omega Technologies might improve your system. Based on our assessment of Gemini's requirements for the most up-to-date hardware available, we recommend that you replace your current Alpha 500 notebooks, which you purchased three years ago, with ten Lightstar 686 notebooks, one for each member of your sales staff. The Lightstar 686, the latest generation of notebooks, can provide your staff with a much more powerful and far more cost-effective way to communicate both in-house and with your customers.

PROBLEM AREAS

Clearly identifies and quantifies the problem and explains why it is important

Consistently demonstrates "you attitude"

Dated technology is costing Gemini time and money. Using their current Alpha 500 notebooks, each of your ten sales staff spends anywhere from 5 to 15 minutes in a client presentation waiting for your product videos to load. Cumulatively, that amounts to over 50 hours a week. Moreover, the lack of memory required to run large files causes the Alpha 500 notebooks to crash often, further wasting your staff's time as they have to enter sales data onto company desktops. Finally, and most important, the playback of your sales videos on the Alpha 500 is often choppy, falling far short of the sharp and professional presentations Gemini has been known for and expects.

FIGURE 13.7 (Continued)

Page 2

ADVANTAGES OF THE LIGHTSTAR 686 NOTEBOOK

The Lightstar 686 will give Gemini state-of-the-art notebook technology by doing the following:

- Providing you with 6 GB of RAM (memory) as opposed to the Alpha 500, which has only 1 GB
- Enabling you to expand Lightstar's memory to 12 GB of RAM as your needs increase
- Running applications three times as fast as the Alpha 500 can, without risk of crashing, even with large files
- Expanding your graphics capabilities with a dedicated graphics card unique to the Lightstar 686
- Improving your networking capabilities and eliminating duplication of efforts by your sales staff in entering data too extensive for their notebooks onto desktops

The Lightstar has received 5-star reviews from business users across the country. According to Felicia Gomez, writing for her widely respected independent blog *TechToday* (**techtoday.com**), the Lightstar is "one of the most advanced laptops of the decade; its extraordinary graphics capabilities make it a powerhouse that will amaze the most particular business user." The description of the Lightstar 686 below points out its many advantages.

ADVANCED TECHNOLOGY

The Lightstar comes equipped with the following advanced features:

- Intel® Core™ i7 processor
- 64-bit operating system
- 6 GB of internal memory (RAM)
- 500 GB 7200 rpm SATA II internal drive
- 450 GB hard drive
- 15" diagonal LED-backlit HD+ WVA anti-glare (1600x900)
- Genuine Windows® Professional 64
- DVD+/-RW optical drive
- 1 GB dedicated GDDR5 video memory
- 8-cell Li-Ion battery
- HD webcam
- Touchpad
- Touch-screen
- Stereo microphone in and headphone/line-out
- Intel Centrino 802.11a/b/g/n and Bluetooth 2.1+EDR

Please visit our website (**www.otech.com**) for even more information about Lightstar's state-of-the-art technical specifications.

Gives a detailed list of features that prove the Lightstar 686 is more efficient than the Alpha 500

Uses parallel verbs for each bulleted point

Provides documented endorsement from outside source

Provides selected specifications for main components and refers reader to website for complete listing

(Continued)

FIGURE 13.7 (Continued)

Page 3

Details Lightstar's many advantages, each clearly and persuasively described under easy-to-follow headings

Enhanced Graphics/Video Capabilities

Unlike your current Alpha laptops, Lightstar comes with a dedicated graphics card as a standard feature enabling your sales force to perform any graphically demanding job they or your customers need. In fact, the Lightstar stores seven times the graphics files you currently have, saving your staff valuable time. Moreover, the Lightstar can play up to 5½ hours of video as well as allow your staff to edit videos. The Lightstar's 15-inch LED-backlit HD+ screen will give you unparalleled crisp HD images, tremendously improving PowerPoint presentations.

Multitasking Functions

You will not have to worry about the crashes, freezes, and long load times of your current laptops, since the Lightstar can process up to 8 streams of information simultaneously, giving your staff the ability to perform many vital jobs faster and far more efficiently. With the Lightstar, they can also download information faster, make a PowerPoint presentation, prepare documents in Word or Excel, and even listen to video reports simultaneously, all in the time it takes to do just one of these tasks with your current Alpha notebooks.

Stresses the Lightstar's energy-saving and ecofriendly features

Energy Efficient

The Lightstar is *Energy Star* qualified, meets or exceeds all EPA guidelines for energy efficiency, and is designed with state-of-the-art heat dissipation. Moreover, the Lightstar keyboard is quiet, eliminating the disturbing clicking sounds heard with other notebooks.

Explains financial benefits of buying Lightstar 686 notebooks

Cost Effective

Your Lightstar 686 will cost 25 percent less than your Alpha notebook did 3 years ago, yet it offers the latest high-tech performance available today. Here are some of the bottom-line savings you can expect:

- includes a top-of-the-line processor at no extra cost
- provides software upgrades at no charge
- includes everything integrated into the system—monitor, graphics card, software

Emphasizes multiple security features

Assured Security

Lightstar offers a unique protection system to ensure the security of all the data sent by or stored on your unit:

- an encryption system requiring a special password to log on to and activate the notebook
- an option to include fingerprint recognition

FIGURE 13.7 (Continued)

Page 4

- a special feature—"Call Home"—that will alert you to unauthorized use
- Business Pro antivirus software free of charge

Efficient Accessibility and Customer Service
Because the Lightstar is compatible with the desktops in your network, your representatives have access to all pertinent records at your main office while making sales calls and doing presentations. By having direct access to this vital documentation, your staff can more efficiently network with you and with clients, as well as fulfill, verify, and store information about customer orders. In brief, the Lightstar makes the virtual office a reality. And because your sales force would be carrying all the resources of your office with them, you will significantly enhance their telecommuting capabilities now and into the future.

Shows how Lightstar is compatible with reader's current technology and is therefore easy to install and use

Power Supply Capabilities
Equipped with an 8-cell lithium-ion battery, the Lightstar 686 operates up to six hours without recharging, enabling your staff to make sales calls at any location for extended periods of time.

Portability/Durability
- weighs only 4.6 pounds—it is ultraportable
- designed for mobile business success
- slips easily into a briefcase
- comes with protective, dent-proof case

Emphasizes convenience and efficiency

TRAINING AND SERVICE
Because all of Omega's software can be preloaded, setup will take less than 30 minutes. Since Lightstar is compatible with Gemini's network, your staff will need minimal training to learn how to use the Lightstar to upload sales orders, graphics, and files, and download product availability information. To help Gemini with any technical needs, our local tech representative, Ory Mawfic, is available online, by phone, on Skype, and for site visits to instruct your staff about the operation and to provide routine maintenance and assessment. Moreover, we offer our customers the latest in remote diagnosis, a service we have been honored to provide many times to Gemini as part of our service agreement.

Stresses varieties of service options after the sale and refers to previous business relationship

(Continued)

FIGURE 13.7 (Continued)

Spells out
costs exactly,
including
options reader
might select

Uses boldface
for bottom-line
figures

Provides a
short history
of Omega's
outstanding
service record
and supplies
references

Closes with a
reminder of
past favorable
business
relationship
to help ensure
future sales

Page 5

COSTS

Below is an estimated price for one Lightstar 686 notebook with upgraded features as itemized on page 2.

Total purchase price:	**$1,766.00**

You might also consider adding one or all of these options:

Windows anytime upgrade	89.00
USB 2-button optical mouse	30.00
Total cost for options:	**$119.00**

OMEGA TECHNOLOGIES' REPUTATION

For the past 16 years, Omega Technologies has provided state-of-the-art computer technology and fast, efficient service after the sale. Last year Omega won an American Business Award for quality service after the sale. A list of our business references is attached. Please feel free to write or call them.

We appreciate your past confidence in choosing Omega Technologies and look forward to providing Gemini with the latest generation of desktops and laptops and the most efficient and convenient support services possible. We are confident that our notebooks will be able to address and provide ample solutions to all of Gemini's technology troubles.

Please call me if you have any questions about the Lightstar or Omega Technologies. I would be happy to bring a few Lightstar 686's to Gemini for you and your sales staff to use for a trial period.

If this proposal is acceptable, please sign and return a copy of the enclosed agreement with this letter.

Sincerely yours,

Marion Copely

Marion Copely

Encl. 2: Business References
　　　Sales Agreement

Being Ethical and Legal

In addition to the guidelines in Chapter 1, here are some ways to make sure your sales proposal follows the highest ethical standards:

- **View your proposal as a contract.** If you omit information, misrepresent claims, or minimize risks, you can be taken to court and sued for damages.
- **Submit a complete, accurate, and fair budget.** Neither underestimate nor inflate costs or charges. Break down all costs in your budget. Indicate if your fees are by the job, weekly, or hourly. Always alert readers to any possible additional charges (e.g., the fees for permits, an increase in the price of materials). Let readers know, too, if you are willing to match a competitor's price, but make sure your price includes the same type and quality of work for the same amount of time.
- **Estimate a realistic timeframe to do the work.** It would be unethical to say a job takes more time than necessary so you can then charge more for your service.
- **Stipulate precisely what your product can (or cannot) do and what a service contract includes and excludes.** Don't make false claims. Always identify exceptions, limitations, and restrictions.
- **Clearly specify any technical problems or risks involved, and suggest ways to eliminate or minimize them** (e.g., by purchasing additional antivirus software, safety equipment, or extra liability insurance or by meeting all codes).
- **Don't misrepresent your competition** by misquoting their prices or maligning their work.

Organization of a Sales Proposal

Most sales proposals include the following elements: introduction, description of the proposed product or service, timetable, costs, qualifications of your company, and conclusion.

Introduction

The introduction to your sales proposal can be a single paragraph in a short sales proposal or several pages in a more complex one. Basically, your introduction should prepare readers for everything that follows in your proposal. The introduction itself may contain the following sections, which are sometimes combined:

1. **Statement of purpose and subject of proposal.** Tell readers why you are writing, and identify the specific subject of your work. If you are responding to an RFP, use specific code numbers or cite application dates. If your proposal is unsolicited, indicate how you learned of the problem, as Figure 13.7 does. Briefly define the solution you propose. Tell your readers exactly what you propose to do for them. Be clear about what your plan covers and, if there could be any doubt, what it does not.

2. **Background of the problem you propose to solve.** In a solicited proposal like Figure 13.5, this section is usually unnecessary, because the potential client has already identified the problem and wants to know how you would solve it. In that

case, just point out how your company would address the problem, mentioning your superiority over your competitors (see the fourth paragraph of Figure 13.5).

In an unsolicited proposal, you need to describe the problem in convincing detail, identifying the specific trouble areas. You may want to focus briefly on the dimensions of the problem—when it was first observed, who or what it most acutely affects, and the specific organizational or community context in which the problem is most troubling.

If it is an external proposal to a current customer, such as the one in Figure 13.7, it would be unwise to point out previous problems your client may have had with your company's service or products. In Figure 13.7, however, note how the problems the sales staff encountered in the field are described mainly as a way to sell the advantages of the Lightstar 686 notebook.

Description of the Proposed Product or Service

This section is the heart of your proposal. Before spending their money, customers will demand hard, factual evidence of what you claim can and should be done. Here are some points that your proposal should cover:

1. Carefully show potential customers that your product or service is right for them. Stress the particular benefits of your product or service most relevant to your reader. Blend sales talk with descriptions of hardware, as the section "Cost Effective" does in Figure 13.7.

2. Describe your work in suitable detail. Specify what the product looks like, what it does, and how consistently and well it will perform in the readers' office, plant, hospital, or agency. You might include a brochure, a picture, a diagram, or, as the writer of the proposal in Figure 13.5 does, a sample of your product for customers to study. See how in Figure 13.7 the reader is given an independent website to consult about the Lightstar's capabilities and endorsement.

3. Stress any special features, maintenance advantages, warranties, or service benefits. Convince readers that your product is up-to-date and energy efficient. Highlight features that show the quality, consistency, or security of your work. For a service, emphasize the procedures you use, the terms of the service, and even the kinds of tools you use, especially any state-of-the-art equipment.

Timetable

A carefully planned timetable shows readers that you know your job and that you can accomplish it in the right amount of time. Your dates should match any listed in the RFP. Provide specific dates to indicate

- when the work will begin
- how the work will be divided into phases or stages
- when you will be finished
- whether any follow-up visits or services are involved

For proposals offering a service, specify how many times—an hour, a week, a month—customers can expect to receive your help; for example, spraying two times a month for an exterminating service.

Costs

Make your budget accurate, complete, and convincing. Don't underestimate costs in the hope that a low bid will win you the job. You may get the job but lose money doing it, because the customer will rightfully hold you to your unrealistic figures. Accepted by both parties, a proposal is a binding legal agreement. Neither should you inflate prices; competitors will beat you in the bidding.

Give customers more than merely the bottom-line cost. Show exactly what readers are getting for their money so they can determine if everything they need is included. Itemize costs for

- specific services
- equipment and materials
- labor (by the hour, by the day, or by the job)
- transportation
- travel
- training

If something is not included or is considered optional, say so—additional hours of training, replacement of parts, upgrades, and the like.

If you anticipate a price increase, let the customer know how long current prices will stay in effect. That information may spur them to act favorably now.

Qualifications of Your Company

Emphasize your company's accomplishments and expertise in providing similar products or services—certificates, accreditations, awards—as the proposal in Figure 13.7 does. Indicate if your firm assigns a project manager to monitor each job it does and works closely with customers to meet their needs and answer questions. Say you place a high priority on cost savings and that you have a reputation for fast response times if clients have questions or concerns. Mention the names of important local firms for whom you have worked that would be able to recommend you. But never misrepresent your qualifications or those of the individuals who work with or for you. Your prospective client may verify if you have in fact worked on similar jobs during the last five or six years.

Conclusion

This is the "call to action" section of your proposal. Encourage your reader to approve your plan by stressing its major benefits. Offer to answer any questions the reader may have. Some proposals end by asking readers to sign and return a copy of a sales agreement, as the proposal in Figure 13.7 does. And build on any previous goodwill or business, as the last paragraph does in Figure 13.7, to further establish trust with your customer or client.

Proposals for Research Reports

You can also expect to write a proposal to your boss seeking her or his authorization to do research, or to an instructor in a course to get a paper or report topic approved, as in Figure 13.8 (pages 631–634).

The principles guiding internal and sales proposals also apply to research proposals. As with internal and sales proposals, you will be writing to convince the reader—your boss or instructor—to approve a major piece of work. Your boss will read your proposal to make sure you write the best possible report to solve a company problem and to increase company profits. Your instructor will look at your report to make sure it meets the course objectives. In drafting any research proposal, make sure of four things:

- that you have chosen a significant topic
- that you have sufficiently restricted the topic
- that you will investigate important sources of information about that topic
- that you can accomplish your work in the specified time

Your proposal will give your reader an opportunity to spot omissions or inconsistencies and to provide helpful suggestions.

To prepare an effective proposal for a research project, you must do some preliminary research. You cannot pick any topic that comes to mind or guess about procedures, sources, or conclusions. As other proposal readers will, your boss or instructor will want convincing and specific evidence about why you are researching this topic and your approach to it. Be prepared to cite key facts to show that you are familiar with the subject and that you are prepared to write about it knowledgeably.

Be very clear about the type of research you will do and what resources you intend to use. For example, you need to do some preliminary research using the following sources before you write a research proposal:

- Internet—list relevant websites you have consulted
- Search engines or databases—for a working bibliography
- Books and articles, online and print—but only those that bear directly on your topic
- Interviews—in person, email, over the telephone, and so on
- Proposed visits to relevant sites—laboratories, salt marshes, health care facilities, plants, offices, and so on

Review Chapter 8 on primary and secondary research methods.

Organization of a Proposal for a Research Report

Your proposal for an in-house report or for a school research project can be a memo or an email divided into six sections, as illustrated in Figure 13.8: *purpose* (introduction), *background of the problem* (why it is important), *questions to be investigated, methods of research, timetable,* and *request for approval.* However, be ready to reorder or expand these sections if your reader wants you to follow a different organizational plan.

Purpose (Introduction)

Keep your introduction short—a paragraph, maybe two, pinpointing the subject and purpose of your work.

> I propose to research and write a report about the "hot knife" laser used in treating port wine stains and other birthmarks.

> I intend to investigate the relationship that exists between office design and our employees' need for "psychological space."

FIGURE 13.8 A Proposal to Write a Research Report for a Class

To: Prof. Marisol Vega
From: Barbara Shoemake
Date: April 20, 2011
Subject: Proposal to write a report on the ethical issues involved in m-commerce

Purpose

To help consumers better understand the implications of using their smartphones and other mobile devices, I propose to research and write a background report on the ethical issues involved in m-commerce, or mobile marketing.

Concisely states purpose

Background of the Problem

The use of smartphones and other mobile devices from Apple, Android, and other vendors has become so widespread in recent years that Sarah Pentoney has claimed, "More people around the globe use iPhones than computers" (74). Consumers in the U.S. alone use more than 150 million smartphones running over 15 billion apps a day. Tim Sherwin, executive vice president of Cardinal Commerce Corporation, declares, "By leveraging the convenience of a consumer's mobile phone, merchants can drive loyalty, brand affinity, and simplify the transaction process through the new product capability" (n.p.). Clearly, m-commerce is a powerful tool worldwide.

Demonstrates importance of topic with statistics and quotation from m-commerce executive

Despite their immense popularity, iPhones, iPads, and other mobile devices present major ethical challenges. Essential to m-commerce technology and use is what Sara Holmes calls "hotspots that trigger a message or ad delivered to your device if you walk or drive into a given area" (7). Each device includes a UDID, or unique device identity, that sends and time stamps data about the consumer's whereabouts and online activities. Cellular service providers can therefore track and store this data indefinitely. M-commerce retailers use a consumer's UDID to alert him or her to a sale or to the arrival of a new product in a nearby store. Marketers can also text a consumer as he or she drives by a restaurant, a mall, an auto dealership, or even a yard sale.

Provides further details about scope and significance of problem to be investigated

(Continued)

FIGURE 13.8 (Continued)

Page 2

References the research of a security expert

But more threatening, a consumer's UDID can be used for purposes other than promoting a particular brand or service. Security experts and members of Congress are concerned that m-commerce strategies violate a consumer's privacy. Granted, consumers have to divulge their location to receive mobile sales alerts. But Eric Smith, whose research is often cited in the literature on m-commerce, still worries about the ethics of tracking an individual's trips, locations, and driving habits, which allows cell phone providers to predict where he or she goes and when. That information, Smith contends, could be "collected and sold to unintended customers such as . . . divorce lawyers, debt collectors, or even industrial spies" (n.p.), not to mention hackers.

Convincingly shows widespread implications of problem

Identifies audience for whom report is intended

Understanding the ethical and legal consequences consumers face in using their smartphones is vital for evaluating the terms of their service agreements and the risks they run in using their devices. To ensure that consumers have the latest information on the privacy risks involved in mobile marketing, I request permission to do research for a background report.

Questions to Be Investigated

At this preliminary stage of my research, I think my report will need to address the following questions related to the ethics of m-commerce:

Formulates specific questions report will address

1. How do strategies used in m-commerce compare with those used in e-marketing?
2. What marketing strategies are unique to m-commerce?
3. What specific risks to privacy do consumers face in using their smartphones?
4. What obligations should cellular service providers have to protect consumers' data and identity?
5. What have providers historically done with the data they collect about consumers' driving and buying habits?
6. What legal rights do consumers have to protect their identity and location when using their smartphones or other mobile devices?
7. What specific changes need to be made to safeguard cell phone alerts for legitimate reasons (e.g., warnings about natural disasters)?

Outlines tentative organization of report

I propose to divide the body of my paper according to three key issues of m-marketing: *Monitoring*, *Consumer Security*, and *Necessary Changes*.

FIGURE 13.8 (Continued)

Page 3

Methods of Research

I will survey recent literature dealing with m-commerce and ethics and interview one or two m-marketing experts and one or two branch managers of local smartphone stores. Judging from the large number of entries found on this topic through Google, Yahoo!, and Academic Search Premier, ethical issues in the mobile marketplace, though a relatively new field, have attracted a great deal of attention.

To restrict my topic, I narrowed my keyword search to concentrate on issues of *privacy*, *security*, *surveillance*, *m-consumer*, and *protection* as they affect consumers using smartphones and other mobile devices. From a preliminary check of documents available at our university library and through online databases, I believe that the following books, articles, and blogs may be most useful in my research:

Claburn, Thomas. "Android, iPhone Apps Pose Privacy Problems." *Information Week*. 5 Oct. 2010. Web. 2 Mar. 2011.

Crowe, Marianne, Marc Rysman, and Joanna Stavins. "Mobile Payments in the United States at Retail Point of Sale: Current Market and Future Prospects." *Research Review* 13 (2010): 6–9. Print.

"First Data Introduces Mobile Voucher Technology for Merchants." *BusinessWire*. 22 Mar. 2011. Web. 30 Mar. 2011.

Holmes, Sara. "Watch for Hotspots When You Use Your Apps." *Springfield Herald* 7 Feb. 2011, B3+. Print.

Ketter, Wolfgang, Han La Poutr, and Norman M. Sadeh, eds. *Agent-Mediated Electronic Commerce and Trading Agent Design and Analysis*. New York: Springer, 2011. Print.

Kortlang, Niels. *The Impact of 3G Technologies on User Acceptance of Mobile Commerce*. Saarbrücken, Germany: VDM, 2010. Print.

Li, Yung-Ming, and Yung-Shao Yeh. "Increasing Trust in Mobile Commerce through Design Aesthetics." *Computers in Human Behavior* 26.4 (2010): 673–84. Print.

Maamar, Zakaria. "Commerce, E-Commerce, and M-Commerce: What Comes Next?" *Communications of the ACM* 46.12 (2003): 251–57. Print.

Mukherjee, Supuntha, and Sagib Ighal Ahmed. "Apple Sued over Apps Privacy Issues: Google May Be Next." *Reuters* 28 Dec. 2010. Print.

Pentoney, Sarah. "Mobile Apps and International Security." *Technology and Risk* 7 (Oct. 2010): 73–101. Print.

Gives detailed lists of primary and secondary sources to be consulted, with rationale

Demonstrates how and why topic is restricted

Uses proper MLA style for documentation

Cites only most current and relevant sources

Includes both print and Web-based research

FIGURE 13.8 (Continued)

Page 4

Perez, Sarah. "Mobile App Marketplace: $17.5 Billion by 2014." *Read Write Web.* 17 Mar. 2010. Web. 19 Mar. 2011.

Robertson, Jordan. "Apple Slammed over iPhone, iPad Location Tracking." *Yahoo! News.* 22 Apr. 2011. Web. 23 Apr. 2011.

Sherwin, Tim. "M-commerce and Brand Loyality." *BusinessWire.* Web. Mar. 2011.

Smith, Eric. "iPhone Applications and Privacy Issues: An Analysis of Application Transmission of iPhone Unique Device Identifiers (UDIDS)." pskl.us. 30 Sept. 2010. Web. 23 Feb. 2011.

Thames, Eleanor. "UDIDs and Consumers' Rights." *Marketing,* 6 July 2009: 17-24. Print.

Vizjak, Andrej, and Max Josef Ringlstetter. *Media Management:* Leveraging Content for Profitable Growth. New York: Springer, 2010. Print.

Wasserman, Todd. "Gap Customers Can Now Get Discounts Via Text Message," *Mashable.com.* 21 Apr. 2011. Web. 13 July 2011.

Zinsmeister, Sean. "Top 5 Mobile Applications Sales Tips." *Business Blogs.* 20 Sept. 2010. Web. 19 Mar. 2011.

Identifies need for additional interviews

I also plan to interview at least two marketing experts in the Springfield area to learn how they assess ethical issues involving smartphones. My first choice is Katarina Kuhn at M-Trade, who has offered several seminars on smartphone security and ethics in the past year. If she is unavailable, I will try to interview human resource directors with impressive credentials in the field, such as Paul Goya at Consumer Advocacy Rights and Jen Holka at Tech Consultants, Inc. Further, I have a list of four or five branch managers at Apple, Android, and other providers and intend to call or visit them as well.

Specifies schedule and how to meet it

Timetable

I hope to complete my research by April 15 and my interviews by April 26. Then I will spend the following two weeks working on a draft, which I will turn in by May 9. After receiving your comments on my draft, I will work on revising the final copy of my background report and submit it by June 7, the date you specified. I will also send to you two progress reports—one when I finish my research and another when I decide on the final organization of my repot.

Politely requests approval and feedback

Request for Approval

Thank you for approving my plan for this background report. I would appreciate any further suggestions on how you think I might best proceed.

© Cengage Learning 2013

Then briefly indicate why the topic or the problem you propose to study is significant. In other words, be prepared to explain why you have chosen that topic and why research on it is relevant or worthwhile for a specific audience or objective. Study how Barbara R. Shoemake in Figure 13.8 states how and why her background report will be useful for consumers.

Supply your boss with a few background details about your topic — for example, the importance of using a laser as opposed to conventional ways of treating birthmarks or why psychological space plays a crucial role in employee productivity and morale. Prove that you have thought carefully about selecting a significant, relevant topic.

Background of the Problem

In this second section, give your reader information about why the problem is important. Cite evidence from your preliminary research and explain key terms or ideas. Show how you propose to break the topic into meaningful units. Tell your reader what specific issues, points, or areas you hope to explore.

Questions to Be Investigated

In this section, formulate a list of questions you intend your research report to answer. The topics included in such questions might later become major sections of the report. Make sure your issues or questions do not overlap and that each relates directly to and supports your restricted topic. See how in Figure 13.8, Shoemake proposes to divide her report on m-commerce into three distinct yet related areas.

Methods of Research

In the fourth section of your proposal inform your boss how you expect to find the answers to the questions you raised in the previous section or how you intend to locate information about your list of subtopics. It's not enough to write, "I will gather appropriate information and analyze it." Specify what data you hope to include, where they are located, and how you intend to retrieve them.

In researching information about the problem, and any proposed solutions, you can expect to use both primary and secondary research methods and tools. Review relevant sections of Chapter 8. Certainly, you will gather data from the Web and from literature published in print sources. (In fact, research reports can be based exclusively on literature searches.) The literature can include

- websites, blogs, webfolios
- books
- encyclopedias or other reference materials, such as statistical data found in manuals or almanacs, online and in print
- articles in professional journals, online and in print
- newspapers and magazines, online and in print
- reviews

Inform your reader what indexes, abstracts, or Internet searches you intend to use as part of your research (review pages 369–375). To document your preliminary work, provide your reader with a list of appropriate titles on your topic, and follow the style of documentation used for an MLA Works Cited page (discussed on pages 391–397) or in another style preferred by your supervisor or employer.

In addition to these online and print materials, you might also collect information from primary research, including lab experiments, field tests, interviews with experts, surveys, or a combination of any of those sources.

Timetable

Indicate when and in what order you expect to complete the different phases of your project. Your boss or instructor needs that information to keep track of your progress and to make sure you will complete your work on time. Specify tentative dates for completing your research, draft(s), revisions, and final copy.

You may have to submit progress reports (see pages 654–661) at regular intervals. If you are asked to do that, indicate when you will submit the progress reports, as Shoemake does in Figure 13.8.

Request for Approval

End your proposal with a request for approval of your topic and a plan of action. You might also invite suggestions from your employer or instructor on how to restrict, research, or organize your topic.

A Final Reminder

This chapter has given you some basic information and specific strategies for writing winning proposals. Keep in mind that a proposal presents a plan to decision makers for their approval. To win that approval, your proposal must be (1) realistic, (2) carefully researched, (3) highly persuasive, (4) ethical, and (5) visually appealing and easy to follow. Those essential characteristics apply to internal proposals in memo or email format written to your employer, more formal sales proposals sent to a potential customer, and research proposals submitted to your instructor.

✓ Revision Checklist

General Guidelines
- ☐ Established and distinguished the roles of the collaborative team members involved in the preparation of the proposal.
- ☐ Researched appropriate sources for RFPs, and followed their instructions.
- ☐ Identified a realistic problem in my proposal—one that is restricted and relevant to my audience's needs.
- ☐ Effectively convinced my audience that the problem exists and that it needs to be solved.

☐ Incorporated quantifiable details demonstrating the scope and importance of the problem.

☐ Persuasively emphasized the benefits of solving the problem according to the proposal, and incorporated the "you attitude" throughout.

☐ Offered a solution that can be realistically implemented—that is, a solution that is both appropriate and feasible for the audience.

☐ Wrote clearly so the audience can understand how and why my proposal would work.

☐ Researched the background of the problem.

☐ Used specific figures and concrete details to show how the proposal saves time and money.

☐ Double-checked the proposal to catch errors, omissions, and inconsistencies.

☐ Avoided exaggerations and underbidding.

☐ Presented information ethically.

☐ Organized the proposal with appropriate headings for clarity and ease of reading.

☐ Used white space, lists, and a professional-looking font to make my proposal visually attractive and reader friendly.

For Internal Proposals

☐ Demonstrated how the proposal benefits my company and my supervisor.

☐ Took into account office politics in describing the problem and offering a solution.

☐ Discussed the proposal with co-workers or supervisors who may be affected.

For Sales Proposals

☐ Related my product or service to the prospective customer's needs and showed a clear understanding of those needs.

☐ Prepared a comprehensive, realistic, and ethical budget; accounted for all expenses; and itemized costs of products and services.

☐ Linked costs to benefits.

☐ Provided a timetable with exact dates for implementing the proposal.

☐ Cited other successful jobs and satisfied clients to show my company's track record.

☐ Concluded the proposal with a summary of the main benefits to readers and a call to action.

For Research Papers/Reports

☐ Proved to my supervisor or instructor that I researched the problem by supplying a list of possible references and sources.

☐ Selected a major problem to investigate in my report.

☐ Supplied relevant and restricted questions my report will answer.

Exercises

1. In two or three paragraphs, identify and document a problem (in services, safety, ecology, communication, traffic, scheduling) that you see in your office or your community. Make sure you give your reader—a civic official or an employer—specific evidence that a problem does exist and that it needs to be corrected.

2. Write a short internal proposal, modeled after Figure 13.6, based on the problem you identified in Exercise 1.

3. As a collaborative group project, write a short internal proposal, similar to that in Figure 13.6, recommending to a company or a college a specific change in procedure, technology, training, transportation, safety, personnel, or policy. Make sure your team provides an appropriate audience (college administrator, department manager, or section chief) with specific evidence about the existence of the problem and your solution of it. Possible topics include these:
   a. providing more and safer parking or lighting
   b. instituting job sharing for mothers
   c. purchasing new office or laboratory equipment or software
   d. hiring more faculty, student workers, or office help
   e. allowing employees to telecommute
   f. changing the lighting or furniture in a student or company lounge or kitchen to make it more eco-friendly
   g. increasing the number of weekend, night, or online classes in your major
   h. adding more health-conscious offerings to the school or company cafeteria menu
   i. altering the programming on a campus or local commercial radio station
   j. installing a new digital projector

4. Write an unsolicited sales proposal, similar to the one in Figure 13.7, on one of the following services or products or on a topic your instructor approves:
   a. providing exterminating service to a store or restaurant
   b. supplying a hospital with automatic hand sanitizers for visitors
   c. designing websites or blogs
   d. providing temporary office help or nursing care
   e. supplying landscaping and lawn care work
   f. testing for noise, air, or water pollution in your community
   g. furnishing transportation for students, employees, or members of a community group
   h. offering technical consulting service to save a company money
   i. digging a well for a small apartment complex
   j. supplying insurance coverage to a small firm (five to ten employees)
   k. cleaning the parking lot and outside walkways at a shopping center
   l. selling a piece of equipment to a business
   m. making a work area safer or greener
   n. preparing an IT seminar or training program for employees
   o. increasing donations to a community or charitable fund
   p. offering discounted memberships at a fitness center

5. Write a solicited proposal for one of the topics listed in Exercise 4 or for a topic that your instructor approves. Do this exercise as a collaborative project. Review Figure 13.5.

6. Write an appropriate proposal—internal, solicited sales, or unsolicited sales—based on the information contained in one of the following three articles. Assume that your or your prospective customer's company or community faces a problem similar to one discussed in one of these articles. Use as much of the information in the article as you need, and add any details of your own that you think are necessary. This exercise can be done as an individual or a collaborative assignment.

a. Multiuse Campuses: A Plan That Works

Gaylord Community School in Gaylord, Michigan, is a bustling center of activity from the first light of dawn to well after dusk. People of all ages come and go until late into the evening for a multitude of activities that include attending classes and meetings, catching up with friends, getting a flu shot, and seeing a play. That's because in addition to a high school, the campus also includes senior and day-care centers, classrooms for adult education, an auditorium for the performing arts, a community health care site, and even a space that can be booked for weddings and other special occasions.

In Big Lake, Minnesota, elementary, middle, and high school buildings are all situated on one centrally located campus that makes up the entire Big Lake School District. Also included in this innovative layout are a state-of-the-art theater, a community resource center, and a multipurpose athletic arena, all of which are used extensively by the entire community.

Both the Gaylord and Big Lake schools are models of a growing movement toward multiuse community campuses that serve as "anchor[s] in the civic life of our nation," according to U.S. Secretary of Education Richard W. Riley. I recently had the pleasure of visiting Secretary Riley in his office. Also present were AARP President Joe Perkins and National Retired Teachers Association Director Annette Norsman, both of whom are involved in many facets of education and lifelong learning.

We discussed many things, including our concerns about the current increase in the number of students caused by the Baby Boom echo (children of the Boomers) and how that population is going to further stress the already crumbling infrastructure of American schools. We also talked about the need for resources—to employ more teachers, bring technology into the classroom, strengthen educational curriculum and opportunities for all ages—and the pressing need to build and renovate schools. That led to a discussion about the necessity and benefit of involving the whole community in the design and use of new school facilities.

I always thought that it was a shame that the majority of schools are used only a third of the day, three fourths of the year, by only a fifth of the population. Considering that there will be more school construction over the next decade than at any time since the 1950s, it just makes sense to consider the intergenerational and community benefits of multiuse spaces, benefits that include everything from establishing better learning environments to getting more bang for the tax buck.

There are many additional bonuses for multiuse educational complexes: They create an exciting community hub, bring life and culture to a central area, and revitalize and nourish the neighborhood in which they are located.

It's a win-win situation for everyone involved.

Source: Horace B. Deets, "Multiuse Campuses: A Plan That Works," *Modern Maturity* (July–August 1999): 72. Reprinted with permission from *Modern Maturity*. Copyright © 1999 by American Association of Retired Persons.

b. Self-Illuminating Exit Signs

The Marine Corps Development and Education Center (MCDEC), Quantico, Virginia, submitted a project recently, to replace incandescent illumination exit signs with self-illuminating exit signs for a cost of $97,238. The first-year savings were anticipated to be about $37,171 with an anticipated payback time of 2.6 years—making this an excellent way to save money.

The primary benefit of these self-illuminating exit signs is that virtually all operation and maintenance expense is eliminated for the life of the device, normally from 10 to 12 years. Power failures or other disturbances will not cause them to go out. In new construction, expensive electrical circuits can be totally eliminated. In retrofits, the release of a dedicated circuit for other use may be of considerable benefit. Initial total cost of installing circuits and conventional devices approximately equals the cost of the self-illuminating signs. Installation labor and expense for the self-illuminating signs is about that of hanging a picture.

The amount of electricity saved varies and depends on whether your existing fixtures are fluorescent (13 to 26 watts) or incandescent (50 to 100 watts). Multiply the number of fixtures × wattage/fixture × hours operated/day × days/year – KWH/year savings. For example, assume:

$$400 \text{ incandescent fixtures}$$
$$\$0.08/\text{KWH } 0.05 \text{ KW/fixture}$$
$$24 \text{ hours/day } 365 \text{ day/year operation}$$
$$400 \times 0.1 \times 365 - 350400 \text{ KWH/year}$$
$$350400 \times 0.08 - \$28,032/\text{year for electricity}$$

Now add in savings achieved from reducing labor to change bulbs; avoiding bulb material, stocking, and storage costs; avoiding transportation costs involved in bulb changes; and reusing existing bulbs.

The above savings can be significant. For the MCDEC Quantico project, estimates of bulb change interval and savings were 700 hours (29 days) and $13,512/ year when all factors were considered.

The cost of a self-illuminating sign depends on whether one or two faces are illuminated primarily and varies between different suppliers. Single-face prices will likely be $100 to $150 while double-face prices may be $250 to $330. The contractor at Quantico found better prices than these ranges indicate. The labor cost should be about $10 per sign.

If you can use an exit-sign system with high dependability, no maintenance, and zero operations cost in your retrofit or new construction projects, try a self-illuminating exit-sign system in your economic analysis today. "Isolite" signs, by Safety Light Corp., are listed as FSC (Fire Safety Code) Group 99, Part IV, Section A, Class 9905 signs and are available through GSA contract. Contact Gerald Harnett, Safety Light Corp., P.O. Box 266, Greenbelt, MD 20070 for more information.

c. Wheelchair-Lift Switch Covers

In order to ensure year-round access to the Springfield Armory National Historic Site (Massachusetts) museum, Michael C. Trebbe designed the cover for switches on wheelchair lifts. During the extreme New England winters, the switch buttons would freeze, thus making the lift inoperable, and would require several hours to thaw. The installation of these covers prevented the freezing of the switch buttons and, therefore, allowed maintenance personnel to attend to matters such as snow removal.

The covers were made of materials found on site, which resulted in the covers being almost cost free. The covers can be quickly built, and they are mounted with the same mounting screws as the switch boxes so as to not destroy any original fabric (in the case of Springfield Armory NHS, brownstone). The materials used included

- 1/8-in. by 4½-in. by 12-in. piece of rubber mat
- 3½-in. by 6½-in. piece of sheet metal
- three aluminum pop rivets
- primer for the sheet metal
- wheelchair icon stencil
- white paint for the icon

The sheet metal is bent to a 90-degree angle at the 5½-inch point. The lowest two holes (see diagram) are drilled to mount the screws on the switch box, which also secure the cover. Triangular cutouts and other holes are drilled for the clearance of the housing screws on the rear of the switch box (see diagram).

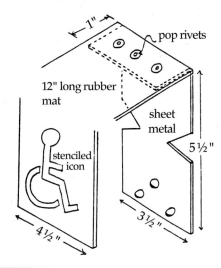

The disability icon is stenciled on the front of the piece of rubber mat using white paint.

---

7. Write a research proposal on which the report on recruiting and retaining multi-national workers in Chapter 15 (pages 702–718) could have been based.

8. Write a research proposal, similar to the one in Figure 13.8, to your instructor seeking approval for a research-based long report. Do the necessary preliminary research to show that you have selected a suitable topic, narrowed it, and identified the sources of information you have to consult. List at least six relevant and recent articles, two books, and several websites or professional blogs pertinent to your topic.

CHAPTER 14

Writing Effective Short Reports

Access chapter-
specific interactive
learning tools,
including quizzes and
more in your English
CourseMate, accessed
through www
.cengagebrain.com.

This chapter shows you how to write short reports and proposals, which are among the most important and frequent types of business communications you may be called upon to prepare. Short, informal reports give up-to-date information (and sometimes what it means and what should be done about it) to help a company or an organization run smoothly, efficiently, and profitably. These reports, which cover a wide range of topics, can help a company fulfill its obligations and plan for its future. Short reports are crucial to day-to-day operations of a company or an organization and are designed primarily for an audience of busy decision makers.

Why Short Reports Are Important

A short report can be defined as an organized presentation of relevant data on any topic—money, travel, sales, technology, personnel, service equipment, weather, the environment—that a company or an agency tracks in its day-to-day operations. Short reports are practical and to the point. They show that work is being done, and they also show your boss that you are alert, professional, and reliable. Short reports are written to co-workers, employers, vendors, and even clients. When they are intended for individuals within your organization, short reports are most often sent as memos or as emails. But for clients, usually you will send your reports as letters.

Businesses cannot function without short written reports. Reports tell whether

- work is being completed
- schedules are being met
- costs have been contained
- sales projections are being met
- trips or conferences have been successful
- locations have been selected
- unexpected problems have been solved

You may write an occasional report in response to a specific question, or you may be required to write a daily or weekly report about routine activities. For example, a short report can update your manager or client about the status of a project, provide feedback on a customer survey, prove you followed the regulations of a state or federal agency, or assess your own or someone else's accomplishments at work.

Types of Short Reports

To give you a sense of some of the topics you may be required to write about, here is a list of various types of short reports common in the business world.

appraisal report	feasibility report	production report
audit report	incident report	progress/activity report
budget report	investigative report	recommendation report
compliance report	laboratory report	research report
construction report	management trainee report	sales report
design report	manager's report	status report
employee activity/	marketing report	survey report
performance report	medicine/treatment	test report
evaluation report	error report	travel report
experiment report	operations report	
inventory report	periodic report	

This chapter concentrates on seven of the most common types of reports you are likely to encounter in your professional work.

1. periodic reports
2. sales reports
3. progress reports
4. employee activity/performance reports
5. trip/travel reports
6. test reports
7. incident reports

Although there are many kinds of short reports, they all are written for readers who need factual information so that they can get a job accomplished. Never think of the reports you write as a series of casual notes jotted down for *your* convenience.

Seven Guidelines for Writing Short Reports

The following seven guidelines will help you write any kind of short report successfully.

1. Anticipate How and Why an Audience Will Use Your Report

Knowing who will read your report, and why, is crucial to your success as a writer. Consider how much your audience knows about your project and what types of information they will need most. Meeting their needs should be one of your major goals. You cannot always assume that your readers have your background and technical knowledge or that they are as directly involved in a project as you are. Consider what they are looking for.

Tech Note

Creating Templates for Short Reports

You can use your word processor's template function to create professional-looking reports. Templates, as we saw on page 523, are predesigned formats for page layouts that specify style elements of a document. They allow you to automate designs for periodic, sales, progress, incident, and other types of reports. An existing template will format information to create a professional-looking report. Using templates, you reduce the risk of errors, omissions, and inconsistencies and ensure that you follow your company's style and format.

To create a specific report template, simply set the format options (such as font, margins, and line spacing) for a new document, type place markers for your text (including headers, footers, and titles), import custom visuals (such as a company logo), and save the document as a template. When you need to create a new report using the template, simply open it, insert the content of your report, and save the document under a new file name as a standard word-processing file. You can customize the formatting of any or all of the following elements:

■ Headers, footers for a company address, titles, and dates
■ List formatting, such as bullet-point style and numbered lists
■ Line spacing and text justification
■ Font style, size, and color
■ Standard graphics, e.g., a flowchart or organizational charts
■ Margin size, paragraph indentation, and columns
■ Standard visual elements such as tables, which can also be filled with the data appropriate to your report
■ Automatic table of contents based on the titles used in your report

But when using templates keep the following precautions in mind:

■ Be sure to follow your company's style when creating templates, particularly those using a company logo.
■ Double-check everything in the template to make sure it is appropriate for the type of report you're writing. Differences between reports will require visual adjustments.
■ When creating a template with place markers to indicate the position of certain textual elements, insert them in boldface for emphasis. Using brackets (for example, **[type title here]** or **[body of text here]**) will highlight and thus emphasize these place holders.

Below are the three basic audiences for whom your reports will be written and a brief overview of their needs.

■ **Managers,** who will constitute the largest audience for your reports, may not always understand or be interested in technical specifications and details, descriptions, and methods. Instead, they want bottom-line information about costs, sales, schedules, personnel, markets, organizational problems, and any delays to make their decision. They are looking for quick, accurate summaries and clear and convincing conclusions and recommendations.

- **Co-workers and team members** may be familiar with your project and understand the technical details, but colleagues in other departments or divisions in your company may not. Gear your report to help other employees carry out their job duties. Find out about their responsibilities and needs and how they will use and benefit from the information you will send them.

- **Audiences** outside your company (clients, media personnel, government officials, community agencies) will likely not be interested in technical information or want long discussions on backgrounds and methods of research. They may not understand jargon, technical abbreviations, or research techniques. You may have to supply examples, comparisons, even definitions, as Mary Fonseca did in Figure 2.7 (page 65). There is no need to spell out every detail for these readers. Rather, they want a concise report that helps them understand your company and how it will work with or serve them.

2. Do the Necessary Research

An effective short report needs the same careful research that goes into other on-the-job writing. Your research may be as simple as instant messaging, emailing, or leaving a voice mail for a colleague or checking a piece of equipment. Or you may have to test or inspect a product or service or assess the relative merits of a group of competing products or services. Some frequent types of research you can expect to do on the job include:

- verifying data in reference manuals or code books
- searching online archives and databases for recent discussions of a problem or procedure
- reading background information in professional and trade journals
- comparing and contrasting competitors' products or services
- pricing equipment
- preparing a budget
- reviewing and updating a client's file
- testing equipment
- performing an experiment or a procedure
- conferring with or interviewing colleagues, managers, vendors, or clients
- visiting, investigating, and describing a site; taking field notes
- attending a conference or workshop

Never trust your memory to keep track of all the details that go into making a successful short report. Take notes, either by hand or on your laptop or tablet computer. Collect all the relevant data you will need—names, model numbers, costs, places, technology, etc.—and organize this information carefully into an outline, which will help you interpret these facts for your readers. (Review Chapter 8 for a variety of research methods used in the world of work.)

3. Be Objective and Ethical

Your readers will expect you to report the facts objectively and impartially—locations, costs, sales, weather conditions, eyewitness accounts, observations, statistics, test measurements, and descriptions. Your reports should be truthful, accurate, and complete. Here are some guidelines to follow:

- Avoid guesswork. If you don't know or have not yet found out, say so and indicate how, where, and when you'll try to find out.
- Do not substitute impressions or unsupported personal opinions for careful research.
- Be ethical. Don't use biased, skewed, or incomplete data. Provide a balanced, straightforward, and honest account; don't exaggerate or minimize. Don't omit key facts. If a project is over budget or late, state so but indicate why and what might be done to correct the problem.
- Make sure your report is relevant, accurate, and reliable. Double-check your details against other sources, and make sure you have sufficient information to reach your conclusions or provide recommendations.

Review the discussion of ethics in business writing in Chapter 1 (pages 29–41).

Tech Note

Using the Web to Write Short Reports

Many government agencies provide the statistical raw data that go into various types of short reports—employment figures, population data, environmental statistics, and so on. Downloading and incorporating information from these and other relevant sites give readers the necessary documentation they need to accept a conclusion or recommendation.

Other businesses also post information that may be relevant to your report for your company. Large corporations such as General Motors or IBM help you see the progress of stocks and mutual funds or obtain other industry-wide information on their websites. Finally, ConsumerReports.org (www.consumerreports.org) evaluates different brands of products so that you don't have to do testing yourself.

Here are some sites that publish appropriate information for short reports:

BizStats, www.bizstats.com
Bureau of Labor Statistics, www.bls.gov
Bureau of Transportation, www.bts.gov
Federal Reserve, www.federalreserve.gov
National Center for Educational Statistics, http://nces.ed.gov
National Center for Health Statistics, www.cdc.gov/nchs
New York Stock Exchange, www.nyse.com
UNESCO Institute for Statistics, www.uis.unesco.org
United Nations Statistics Division, http://unstats.un.org/unsd
U.S. Census Bureau, www.census.gov

4. Organize Carefully

Organizing a short report effectively means that you include the right amount of information in the most appropriate places for your audience. Make your report easy to read and to follow. Many times a simple chronological or sequential organization will be acceptable for your readers. Your headings will show readers how you organized your report. Readers will expect your report to contain information on such topics as purpose, findings, conclusions, and, in many reports, recommendations, as described in the following sections.

Purpose

Always begin by telling readers why you are writing (your purpose) and by alerting them to what you will discuss and why it is significant. Give your readers a summary of key events and details at the beginning to help them follow the remainder of the report quickly. Essential background information alerts readers to the importance of your report. When you establish the scope (or limits) of your report, you help readers zero in on specific times, costs, places, procedures, or problems.

Findings

This should be the longest part of your report and contain the data (the results) you have collected—facts about prices, personnel, equipment, events, locations, incidents, or tests. Gather the data from your research, interviews, or conversations with co-workers, employers, or clients. Use statistical sources, as on pages 366–368. Again, remember to choose only those details that have the greatest importance and relevance to your reader. Separate major points from minor ones.

Conclusion

Your conclusion tells readers what your data mean. A conclusion can summarize what has happened; review what actions were taken; or explain the outcome or results of a test, a visit, or a program. Be aware, though, that readers are skeptical and may ask why you didn't reach a different conclusion. Anticipate possible objections and explain why other conclusions are unworkable.

Recommendations

A recommendation informs readers what specific actions you think your company or client should take—market a new product, hire more staff, institute safety measures, select among alternative plans or procedures, and so on. Recommendations must be based on the data you collected, the resources (budget) and schedule that your company or department follows, and the conclusions you have reached. They need to show persuasively how all the pieces fit together.

Note how the periodic report in Figure 14.1 (page 651) fails to help readers see and understand the organization and importance of the information. The revised version of the report, Figure 14.2 (pages 652–653), clearly illustrates effective report writing. The reader is provided with information at regularly scheduled intervals—daily, weekly, bimonthly (twice a month).

Note that because of the daily deadlines executives face, some companies prefer that recommendations come at the beginning of the report (as in Figure 14.9), followed by supporting documentation.

5. Write Clearly and Concisely

Writing clearly and concisely is essential in all business reports. Ask your boss or experienced co-workers about appropriate style for your company. Also look at previous, similar reports to get a sense of your company's style and tone.

Here are a few guidelines to help you write clearly and concisely.

- **Use an informative title or subject line that gets to the point right away.** "Software Options" is not as clear as "Most Economical Options for Spreadsheet Software."
- **Write in plain English.** Make every word count, avoid jargon, and keep your writing simple and straightforward. Prune business clichés such as "at the end of the day" or "to put a fine point on it."
- **For global readers, make sure you use international English.** Keep your sentences short, and write in the active voice. Do not use U.S. idioms, slang, or abbreviations. (See pages 187–188.)
- **Adopt a professional yet personal tone.** Avoid being overly formal or too casual—strike a balance between these two extremes. Don't sound arrogant by adopting a tone that suggests you alone have the final authority.
- **Keep your report as concise as possible while still giving readers the essential information they need.** Don't burden them with lengthy project histories when all they ask for is a quick update on a project, and don't pad the report with unnecessary details to sound important. A short report should be no longer than two to three pages.

6. Create a Reader-Centered Design

The appearance of your report will influence how your readers will respond to it and to you. Here are some useful guidelines. (You may also want to review Chapters 10 and 11 on visuals and document design.)

- **Help readers locate and digest information quickly.** Use headings, subheadings, bullets, and numbered lists to show how you have organized your report. Doing this, you break large portions of text into easy-to-read parts. Your headings and subheadings give readers the big picture at a glance. Many reports in this chapter demonstrate how headings and bulleted or numbered lists assist readers. For instance, see Figures 14.2, 14.3, 14.6 and 14.11. If you are submitting your report as an electronic document, use hyperlinks to help readers find information quickly.
- **Make your report look professional, readable, and easy to follow.** Don't flood your report with color. (See the Tech Note on page 644.) Avoid using flashy color or fancy fonts that are hard to read. Also, don't try to

squeeze too much text onto the page. Always leave comfortable margins. Design your report to assure your reader that it is well thought out and self-explanatory.

- **Be consistent in your design and format.** Use the same font throughout the text of your report and a consistent typeface for headings and subheadings.
- **Include only the most essential visuals.** Use visuals only if they make the reader's job easier, reinforcing or summarizing key data quickly, as the table in Figure 14.2 and the map in Figure 14.9 do. Keep visuals simple and relevant—a picture or drawing to illustrate a point. Do not get carried away with inserting numerous, and unnecessary, visuals for a short report.
- **Place visuals in the most appropriate place in your report.** Design or import visuals so that they are easy to read, and place them close to the text that they help illustrate or explain.

7. Choose the Most Appropriate Format

Depending on your audience, you can send your short report as an email, a memo, or a letter. For routine reports to your boss or others inside your company, you will likely use a memo format, as in Figures 14.2, 14.3, and 14.9. Note that with a memo format your readers will not expect you to include an inside address or formal salutation and complimentary close. Depending on your company's policy, you might also send a short report in the body of an email, as in Figures 14.4 and 14.7, or as an attachment to an email. Incident reports, however, are often submitted as hard-copy memos for legal reasons; they can also be written as a memo, as in Figure 14.13, or you may be required to complete a special form. When writing to clients and other readers outside your company or organization, it is best to send your report as a formal letter (including a salutation and complimentary close, as in Figure 14.6).

Periodic Reports

Periodic reports, as their name signifies, provide readers with information at regularly scheduled intervals—daily, weekly, bimonthly (twice a month), monthly, or quarterly. They help a company or an agency keep track of the quantity and quality of the services it provides and the amount and types of work done by employees. Information in periodic reports helps managers make schedules; order materials; hire, train, or assign personnel; budget funds; and, generally speaking, meet corporate needs and fulfill a corporate mission.

Figure 14.2 provides essential police information clearly and concisely. It summarizes, organizes, and interprets the data collected over a three-month period from individual officers' activity logs. Because of this report, Captain Martin will be better able to plan future protection for Springdale and to recommend changes in police services.

A Poor and an Effective Short Report

Sergeants Daniel Huxley, Jennifer Chavez, and Ivor Paz of the Springdale Police Department were responsible for writing a monthly status report for the second quarter of 2011. Confronted with a mass of data about various crimes and misdemeanors, they had to organize, compare, and contrast this data as well as draw conclusions and make recommendations. Figure 14.1, an early draft of their report, does not follow the guidelines on pages 643–649. But Figure 14.2, a revised version of Figure 14.1, does. Read through both reports, keeping in mind the differences below:

- **Research.** Although Figure 14.1 includes department statistics, it does not explain or provide recommendations based on them. Figure 14.2, however, provides explanations, supplies more detail, and gives recommendations. In addition, this figure includes other relevant sources rather than just presenting the reader with "uncrunched" numbers.
- **Audience analysis.** Instead of providing the most important information first, analyzing the statistics, or offering recommendations, the report in Figure 14.1 simply throws facts at the reader with only a vague consideration of why they are important. Figure 14.2 consistently takes the reader's needs into account by focusing on how Captain Martin will use the statistical information about the crime rate during the second quarter of the year. To help Captain Martin, Figure 14.2 presents the most important information first, then explains and analyzes the numbers for her and provides realistic and direct recommendations.
- **Objectivity/ethics.** In boasting, "Here we'll let the facts speak for themselves," Figure 14.1 does not give Captain Martin the full picture. In addition, since details have not been checked against any other sources, the report is incomplete and possibly inaccurate. Figure 14.2 eliminates the guesswork and, more ethically, offers solutions, careful analysis, and a variety of relevant sources.
- **Organization.** Figure 14.1 makes no attempt to organize the report in a reader-friendly manner. The purpose statement is vague, the findings are thrown at the reader in three dense and disorganized paragraphs, and the conclusion does not summarize the facts. Figure 14.2 supplies a clear purpose statement, organizes and summarizes the facts concisely, and helpfully groups recommendations.
- **Writing style and tone.** Figure 14.1 is just an accumulation of numbers, making it hard for the reader to access or understand their importance. The tone is smug, arrogant, and self-congratulatory. Figure 14.2 is easy to follow and to understand. It uses helpful connective words and phrases ("compared to last quarter," "however," "overall," "yet") and includes important contexts ("were less than last quarter"). This report is both professional and reader-friendly at the same time.
- **Format and visuals.** Figure 14.1 does not include headers, bullets, visuals, or consistent paragraph indentation. Figure 14.2 instead supplies clear heads, breaks the text beneath into easy-to-digest subheads, places numerical data in visual form (Table 1), and uses bulleted lists to help Captain Martin better understand the significance and scope of the report.

FIGURE 14.1 An Example of a Poorly Written, Poorly Organized, and Poorly Formatted Short Report

Springdale Police Department

Emergency 555-1000   **Administration** 555-1001   **Traffic** 555-1002

www.springdalepd.gov

TO: Captain Alice Martin
FROM: Sergeants Daniel Huxley, Jennifer Chavez,
             and Ivor Paz
SUBJECT: Crimes
DATE: July 12, 2011

This report will let you know what happened this quarter as opposed to what happened last quarter as far as crimes are concerned in Springdale. This **report is based on statistics** the department has given us over the quarter.

   Here we'll let the facts speak for themselves. From Jan.–Mar. we saw 126 robberies while from Apr.–June we had 106. Home burglaries for this period: 43; last period: 36. 33 cars were stolen in the period before this one; now we have 40. Interestingly enough, **last year at this time we had only 27** thefts. Four of them involved heirlooms.

Homicides were 4 this time versus 5 last quarter; assault and battery charges were 92 this time, 77 last time. Carrying a concealed weapon 11 (10 last quarter). We had 47 arrests (55 last quarter) for charges of possession of **a controlled substance**. Rape charges were 8, **1 less than last quarter**. 319 citations this time for moving violations: **speeding** 158/98, and failing to observe the signals 165/102 last quarter. DUIs this quarter—only 45, or 23 fewer than last quarter.

   Misdemeanors this quarter: disturbing the peace 53; vagrancy/public drunkenness 8; violating leash laws 32; violating city codes 39, including **dumping trash**. Last quarter the figures were **48, 59, 21, 43**.

We believe this report is **complete and up to date**. We further hope that this report has given you all the facts you will need.

Vague subject line

Introduction doesn't give overall picture

Throws facts out without any sense of reader's needs

Includes irrelevant detail

No analysis or commentary

Gives undigested numbers

Poor, inconsistent format

Hard-to-follow comparisons and contrasts

Conclusion provides no summary or recom-mendations

FIGURE 14.2 A Well-Prepared Report, Revised from Figure 14.1

 Springdale Police Department

Emergency 555-1000   **Administration** 555-1001   **Traffic** 555-1002

www.springdalepd.gov

TO: Captain Alice Martin
FROM: Sergeants Daniel Huxley, Jennifer Chavez, and Ivor Paz
SUBJECT: Crime rate for the second quarter of 2011
DATE: July 12, 2011

From April 1 to June 30, 852 crimes were reported in Springdale, representing a 5 percent increase over the 815 crimes recorded during the previous quarter.

TYPES OF CRIMES

The following report, based on the table below, discusses the specific types of crimes, organized into four categories: **robberies and theft**, **felonies**, **traffic**, and **misdemeanors**.

Table 1. Comparison of the 1st and 2nd Quarter Crime Rates in Springdale

Category	1st Quarter	2nd Quarter
ROBBERIES AND THEFT		
Commercial	63	75
Domestic	36	43
Auto	33	40
FELONIES		
Homicide	5	4
Assault and battery	77	92
Carrying a concealed weapon	10	11
Poss. of a controlled substance	55	47
Rape	9	8
TRAFFIC		
Speeding	165	197
Failure to observe signals	102	118
DUI	78	65
MISDEMEANORS		
Disturbing the peace	48	53
Vagrancy	40	48
Public drunkenness	19	40
Leash law violations	21	32
Dumping trash	43	39
Other	8	12

Robberies and Theft

The greatest increase in crime was in robberies, 20 percent more than last quarter. Downtown merchants reported 75 burglaries, exceeding $985,000. The biggest theft

Precise subject line

Begins with concise overview of report

Organizes crimes into categories

Supplies easy-to-follow visual

Table is boxed, making it easier to read

Uses clear headings to show organization of report

© Cengage Learning 2013

FIGURE 14.2 (Continued)

Page 2

occurred on May 21 at Weisenfarth's Jewelers, when three armed robbers stole more than $217,000 in merchandise. (Suspects were apprehended two days later.) Home burglaries accounted for 43 crimes, though the thefts were not confined to any one residential area. We also had 40 car thefts reported and investigated.

Felonies

Homicides decreased slightly from last quarter—from 16 to 13. Charges for battery, however, increased—15 more than we had last quarter. Arrests for carrying a concealed weapon were nearly identical this quarter to last quarter's total. But the 47 arrests for possession of a controlled substance were appreciably down from the first quarter. Arrests for rape for this quarter also were less than last quarter's. Three of those rapes happened within one week (May 6–12) and have been attributed to the same suspect, now in custody.

Traffic

Traffic violations for this period were lower than last quarter's figures, yet this quarter's citations for moving violations (362) represent a 5 percent increase over last quarter's (345). Most of the citations were issued for speeding (158) or for failing to observe signals (98). Officers issued 45 citations to motorists for DUIs, a significant decrease from the 78 DUIs issued last quarter. The new state penalty of withholding a driver's license for six months of anyone convicted of driving while under the influence appears to have been an effective deterrent.

Misdemeanors

The largest number of arrests in this category were for disturbing the peace—53. Compared to last quarter, this is an increase of 10 percent. There were 88 arrests for vagrancy and public drunkenness, an increase from the 59 charges made last quarter. We issued 32 citations for violations of leash laws, which represents a sizable increase over last quarter's 21 citations. Thirty-seven citations were issued for dumping trash at the Mason Reservoir.

CONCLUSION

Overall, while the crime rate has decreased for traffic violations (especially DUIs) and possession of controlled substances this quarter, we have seen a marked increase in arrests for robberies and battery.

RECOMMENDATIONS

To help deter robberies in the downtown area, we recommend the following:

- increasing surveillance units to 15 rather than the 10 now in the area
- offering businesses our workshop on safety and security precautions, as we did during the first quarter

Historically, battery arrests have risen during the second quarter. Our recommendations to counter this trend include:

- continuing to work closely with Neighborhood Watch Groups
- providing more foot and bicycle patrols in the neighborhoods with the highest incidence of battery complaints

Documents effective actions

Provides essential background and statistical information and comparative analyses

Easy-to-read sentences

Draws logical conclusion

Includes only data reader needs

Summarizes findings of report

Offers specific actions/ changes based on conclusion of report in bulleted lists

Sales Reports

Sales reports provide businesses with a necessary and ongoing record of accounts, online and mail purchases, losses, and profits over a specified period of time. They help businesses assess past performance and plan for the future. In doing that, they fulfill two functions: *financial* and *managerial*. As a financial record, sales reports list costs per unit, discounts or special reductions, and subtotals and totals. Like a spreadsheet, sales reports show gains and losses. They may also provide statistics for comparing two quarters' sales.

Sales reports are also a managerial tool because they help businesses make both short- and long-range plans. The restaurant manager's sales report illustrated in Figure 14.3 guides the owners to decide which popular entrées to highlight and which unpopular ones to modify or remove from the menu. Note how the recommendations follow logically from the figures Sam Jelinek gives to Gina Smeltzer and Alfonso Zapatta, the owners of the Oaks. Note, too, that because the readers are familiar with the subject of the report, the writer did not have to supply background information on the entire offerings.

Progress Reports

A progress report, such as those in Figures 14.4, 14.5, and 14.6 (pages 657–659), informs readers about the status of an ongoing project. It lets them know how much and what type of work has been done by a particular date, by whom, how well, and how close the entire job is to being completed. A progress report reveals whether you are

- specifying what work has been done
- keeping on your schedule
- staying within your budget
- using the proper technology or equipment
- making the right assignments
- identifying an unexpected problem
- providing adjustments in schedules, personnel, and so on
- indicating what work remains to be done
- completing the job efficiently, correctly, and according to codes

Almost any kind of ongoing work can be described in a progress report—research for a paper, construction of an apartment complex, preparation of a website, documentation of a patient's rehabilitation. Progress reports are often prepared at key phases, or milestones, in a project.

Audience for Progress Reports

A progress report is intended for people who generally are not working alongside you but who need a record of your activities to coordinate them with other individuals' efforts and to learn about problems or changes in plans. For example, supervisors who do not work in a field or branch office (and non-native speakers of English who manage overseas offices) will rely on your progress reports for much

FIGURE 14.3 **A Sales Report**

Dayton, OH 43210 ● (813) 555-4000 ● (813) 555-4100 fax
www.theoaks.com

THE OAKS

TO:	Gina Smeltzer	DATE:	June 28, 2011
	Alfonso Zapatta, Owners		
FROM:	Sam Jelinek *S J*	SUBJECT:	Analysis of entrée sales,
	Manager		June 10–16 and June 17–23

As we agreed at our monthly meeting on June 3, here is my analysis of entrée sales for two weeks to assist us in our menu planning. Below is a record of entrée sales for the weeks of June 10–16 and June 17–23 that I have compiled and put in a table for easier comparisons.

	Portion Size	June 10–16		June 17–23		Both Weeks	
		Amount	Percentage	Amount	Percentage	Amount	Percentage
Cornish Hen	6 oz.	238	17	307	17	545	17
Stuffed Young Turkey	8 oz.	112	8	182	10	294	12
Broiled Salmon Steak	8 oz.	154	11	217	12	371	13
Brook Trout	12 oz.	182	13	252	14	434	9
Prime Rib	10 oz.	168	12	198	11	366	11
Lobster Tails	2–4 oz.	147	10	161	9	308	10
Delmonico Steak	10 oz.	56	4	70	4	126	4
Moroccan Chicken	6 oz.	343	25	413	23	756	24
		1,400	**100**	**1,800**	**100**	**3,200**	**100**

Recommendations

Based on the figures in the table above, I recommend that we do the following:

1. Order at least 100 more pounds of prime rib each two-week period to be eligible for further quantity discounts from the Northern Meat Company.
2. Remove the Delmonico Steak entrée because of its low acceptance.
3. Introduce a new chicken or fish entrée to take the place of the Delmonico steak; I would suggest grilled lemon chicken to accommodate those patrons interested in a tasty, low-fat, lower-cholesterol entrée.

Please give me your responses within the next week. It shouldn't take more than a few days to implement these changes. Thank you.

Restricted subject line

Begins with purpose and scope of report

Organizes findings of the report in helpful table

Boldfaces totals

Offers precise and relevant recommendations

Requests authorization to implement recommendations

of their information. Customers, such as a contractor's clients, expect reports on how carefully their money is being spent, if schedules are being met, and whether there is a risk of going over the budget.

The length of a progress report will depend on your audience and on the complexity of the project. A short email to a supervisor about organizing a time management workshop, such as that in Figure 14.4, might be all that is necessary. A report to an instructor about the progress a student is making on a research report easily could be handled in a memo, such as Barbara Shoemake's progress report in Figure 14.5 on the research report described in her proposal in Chapter 13 (Figure 13.8, pages 631–634). Dale Brandt's assessment of the progress his construction company is making in renovating Dr. Burke's office is given in a letter in Figure 14.6.

Frequency of Progress Reports

Progress reports can be written at any regular interval, even annually. Your specific job and your employer's needs will dictate how often you have to keep others informed of your progress. A single progress report is sufficient for Philip Javon's purpose in Figure 14.4. Barbara Shoemake was asked to submit two progress reports, the first of which is found in Figure 14.5. Contractor Dale Brandt determined that three reports, spaced four to six weeks apart, would be necessary to keep Dr. Burke posted; Figure 14.6 is the second of those reports.

Parts of a Progress Report

Progress reports should contain information on (1) the work you have done, (2) the work you are currently doing, and (3) the work you will do.

How to Begin a Progress Report
In a brief introduction,

- indicate why you are writing the report
- provide any necessary project titles or codes and specify dates
- help readers recall the job you are doing for them

If you are writing an initial progress report, supply brief background information in the opening. Philip Javon's first sentence in Figure 14.4, for example, quickly establishes his purpose by reminding Thad Sands, his boss, of their discussion last week. Similarly, Barbara Shoemake states the purpose and scope of her work in the first paragraph in Figure 14.5.

If you are submitting a subsequent progress report, inform your reader about where your previous report left off and where the current one begins. Make sure you clearly specify the period covered by each report. Note how Dale Brandt's first paragraph in Figure 14.6 calls attention to the continuity of his work.

How to Write the Body of a Progress Report
The body of the report should provide significant details about costs, materials, personnel, and times for the major stages of the project.

FIGURE 14.4 A One-Time Progress Report Sent as an Email

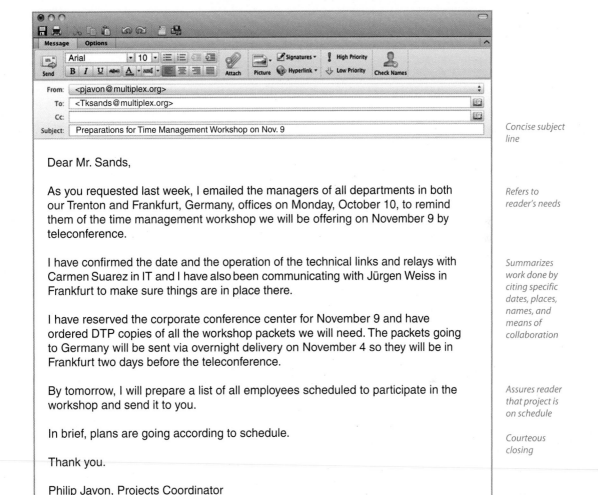

Concise subject line

Refers to reader's needs

Summarizes work done by citing specific dates, places, names, and means of collaboration

Assures reader that project is on schedule

Courteous closing

Includes signature block

- Emphasize completed tasks, not false starts. If you report that the carpentry work or painting is finished, readers do not need an explanation of paint viscosity or geometrical patterns.
- Omit routine or well-known details ("I had to use the library when I wanted to read the back issues of *Safety News* that were not archived on the Web").

FIGURE 14.5 A Progress Report Regarding a Student's Research

TO: Professor Marisol Vega
FROM: Barbara R. Shoemake
DATE: April 25, 2011
SUBJECT: First Progress Report on Research Report

Clearly states purpose of report

This is the first of two progress reports that you asked me to submit about my background report on the ethical issues over using m-commerce in the mobile marketplace.

Provides updates by citing individual references and actions taken

From March 30 until April 14, I gathered information from library holdings, the Internet, and one interview. Of the seventeen references listed in my proposal, I've so far found only thirteen. The text by Niels Kortlang (*The Impact of 3G Technologies on User Acceptance of Mobile Commerce*) and the articles by Zakara Maamar ("Commerce, E-Commerce, and M-Commerce: What Comes Next?") and Eleanor Thames ("UDIDs and Consumers' Rights") are not available at our campus library, and I'm working with Document Delivery to receive these texts as soon as possible. One of my Internet sources, *Business Blogs,* has been down for the past several days, but according to the Webmaster, service should be restored in the next 4–5 days.

Details results of primary research

On April 21, I had an extended interview (1½ hours) with Katrina Kuhn, a marketing specialist and m-commerce consultant. She gave me some helpful handouts from her m-commerce seminars, as well as a copy of a conference presentation she'd recently delivered on how consumers can better safeguard their privacy when using social media — which I hope to include in the research for my report.

Gives detailed descriptions of steps in the research process

Because of an extended trip to Denver, Paul Goya could not meet with me. At his suggestion, though, I am trying to schedule an interview with Robert Sims, the manager at Mobile General here in Springfield. Sims has made numerous sales presentations and in-house briefings on the mobile devices his company developed to improve customer service and provide better monitoring. Even if Mr. Goya cannot meet with me, Ms. Kuhn gave me a great deal of information about the legal and ethical foundations of m-commerce. Also, there is a chance I might interview Jen Holka at Tech Consultants in early May when her schedule clears. Not currently having the secondary sources listed above may temporarily slow, but not stop, my work.

Anticipates problem

Concludes with next steps

Starting tomorrow, I will begin my report and will submit my final copy by June 7. You will receive my second progress report by May 15.

FIGURE 14.6 The Second of Three Progress Reports from a Contractor to a Customer

Brandt **C**onstruction **C**ompany   *"Building a Greener Tomorrow"*

Halsted at Roosevelt, Chicago, Illinois 60608-0999 • 312-555-3700 • Fax: 312-555-1731
www.brandtcon.com

April 27, 2012

Dr. Pamela Burke
1439 Grand Avenue
Mount Prospect, IL 60045-1003

Dear Dr. Burke:

Here is my second progress report about the renovation work being done at your new clinic at Hacienda and Donohue. I am pleased to report that work proceeded satisfactorily in April according to the plans you had approved in March.

Review of Work Completed in March
As I informed you in my first progress report on March 30, we tore down the walls, pulled the old wiring, and removed existing plumbing lines. All the gutting work was finished in March.

Work Completed During April
By April 6, we had laid the new pipes and connected them to the main septic line. We also installed the two commodes, four standard sinks, and a utility basin. The heating and air-conditioning ducts were installed by April 13. From April 16–20, we erected soundproof walls in the four examination rooms, the reception area, your office, and the laboratory. Throughout your clinic we used environmentally safe (green) materials. We also installed solar panels on the roof and made the first examination room larger by five feet, as requested.

Problems with the Electrical System
We had difficulty with the electrical work, however. The outlets and the generator for the laboratory equipment required extra-duty power lines that Con Edison and Cook County inspectors had to approve, which slowed us down by three days. Also, Midtown Electric failed to deliver the recessed lighting fixtures by April 26. Those fixtures and the generator are now installed. Nevertheless, the cost of those fixtures increased the material budget by **$5,288.00**. But the cost for labor remains as we had projected— **$94,550**.

Work Remaining
The finishing work is scheduled for May. By May 11, the floors in the examination rooms, laboratory, washrooms, and hallways should be tiled and the reception area and your office carpeted. By May 14, the reception area and your office should be paneled and painted. If everything stays on schedule, touch-up work is planned for May 14–18. You should be able to move into your new clinic by May 21.

You will receive a third and final progress report by May 14. Thank you again for your business and for the confidence you have placed in our company.

Sincerely yours,

Dale Brandt

Dale Brandt

Professional-looking letterhead

Begins with key information: project status

Recaps activities for background

Uses boldface headings

Summarizes current accomplishments

Attention to greening the building

Identifies problems, how they were solved

Specifies work remaining

Projects successful completion

Promises to keep reader informed

RSS Feeds

Really simple syndication (RSS) feeds help readers keep up with constantly changing information on the Internet. Available through a subscription service, RSS feeds track updates from a variety of sources, including blogs, podcasts, news sites, weather reports, financial information, stock quotations, sales trends, product information, medical websites, or even your local library. RSS feeds are distributed in "streams" or "channels" and are collected by an RSS "aggregator," a software program that allows users to identify, gather, organize, and read updated information on a regular basis in areas most relevant to their job.

Most feeds generally consist of a headline, a summary/abstract, and a link to the original website, as in the feed below:

Nutrition Labels: How Precise Are They?
The FDA allows food to have 20 percent more diet-damaging nutrients than the labels state.
http://abcnews.go.com

RSS feeds may also contain photographs, videos, sound, or other media content. Often, too, the user's RSS aggregator will offer several helpful features and tools for sharing email, blog posts, and the like and will indicate that the user has already seen or read a particular post.

RSS feeds benefit report writers by:

■ listing the latest headlines in an area of interest
■ providing summaries and time stamps so readers do not have to visit large numbers of websites each day to see if anything new has been added
■ helping users quickly determine if they need to visit the updated websites themselves
■ offering custom searches saved as RSS feeds
■ organizing feeds for convenient browsing

RSS feeds are especially helpful to writers who have to generate short reports based on assimilating data from a wide variety of sources. For instance, when an advertising accounts manager has to create a weekly report about automotive industry data to determine what advertisements should be shown in a particular market, she can use an RSS aggregator to learn about trends and recent industry news from a variety of relevant sources. An aggregator helps the manager group feeds by subject, date, or by another area relevant to her report, making it a relatively simple task to quickly find, scan, and extract the data she needs for her report.

Source: Steven Turner.

- Describe in the body of your report any snags you encountered that may affect the work in progress. See Dale Brandt's section on electrical problems in Figure 14.6. Ethically, the reader needs to know about trouble early in the project, so appropriate changes or corrections can be made.

How to End a Progress Report

The conclusion should give a timetable for the completion of duties or the submission of the next progress report. Give the date by which you expect work to be completed. Be realistic; do not promise to have a job done in less time than you know it will take. Readers will not expect miracles, only informed estimates. Even so, any conclusion must be tentative. Note that the good news Dale Brandt gives Dr. Burke about moving into her new clinic is qualified by the words "If everything stays on schedule." He also employs the "you attitude" by thanking Dr. Burke again for her business.

Employee Activity/Performance Reports

An employee activity/performance report informs employers about what you did during a specified period (weekly, monthly, quarterly). Your supervisor will expect you to explain how you managed your time and fulfilled the requirements of your job. Accountability is a major objective in the world of work; employers closely monitor employees, even online. An activity report is a vital indication of an employee's performance. Your supervisor or manager will want to know about the specific tasks you accomplished, how many of them, when, and why, as well as any ongoing projects in which you are involved. Activity reports will play a role in assessing your annual job performance and determining whether you are promoted.

Figure 14.7 shows an employee activity report written by Carey Lewis, an administrative assistant, for his supervisor, Beth-Anne Prohaska. Note how he classifies his accomplishments into four major categories and provides a concise explanation of what he has done and why. Refer to Figure 14.7 as you study the guidelines below for writing an activity report.

Guidelines for Writing an Activity Report

To write a reliable and efficient activity report, you need to document what you have done. Keep track of your accomplishments by using a log, recording and saving job-related duties on your workplace calendar, and correlating the hours you spent on a project with your timesheets. Here are some guidelines to follow when you write an activity/performance report:

- **Use the format that your employer or agency dictates.** This might be a special form, a memo, an email, or an email with an attachment, such as an annotated day-to-day calendar.

FIGURE 14.7 An Employee Activity Report Sent as an Email with Attachments

Includes attachment

Gives background and briefly summarizes purpose of report

Uses boldfaced headings to group activities

Uses strong verbs to convey type of work

Provides key dates

Describes job duties clearly

References attachment

Specifies length and topic of presentation

Acknowledges collaboration

Promises continuity and requests feedback

From: Carey Lewis (clewis@wdynamics.com)
To: Beth-Anne Prohaska (baprohaska@wdynamics.com)
Cc: Gloria Arrelo (garrelo@starinstruments.com)
Subject: Monthly Activity Report for August
Purchase Orders   Workflow

Dear Ms. Prohaska:

During the past pay cycle (August 1–August 31), I worked on several projects that I believe helped to ensure, and even increase, the efficiency of our department. Below I have categorized my accomplishments (all of which are specified in my job description), included completion dates for major tasks, and attached relevant documentation.

Oversaw Day-to-Day Management of Office
• Maintained adequate office supplies (see attachment: "Purchase Orders")
• Researched need for new scanner and priced models for purchase
• Submitted recommendation for model scanner (August 9)
• Trained two new interns (Loretta Bauer, Scott Chu)

Prepared and Delivered Documents
• Edited, printed, and mailed the monthly newsletter (August 17)
• Compiled, printed, and distributed monthly sales report (August 24)
• Updated, stored, and retrieved records

Planned Schedules
• Maintained office calendar of events and meetings and posted it to the website
• Arranged travel plans for four staff on sales visits; two of these were overseas trips
• Coordinated workflow charts (see attachment: "Workflow")
• Logged staff timesheets in master file

Organized Meetings
• Presented short report (15 minutes) at HR meeting on August 8 about our department's successes in marketing new products
• Created—with Richard Fleming, assistant manager—and distributed agenda for monthly staff meeting (August 27)
• Took minutes for monthly meeting and shared them on company intranet (August 29)

For the coming month I will fulfill my ongoing responsibilities as well as meet our department's goals for any new assignments. I look forward to any comments you may have about my past or current performance.

Sincerely,

Carey Lewis
Administrative Assistant

- **Make sure you are honest, objective, and accurate.** It is unethical to misrepresent what you did—for example, saying a job took longer than it did or exaggerating the number of tasks you actually performed. Do not claim credit for tasks you did not accomplish or finish. But do give credit to coworkers with whom you collaborated, as Carey Lewis does in Figure 14.7.
- **Describe your major accomplishments.** Quantify and give dates, if necessary. But be careful that you do not dwell on smaller duties ("I answered the phone every day") at the expense of omitting tasks that demonstrate your writing, organizational, and collaborative skills at work.
- **Be sure that each of your accomplishments squares with your job description.** Relate what you accomplished to what you were hired to do. Indicate how your performance meets the goals your employer set for you. Specify how your tasks assist, improve, or contribute to your department or division.
- **Include training sessions or workshops you attended, licensure/certification updates, committee memberships, and presentations you made.** List and describe any meetings, classes, or CEU-giving sessions that help you to do your job more efficiently, accurately, or promptly.
- **Stress how your job accomplishments benefited the company, your department, or your community.** Link what you have done to the goals your company or department hopes to meet, such as sharing minutes in Figures 14.7.
- **Be prepared to verify your activities with copies of relevant documents, emails, and even IMs.** Maintain a folder of all these documents so you can compile them for yearly evaluations.

Trip/Travel Reports

Reporting on the trips you take is an important professional responsibility in the world of work. Basically, you are on a fact-finding mission. In documenting what you did and saw, trip reports keep readers informed about your efforts and how they affect ongoing or future business. Moreover, such reports help you better understand your job and develop your networking skills. Trip reports also should be written after you attend a convention or sales meeting or call on customers.

Questions Your Trip/Travel Report Needs to Answer

Specifically, a trip report should answer the following questions for your readers:

- Where did you go?
- When did you go?
- Why did you go?
- What did you see?
- Whom did you see?
- What did they tell you?
- What did you do about it?

For a business trip, you are also likely to have to inform readers how much it cost and to supply them with receipts for all of your business expenses.

Common Types of Trip/Travel Reports

Travel reports can cover a wide range of activities and are called by different names to characterize those activities. Most likely, you will encounter the following four types of travel reports.

  1. **Field trip reports.** These reports, often assigned in a course, are written after a visit to a laboratory, hospital, detention center, or other location to show what you have learned about the operation of a facility. (Such visits might even be done via "virtual tours" on the Internet.) You will be expected to describe how an institution is organized, the technical procedures and equipment it uses, pertinent ecological conditions, or the ratio of one group to another. The emphasis in such reports is on the educational value of the trip, as nursing student Mark Tourneur's report in Figure 14.8 demonstrates.

  2. **Site inspection reports.** These reports inform managers about conditions at a branch office, plant, or customer's business, or report on the advisability of relocating an office or other facility. After visiting the site, you will determine whether it meets your employer's (or customer's) needs. Site inspection reports can provide information about the physical plant, the environment (air, soil, water, vegetation), safety standards, or IT or financial operations.

  Figure 14.9 (page 667), which begins with a recommendation, is a report written to a district manager interested in acquiring a new site for a fast-food restaurant chain.

  3. **Home health or social work visit reports.** Nurses, social workers, and probation officers, for instance, report routinely on their visits to patients and clients. Their reports describe clients' lifestyles, assess needs, and make recommendations based on a variety of sources—clients, health care professionals, charitable organizations, and the like. These reports are often divided into Purpose of the Visit, Description of the Visit, and Action Taken as a Result of the Visit, as in Figure 14.10 (page 669), a report from a social worker to a county family services agency.

  4. **Sales/customer visit reports.** Visiting current or potential customers can result in further business opportunities for you and your company. Note how a routine service visit leads to a proposal for an important sale of upgraded notebooks in Figure 13.7 (pages 622–626). Your supervisor will need to know the following about your sales call: (a) What are the customer's needs and history of transactions with your company? (b) Which products or services that your company provides best meet those needs? (c) What types of demonstrations, samples, or written materials (brochures) did you give the customer? (d) What questions or reservations does the customer have about your products or services? (e) Who are your company's chief competitors? (f) How does your firm stand in terms of landing a contract? (g) What kinds of follow-up visits, calls, or information might be necessary to clinch the deal? Make sure you spell out precisely how you successfully represented your company and its products or services.

FIGURE 14.8 A Field Trip Report

TO: Katherine Holmes, RN, MSN
           Director, RN Program
FROM: Mark Tourneur M.T.
           RN Student
DATE: November 10, 2011
SUBJECT: Field Trip to Water Valley Extended Care Center

On Monday, November 7, I visited the Water Valley Extended Care
Center, 1400 Medford Boulevard, in preparation for my internship in an
extended care facility next semester.

Philosophy and Organization
Before my tour started, the director, Sue LaFrance, explained the holistic
philosophy of health care at Water Valley and emphasized the diverse
kinds of nursing practiced there. She stressed that the agency is not
restricted to geriatric clients but admits anyone requiring extended care.
She pointed out that Water Valley is a medium-sized facility (150 beds)
and contains three wings: (1) the Infirmary, (2) the General Nursing Unit,
and (3) the Ambulatory Unit.

Primary Client Services
My tour began with the Infirmary, staffed by one RN and two LPNs, where
I observed a number of life-support systems in operation:

- IVs
- oxygen setups
- feeding tubes
- cardiac monitors

Then I was shown the General Nursing Unit (40 beds), staffed by three
LPNs and four aides. Clients can have private or semiprivate rooms;
bathrooms have wide doors and lowered sinks for patients using
wheelchairs or walkers. The Ambulatory Unit serves 90 clients who can
provide their own daily care.

Additional Client Services
Dietetics
Before lunch in the main dining room, I was introduced to Jack Isoke, the
dietitian, who explained the different menus he coordinates. The most
common are low-sodium and ADA (American Diabetic Association)
restricted-calorie. Staff members eat with the clients, reinforcing the
holistic focus of the agency.

Pharmacy
After lunch, Kendra Tishner, the pharmacist, discussed the agency's
procedures for ordering and delivering medications. She also described
the client teaching she does and the in-service workshops she conducts.

*Explains
purpose of trip*

*Describes the
mission and
organization of
the facility*

*Summarizes
what he
observed,
whom he met,
and the various
units of the
facility*

*Usefully
provides
subheads to
organize his
report*

*Documents
conferences
with staff*

(Continued)

FIGURE 14.8 (Continued)

Describes the function of various locations in the facility

Physical/Spiritual Therapy
I then observed clients in both recreational and physical therapy. Water Valley's full-time physical therapist, Tracy Cook, works with stroke and arthritic clients and helps those with broken bones regain the use of their limbs. In addition to a weight room, Water Valley has a small sauna that most of the clients use at least twice a week.

The clients' spiritual needs are not neglected, either. A small interdenominational chapel is located just south of the Ambulatory Unit.

Concludes with the importance of the trip

Benefits for My Internship
From my visit to Water Valley, I learned a great deal about the health care delivery system at an extended care facility. I was especially pleased to have been given so much information on emergency procedures, medication orders, and physical therapy programs. My forthcoming internship should be even more productive, since I now have firsthand knowledge about these various services.

How to Gather Information for a Trip/Travel Report

Regardless of the kind of trip report you have to write, your assignment will be easier and your report better organized if you follow these suggestions.

1. Before you leave on the field trip, site inspection, or visit, be sure you are prepared:
   a. Obtain all necessary names; street, email, and website addresses; and relevant telephone, cell, and fax numbers.
   b. Check files for previous correspondence, case studies, warranties, or terms of contracts or agreements.
   c. Download work orders, instructions, or other documents pertinent to your visit, for example, websites and ads.
   d. Bring a laptop or notebook with you. Keep a journal of what you saw and heard.
   e. Locate a map of the area and get the directions you'll need beforehand. (Both maps and directions can be obtained easily at www.mapquest.com, http://maps.yahoo.com, and http://maps.google.com).
   f. Make a record of appointment times and locations as well as the job titles of the people whom you expect to meet.
   g. Save all receipts.
   h. Bring a camcorder, audio recorder, smartphone, or camera, if necessary, to record important data. You may be asked to post photographs from your trip on your company's website or intranet site.

FIGURE 14.9 A Site Inspection Report Using a Map

VAIL's
Chicken House

TO: Pretha Bandi   DATE: March 30, 2012
FROM: Beth Armando-Ruiz *B.A.R.* SUBJECT: New Site for Vail's #7
        Development Department

Recommendation

To follow up on our discussions earlier this month, I think the best location for the new Vail's Chicken House is the vacant Dairy World restaurant at the northeast corner of Smith and Fairfax avenues—1701 Fairfax. I inspected this property on March 19 and 20 and also talked to Kim Shao, the broker at Crescent Realty representing the Dairy World Company—**kims@crescent.org.** The location, parking facilities, and building at the Dairy World site all present the best opportunity for future growth and increased sales for Vail's.

Begins with most important details about the writer's recommendation

Gives essential contact information

The Location

Please refer to the map below. Located at the intersection of the two busiest streets on the southeast side of the city, the property will allow us to take advantage of the traffic flow to attract customers. Being only one block west of the Cloverleaf Mall should also help attract business.

Provides necessary background details

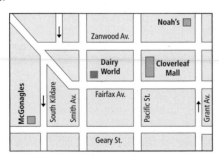

Includes map and traffic flow information essential for reader's purpose

Another benefit is that customers will have easy access to our location. They can enter or exit the Dairy World site from either Smith or Fairfax. Left turns onto Smith are prohibited from 7 a.m. to 9 a.m., but since most of our business is done after 11 a.m., the restriction poses few problems.

Denver, CO 87123   (303) 555-7200   www.vails.com

(Continued)

FIGURE 14.9 (Continued)

Pretha Bandi
March 30, 2012
Page 2

Area Competition

Assesses the location, in light of the competition

Only two other fast-food establishments are within a one-mile vicinity. McGonagles, 1534 South Kildare, specializes in hamburgers; Noah's, 703 Grant, serves primarily seafood entrées. Their offerings will not directly compete with ours. The closest fast-food restaurant serving chicken is Johnson's, 1.8 miles away.

Parking Facilities

Gives only the most essential facts audience needs on parking, seating capacity, and alterations

The parking lot has space for 25 cars, and the area at the south end of the property (38 feet × 37 feet) can accommodate 14–15 more vehicles. The driveways and parking lot were paved with asphalt last July and appear to be in excellent condition. We will also be able to make use of the drive-up window on the north side of the building.

The Building

Writer has done research on equipment

The building has 3,993 square feet of heated and cooled space. The air-conditioning units and heating units were installed within the last fifteen months and seem to be in good working order; nine more months of transferable warranty remain on all these units.

Paragraphs are easy to follow with precise topic sentences

The only major changes we must make are in the kitchen. To prepare items on the Vail's menu, we need to add three more exhaust fans (there is only one now) and expand the grill and cooking areas. The kitchen also has three relatively new sinks and offers ample storage space in the 16 cabinets.

Does not overwhelm reader with petty details

The restaurant has a seating capacity of up to 34 persons; 10 booths are covered with red vinyl and are comfortably padded. A color-coordinated serving counter could seat 8 to 10 patrons. The floor does not need to be retiled, but the walls will have to be painted to match Vail's decor.

FIGURE 14.10 A Social Worker's Visit Report

GREEN COUNTY FAMILY SERVICES

Randall, VA 21032
703-555-4000
fax 703-555-4210
greenfam@msn.net

TO: Margaret S. Walker, Director
            Green County Family Services
FROM: Jeff Bowman, Social Worker
SUBJECT: Visit to Mr. Lee Scanlon
DATE: October 10, 2011

PURPOSE OF VISIT

At the request of the Green County Home Health Office, I visited Mr. Lee Scanlon at his home at 113 West Diversy Drive in Randall on Wednesday, October 3. Mr. Scanlon and his three children (ages six, eight, and eleven) live in a two-bedroom apartment above a garage. Last week, Mr. Scanlon was discharged from Lutheran General Hospital after leg surgery, and he has asked for financial assistance.

Starts with why, where, when, and with whom the visit was made

DESCRIPTION OF VISIT

Mr. Scanlon is a widower with no means of support except unemployment compensation of $1,084 a month. He lost his job at Beaumont Industries when the company went out of business four weeks ago and wants to go back to work, but Dr. Marilyn Canning-Smith advised against it for six to seven weeks. His oldest child is diabetic and the six-year-old daughter must have a tonsillectomy. Mr. Scanlon also told me the problems he is having with his refrigerator; it "is off more than it is on," he said.

Gives relevant background information about client

Mrs. Alice Gordon, the owner of the garage, informed me that Mr. Scanlon had paid last month's rent but not this month's. She also stressed how much the Scanlons need a new refrigerator and that she has often let them use hers to store their food.

Lists information ethically and legally obtained from client, doctor, and owner

Here is a breakdown of Mr. Scanlon's monthly expenses and income:

Expenses		Income
$ 550	rent	$1,084 unemployment compensation
150	utilities	
450	food	
120	medicines	
80	transportation	
$ 1,350		

Itemizes information to make it easy to follow and compare

(Continued)

FIGURE 14.10 (Continued)

Margaret S. Walker
October 10, 2011
Page 2

ACTION TAKEN AS A RESULT OF THE VISIT
To assist Mr. Scanlon, I have done the following:

1. Set up an appointment (10/19/11) for him to apply for food stamps.
2. Conferred with Blanche Derringo regarding Medicaid assistance.
3. Requested that the State Employment Commission help him to find a job as soon as he is well enough to work. My contact person is Wesley Sahara; his email address is **wsahara@empcom.gov.**
4. Visited Robert Hong at the office of the Council of Churches to obtain food and money for utilities until federal assistance is available; he also will try to find the Scanlons another refrigerator.
5. Telephoned Sharon Muñoz at the Green County Health Department (555-1400) to have Mr. Scanlon's diabetic daughter receive insulin and syringes gratis.

Documents activities done to assist client rather than giving reader recommendations to follow

© Cengage Learning 2013

2. When you return from your trip, keep the following hints in mind as you compile your report:
   a. Write your report promptly. If you put it off, you may forget important details.
   b. When a trip takes you to two or more widely separated places, note in your report when you arrived at each place and how long you stayed.
   c. Exclude irrelevant details, such as whether the trip was enjoyable, what you ate, or how delighted you were to meet people. Concentrate on the information your reader needs to make a decision, meet a goal, or receive a timely update.
   d. Be objective about what you saw and heard. Indicate when you quote someone as part of an interview.
   e. Offer to answer any questions your reader may have about the trip and its outcomes.
   f. Double-check to make sure you have spelled names and calculated figures correctly. Mistakes in math make you look bad.

Test Reports

Much physical research (the discovery and documentation of facts) is communicated through short reports variously called *experiment*, *investigation*, *laboratory*, or *operations reports*. They all record the results of tests, whether the tests were conducted in a forest, computer center, laboratory, shopping mall, or soybean field. You may be asked to test an existing or a new product or procedure or verify certain physical or environmental conditions for a class or an employer. Figures 14.11 and 14.12 contain two test reports.

Objectivity and accuracy are essential ingredients in a test report. Readers want to know about your empirical research (the facts), not about your feelings. Record your observations without bias or guesswork in a laboratory journal or log book and always document the results with precise measurements using the standard symbols and abbreviations of your profession.

Case Study

Two Sample Test Reports

Figure 14.11 is a relatively simple and short test report in memo format regarding sanitary conditions at a hospital psychiatric unit. The report follows a direct and useful pattern of organization:

- statement of purpose—*why?*
- findings—*what happened?*
- recommendations—*what next?*

Submitted by an infection control officer, the report does not provide elaborate details about the particular laboratory procedures used to determine whether bacteria were present, nor does it describe the pathogenic (disease-causing) properties of the bacteria. Such descriptions are unnecessary for the audience (the housekeeping department) to do its job.

A more complex example of a short test report is found in Figure 14.12, which studies the effects of four light periods on the growth of paulownia seedlings (a flowering tree cultivated in China). The report, published in a scientific journal, is addressed to specialists in forestry. Such a test report follows a different, more detailed pattern of organization than the report in Figure 14.11 and includes an informative abstract, an introduction, a materials and methods section, a results and discussion section, the conclusion, and a list of references cited in the study.

To meet the needs of an expert audience, the writers of the report in Figure 14.12 had to include much more information than did Janeen Cufaude, the infection control officer who wrote the report in Figure 14.11, about the way the test was conducted and the types of scientific data the audience expects and needs. The researchers did not have to define technical terms for their audience, and they could confidently use scientific symbols and formulas as well.

FIGURE 14.11 A Short, Informal Test Report

Charleston General
HOSPITAL

Charleston, WV 25324-0114 / (304) 555-1800 / www.charlestongeneral.com

TO: James Dill, Supervisor   FROM: Janeen Cufaude
       Housekeeping   Infection Control Officer

DATE: December 12, 2011   SUBJECT: Routine sanitation inspection,
                                              Dec. 8

States why tests were performed and how

As part of the monthly check of the psychiatric unit (11A) on December 8, five areas were swabbed and tested for bacterial growth. The results of the lab tests of these samples are as follows:

Gives results of tests conducted in different locations

AREA	FINDINGS
1. cabinet in patients' kitchen	1. positive for 2 colonies of strep germs
2. rug in eating area	2. positive for food particles and yeasts and molds
3. baseboard in dayroom	3. positive for particles of dust
4. medicine counter in nurses' station	4. negative for bacteria—no growth after 48 hours
5. corridor by south elevator	5. positive for 4 colonies of staph germs isolated

ACTIONS TO BE TAKEN AT ONCE

Provides detailed directions based on results

1. Clean the kitchen cabinets with K-504 liquid daily, 3:1 dilution.
2. Shampoo rug areas bimonthly with heavy-duty shampoo, and clean visibly soiled areas with Guard-Pruf as often as needed.
3. Wipe all baseboards weekly with K-12 spray cleanser.
4. Mop heavily traveled corridors and access areas with K-504 cleanser daily, 1:1 dilution.

FIGURE 14.12 A Test Report Published in a Scientific Journal

Paulownia Seedlings Respond to Increased Daylength

M. J. Immel, E. M. Tackett, and S. B. Carpenter

Abstract

Paulownia seedlings grown under four photoperiods were evaluated after a growing period of 97 days. Height growth and total dry weight production were both significantly increased in the 16- and 24-hour photoperiods.

Begins with informative abstract

Introduction

Paulownia (*Paulownia tomentosa* [Thunb.] Steud.), a native of China, is a little known species in the United States. Recently, however, there has been increased interest in this species for surface mine reclamation (*1*).* Paulownia seems to be especially well adapted to harsh micro-climates of surface mines; it grows very rapidly and appears to be drought-resistant. In Kentucky and surrounding states, paulownia wood is actively sought by Japanese buyers and has brought prices comparable to black walnut (*2*).

This increased interest in paulownia has resulted in several attempts to direct seed it on surface mines, but little success has been achieved. The high light requirements and the extremely small size of paulownia seed (approximately 6,000 per gram) may be the limiting factors. Planting paulownia seedlings is preferred; but, because of their succulent nature, seedlings are usually produced and outplanted as container stock rather than bareroot seedlings. Daylength is an important factor in the production of vigorous container plants (*5*).

Our study compares the effects that four photoperiods—8, 12, 16, and 24 hours—had on the early growth of container-grown paulownia seedlings over a period of 97 days.

Gives background, purpose, and scope of study

Materials and Methods

Seeds used in this study were stratified in a 1:1 mixture of peat moss and sand at 4°C for 2 years. Following cold storage, seeds were placed on a 1:1 potting soil–sand mix and mulched with cheesecloth. They were then placed under continuous light until germination occurred. Germination percentages were high, indicating paulownia seeds can survive long periods of storage with little loss of viability (*3*).

Describes steps taken: procedures, conditions, and equipment used

Thirty days after germination, 3- to 4-centimeter seedlings were transplanted into 8-quart plastic pots filled with an equal mixture of potting soil, sand, and peat moss. Seventy-five seedlings were randomly assigned to each of the four treatments. Treatments were for 4 photoperiods—8, 12, 16, and 24 hours—and were replicated three times in 12 light chambers. Each chamber was 1.2- by 1.2-meters with an artificial light source 71 centimeters above the chamber floor.

The light source consisted of eight fluorescent lights: four 40-watt plant growth lamps alternated with four 40-watt cool white lamps. Light intensity averaged 550 foot-candles (1340μ einsteins/m²/s) at the top of each pot and the temperature averaged 550 foot-candles (1340 einsteins/m²/s) at the top of each pot and the temperature averaged 23°C (+2°C).

Uses technical terms and symbols audience expects

Seedlings were watered and fertilized after transplanting with a 6-gram 14-4-6 agriform container tablet. Beginning 1 month after transplanting, two seedlings were randomly selected and harvested from each chamber for a total of 24 trees.

© Cengage Learning 2013

(Continued)

*To save space, the references section has been omitted.

FIGURE 14.12 (Continued)

Height, root collar diameter, length of longest root, and oven-dry weight (at 65°C) were determined for each seedling. Harvests continued every week for 5 additional weeks.

Results and Discussion

Results indicate that early growth of paulownia is influenced by photoperiod, as shown in Table 1.

Includes a visual to summarize results and then explains what happened

TABLE 1. Height Diameter, Root Length, Total Dry Weight, and R/S Ratio for Paulownia Seedlings Grown Under Four Photoperiods After 97 Days.

Photo period (hrs.)	Height (cm)	Diameter (cm)	Root length (cm)	Total dry weight (gm)	R/S ratio
8	13.1	0.48	16.0	1.65	0.18
12	17.8	0.67	34.7	7.27	0.32
16	27.3	0.93	31.1	15.92	0.39
24	29.2	0.90	43.9	18.56	0.33

Expanding the photoperiod from 8 to either 16 or 24 hours increased height growth by 100 percent. Height growth in the 12-hour treatment also increased, but did not differ significantly from the 8-hour treatment. Heights under photoperiods of 8, 12, 16, and 24 hours were 13.1, 17.8, 27.3, and 29.2 centimeters, respectively.

Cites related studies

Previous studies have also shown that photoperiod affects the growth of paulownia seedlings (*4, 6*). Sanderson (*6*), for example, found that paulownia seedlings grown under continuous light averaged 27.2 centimeters in height after 101 days compared with 29.2 centimeters for our 24-hour seedlings. Other corresponding photoperiods were equally comparable. Downs and Borthwick (*4*) also concluded that height growth of paulownia was affected by extending the photoperiod.

The great treatment differences were shown in total dry weight production. Refer again to Table 1. The mean weight of 1.65 grams for seedlings in the 8-hour treatment was significantly less than that of any of the other photoperiods. The 16- and 24-hour treatments did not differ significantly. In fact, they more than doubled the average weight for seedlings in the 12-hour treatment.

Provides accurate measurements in a clear, objective tone

Root-to-shoot ratio (R/S) indicates the relative proportion of growth allocated to roots versus shoots for the seedlings in each photoperiod. In this study, shoots were developing at nearly three times the rate of the roots for seedlings in the 12-, 16-, and 24-hour treatments.

The 0.18 R/S ratio for seedlings in the 8-hour treatment was much lower, indicating that relative growth of the shoot is approximately five times that of the root. The shorter photoperiod therefore decreased root development relative to shoot development as well as significantly reduced total dry weight production.

Although root collar diameter and root length did not significantly differ under the different photoperiods after 97 days, there was a trend for greater diameter and root growth with longer photoperiods.

FIGURE 14.12 (Continued)

Page 3

Conclusions

Results indicate that the growth of paulownia seedlings is affected by changes in the photoperiod. Increasing the photoperiod significantly increased height growth and total dry matter production. The distribution of dry matter (R/S ratio) was altered by increasing the photoperiod; the ratio was larger in the longer photoperiods. In contrast to earlier studies (4), we found paulownia seedlings subjected to extended photoperiods were still growing after 97 days.

Interprets the significance of the results

Questions Your Test Report Needs to Answer

Readers will expect your test report to supply the following information:

- why you performed the test—an explanation of the reasons, your goals, and who authorized you to perform the test
- how you performed the test—under what circumstances or controls you conducted the test, what procedures and equipment you used, etc.
- what the outcomes were—your conclusions
- what implications or recommendations follow from your test—what you learned, discovered, confirmed, or disproved or rejected

When you sign the final copy of your report, you certify that things happened exactly when, how, and why you say they did.

Incident Reports

The short reports discussed thus far in this chapter have dealt with routine work. They have described events that were anticipated or supervised. But every business or agency runs into unexpected trouble that delays routine work, damages equipment or property, or results in personal injury. These circumstances need to be documented in an incident (or accident) report. The audience for an incident report can be within your organization or outside it. Employers use incident reports to make changes so that the problem does not occur again or so that a job can be done more effectively and safely. On some occasions government inspectors, insurance agents, and attorneys must be informed about those events that have interfered with or threatened normal, safe operations.

When to Submit an Incident Report

An incident report is submitted when there is, for example,

- an accident—fire, automobile, physical injury
- a law enforcement offense
- an environmental danger
- a computer virus
- a machine breakdown
- a delivery delay
- a cost overrun
- a production slowdown

Figure 14.13 is an incident report about a train derailment submitted by the engineer on duty. This report is in memo format, but some companies or agencies require you to fill out a special form. Because it can contain legally sensitive information needed in hard copy, an incident report should not be sent as an email.

Parts of an Incident Report

Include the following information in your incident report. Note how Figure 14.13 includes precise and accurate information for each of these parts.

1. **Identification details.** Indicate who and what was involved, and gather all relevant data—names, contact information, model/serial numbers, and so on. Record titles, department, and employment identification numbers. Indicate if you or your fellow employees were working alone. For customers or victims, record home addresses, phone numbers, and places of employment. Insurance companies will also require policy numbers.

2. **Type of incident.** Briefly identify the incident—personal injury, fire, burglary, equipment failure. Identify any part(s) of the body precisely. "Eye injury" is not enough; "injury to the right eye, causing bleeding" is better. "Dislocated right shoulder" or "punctured left forearm" is descriptive and exact. A report on damaged equipment should list make and model numbers.

3. **Time and location of the incident.** Include precise date (not "Thursday") and time (a.m. or p.m.).

4. **Description of what happened.** This section is the longest part of the report. Let readers know exactly what happened and why, how it occurred, and what led up to the incident.

5. **What was done after the incident.** Describe the action you took to correct conditions, how things got back to normal, and what was done to treat the injured, make the environment safer, speed a delivery, or repair damaged equipment.

FIGURE 14.13 An Incident Report in Memo Format

THE GREAT HARVESTER RAILROAD
Des Moines, IA 50306-4005
www.ghrr.com

TO: Angela O'Brien, District Manager
           James Hwang, Safety Inspector
FROM: Nick Roane, Engineer *Nick Roane*
DATE: May 5, 2011
SUBJECT: Derailment of Train 26 on May 5, 2011

Signs report to verify account of incident

Type of Incident
Two grain cars went off the track while I was driving Engine 457 of Train 26 on May 5, 2011. There were no injuries to the crew.

Begins with most important details

Description of Incident
At 7:20 a.m., I was traveling north at a speed of 52 miles an hour on the single main-line track four miles east of Ridgeville, Illinois. Weather conditions and visibility were excellent. Suddenly, the last two grain cars, 3022 and 3053, jumped the track. The train automatically went into emergency braking and came to a stop. But it did not stop before both grain cars turned at a 45° angle. After checking these cars, I found that half the contents of their loads had spilled. The train was not carrying any hazardous chemical shipments.

Gives precise time, location

Describes what happened

I notified Supervisor Bill Purvis at 7:40 a.m., and within 45 minutes he and a section crew arrived at the scene with rerailing equipment. The section crew removed the two grain cars from the track, put in new ties, and made the main-line track passable by 9:25 a.m. At 9:45 a.m. a vacuum car arrived with Engine 372 from Hazlehurst, Illinois, and its crew proceeded with the clean-up operation. By 11:25 a.m. all the spilled grain was loaded onto the cars brought by the Hazlehurst train. Bill Purvis notified Barnwell Granary that their shipment would be at least four hours late.

Explains what was done

Causes of Incident
Supervisor Purvis and I checked the stretch of train track where the cars derailed and found it to be heavily worn. We believe that a fisher joint slipped when the grain cars hit it, and the track broke. You can see the location of the cracked fisher joint in the graphic below.

Determines likely cause

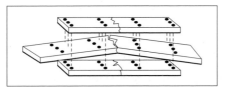

Supplies easy-to-follow exploded visual

(Continued)

FIGURE 14.13 (Continued)

Page 2

Recommendations
We made the following recommendations to the switch yard in Hazlehurst to be carried out immediately.

1. Check the section of track for 10 miles on either side of Ridgeville for any signs of defective fisher joints.
2. Repair any defective joints at once.
3. Instruct all engineers to slow down to 5 to 10 mph over this section of the track until the rail check is completed.

Offers precise recommendations to prevent problem from recurring

© Cengage Learning 2013

6. **What caused the incident.** Make sure your explanation is consistent with your description of what happened. Pinpoint the trouble. In Figure 14.13, for example, the defective fisher joint is listed under the heading "Causes of Incident."

7. **Recommendations.** Recommendations about preventing the problem from recurring may involve repairing any broken parts (as in Figure 14.13), calling a special safety meeting, asking for further training, adapting existing equipment, doing emergency planning, or modifying schedules. Make sure all of your recommendations are consistent with what actually happened.

Protecting Yourself Legally

An incident report can be used as legal evidence, and it then becomes part of a permanent legal record that can be used by law enforcement and attorneys in court to establish negligence and liability on your and your company's part. It can also be used by an employer to determine employee responsibility. An incident report frequently concerns the two topics over which powerful legal battles are waged— health and property. You could lose a case in court if your report is not written clearly, accurately, and completely.

You have to be very careful about collecting and recording details. Make sure your report is not sketchy, confusing, or incomplete. To avoid these errors, you may have to interview employees or bystanders; travel to the incident site; check manuals, code books, or other guides; consult safety experts; collect and describe evidence; or research records/archives.

To ensure that what you write is legally proper, follow these guidelines:

1. **Submit your report promptly, and sign or initial it.** Any delay might be seen as a cover-up. Send your report to the appropriate parties immediately after you

have gathered the necessary information and had it reviewed by your supervisor. You may have to post photographs as well.

2. Double-check your spelling (individuals' names, pieces of equipment, etc.) and your punctuation. An error here calls the accuracy and validity of your whole report into question.

3. Be accurate, objective, and complete. Recount clearly what happened in the order it took place. Give sufficient information for readers to know exactly what happened. Never omit or distort facts; the information may surface later, and you could be accused of a cover-up. Do not just write "I do not know" for an answer. If you are not sure, state why. Also be careful that there are no discrepancies in your report.

4. Give facts, not opinions. Provide a factual account of what actually happened, not a biased interpretation of events or one based on speculation or hearsay. Vague terms such as "I guess," "I wonder," "apparently," "perhaps," or "possibly" weaken your objectivity. Indicate who discovered, reported, or witnessed the incident. But stick to details you witnessed or that were seen by eyewitnesses. Identify witnesses or victims by giving complete names, addresses, places of employment, and so on. Keep in mind that stating what someone else saw is regarded as hearsay and therefore is not admissible in a court of law. State only what *you* saw or heard. When you describe what happened, avoid drawing uncalled-for conclusions. Consider the following statements of opinion versus fact:

> Opinion: The patient seemed confused and caught himself in his IV tubing.
> Fact: The patient caught himself in his IV tubing.
>
> Opinion: The equipment was defective.
> Fact: The bolt was loose.

Be careful, too, about blaming someone. Statements such as "Baxter was incompetent" or "The company knew of the problem but did nothing about it" are libelous remarks.

5. Do not exceed your professional responsibilities. Answer only those questions you are qualified to answer. Do not presume to speak as a first responder, a detective, an inspector, a physician, a supervisor, or a judge. Do not represent yourself as an attorney or a claims adjuster in writing the report. And don't take sides.

Short Reports: Some Final Thoughts

To prepare successful short reports, keep in mind the rules of short report writing discussed in this chapter. Always take into account your readers' needs and expectations at every stage of your writing, document carefully what you write about, take accurate and complete notes, write objectively and ethically, present complicated data clearly and concisely, provide background and context where necessary, and include specific recommendations, where called for, based upon the facts. Remembering these basic rules will earn you praise and possibly promotions at work.

✓ Revision Checklist

- [ ] Understood why short reports are important for my company or organization.
- [ ] Recognized the type of report (e.g., periodic, progress, employee activity, etc.) I need to write.
- [ ] Chose the report format that is most appropriate for my audience (memos or emails for supervisor or co-workers; letters for clients).
- [ ] Had a clear sense of how my readers will use my report.
- [ ] Did appropriate research to give my readers enough information to help them make careful decisions.
- [ ] Provided significant information about costs, materials, personnel, and times so readers will know that my work consists of facts, not impressions.
- [ ] Double-checked all data—names, costs, figures, dates, places, and equipment models and numbers.
- [ ] Verified necessary statistics and trends so that my periodic and sales reports are thorough and accurate.
- [ ] Made sure all my comments and recommendations were ethical.
- [ ] Followed all organization guidelines.
- [ ] Adhered to all legal requirements.
- [ ] Eliminated unnecessary details or those too technical for my audience.
- [ ] Kept report concise and to the point.
- [ ] Used headings whenever feasible to organize and categorize information.
- [ ] Supplied relevant visuals to help readers understand my message and crunch any numbers.
- [ ] Employed underlining, boldface, or italics to set headings apart or to emphasize key ideas.
- [ ] Began the report with a statement of purpose that clearly described the scope and significance of my work.
- [ ] Incorporated tables and other pertinent visuals to display data whenever appropriate.
- [ ] Explained clearly in a conclusion section what the data mean.
- [ ] Determined that recommendations logically follow from the data and that recommendations are relevant and realistic.

Periodic/Sales Reports
- [ ] Submitted my report at scheduled intervals.
- [ ] Provided accurate data on quantities sold or services provided.
- [ ] Documented, organized, described, and explained the importance of the data, including sales figures.

Progress Reports
- [ ] Reported at key stages of the project.
- [ ] Gave readers bottom-line information on costs, equipment, personnel, and schedules.

☐ Described the work or projects to be completed.

☐ Alerted readers about any problems—delays or cost overruns.

Employee Activity/Performance Reports

☐ Accounted for how I spent my time at work for a given period.

☐ Categorized types of projects and tasks accomplished.

☐ Indicated when a job was a collaborative effort—with whom I worked and when.

☐ Identified projects and tasks that continue into the next scheduled evaluation period (monthly, quarterly, etc.).

Trip/Travel Reports

☐ Explained the purpose of the trip—where and when I went and whom I visited, interviewed, made a presentation for, or saw at a conference.

☐ Recorded all names, places, dates, costs, etc., accurately.

☐ Supplied relevant visuals, such as photographs and charts.

☐ Included findings and recommendations.

☐ Indicated I am available for questions or clarifications.

Test Reports

☐ Recorded data accurately and objectively.

☐ Specified why the test was performed and which procedures and equipment were used.

☐ Explained the significance of results and outcomes, emphasizing how the information can be applied to meet my company's needs.

Incident Reports

☐ Identified the type of incident, time, place, victims and eyewitnesses, injuries, and so on.

☐ Provided a factual, objective, and accurate account of the incident in a logical or chronological order.

☐ Stated without bias what may have caused the incident; did not draw unjustified conclusions.

☐ Explained what has been or can be done to correct the problem and prevent it from recurring.

Exercises

1. Bring to class an example of a periodic report from your present or previous job or from any community, religious, or social organization to which you belong. In an accompanying memo to your instructor, indicate who the audience is and why such a report is necessary, stressing how it is organized, what kinds of factual data it contains, what visuals were used, and how it might be improved in content, organization, style, and design.

2. Assume that you are the manager of an apartment complex of 200 units. Write a periodic report based on the following information: 26 units are vacant, 38 soon will be vacant, and 27 will be leased by June 1. Also add a section of recommendations to your supervisor (the head of the real-estate management company for which you work) on how vacant apartments might be leased more quickly and perhaps at increased rents. Consider such important information as decorating, advertising, and installing a new security system.

3. You work for a household appliance store and your manager has asked you to prepare a sales report based on the information contained in the following table. Include a recommendation section for your manager.

Product	Number Sold	
	October	November
Kitchen Appliances		
Refrigerators	72	103
Dishwashers	27	14
Freezers	10	36
Electric ranges	26	26
Gas ranges	10	3
Microwaves	31	46
Laundry Appliances		
Washers	50	75
Dryers	24	36
Air Treatments		
Room air conditioners	41	69
Dehumidifiers	7	2

4. Write a progress report on the wins, losses, and ties of your favorite sports team for the past season. Address the report to the director of publicity for the team and stress how the director might use those facts for future publicity. As part of your report, indicate what might be an effective lead for a press release about the team's efforts.

5. Submit a progress report to your writing teacher on what you have learned in his or her course so far this term, which writing skills you want to develop further, and how you propose doing so. Mention specific memos, emails, letters, instructions, reports, and proposals you have written or will soon write.

6. Compose a site inspection report on any part of the college campus or on the plant, office, or store in which you work that might need remodeling, expansion, rewiring for computer use, or new or additional air-conditioning or heating work.

7. You and your collaborative team have been asked to write a short preliminary inspection report on the condition of a historic building for your state historical society. Inspecting the building, the home of a famous late-nineteenth-century governor, you discover the problems listed below. Include all these details in your report. Also supply recommendations for your readers: a director of the state historical society, a state architect, and four representatives of the subcommittee on finance from your state legislature. Design two appropriate visuals to include in your report.

   - The eight front columns are all in need of repair; two of them may have to be replaced.
   - The area below each bottom window casement needs to be excavated for waterproofing.
   - The slate tile on the roof has deteriorated and needs immediate replacement.
   - The front stairs show signs of mortar leaching and require attention at once.
   - Sections of gutter on the northwest and northeast sides of the house must be changed; other gutters are in fair shape.
   - Wood shutters need to be repainted; four of the twelve may even need to be replaced.
   - All trees around the house need pruning; an old elm in the backyard shows signs of decay.
   - The siding is in desperate need of preparation and painting.
   - The brick near the front entrance is dirty and moss-covered.

8. Write a report to an instructor in your major about a field trip you took recently—to a museum, laboratory, health care agency, correctional facility, radio or television station, agricultural station, or office. Indicate why you took the trip, name the individuals you met on the trip, and stress what you learned and how that information will help you in course work or on your job.

9. Write an employee activity/performance report on what you accomplished at your job, full or part time, during the last month. In your introduction, indicate how the work you did helps your employer meet his or her goals. In your conclusion, forecast how the work you expect to do next month illustrates your dedication to doing an effective job.

10. Submit a test report on the purpose, procedures, results, conclusions, and recommendations of an experiment you conducted on one of the following subjects:
    a. soil
    b. Internet access
    c. water
    d. automobiles
    e. textiles/clothing

    f. animals
    g. recreational facilities
    h. computer hardware or software
    i. forests
    j. food
    k. air quality
    l. transportation
    m. blood
    n. noise levels

11. Write an incident report about a problem you encountered in your work or at home in the last year. Document the problem and provide a solution. Use the memo format shown in Figure 14.13.

12. Write an incident report about one of the following problems. Assume that it happened to you. Supply relevant details and visuals in your report. Identify the audience for whom you are writing and the agency you are representing or trying to reach.

    a. After hydroplaning, your company car hit a tree and suffered damage to the front fender.

    b. You were the victim of an electrical shock because an electrical tool was not grounded.

    c. You strained your back lifting a bulky package in the office or plant.

    d. A virus infected your company's intranet, and it will have to be shut down for 12 hours to debug it.

    e. The crane you were operating broke down, and you lost a half-day's work.

    f. The vendor shipped the wrong replacement part for your computer, and you cannot complete a job until the correct part arrives in three days.

    g. An electrical storm knocked out your computer; you lost 1,000 mailing addresses and will have to hire additional help to complete a mandatory mailing by the end of the week.

    h. A scammer stole some sensitive files about a new product or service your company had hoped to launch next month.

    i. An irate customer threatened one of your sales clerks, but there was no physical violence.

13. Choose one of the following descriptions of an incident, and write a report based on it. The descriptions contain unnecessary details, vague words, insufficient information, unclear cause-and-effect relationships, or a combination of those errors. In writing your report, correct the errors by adding or deleting whatever information you believe is necessary. You may also want to rearrange the order in which information is listed. Use a memo format, like that in Figure 14.13, to write the report.

    a. After sliding across the slippery road late at night, my car ran into another vehicle, one of those fancy imported cars. The driver of that car must have been asleep at the wheel. The paint and glass chips were all over. I was driving back from our regional meeting and wanted to report to the home office the next day. The accident will slow me down.

b. Whoever packed the glass mugs did not know what he or she was doing. The string was not the right type, nor was it tied correctly. The carton was too flimsy as well. It could have been better packed to hold all those mugs. Moreover, since the bus had to travel across some pretty rough country, the package would have broken anyhow. The best way to ship these kinds of goods is in specially marked and packed boxes. The value of the box contents was listed at $575.

CHAPTER 15

Writing Careful Long Reports

Access chapter-specific interactive learning tools, including quizzes and more in your English CourseMate, accessed through www .cengagebrain.com.

This chapter introduces you to long reports—how they differ from short reports, how they are written, and how they are organized. It is appropriate to discuss long reports in one of the last chapters of *Successful Writing at Work*. A major project in the world of work, the long report gives you an opportunity to use and combine many of the writing skills and research strategies you have already learned. In business, a long report is the culmination of many weeks or months of hard work on an important company project.

The following skills will be most helpful to you as you study long reports; appropriate page numbers appear for the topics that have already been discussed:

- functioning as a member of a collaborative writing team (pages 84–86)
- respecting corporate image and mission (pages 31–32) and adhering to corporate guidelines on preparing and routing reports
- assessing and meeting your audience's multiple needs (pages 12–17)
- gathering and summarizing information from print and online sources, and conducting interviews (pages 324–383)
- generating, drafting, revising, and editing your ideas and those of your team (pages 50–78)
- reporting the results of your research accurately and concisely (pages 327–328)
- creating and introducing visuals and designing documents (pages 517–536)
- using an appropriate method of documentation (pages 384–385)
- preparing an informative abstract (pages 443–444)

Having developed these skills, you should be ready to write a successful long report. The collaborative team of writers of the model long report in Figure 15.3 on pages 701–718 uses all of the above skills in their work for RPM Technologies.

How a Long Report Differs from a Short Report

Both long and short reports are invaluable tools in the world of work. Basic differences exist, though, between the two types of reports. A short report is not a watered-down version of a long report; nor is a long report simply an expanded version of a short one. The two types of reports differ in scope, research, format, timetable, audience, and collaborative effort.

686

The following sections explain some of the key differences between these two reports. By understanding these differences, you will be better able to follow the rest of this chapter as it covers the process of writing a long report and the organization and parts of such a report. Study the long report at the end of this chapter to see how these two types of documents differ.

Scope

A long report is a major study that provides an in-depth view of a key problem or idea. It might be eight to twenty pages long or even much longer, depending on the scope of the subject. The implications of a long report are wide-ranging for a business or an industry—relocating a plant, adding a new network, changing a programming operation, or adapting the workplace for multinational employees, as in Figure 15.3.

Long reports examine a major problem in detail, while short reports cover just one part of the problem. Unlike a short report, a long report may discuss not just one or two current events but rather the continuing history of a problem or an idea (and the background information necessary to understand it in perspective). For example, the short test report on paulownia in Figure 14.12 (pages 674–676) would be used with many other test reports in a long report for a group of environmentalists or a government agency on the value of planting those trees to prevent soil erosion.

The titles of some typical long reports further suggest their extensive (and in some cases exhaustive) coverage:

- A Master Plan for the Recreation Needs of Dover Plains, New York
- The Transportation Problems in Kingford, Oregon, and the Use of Monorails
- Promoting More Effective E-Commerce and E-Tailing at TechWorld, Inc.
- The Use of Virtual Reality Attractions in Theme Parks in Jersey City, New Jersey
- Public Policy Implications of Expanding Health Care Delivery Systems in Tate County
- Internet Medicine in Providing Health Care in Rural Areas: Ways to Serve Southern Montana

Research

A long, comprehensive report requires much more extensive research than a short report does. Information can be gathered over time from primary and secondary research—databases, Internet searches, surveys, books, articles, blogs, statistical sources, laboratory experiments, site visits and tests, interviews, and the writer's own observations. For a long report, you will have to do much more research than for a short report, including possibly interviews to track down the relevant background information and to discover what experts have said about the subject and what they propose should be done. Note how much research went into the business report in Chapter 8 and the long report in Figure 15.3.

Information gathered for many short reports can also help you prepare a long report. In fact, as the example of paulownia in Figure 14.12 again shows, a long report can use the experimental data from a short report to arrive at a conclusion.

Also for a long report, writers often supply one or more progress reports (one type of short report).

No matter what type of research you do for a report or a proposal, you will have to evaluate the sources you use. That means verifying their accuracy, making sure they are current, and gathering necessary and relevant information from secondary sources or through interviews and surveys to justify your conclusions.

Finally, preparing a proposal can lead to writing a long report. You might suggest a change to an employer, who then would ask you to write a long report containing the research necessary to implement that change.

Tech Note

Using Government-Sponsored Research

The U.S. government conducts or sponsors a great deal of research that you might find relevant for your reports. Government research is at the heart of science, technology, health care, urban planning, and so on. Following is a short list of some major government and other relevant websites.

- **www.epa.gov** will lead you to press releases, test guidelines, and information about grants, contracts, and job opportunities at the Environmental Protection Agency.
- **www.osha.gov**, the website for the Occupational Safety and Health Administration, supplies information about OSHA standards, news releases and fact sheets, publications, technical information, and safety links.
- **www.sba.gov** leads to the Small Business Administration's website, where you'll find guides on beginning a small business, the opportunity to include your business in the national register, and contact information for local SBA offices.
- **www.bls.gov** is sponsored by the Bureau of Labor Statistics. This organization compiles and maintains financial information such as charts on U.S. dollar inflation, growth or decline in the number of businesses, and wage increases.
- **www.gpo.gov/fdsys** is the next generation for federal government online information through the Government Printing Office's Federal Digital System. It provides free electronic access to a variety of federal documents, including legislative, judicial, and executive resources.
- **www.archives.gov** is the address for the National Archives and Records Administration, an independent federal agency that oversees the management of all federal records.
- **www.ars.usda.gov** is the portal to the Agricultural Research Service (ARS), the U.S. Department of Agriculture's chief scientific research agency. ARS conducts research to ensure high-quality, safe food; assess the nutritional needs of Americans; sustain a competitive agricultural economy; and enhance the environment.
- **www.rita.dot.gov** is the site for the Research and Innovative Technology Administration (RITA), which coordinates the U.S. Department of Transportation's research and education programs. RITA also offers vital transportation statistics and analysis for aviation, highway traffic safety, maritime travel, hazardous materials transportation, and so on.

Format

A long report is too detailed and complex to be adequately organized in a memo or letter format. The product of thorough research and analysis, the long report gives readers detailed discussions and interpretations of large quantities of data. To present the information in a logical and orderly fashion, the long report contains more parts, sections, headings, subheadings, documentation, and supplements (appendices) than would ever be included in a short report. Look at the Table of Contents for Figure 15.3 (page 703) to see at a glance the various divisions and parts of a long report.

Timetable

The two types of reports differ in the time it takes to prepare them. Writers of these two reports work under different expectations from their readers and under different deadlines. A long report explores with extensive documentation a subject involving personnel, locations, costs, safety, or equipment. Many times a long report is required by law, for example, investigating the feasibility of a project that will affect the ecosystem. A short report is often written as a matter of routine duty; the writer is sometimes given little or no advance notice. The long report, however, may take weeks or even months to write. Below is a timetable for preparing a typical long report.

Do Research and Conduct Interviews	Outline or Draft	Conferences for Revisions and Further Research	Revise and Prepare Figures	Proof-read and Polish	Submit Long Report
4 weeks	3 weeks	1 week	3 weeks	1 week	Due Date

Audience

The audience for a long report is generally broader—and includes individuals higher up in an organization's hierarchy—than that for a short report. Your short report may be read by co-workers, a first-level supervisor, and possibly that person's immediate boss, but a long report is primarily intended for people in the top levels of management—presidents, vice presidents, superintendents, directors—who make executive, financial, and organizational decisions. These individuals are responsible for long-range planning and seeing the big picture, so to speak.

Keep in mind, too, that other individuals may consult your report or at least read the parts of it that are most relevant for their needs. Human resources personnel may examine your report to make sure it follows a company's policies and adheres to all government guidelines regarding the privacy and protection of individuals being surveyed or interviewed. Specialists in different fields may also study the sections of your report on protocols, experiments, or other technical matters upon which your recommendations are based.

Collaborative Effort

Unlike many short reports, the long report in the world of business may not always be the work of one employee. Rather, it may be a collaborative effort, the product of a committee or team whose work is reviewed by a main editor to make sure that the final copy is consistently and accurately written. Individuals in many departments within a company—IT, document design, engineering, sales, transportation, legal affairs, public relations, safety—may cooperate in planning, researching, drafting, revising, and editing a long report.

The team should estimate a realistic timeframe necessary to complete the various stages of their work—when drafts are due and when editing must be concluded, for example. A project schedule based on that estimate should then guide a writing team's work. But remember: Projects almost always take longer than initially planned. Prepare for a possible delay at any one stage. The team may have to submit written progress reports (see pages 654–661) to its members, as well as to management.

To be successful, a collaborative writing team should observe the guidelines and procedures for collaborative writing (review Chapter 3).

The Process of Writing a Long Report

As we just saw, writing a long report requires much time and effort. Since your work will be spread over many weeks, you need to see your report not as a series of static or isolated tasks but as an evolving project. Before you embark on that project, review the information on the writing process in Chapter 2 (pages 51–78). You may also want to study the flow chart in Figure 10.14, which illustrates the different stages in writing a research paper or report. The following guidelines will also help you to plan and write your long report:

1. **Identify a significant topic.** You'll have to do preliminary research—reading, online searching, compiling a questionnaire or survey, conferring with and interviewing experts—to get an overview of key ideas, individuals involved, and the implications for your company, organization, or community. Note the kinds of research Terri Smith Ruckel and her team did for the long report in Figure 15.3. As they did, expect to search a variety of print and Internet sources, to read and evaluate them, and to incorporate them in your work.

2. **Expect to confer regularly with your supervisor and team members.** In these meetings, be prepared to ask pertinent and researched questions to pin down exactly what your boss wants and how your writing team can accomplish this goal. Focus your questions on the company's use of your report, how it wants you to express certain ideas, and the amount of information it needs. Your supervisor may want you to submit an outline before you draft the report and may expect several more drafts for approval before you write the final version.

3. **Revise your work often.** Be prepared to work on several outlines and drafts. Your revisions may sometimes be extensive, depending on what your boss or collaborative team recommends. You may have to consult new sources and arrive at a

new interpretation of those sources. As you narrow your purpose and scope, you may find yourself deleting information or modifying its place in your work. This shifting around as well as adding and deleting information will help you to arrive at a carefully organized report. At the later stages of your work, you will be revising and editing your words, sentences, and paragraphs.

4. **Keep the order flexible at first.** As you work on your drafts and revisions, remember that a long report is not written in the order in which the parts will finally be assembled. You cannot write in "final" order—abstract to recommendations. Instead, expect to write in "loose" order to reflect the process in which you gathered information and organized it for the final copy of the report. Usually, the body is written before the introduction so the authors can make sure they have not left anything out. The abstract, which appears very early in the report, is always written after all the facts have been recorded and the recommendations made or the conclusions drawn. The title page and the table of contents are always prepared last.

5. **Prepare both a day-to-day calendar and a checklist.** Keep both posted where you do your work—above your desk or computer, or use your computer's calendar—so you can track your progress. Make sure your collaborative team is following the same calendar and using the same checklist. The calendar should mark *milestones*—that is, dates by which each stage of your work must be completed. Match the dates on your calendar with any dates your employer gave you to submit an outline, progress reports, and the final copy. Your checklist should list the major parts of your report. As you complete each section, check it off. Before assembling the final copy of your report, use the checklist to make sure you have not omitted something.

Parts of a Long Report

A long report may include some or all of the following twelve parts, which form three categories: *front matter* (letter of transmittal, title page, table of contents, list of illustrations, abstract); *report text* (introduction, body, conclusion, recommendations); and *back matter* (glossary, citations list, appendix).

Numbering the Pages of a Long Report

You will use two sets of numbers for the pages of your long report, one for the front matter and another for the text and back matter of your report. See Figure 15.3 for an example of proper pagination in a long report. Use lowercase Roman numerals (e.g., i, ii, iii, iv) for the front matter. The title page counts as page i, but do not number it. Instead, start with the table of contents as page ii. Then number the list of illustrations as page iii and the abstract, if it appears on a separate page as in Figure 15.3, as page iv. Format all front matter page numbers as footers. For the text and back matter of your report, use consecutive Arabic numerals in the headers, that is, in the upper left-hand side of each page. Keep in mind, though, that your instructor or employer may prefer a different placement for the page numbers—for example, putting all report page numbers in footers or in headers.

Front Matter

As the name implies, the front matter of a long report consists of everything that precedes the actual text of the report. Such elements introduce, explain, and summarize to help the reader locate various parts of the report.

Letter of Transmittal

This short (usually only one-page) letter states the purpose, scope, and major recommendation of the report in three or four paragraphs. Figure 15.1 shows a sample letter of transmittal for a business report. It highlights the main points of the report that a decision maker would be most interested in reading. In the first paragraph of the letter, the writers indicate why they are sending the report and point out why it is important. The second paragraph summarizes and justifies the recommendations they make in the report. In the final paragraph, the writers thank the reader for the opportunity to prepare the report and offer to answer any questions or discuss or clarify any parts of the report.

Title Page

Since MLA and APA have different formats for title pages, find out what your employer prefers. Basically, though, your title page should contain the following:

- the full title of your report. Tell readers what your topic is and how you have restricted it in time, space, or method. Your title should suggest the scope and research behind your work. Avoid titles that are vague, too short, or too long.

> Vague Title: A Report on the Internet: Some Findings
> Too Short: The Internet
> Too Long: A Report on the Internet: A Study of Dot-Com Companies, Their History, Appeal, Scope, Liabilities, and Relationship to Ongoing Work Dealing with Consumer Preferences and Protection Within the Last Five Years in the Midwest

- the name of the company or agency preparing the report
- the name(s) of the report writer(s)
- the date the report was submitted
- any agency, order, or grant number
- the name of the firm or organization for which the report was prepared

Make sure that your title page looks professional. Center your title and graphically subordinate any subtitles. Do not use abbreviations (e.g., *bldgs.* for *buildings*, *gov't* for *government*, *bus.* for *business*) or acronyms (e.g., *AMS* for *Association of Marketing Students*; *PTAs* for *Physical Therapist Assistants*).

Table of Contents

The table of contents lists the major headings and subheadings of your report and tells readers on which pages they can be found. Make sure that your table of contents exactly matches the order and wording of your main headings and subheadings. It reveals the scope of your report and helps readers identify the parts of your report that are of most interest to them.

FIGURE 15.1 A Letter of Transmittal for a Long Report

ALPHA CONSULTANTS

■ 1400 Ridge ■ Evanston, California 97214-1005 ■

■ 805-555-9200 ■ FAX 805-555-0221 ■

www.alphaconsultants.com

August 15, 2011

Dr. K. G. Lowry, President
Coastal College
San Diego, CA 93219-2619

Dear Dr. Lowry:

We are happy to offer you the enclosed report, **A Study to Determine New Directions in Women's Athletics at Coastal College**, which you commissioned. Our report contains our recommendations about strengthening existing sports programs and creating new ones at Coastal College.

We recommend that Coastal should engage in more aggressive recruitment to establish a more competitive women's baseball team, should offer additional athletic activities in women's track and field by August 2012, and should create a new interdisciplinary program between the Athletic Department and the Women's Studies Program. Based on the findings in our report, we believe that these recommendations are cost-effective, timely, and consistent with the mission of Coastal College.

It has been an honor to prepare this report for you. We hope that you find it helpful in meeting students' needs at Coastal College. If you have any questions or if you would like to discuss any of our recommendations, please call us.

Sincerely yours,

Barbara Gilchrist

Barbara Gilchrist

Lee T. Sidell

Lee T. Sidell

Encl. Report

Indicates why report was done

Gives three key recommendations

Provides justification for recommendations

Encourages response and questions

Essentially, a table of contents shows how you organized your report. In Figure 15.2, for example, the reader can see how the report "A Study to Determine New Directions in Women's Athletics at Coastal College" is divided into four chief parts (Introduction, Discussion, Conclusion, and Recommendations). A table of contents emerges from many outlines and drafts. Your outlines may undergo considerable change until you decide on the formal divisions and subdivisions of your report.

Include front matter components in your table of contents, but never list the contents page itself, the letter of transmittal, or the title page. Also, never include just one subheading under a heading.

Incorrect: EXPANDING THE SPORTS PROGRAM
                 Basketball
        BUILDING A NEW ARENA
                 The West Side Location

Correct: EXPANDING THE SPORTS PROGRAM
                 Basketball
                 Track and Field
        BUILDING A NEW ARENA
                 The West Side Location
                 Costs

Tech Note

Automatically Formatting the Long Report

You can save a considerable amount of time when writing your long report by using your word processor's automatic formatting features. You can use the automatic formatting feature either as you keyboard or after you've completed your report, by turning on the automatic formatting options and selecting the types of formats you'd like to apply to the report. You can

- apply bulleted or numbered lists
- automatically format the heading structure of your report by using your word processor's tagging function
- create a table of contents via your tagged headers
- produce an index by tagging individual words

It will take only a short time to familiarize yourself with your word processor's autoformatting features, and by doing so you can reduce the cumbersome process of formatting each item individually. You should always carefully proofread your report to make sure that the autoformatting has been correctly applied.

FIGURE 15.2   A Table of Contents for a Long Report

TABLE OF CONTENTS

Uses Roman numerals to number pages of front matter

Records key sections and subsections of report

Indents subheads under major section headings

Uses at least two subdivisions under a heading

Boldfaces major sections

Lists references on separate pages in report

Provides titles for documents included in appendices

Uses Roman numeral as footer to number table of contents page

ii

List of Illustrations
A list of all the visuals indicates where they can be found in your report. Note the variety of visuals found in Figure 15.3.

Abstract
As discussed in Chapter 9, an abstract summarizes the report, presenting a brief overview of the problem and conclusions. An informative abstract is far more helpful to readers of a report than is a descriptive one, which gives no conclusions or results. Abstracts may be placed at various points in long reports—on the title page, on a separate page preceding or following the table of contents, or as the first page of the report text.

Not every member of your audience will read your entire report, but almost everyone will read the abstract. For example, the president of the corporation or the director of an agency may use the abstract as the basis for approving the report and passing it on for distribution. Thus, the abstract may be the most important part of your report. See how the abstract Terri Smith Ruckel and her team prepared for their readers on page 704 succinctly gives the reporter's conclusions, or findings.

Begin your abstract with a sentence that identifies the subject, purpose, scope, and importance of your report. Then concentrate on the main points your report covers, briefly comment on the results and outcomes you reached, and clearly pinpoint your major recommendations. Your abstract needs to emphasize the major sections of your report.

Some information does not belong in an abstract. Do not give readers a detailed description of the methods you used or try to squeeze in every minor point. Don't worry about describing the visuals accompanying your report, such as maps or graphs. It is also unnecessary to repeat who commissioned the report and why. And never include information in your abstract that is not found in your report.

When you write your abstract, use complete sentences with keywords. Appropriate keywords make it easy for search engines to find your report if it is posted on the Web or if it will be archived on, say, a company intranet. (Review pages 371–374.)

Text of the Report
The text of a long report consists of an introduction, the body, a conclusion, and sometimes recommendations.

Introduction
The introduction may constitute as much as 10 or 15 percent of your report, but usually it is not any longer. If it were, the introduction would be disproportionate to the rest of your work, especially the body. The introduction is essential because it tells readers why your report was written and thus helps them to understand and interpret everything that follows. See how Terri Smith Ruckel and her team emphasize the importance of their research for their employer, RPM Technologies, in the

introduction to Figure 15.3 on pages 705–707. But make sure you do not put your findings, conclusion, or recommendations in your introduction.

Do not regard the introduction as one undivided block of information. It includes the following related parts, which should be labeled with subheadings. Keep in mind, though, that your employer may ask you to list these parts in a different order.

   1. Background. To understand why your topic is significant and hence worthy of study, readers need to know about its history. This history may include information on such topics as who was originally involved, when, and where; how someone was affected by the issue; what opinions have been expressed on the issue; and what the implications of your study are. Note how the long report in Figure 15.3 provides useful background information on when, where, how, and why multinational employees entered the U.S. workforce.

   2. Problem. Identify the problem or issue that led you to write the report. Your problem needs to be significant enough to warrant a long report. Because the problem or topic you investigated will determine everything you write about in the report, you need to state it clearly and precisely. That statement may be restricted to a few sentences. Here is a problem statement from a report on how construction designs have not taken into account the requirements of Americans with disabilities:

> The construction industry has not sufficiently met the needs for accessible workplaces and homes for all age and physical ability groups. The industry has relied on expensive and specialized plans to modify existing structures rather than creating universally designed spaces that are accessible to everyone.

   3. Purpose statement. The purpose statement, crucial to the success of the report, tells readers why you wrote the report and what you hope to accomplish or prove. It expresses the goal of all your research. In explaining why you gathered information about a particular problem or topic, indicate how such information might be useful to a specific audience, company, or group. Like the problem statement, the purpose statement does not have to be long or complex. A sentence or two will suffice. You might begin simply by saying, "The purpose of this report is. . . ."

   4. Scope. This section informs readers about the specific limits—number and type of issues, time, money, locations, personnel, and so forth—you have placed on your investigation. You inform readers about what they will find in your report or what they won't through your statement about the scope of your work. The long report in Figure 15.3 concentrates on adapting the U.S. workplace to meet the communication and cultural needs of a workforce of multinational employees, not on trends in the international employment market—two completely different topics.

Body

Also called the *discussion*, the body is the longest section of a report, possibly making up as much as 70 percent or more of the document. Everything in this and all the other sections of your report grows out of your purpose and how you have limited your scope. The body contains statistical information, details about the environment, and physical descriptions, as well as the various interpretations and comments of the authorities whose work you consulted or the individuals whom you have interviewed as part of your research. The body can also identify and describe the range of options you surveyed and earmark the most appropriate, as the business report in Chapter 8 does in listing various print and online publications in which to advertise the new apartment complex.

What to Include in the Body of a Report   The body of your report should

- be carefully organized to reveal a coherent and well-defined plan
- separate material into meaningful parts to identify the major issues as well as subissues in your report
- clearly relate the parts to one another
- use headings to help your reader identify major sections more quickly

Headings   Your organization is reflected in the different headings and subheadings included in your report. Use them to make your report easy to follow. Organizational headings will also enable someone skimming the report to find specific information quickly. The headings, of course, will be included in the table of contents. (Note how Figure 15.3 is carefully organized into sections.)

Transitions   In addition to headings, use transitions to reveal the organization of the body of your report. At the beginning of each major section of the body, tell readers what they will find in that section and why. Summary sentences at the end of a section will tell readers where they have been and prepare them for any subsequent discussions. The report in Figure 15.3 does an effective job of providing internal summaries.

Conclusion

The conclusion should tie everything together for readers by presenting the findings of your report. Findings, of course, will vary depending on the type of research you do. For a research report based on a study of sources located through various reference searches, the conclusion should summarize the main viewpoints of the authorities whose works you have cited. For a report done for a business, you must spell out the implications for your readers in terms of costs, personnel, products, location, and so forth.

Regardless of the type of research you do, your conclusions should

- be based on the information and documentation in the body of the report
- corroborate the evidence/information you gave in the body of your report
- grow out of the work you describe in the body of the report
- stick to the areas that your report covers, and not stray into areas it does not

In essence, to write an effective conclusion, you will have to summarize a great deal of information accurately and concisely. Notice how the following conclusion, from a long report on the Japanese tuna market, clearly summarizes the market opportunities explained in the report.

Conclusion

The U.S. tuna industry has great potential to expand its role into the Japanese market. This market, currently 400,000 tons a year and growing rapidly, is already being supplied by imports that account for 35 percent of all sales. Our report finds that not only will the Japanese market expand, but its share of imports will continue to grow. The trend is alarming to Japanese tuna industry leaders because the market, close to a billion dollars a year, is increasingly subject to the influence of foreign imports. Decreasing catches by Japan's own tuna fleet as well as Japanese consumers' increased preference for tuna have contributed significantly to this trend.

Recommendations

The most important part of the report, after the abstract, is the recommendations section. Clearly, your recommendations must be based on the findings and conclusions you reached in your report. They should tell readers exactly what plan of action you want them to follow. Make sure your recommendations are specific and consistent with your company's or organization's goals, resources, and ethical standards.

Here are some things that recommendations can tell readers:

- How to solve the problem your report has focused on, as in Figure 15.3
- What new technology, equipment, and so on a company should purchase and why
- When and where to expand into a new market
- Whom to recruit and retain, as in Figure 15.3
- How to create a more profitable Web presence, as in the long report in Chapter 8 (pages 399–413)

The report on the Japanese tuna market discussed above uses a numbered list to make its recommendations easy to see and to follow.

Recommendations

Based on our analysis of the Japanese tuna market, we recommend five marketing strategies for the U.S. tuna industry:

1. Farm greater supplies of bluefin tuna to export.
2. Market our own value-added products.
3. Sell fresh tuna directly to the Tokyo Central Wholesale Market.
4. Sell wholesale to other Japanese markets.
5. Advertise and supply to Japanese supermarket chains.

Back Matter

Included in the back matter of a long report are all the supporting data that, if included in the text of the report, would bog the reader down in details and make the main points harder to find and to grasp.

Glossary

The glossary is an alphabetical list of the specialized vocabulary used in the report and the definitions. A glossary might be unnecessary if your report does not use a highly technical vocabulary or if *all* members of your audience are familiar with the specialized terms you do use.

Citations List

Any sources cited in your report—websites, books, articles, television programs, interviews, reviews, podcasts, webinars—are usually listed in this section (see Chapter 8 on preparing a Works Cited or References list). Also, ask your employer how he or she wants information to be documented. Sometimes employers prefer all information to be documented in footnotes or cited parenthetically in the text. Note that the long report in Figure 15.3 follows the American Psychological Association (APA) system of documentation. Although APA no longer recommends a table of contents or a list of illustrations, these pages are supplied here as models for students who are asked to use them.

Appendix

An appendix contains supporting materials for the report—tables and charts too long and technical to include in the discussion, sample questionnaires, budgets and cost estimates, correspondence about the preparation of the report, case histories, transcripts of telephone conversations. Group such items under the heading "Appendices," as in Figure 15.2.

A Model Long Report

The long report in Figure 15.3 was written by a senior training specialist, Terri Smith Ruckel, and her collaborative team for the vice president of human resources who commissioned it. Note that only Ruckel's name appears on the report, according to her company's policy. The main task facing Ruckel and her team was to demonstrate what RPM Technologies needed to do to meet the needs of multinational workers and thus promote diversity in the workplace. They gathered relevant data from both primary and secondary research, including reports, journal articles, websites, interviews, government documents, and even RPM in-house publications.

Figure 15.3 contains all the parts of a long report discussed in this chapter except a glossary and an appendix. The report does not contain the technical terms and data that would require a glossary and an appendix. Note how the cover letter introduces the report and its significance for RPM Technologies and how the abstract succinctly identifies the main points of the report.

FIGURE 15.3   A Long Report

RPMTechnologies

4500 Florissant Drive   St. Louis, MO 63174

314.555.2121   **www.rpmtech.com**

May 5, 2011

Jesse Butler
Vice President, Human Resources
RPM Technologies

Dear Vice President Butler:

With this letter I am enclosing the report my team and I prepared on effective ways to recruit and retain a multinational workforce for RPM Technologies, which you requested we submit by early May. The report argues for the necessity of adapting the RPM workplace to meet the needs of multinational employees, including promoting cultural sensitivity and making our written business communications more understandable for this audience.

Multinational workers undoubtedly will continue to play a major role in U.S. businesses and at RPM as well. With their technical skills and homeland contacts, these employees can help RPM Technologies successfully compete in today's global marketplace.

But businesses like RPM need to recruit qualified multinational workers more aggressively and then provide equal opportunities for them in the workplace. We must also be sensitive to cultural diversity and communication demands of such an international workforce. By including cross-cultural training—for native and non-native English-speaking employees alike—RPM can more effectively promote cultural sensitivity. In-house language programs and plain English or translated versions of key corporate documents can further improve the workplace environment for our multinational employees.

I hope you find this report helpful in recruiting and retaining additional multinational employees for RPM Technologies. If you have any questions or want to discuss any of our recommendations or research findings, please call me at extension 5406 or email me. I look forward to receiving your suggestions.

Sincerely yours,

Terri Smith Ruckel

Terri Smith Ruckel
Senior Training Specialist

Enclosure: Report

Begins with major recommendation of report

Presents findings of report

Alerts reader to major ways to solve problems

Offers to answer questions

Enclosure notation specifies report is attached

(Continued)

FIGURE 15.3 (Continued)

Title page is carefully formatted and uses boldface

Adapting the RPM Workplace for Multinational Employees

Identifies writer and job title

Terri Smith Ruckel

Ruckel presents report from entire staff— writing for another's signature

Senior Training Specialist RPM Technologies

Prepared for

RPM executive who assigned the report

Jesse Butler Vice President, Human Resources

Date submitted

May 5, 2011

Title page is not numbered

FIGURE 15.3 (Continued)

Table of Contents

While APA does not include a table of contents, individual employers such as RPM may require one

Major divisions of report in all capital letters

Subheadings indicated by indentations and italics

Page numbers included for major sections of report

No subsections needed here

ii

(Continued)

FIGURE 15.3 (Continued)

*Identifies
each figure
by number,
title, and page
number*

*Provides a title
for each visual*

iii

© Cengage Learning 2013

Abstract

This report investigates how U.S. businesses such as RPM must gain a competitive advantage in today's global marketplace by recruiting and retaining a multinational workforce. This current wave of immigrants is in great demand for their technical skills and economic ties to their homeland. Yet many companies like ours still operate by policies designed exclusively for native speakers of English. Instead, we need to adapt RPM's company policies and workplace environment to meet the cultural, religious, social, and communication needs of these multinational workers. To do this, we need to promote cultural sensitivity training, both for multinationals and employees who are native speakers of English. Additionally, as other U.S. firms have successfully done, RPM should adapt vacation schedules and daycare facilities for an increasing multicultural workforce. Equally important, RPM needs to ensure, either through translations or plain-English versions, that all company documents can be easily understood by multinational workers. RPM might also offer non-native speakers of English in-house language instruction while providing foreign language training for employees who are native speakers of English.

*Concise,
informative
abstract that
states purpose
of report
and why it is
important for
audience*

*Uses helpful
transitional
terms such as
"additionally"
and "equally
important"*

*Footer uses
Roman
numerals for
front matter
pages*

iv

© Cengage Learning 2013

FIGURE 15.3 (Continued)

1

Introduction

Background

The U.S. workforce has been undergoing a remarkable revolution. The U.S. Bureau of Labor Statistics predicts that by 2015 the labor force in the United States will comprise 162 million workers who must fill 167 million jobs (2009). According to the U.S. Chamber of Commerce, a shortage of skilled workers is sure to increase "even more heavily in the future when many of the baby boomers begin to retire" (2010).

The most dramatic effect of filling this labor shortage will be in hiring greater numbers of highly skilled multinational employees, including those joining RPM. Currently, "one of every five IT specialists [and] one of every six persons in engineering or science occupations . . . is foreign born" (Kaushal & Fix, 2006). This new wave of immigrants will make up 37 percent of the labor force by 2015 and continue to soar afterward. By 2025 the number of international residents in the United States will rise from 26 million to 42 million, according to the U.S. Chamber of Commerce (2010). The Congressional Budget Office claims that "over the next decade net immigration will average between 500,000 and 1.5 million people annually" (2010). As Alexa Quincy aptly put it, "The United States is becoming the most multiculturally diverse country in the global economy" (2010).

The following commissioned report explores the impact that this new multinational workforce will have on RPM Technologies, and what our company must do to recruit and to retain these workers.

Global Immigration

Unlike earlier generations, today's immigrants actively maintain ties with their native countries. These new immigrants travel back and forth so regularly they have become global citizens, exercising an enormous influence on the success of a business like RPM. Several years ago, demographers Crane and Boaz declared, "Immigration [will] give America an economic edge in the global economy. . . most notably in the Silicon Valley and other high-tech centers. Several major ethnic groups have immigrated to New York City in 2009 alone. They provide business contacts with other markets, enhancing [a company's] ability to trade and invest profitably abroad" (2005). Figure 1 on page 2 identifies these major groups.

Today's High-Tech Immigrants

Undeniably, many immigrants today often possess advanced levels of technical expertise. A report by the Kaiser Foundation found that California's Silicon Valley had significantly benefited from the immigrants who have arrived with much needed technical training. East Asian, Indian, Pakistani, and Middle

APA requires the first line of every paragraph to be indented

Gives convincing statistical evidence about the importance of topic

APA cites year of publication

Cites various sources to validate projections

Explains why the report was written

Uses ellipses and brackets to show deletions/ additions in quotes

Subheads listed in italics

(Continued)

FIGURE 15.3 (Continued)

*Provides
number and
title for figure*

2

Figure 1
**Major Ethnic Groups
Immigrating to New York City in 2009**

*3-D pie
chart reveals
differences
in immigrant
workers*

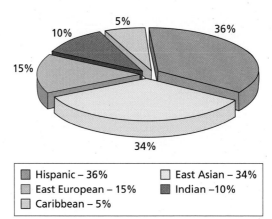

■ Hispanic – 36%	□ East Asian – 34%
▨ East European – 15%	▨ Indian –10%
▤ Caribbean – 5%	

*Provides key for
visual*

*Cites source for
visual*

Source: Brown, P. (2010, February). *History of U.S. immigration.*
Retrieved from http://immigration.ucn.edu

Eastern scientists and engineers, who have relocated from a number of countries, now hold more than 40 percent of the region's technical positions ("Immigrants Find," 2010). Figure 2 on page 3 indicates the leading countries of origin for Silicon Valley's immigrants in 2009 and records the percentage for each nationality.

Information technology (IT) employers are always enthusiastically searching for highly skilled international workers. Murali Krishna Devarakonda, president of the Immigrants Support Network in Silicon Valley, states why: "We contribute significantly to the research and development of a company's products and services" (personal communication, April 3, 2010).

*Email not
included in APA
Reference list*

Problem

*Identifies a
major problem
and explains
why it exists*

RPM, like other e-companies, is experiencing a critical talent shortage of IT professionals, making the recruitment of a multinational workforce a vital priority for us. Meeting the cultural and communication demands of these workers, however, poses serious challenges for RPM. The traditional workplace has to be transformed to respect the ways multinational employees communicate about business. Native English-speaking employees will also have to be better prepared to understand and to appreciate their international co-workers. Unfortunately, many corporate policies and programs at RPM, and at other U.S. companies as well, have been created for native-born, English-speaking employees (Morales, 2009; Reynolds, 2010). Rather than rewarding multinational workers, such policies unintentionally punish them.

*Two separate,
corroborating
sources*

FIGURE 15.3 (Continued)

3

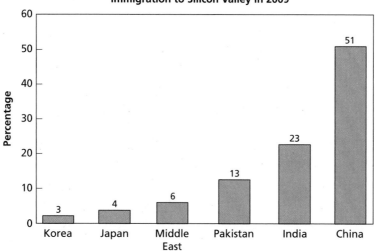

Figure 2
Immigration to Silicon Valley in 2009

Source: Immigrants find the American dream in California's Silicon Valley. (2010, March). *Silicon Valley News*, p. 37.

Bar chart identifies and quantifies major groups of immigrants

Relevant visual in appropriate place in text

Purpose

The purpose of this report is to show that because of the need to increase the number of multicultural employees in the workplace, RPM must adapt its business environment to recruit and retain this essential and diverse labor force.

Concisely states why the report was written

Scope

This report explores cultural diversity in the current U.S. workplace and suggests ways for RPM to compete successfully in the global marketplace by providing equal employment opportunities for multinational workers. Doing so, we will foster cross-cultural literacy, and improve training in intercultural communication at our firm.

Informs reader that report will focus directly on RPM's needs

Discussion

Providing Equal Workplace Opportunities for Multinational Employees

Aggressive Recruitment of IT Professionals from Diverse Cultures

A multilingual workforce is essential if RPM wants to compete in a culturally diverse global market. But firms such as ours must be prepared to adapt or modify hiring policies and procedures to attract these multinational employees, beginning with rethinking our recruitment and retention policies. Routine visits to U.S.

Discussion is organized into three main sections, each with subsections

(Continued)

FIGURE 15.3 (Continued)

4

campuses by company recruiters or "specialized international recruiters" can help us to identify and to hire highly qualified multinational job candidates (Hamilton, 2010).

Emphasizes recruiting multinational workers and suggests how to do so

Moreover RPM should visit universities abroad with distinguished IT programs to attract talented multinational employees. We should encourage students and recent graduates from these universities to apply for a 1-J visa to learn more about RPM through an internship program here. As Catherine Bolgar reported, "Boeing went to Russia for specialist software engineers it couldn't find in the U.S." (2007). These searches, along with articles on our website and executive blogs, should emphasize RPM's commitment to globalization. Lobbying more actively to increase the number of H1-B visas for skilled workers will also help RPM. "In 2008 alone, US companies submitted 163,000 applications for the 65,000 H1-B visa slots. Google, for instance, applied for 300 of them, but 90 were denied" (Richtel, 2009).

Identifies specific benefits for RPM

Capitalizing on a diverse workforce, RPM can more effectively increase its multicultural customer base worldwide. Logically, customers buy from individuals they can relate to culturally. RPM might take a lead from Union Bank of California, a business serving a diverse population, especially its Asian and Hispanic customers. The bank has a successful recruitment history of hiring employees with language skills in Hindi, Vietnamese, Korean, and Spanish. In fact, Union Bank ranked fourth as an employer of minorities (Union Bank of California, 2008). Figure 3 below charts the increase in multinational employees hired by Union Bank over an 8-year period.

Tracks key information in a clear and concise graph

Supplies necessary legend

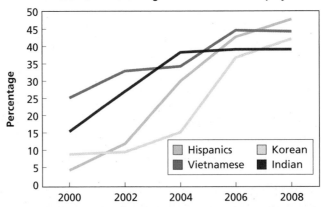

Figure 3
Union Bank of California:
Growth in Percentage of Multinational Employees

Legend: Hispanics, Korean, Vietnamese, Indian

Provides source

Source: Hamilton, B. E. (2010, April 15). Diversity is the answer for today's work force. *Journal of Business Diversity, 10* (38), 35–37.

FIGURE 15.3 (Continued)

5

Another highly competitive business, Darden Restaurants, Inc., selected Richard Rivera, a Hispanic, to serve as president of Red Lobster, the nation's largest full-service seafood chain. Overseeing 680 restaurants nationwide, Rivera was the "most powerful minority CEO in the restaurant industry" (R. Jackson, personal communication, January 15, 2010).

Under Rivera's leadership, Red Lobster hired more international employees—totaling more than 35 percent of its workforce—than it had in previous years. Rivera believes that management cannot properly respond to its diverse customers if the majority of employees are native speakers of English. Many top Fortune 500 companies, such as Cisco and Intel, can also claim that 40% or more of their workforce is comprised of multinationals ("100 Best Companies," 2010). Closer to RPM in St. Louis, Whitney Abernathy—manager of Netshop, Inc.—found that contracts from Indonesia increased by 17 percent after she hired Jakarta native Safja Jacoef (personal communication, April 2, 2010).

Commitment to Ethnic Representation

Many companies have mission statements on diversity and multinational employees in the workplace. G.E., American Airlines, IBM, and WalMart, etc., promote multinationals as mentors and interpreters. Kodak, which is committed to an all-inclusive workplace, has eight employee cultural network groups, including the Hispanic Organization for Leadership and Advocacy, or HOLA, which "helps all Hispanic employees reach their full potential while developing a rewarding career" ("Careers at Kodak," 2008). Such a proactive program, which we might incorporate at RPM, recognizes the leadership abilities of multinational employees. Moreover, "glass ceilings," which in the past have prevented women and ethnic employees from moving up the corporate ladder, are being shattered. The recruitment and promotion of non-native speakers of English is vital for corporate success. Tesfaye Aklilu, vice president at United Technologies, astutely observes:

> In a global business environment, diversity is a business imperative. Diversity of cultures, ideas, perspectives, and values is the norm of today's international companies. The exchange of ideas from different cultural perspectives gives a business additional, valuable information. Every employee can see his/her position from a global vantage point. (Aklilu, 2010)

RPM would also do well to follow the lead of one of our chief competitors, Ablex Plastics. Ablex created the position of Vice President for Diversity two years ago and won an award from the International Business Foundation for hiring more Asian American women managers ("Ablex Appoints," 2008).

Promoting and Incorporating Cultural Awareness Within the Company

Cross-cultural Training

Many of RPM's competitors are creating cultural awareness programs for international employees as well as native speakers. Committed to diversity, Aetna

---

Personal communication supports the need for recruitment but is not cited in references

Personal interview not included in APA References list

Stresses other business precedents that encourage recruiting these employees

Indents quotation of forty or more words

Second major section

Cites business incentive to adapt as competitors did

(Continued)

FIGURE 15.3 (Continued)

6

offers online courses on ethnicity (e.g., "A Bridge to Asia") to "promote an atmosphere of openness and trust" (Aetna, 2008), while Johnson & Johnson conducts Diversity University "to help employees understand and value differences and the benefits of working collaboratively . . . to meet business goals" (2010). Employees find it easier to work with someone whose values and beliefs they understand, while employers benefit from collaboration. Such a program could have prevented the problem RPM confronted when a non-native English-speaking employee a was offended by a cultural misunderstanding ("RPM First Quarter," 2010). We may want to model our programs after those at American Express, which has a workforce representing 40 nations, or those at Extel Communications with its large percentage of Hispanic and Vietnamese employees. United Parcel Service (UPS) profitably pairs a native English-speaking employee with someone from another cultural group to improve on-the-job problem solving and communication. Jamie Allen, a UPS employee since 1998, for instance, found her work with Lekha Nfara-Kahn to be one of the most rewarding experiences of her job (Johnson, 2008).

Although they need to encourage cultural sensitivity training, U.S. firms like RPM also need to be cautious about severing international workers' cultural ties—a delicate balance. When management actively promotes bonds among employees from similar cultures, workers are less fearful about losing their identity and becoming "token" employees. The Lorenzo Lleras law firm, for example, recommends that new multinationals join a "buddy system" to assist their transition to American culture while retaining their cultural identity (U.S. Visa News, 2010). Encouraging such contacts, Globe Citizens Bank has long mobilized culturally similar groups by asking workers of shared ethnic heritages to network with each other (Gordon & Rao, 2009). Employees of Turkish ancestry from Globe's main New York office go to lunch twice a month with Turkish-born employees from the Newark branches. Globe hosts these luncheons and in return receives a bimonthly evaluation of the bank's Turkish and Middle Eastern policies (Hamilton, 2010).

Cultural education must go both ways, though. Both sides have things to learn about doing business in a global environment. The U.S. business culture has conventions, too, and few international employees would want to ignore them, but they need to know what those conventions are (Johnson, 2008). A frequent problem with U.S. corporations such as RPM is that we assume everyone knows how we do things and how we think—it never occurs to us to explain ourselves. Problems often stem from easy-to-correct misunderstandings about business etiquette. For example, native speakers of English are typically comfortable within a space of 1.5 to 2 feet for general personal interactions in business. But workers from Taiwan or Japan, who prefer a greater conversational distance, feel uncomfortable if their desks are less than a few feet away from another employee's workspace (Johnson, 2008; "Taiwanese Business Culture," 2008).

References internal document from company intranet

Transition to new subdivision— networking of employees with similar cultural backgrounds

Relevant source on topic of immigration

Offers two examples RPM could follow

Another clear transitional sentence

Identifies key RPM problem

Gives cultural example

FIGURE 15.3 (Continued)

7

Promotion of Cultural Sensitivity

Corporate efforts to validate diverse cultures might also include the recognition of an ethnic group's holidays. RPM has just begun to do this by hosting cultural events, including Cinco de Mayo and Chinese New Year celebrations. Many companies honor National Hispanic Heritage Month in September, coinciding with the independence celebrations of five Latin American countries (U.S. Equal Employment Opportunity Commission, 2008). Emphasizing their "longstanding relationship with the Chinese community," Wells Fargo participates "every year in San Francisco's Chinese New Year parade" by featuring "a 200-foot-long Golden Dragon" (Babal, 2008). GRT Systems sends New Year's greetings at Waisak (the Buddhist Day of Enlightenment) to its employees who are Buddhist. Chemeka Taylor, GRT's operations manager, points out: "We send native-born employees Christmas cards; why shouldn't we honor our international workforce, too?" (personal communication, April 10, 2010). Techsure, Inc., allows Muslim employees to alter their schedules during Ramadan, (M. Saradayan, personal communication, February 28, 2010). Figure 4 (see page 8) provides a helpful multicultural calender that RPM needs to follow in developing cultural sensitivity policies.

Successful U.S. firms have been sensitive to the needs of their English-speaking employees for decades. Flexible scheduling, telecommuting options, daycare, and preventive health programs have become part of corporate benefit plans. Many of these options and benefits have already been in place at RPM. But an international workforce presents additional cultural opportunities for RPM management to respond to with sensitivity. For example, our company cafeterias might easily accommodate the dietary restrictions of vegetarian workers or those who abstain from certain foods, such as dairy products. At GlobTech, soybean and fish entrees are always available (Reynolds, 2010). Adding ethnic items at RPM would express our cultural awareness and respect for multinational employees.

Daycare remains a key issue in hiring and retaining skilled employees, whether they are native or non-native speakers of English. RPM's child care facilities at our facility in San Luis Obispo have brought us much positive publicity over the past eight years ("RPM Daycare Facilities," 2010). But by modifying child care that reflects our workers' culturally diverse needs, RPM can give a multinational workforce greater peace of mind and better enable them to do their jobs. A pacesetter in this field is DEJ Computers, which insists that at least two or three of its daycare workers must be fluent in Korean or Hindi (Parker, 2009). Another culturally sensitive employer, ITCorp assists its Hispanic workforce by hiring bilingual day care workers and by serving foods the children customarily eat at home (Gordon & Rao, 2009).

Spells out precise ways RPM can incorporate cultural sensitivity into the workplace

Includes valuable information from official corporate blogs

APA lists blogs in references

Includes appropriate calendar of ethnic holidays

Identifies current RPM programs and how they could be easily modified to assist multinational workers

Cites company publication showing research within the organization

© Cengage Learning 2013

(Continued)

FIGURE 15.3 (Continued)

8

Figure 4
A Multicultural Calendar

December 2008

	1	2	3	4	5	6 Feast of St. Nicholas (some European countries)
7 Advent begins (Christian)	8 Bodhi Day (Rohatsu- Buddhist)	9 Eid al-Adha (Islam) Gita Jayanti (Hindu)	10	11	12 Datta Jayanti (Hindu); Feast of Our Lady of Guadalupe (Catholic)	13
14	15	16 Posada begins (Hispanic) (ends Dec. 24)	17	18	19	20
21	22 Ekadashi (Hindu); Hanukkah begins (Jewish) (ends Dec. 29)	23	24 Nocha Buena (Hispanic)	25 Christmas (Christian)	26 Kwanzaa begins (Interfaith) (ends Jan. 1)	27
28	29 Muharram (Islam)	30	31 New Year's Eve (Western)			

Some Information About December's Holidays

Bodhi Day: Buddhists celebrate Prince Gautama's vow to attain enlightenment.
Christmas: Christians celebrate the birth of Jesus Christ.
Datta Jayanti: Celebration of the birth of Lord Datta.
Eid al-Adha: Feast of Sacrifice, commemorating the prophet Abraham's sacrifice of his son Ishmael to Allah.
Ekadashi: Twice monthly purifying fast.
Feast of Our Lady of Guadalupe: Roman Catholic holiday honoring the appearance of the Virgin Mary near Mexico City in 1531.
Feast of Saint Nicholas: Christmas celebration in Austria, Belgium, France, Germany, the Netherlands, Czech Republic, Poland, Russia, and Switzerland.
Gita Jayanti: Celebration of the anniversary of the Bhagavad Gita, the Hindu scripture.
Hanukkah: Eight-day celebration commemorating the rededication of the Temple of Jerusalem.
Kwanzaa: African American and Pan-African celebration of family, community, and culture. Seven life virtues are presented.
Muharram: Islamic New Year.
Nocha Buena: The final day of the Posada celebration.
Posada: Nine-day Latin American Christian celebration, symbolizing Joseph and Mary's journey to find an inn.

Source: Johnson, V. M. (2008). Growing multinational diversity in business sparks changes. *Business Across the Nation, 23*(7), 43–48.

Major visual with a great deal of detail merits a full page to make it readable

Pays attention to major world holidays

Visual helps to convince RPM to adopt similar policy

Provides descriptions of cultural significance of dates

Lists source

FIGURE 15.3 (Continued)

9

Making Business Communication More Understandable for Multinational Employees

Translation of Written Communications

All employees must be able to understand business communications affecting them. Among the essential documents causing trouble for multicultural readers are company handbooks, insurance and health care obligations, policy changes, and OSHA and EPA regulations (Hamilton, 2010). To ensure maximum understanding of these documents by a multinational workforce, RPM needs to provide a translation or at least a plain English version of them. To do this, RPM could solicit the help of employees who are fluent in the non-native English speakers' languages as well as contract with professional translators to prepare appropriate work-related documents.

Workplace signs in particular, especially safety messages, must consider the language needs of international workers. In the best interest of corporate safety, RPM needs to have these signs translated into the languages represented by multinationals in the workplace and/or post signs that use global symbols. Unquestionably, we need to avoid signs that workers might find hard or even impossible to decipher. For example, a capital **P** for "parking" or an **H** for "hospital" might be unfamiliar to non-native speakers of English (Parker, 2009).

Exchange of Language Learning: Some Options

Providing English language instruction for employees also has obvious advantages for the RPM workplace. Should RPM Technologies elect not to provide in-house language training, the company should consider reimbursing employees for necessary courses, books, and software. Moreover, we might make available an audio library of specific vocabulary and phrases commonly used on the job as well as multilingual dictionaries with relevant technical terms. Hiromi Naguchi, an e-marketing analyst at PowerUsers Networking, praised the English language training she received on her job. Although she had ten years of English in the Osaka (Japan) schools, Naguchi developed her listening and speaking skills on the job, benefiting her employers, co-workers, and customers. As a result of her diligence, Naguchi was recognized as her company's "Most Valuable Employee for 2010" for her interpersonal skills (H. Naguchi, personal communication, January 30, 2011).

A recent international survey of executive recruiters showed that being bilingual is critical to success in business ("Developing Foreign Language Skills," 2008). According to this survey, 85 percent of European recruiters, 88 percent of Asian recruiters, and 95 percent of those in Latin America stressed the importance of speaking at least two languages in today's global workplace. As Jennifer Torres, CEO of Pacifica Corp., emphasized: "Although English remains the dominant

Third major section

Turns to written communication and multinational workers

Offers practical solution

Gives examples of what to avoid and why

Outlines specific benefits for RPM

Points to apt example

Cites business survey to confirm the necessity of change at RPM

(Continued)

FIGURE 15.3 (Continued)

10

Identifies quote within a source

language of international business, multilingual executives clearly have the competitive advantage. This advantage will only increase with the continued globalization of commerce" (as cited in "Developing Foreign Language Skills," 2008). Stockard Plastics, one of RPM's major customers, boasts that its employees collectively know more than 45 languages, everything from Albanian to Yoruba.

Language Training Must Be Reciprocal

Argues that reader must consider both sides

But language training has to be reciprocal—for native as well as non-native speakers—if communication is to succeed. Sadly, even though foreign language instruction is on the rise, "fewer than 1 in 8 students at U.S. colleges major in [a] foreign language" (Cicorone, 2010). Unfortunately, this is the case with RPM's native-speaking employees. However, many of the international workers RPM needs to recruit are bi- or even trilingual. In India, Israel, or South Africa, for example, the average worker speaks two or more languages every day to conduct business.

Since RPM needs to recruit such workers, we have to learn more about the cultures and languages of these global employees. RPM management should consider contracting with one of the companies specializing in language instruction for businesspeople (www.selfgrowth.com/foreignlanguage.html). We also need to network with international employee groups to solicit their help and advice.

Conclusion

Conclusion concisely summarizes the highlights of the report without repeating the documentation

To compete in the global marketplace, RPM must emphasize cultural diversity much more in its corporate mission and throughout the workplace. Through its policies and programs, RPM must aggressively recruit and retain an increasing number of technologically educated and experienced multinational workers. Such workers are in great demand today and will be even more so over the next ten to twenty years. They can help RPM increase our international customer base and advance the state of our technology. But we need to ensure that the workplace is sensitive to their cultural, religious, and communication needs. Providing equal opportunities, diversity training and networking, and easy-to-understand business documents will keep RPM globally competitive in recruiting and retaining these essential employees. We can expand in the global marketplace only by making a commitment to learn about and respect other cultures and their ways of doing business.

Forecasts continuing benefits for RPM

FIGURE 15.3 (Continued)

11

Recommendations

By implementing the following recommendations, based on the conclusions reached in this report, RPM Technologies can succeed in hiring and promoting the IT multinational professionals our company needs for future success in today's global economy.

1. Recruit multinational workers more effectively through our website, international hiring specialists, and visits to college and university campuses here and abroad while working more closely with the Immigration and Naturalization Service (INS) to retain multinationals.
2. Establish a mentoring program to identify and foster leadership abilities in multinational employees, resulting in retaining and promoting these workers.
3. Promote cultural sensitivity and networking groups comprising both multinationals and native English-speaking employees.
4. Encourage a group's cultural ties by actively supporting such work-related organizations as the Hispanic Organization for Advocacy and Leadership (HOLA).
5. Develop educational materials for employees who are native speakers of English about the cultural traditions of their multinational co-workers.
6. Reassess and adapt RPM's day care facilities to more effectively meet the needs of children of multinational employees.
7. Supply relevant translations and plain-English versions of company hand-books, manuals, new regulations, insurance policies, safety codes, and other human resource documents.
8. Develop second-language training programs to enhance communication and collaboration between multinational and native speaker employees at RPM.

Provides specific, relevant recommendations, based on conclusions, to solve the problem at RPM

Numbered list format is easy for busy executives to read

Uses strong persuasive verbs to introduce each recommendation

(Continued)

FIGURE 15.3 (Continued)

12

References

Includes only sources actually cited in report

100 best companies to work for, 2010. (2010, February 8). *Fortune 500*. Retrieved from http://money.cnn.com/magazines/fortune/bestcompanies/2010/minorities/

Ablex appoints its first vice president for diversity. (2008, February). Retrieved from http://www.ablexinter.org

Double-spaces between entries

Aetna, Inc. (2008, March). *Diversity annual report: The strength of diversity*. Retrieved from http://www.aetna.com/about/aetna/diversity/data/AetnaEnglish_2008.pdf

Aklilu, T. (2010). *Diversity at UTC*. Retrieved from http://www.utc.com/careers/diversity/index4.htm

Arranges all entries by author's last name or (if no author) by first word of title

Babal, M. (2008, January 26). The year of the ox [Web log post]. Retrieved from http://blog.wellsfargo.com/wachovia/2009/01/the_year_of_the_ox.html

Bolgar, C. (2007, June 18). Corporations need a global mindset to succeed in today's multipolar business world. *Wall Street Journal Online*. Retrieved from http://www.accenture.com/global/highperformancebusiness_business/multipolarbusinessworld.htm

Specifies date of publication for every entry after author's name (or title, if author's name not given)

Brown, P. (2010, February). *History of U.S. immigration*. Retrieved from http://immigration.ucn.edu

Careers at Kodak. (2008, February). Retrieved from http://www.kodak.com/US/employment/corp/career/whyemploymentworks.jhtm

Cicorone, M. (2010, November). The importance of foreign languages to success in the world of work. *Language Instruction, 51*(3), 18–27.

Congressional Budget Office. (2010, May). *Projections of net migration to the United States*. Retrieved from http://www.cbo.gov/doc.cfm?index=7249

Capitalizes only first word and proper nouns in title

Crane, E. H., & Boaz, D. (Eds.). (2005). *Cato handbook for policymakers* (6th ed.). Washington, DC: Cato.

Developing foreign language skills is good business. (2008, February 3). *Business World*. Retrieved from http://www.businessworld.ca/article.cfm/newsID/7109.cfm

Provides page numbers for print sources

Gordon, T., & Rao, P. (2009, April). Challenges ahead for American companies. *National Economics Review, 11*(3), 38–42, 56.

FIGURE 15.3 (Continued)

13

Hamilton, B. E. (2010, April 15). Diversity is the answer for today's work force. *Journal of Business Diversity, 10*(38), 35–37.

Immigrants find the American dream in California's Silicon Valley. (2010, March). *Silicon Valley News*, p. 37.

Johnson & Johnson. (2010). *Programs and activities*. Retrieved from http://www.jnj.com/connect/about-jnj/diversity/programs

Kaushal, N., & Fix, M. (2006). The contributions of high-skilled immigrants. In M. Fix (Ed.), *Contributions of immigrants to the United States*. Also published as *Insight*, July 2006, No. 16. Washington, DC: Migration Policy Institute.

Morales, J. (2009). *Immigration News & Notes*. Retrieved from http://www.immigrationnewsandnotes.com

Parker, M. (2009, May 26). Multinational hires—advice and advocacy [Web log post]. Retrieved from http://parkeronimmigration.blogspot.com/2009/05/26/multinational_hires-advice_and_advocacy.php

Quincy, A. (2010, March 3). Multiculturalism makes for a good business. *Workforce, Inc., 14*(2), 5–8.

Reynolds, P. (2010, February 3). Serving up culture [Web log post]. Retrieved from http://www.culture.org/2010/02/serving_up_culture

Richtel, M. (2009, April 12). Tech recruiting clashes with immigration rules. *New York Times*. Retrieved from http://www.nytimes.com/2009/04/12/business/12immig.html

RPM daycare facilities rated high. (2010, January 15). *RPM News*. Retrieved from http://www.rpm.com/rpmnews/01_15_2010/rpm_daycare_facilities_rated_high

RPM first quarter activity report. (2010). *RPM Internal Reports*. Retrieved from http://www.rpm.com/internalreports/2010_firstquarter

Taiwanese business culture. (2008). *Executive Planet*. Retrieved from http://www.executiveplanet.com/business-culture-in/132438266669.html

Indents second and subsequent lines 5 spaces or ½ inch

References blog posts with proper APA citations

Gives full Web addresses for verification and to make source easy to find

Cites material available on company intranet

(Continued)

© Cengage Learning 2013

FIGURE 15.3 (Continued)

14

Article about a company included on company website

Union Bank of California. (2008, June 23). Federation magazine ranks Union Bank one of the best companies for minorities. Retrieved from http://www.uboc.com /about/main/0,3250,2485_11256_502261585,00.html

Italicizes title of government report

U.S. Bureau of Labor Statistics. (2009). *Overview of the 2008–18 projections.* Retrieved from http://www.bls.gov/oco/oco2003.htm

U.S. Chamber of Commerce. (2010). *Immigration issues.* Retrieved from http://www.uschamber.com/issues/index/immigration/default

U.S. Citizenship and Immigration Service. (2009, December). USCIS reaches FY 2010 H-1B cap. *USCIS.gov.* Retrieved from http://www.uscis.gov/pressroom

Government documents provide valuable statistics

U.S. Equal Employment Opportunity Commission, Federal Hispanic Work Group. (2008). Report on the *Hispanic employment challenge in the federal government.* Retrieved from http://www.eeoc.gov/federal/reports/hwg

U.S. Visa News. (2010). *Lorenzo M. Lleras immigration attorney.* Retrieved from http://www.usvisanews.com

Final Words of Advice About Long Reports

Perhaps no piece of writing you do on the job carries more weight than the long report. You can simplify your job and increase your chances for success by following these guidelines for scheduling, researching, and collaborating:

1. Plan and work early. Do not postpone work until a deadline draws near.
2. Confer often and carefully with others in your group office.
3. Do a thorough search among Internet, print, and other resources.
4. Consult with specialists in other fields both in your company and in other organizations, including government officials.
5. Divide your workload into meaningful units. Reassure yourself that you do not have to write the report or even an entire section of the report in one day.
6. Set up mini-deadlines for each phase of your work, and then meet them.

✓ Revision Checklist

☐ Concentrated on a major problem—one with significant implications for my school, neighborhood, city, or employer.

☐ Identified, justified, and described the significance of the main problem.

☐ Did sufficient research—in the library, on the Internet, or through interviewing, personal observation, or testing.

☐ Became familiar with key terms and major trends in the field.

☐ Anticipated how various managers and other decision makers will use and profit from my report for their long-range planning.

☐ Made sure I understood what my readers are looking for.

☐ Adhered to all specified schedules for completing various stages of the long report.

☐ Divided and labeled the parts of the long report to make it easy for readers to follow and to show a careful plan of organization.

☐ Supplied an informative abstract that leaves no doubt in readers' minds about what the report deals with and why.

☐ Designed an attractive title page that contains all the basic information—title, date, for whom the report is written, my name.

☐ Gave my readers all the necessary introductory information about background, problem, purpose of report, and scope.

☐ Included in the body of the report the weight of all my research—the facts, statistics, interview comments, and descriptions—that my readers need.

☐ Included subheadings to reflect the major divisions into which I have organized the research that forms the nucleus of the text.

☐ Wrapped up the report in a succinct conclusion. Told readers what the findings of my research are and accurately interpreted all data.

☐ Supplied a recommendations section (if required) that tells readers concretely how they can respond to the problem using the data. Offered recommendations that are realistic and practical and related directly to the research and topic.

☐ Included in the final copy of the report all the parts listed in the table of contents.

☐ Supplied a one-page transmittal letter informing readers why the report was written and describing its scope and findings.

Exercises

1. Send an email to your instructor describing how one of the short reports in Chapter 14 could be useful to someone who has to write a long report.

2. Using the information contained in Figures 15.1 and 15.2, draft an introduction for the report "A Study to Determine New Directions in Women's Athletics at Coastal College." Add any details you think will be relevant.

3. What kinds of research did Terri Smith Ruckel and her team need to do to write the long report in Figure 15.3? As part of your answer, include the titles of any specific reference works you think the writer may have consulted.

4. Study Figure 15.3 and answer the following questions based on it:
   a. Why can the abstract be termed informative rather than descriptive?
   b. How have Ruckel and her team successfully limited the scope of the report?
   c. Where have Ruckel and her team used internal summaries especially well?
   d. Where and how have the writers adapted their technical information for their audience (a general reader)?
   e. What visual devices do the writers use to separate parts of the report?
   f. How do the writers introduce, summarize, and draw conclusions from the expert opinions in order to substantiate the main points?
   g. How have the writers documented information?
   h. What functions does the conclusion serve for readers?
   i. How do the recommendations follow from the material presented in the report? How are they both distinct and interrelated?

5. Come to class prepared to discuss a major community problem suitable for a long report (e.g., traffic, crime, air and water pollution, housing, transportation). Then write a letter to an appropriate agency or business requesting a study of the problem and a report.

6. Write a report outline for the problem you selected in Exercise 5. Use major headings. Include a cover letter with your outline.

7. Have your instructor approve the outline you prepared for Exercise 6. Then write a long report based on the outline, either on your own or as part of a collaborative writing team.

CHAPTER **16**

Making Successful Presentations at Work

Almost every job requires employees to have and to use carefully developed speaking skills. In fact, to get hired, you have to be a persuasive speaker at your job interview. And to advance up the corporate ladder, you must continue to be a confident, well-prepared, and persuasive speaker. Some jobs may require you to deliver as many as three or four PowerPoint presentations a week.

The world of work receives and shares much of its information through informal briefings, collaborative discussions, PowerPoint presentations, videoconferencing (see pages 120–121), and formal oral reports. Your employer will expect your oral communication skills in all these situations to be as effective and professional as your writing skills are in preparing correspondence, reports, and proposals. The goal of this chapter is to help you be a more successful speaker on the job.

Access chapter-specific interactive learning tools, including quizzes and more in your English CourseMate, accessed through www.cengagebrain.com.

Writing a Document versus Making a Presentation

Writing a document and delivering a report both require you to (1) research your topic, (2) plan your organization, and (3) choose your language and visuals carefully. There are nevertheless some fundamental differences between these two ways of communicating in the world of work. When you make a presentation, you must focus on these additional items:

- Your appearance—how you dress, stand, move, and gesture
- Your delivery—whether you can be heard, your tone of voice, whether you sound confident or nervous
- The complexity of your subject—your talk must be informative yet concise and easily understood the first (and only) time the audience hears it
- The amount of time you have to speak—you cannot exceed your allotted time
- Your audience's attention span—usually not more than fifteen or twenty minutes
- Your introduction—who you are and what you will talk about
- The layout of the room—lighting, capacity, acoustics, etc.
- The technical equipment necessary for your presentation—screen, monitors, whiteboards

- Your visuals—they must be clear and easily seen, even from the back of the room
- Your interaction with the audience—questions, comments, and nonverbal responses including laughs, frowns, puzzled looks

Evaluating a Speaker

As the list of differences between making a presentation and writing a paper or report shows, there are essential differences you have to take into account to be an effective speaker. One of the ways to be a successful speaker is knowing what *not* to do. Table 16.1 identifies the characteristics of a poor speaker that you need to avoid. Figure 16.3 at the end of the chapter presents an evaluation form used to assess a speaker's delivery, presentation, and use of visuals. Study both Table 16.1 and Figure 16.3, which bookend this chapter, to learn how to become a proficient and persuasive speaker.

Types of Presentations

You will make numerous presentations on the job that will vary in the amount of preparation they require, the time they last, and the audience and occasion for which they are intended. Here are some common types of presentations you can expect to make before different audiences in the world of work.

For Your Customers or Clients

- sales appeals stressing how and why your company's products or services meet your listeners' needs
- scenarios about why your company is better qualified and more successful than the competition
- demonstrations of your products or services
- a persuasive overview of your company's or organization's contributions to technology, the community, the environment, the profession at large

For Your Boss

- progress or status reports on how a project is going
- an assessment of your job accomplishments
- a justification of a budget, your own position, or your department or division
- an evaluation of job candidates, proposals, possible sites for a branch location, etc.
- a summary of a conference or meeting you attended and how your company can benefit from the information you gathered

For Your Co-workers

- an end-of-shift report, such as those made by police officers and nurses
- an explanation of a new or revised company policy

TABLE 16.1   Characteristics of a Poor Speaker

Presentation	Appearance	Voice
Does not preview purpose or main ideas Does not provide an introduction, transitions, or a conclusion Loses his or her place Includes too much or too little information Fails to come to the point Does not identify subdivisions Does not give examples or case studies Is disorganized; does not follow a logical sequence	Does not dress professionally Wears clothes or jewelry that call attention to themselves and detract attention from the presentation Does not appear well groomed Disregards company's dress code	Mumbles, making it hard to understand his or her words Speaks in monotone like a robot Talks too softly or too loudly Mispronounces words Injects fillers such as "uh," "right," "like," "you know" Sounds frightened, not confident Laughs nervously or inappropriately

Posture	Interaction with Audience	Visuals
Slumps or slouches Grips lectern fearfully Jingles keys or coins Sways or paces Fumbles with notes, equipment, or visuals Blocks view of visuals by standing in front of them Looks stiff and unnatural	Does not make eye contact Keeps his or her head down looking at notes Apologizes for lack of information or for inability to discuss topic Criticizes room, equipment, time of presentation Finds fault with coworkers or company Keeps checking wristwatch or mobile device Stares at floor or out of window Is condescending or disrespectful to international audience Refuses to or cannot answer audience's questions Argues with questioner Does not thank listeners for their time	Does not provide any visuals Shows too many visuals Just reads visuals without interpreting them or relating them to the topic Shows visuals out of sequence Uses visuals that • are too small to see or are too busy • contain too much writing • contain typos, misspellings, or errors in math • are not relevant to content or context • are too technical or too hard to understand • look unprofessional, are not proportional, or do not look well designed • Does not acknowledge source of visual

- a training session on job safety, operating a piece of equipment, using new software
- a briefing on new job assignments and tasks

For Community Leaders or Groups

- appeals before elected officials
- an explanation of your company's decision or activity
- an update on completing a public works project
- tours of a company facility

Four Ways to Make a Presentation

Your effectiveness depends directly on how you make your presentation. Here are four ways to do that, along with their benefits and drawbacks.

1. Speaking impromptu, or "off-the-cuff." When your boss asks you to "say a few words" about a job you are doing or about the status of a sales campaign, you will have to speak impromptu. In such a situation, where you have no chance to prepare, you have to know your subject very well so that you do not get caught off guard. The impromptu approach works well only for short presentations before a small group of co-workers. But it is lethal for other types of presentations. In fact, the worst way to make a presentation is to speak without any preparation. You may think your experience qualifies you for an impromptu performance, but you only fool yourself if you believe you have all the necessary details and explanations in the back of your head. It is equally dangerous to believe that once you start talking, everything will fall into place smoothly. The "everything works out for the best" philosophy, unaided by a lot of hard work, is a recipe for disaster. Without preparation, you are likely to confuse important points or forget them entirely. Mark Twain's advice is apt here: "It takes three weeks to prepare a good impromptu speech."

2. Memorizing a presentation. The exact opposite of the off-the-cuff approach, a memorized speech does have some advantages for certain individuals—tour bus drivers, museum guides, or salespeople—who must deliver the same speech verbatim many times over. But for the individual who has to make a formal presentation, a memorized speech contains pitfalls:

- The hours you must spend memorizing exact words and sentences would be better devoted to researching and organizing your presentation.
- If you forget a single word or sentence, you may lose the rest of the speech.
- A memorized delivery can make you appear stiff and mechanical. Rather than adjusting to an audience's reactions, you are obligated to speak the exact words you wrote before you ever saw your audience.

3. Scripting a presentation. Scripting, or reading a speech word for word, may be appropriate if you are presenting technical information on a company policy or legal issues on which there can be no deviation from the printed word or if you are making a formal presentation at a professional conference. Most presentations, however, will not require this rigid adherence to a text. You run the risk of not interacting with your audience. Moreover, reading a speech is also likely to bore your audience. With your head buried in your script, you set up a barrier between you and the audience by not establishing eye contact for fear of losing your place.

   4. Delivering a presentation extemporaneously. An extemporaneous delivery is the most widely used and preferred way to share information in the world of work. PowerPoint presentations are examples of an extemporaneous delivery. Unlike memorizing or reading your comments, you do not come before your audience with the entire speech in hand. By no means, though, is an extemporaneous delivery the same as an impromptu performance. It requires a great deal of preparation. What you prepare, though, is an outline of or notes about the major points you want to make. You need to rehearse using your outline or notes, but you won't have the actual words you will use in your speech. Instead, you will talk to your audience while referring to your outline to remember your major points. The best extemporaneous presentations allow you to establish rapport with your audience by looking at them and responding to their reactions. Your goal is to sound spontaneous, conversational, yet very well prepared. Be careful, though, that you do not leave out important points or go over your allotted time.

Informal Briefings

If you have ever given a book report or explained laboratory results in front of a class, you have given an informal briefing. Such reports are a routine part of many jobs. They usually last between ten and twenty minutes and are given to a small group of co-workers and possibly your boss. A briefing may be scheduled for a certain time every week—"Monday Briefing for the Week Ahead"—or called for a special purpose, such as explaining a new procedure or company policy. Other topics for such briefings often focus on training, motivating a workforce, reviewing sales activities, and so on. Whatever the topic, informal briefings have one thing in common: They bring people up-to-date by supplying them with key information. Figure 16.1 (page 726) contains an outline for an informal briefing.

   When you have to make an informal briefing, follow these guidelines:

1. **Prepare.** Confer with your boss, individuals in human resources, collaborative team members. Never speak off-the-cuff.

2. **Decide what main points you want to cover.** Write down a few points you want to cover, and keep this list or outline before you as you speak. Highlight key names, terms, dates, or places you want to stress.

3. **Watch out for information overload.** Do not crowd too many points into one short briefing.

4. **Arrange your points in a logical sequence.** Put the main point first, or try using a cause-and-effect or chronological organization.

5. **Be clear.** Don't use unfamiliar terms or complex explanations. To illustrate main points, include concrete examples that your audience will easily understand and apply to solve a problem or clarify a work-related issue.

FIGURE 16.1 An Outline of Speaking Points for an Informal Briefing

Clearly focused topic

Uses short outline of main points to stress at briefing

Crime Alert Briefing: People's National Bank, Millersville, May 9, 2012

Topic of briefing: Crime alert about the increasing number of counterfeit $20 and $50 bills and cashier's checks in the Millersville area.

I. **Introduction**
 A. Area banks have been asked to watch for the growing number of counterfeit $20 and $50 bills and cashier's checks in circulation
 B. Our customers trust in us to protect their financial security
 C. We have a responsibility as tellers/bank officers to identify counterfeit bills/cashier's checks and to keep them out of circulation

II. **Identifying Counterfeit Bills**
 A. Most of the counterfeit bills are in denominations of $20 or $50
 B. Make sure you compare any suspected fraudulent bills with real currency
 C. Here are some telltale signs to look for:
  1. No watermark is visible when the bill is held up to the light
  2. The portrait is not off centered as in legitimate $20 and $50 bills
   SHOW SLIDE 1 (enlarged photo of counterfeit $20.00 bill)
  3. There are blurred lines and fuzzy scrolls along the border (edges) of each counterfeit bill

Includes two relevant slides to help tellers identify counterfeit currency and cashier's checks

III. **Identifying Bogus Cashier's Checks**
 A. The majority of these bills and cashier's checks carry the Newtown State Bank name and logo
 B. Almost all of these fraudulent checks end with serial numbers between 7632 and 7645
 C. Here are some other things to look for:
  1. Lighter than usual paper stock
  2. The words "Pay to the Order" are streaked in grey
   SHOW SLIDE 2 (enlarged image of fraudulent cashier's check)
  3. The edges of the check are clean cut instead of being jagged

Gives precise directions

IV. **Conclusion**
 A. Thank you for being on the alert
 B. If you have any doubts, call a bank officer
 C. Review bank policy on reporting counterfeit currency
  DISTRIBUTE HANDOUT (list of bank procedures for reporting counterfeit currency and checks)

Supplies handout for quick reference about bank policy

6. **Be concise and to the point.** An informal briefing is not the place to make a lengthy presentation. Your audience may be on a strict timetable and have obligations to fulfill later in the day.

7. **Don't rush your delivery.** Speak calmly and deliberately so that your audience can clearly absorb your message or even take notes.

8. **Stay positive.** Even when you have unpleasant news to impart (e.g., an increase in customer complaints, a project delay), resist blaming or lecturing your audience. Instead, focus on the positive steps needed to resolve the problem.

9. **Use appropriate visuals to clarify a point.** Do not overwhelm your listeners with numerous graphs, charts, and photos, however. Be selective and target only two or three points to illustrate.

10. **End on time.** Be sure to allow time at the end of your briefing for questions, comments, and suggestions.

11. **Always thank listeners for their time and cooperation.**

Note how the speaker's outline in Figure 16.1 identifies key points, illustrates them with examples, inserts appropriate visuals in the most effective places, and makes sure the briefing is audience-focused and runs on schedule.

Formal Presentations

Whereas an informal briefing is likely to be short, generally conversational, and intended for a limited number of people, a formal presentation is much longer, far less conversational, and perhaps intended for a wider audience. It involves much more preparation.

Expect to spend several days preparing your presentation. You cannot just dash it off. Just as you did with an informal briefing you will have to do the following; only more extensively and over a longer period of time:

- research the subject
- interview key resource individuals
- prepare, time, and sequence visuals
- coordinate your talk with presentations by your co-workers or boss
- rehearse your presentation

Many of us are uncomfortable in front of an audience because we feel frightened or embarrassed. Much of that anxiety can be eased if you know what to expect. The two areas you should investigate thoroughly before you begin to prepare your presentation are (1) who will be in your audience and (2) why they are there.

Analyzing Your Audience

The more you learn about your audience, the better prepared you will be to give your listeners what they need. Just as you do for your written work, for your oral presentation you will have to do some research about the audience, emphasizing the "you attitude" and establishing your own credibility.

Consider Your Audience as a Group of Listeners, Not Readers

While audience analysis pertains both to readers of your work and to listeners of your presentation, there are several fundamental differences between these two groups. Unlike a reader of your report, the audience for your presentation

- is a captive audience
- may have only one chance to get your message
- has less time to digest what you say
- has a shorter attention span
- can't always go back to review what you said or jump ahead to get a preview
- is more easily distracted—by interruptions, chairs being moved, people coughing, outside noise, and so on
- cannot absorb as many of the technical details as you would include in a written report

Take all of these differences into account as you plan your presentation and assess who constitutes your audience.

Guidelines for Analyzing Your Audience

Here are six key questions to ask when analyzing your audience.

1. **How much do they know about your topic?**
   - consumers with little or no technical knowledge
   - technical individuals who understand terms, jargon, and background
   - business managers looking only for the bottom line

2. **What unites them as a group?**
   - members of the same profession
   - customers using the same products
   - employees of the company you work for

3. **What do they want to receive from your presentation?**
   - a quick overview
   - bottom-line financial details
   - technical details on materials, methods, and conclusions (results)

4. **What is their interest level or stake in your topic?**
   - friendly and interested
   - highly motivated
   - neutral—waiting to be informed, entertained, or persuaded
   - uninterested in your topic—only there because attendance is mandatory
   - uncooperative, antagonistic, or likely to challenge you

5. **What do you want them to do after hearing your presentation?**
   - buy a product or service
   - adopt a plan
   - change a schedule
   - learn more about your topic

- sign a petition
- accept a new policy (e.g., safety procedures)

6. **What questions are they likely to raise?**
   - about money, profits, expenses, salary
   - about personnel, hiring, training
   - about transfers, mergers, promotions
   - about new job responsibilities, accountability
   - about locations, new, remodeled, domestic, overseas
   - about schedules/timetables, effect on quotes, salary

Special Considerations for a Multinational Audience

Given the international makeup of audiences at many business presentations, you may have to address a group of listeners (international co-workers or customers) whose native language is not English or even make a presentation before individuals in a country other than your own. Consider your audience's particular cultural taboos and protocols. Will they expect you to thank them for listening before you present your information? Will they expect you to stand in one place, or will they be comfortable if you move about the room while you speak?

As you prepare a talk before a multinational audience, keep the following points in mind:

1. Brush up on your audience's culture, especially accepted ways they communicate with one another (see pages 183–189).
2. Find out what constitutes an appropriate length for a talk before your audience. (German listeners might be accustomed to hearing someone read a thirty- to forty-page paper, while members from another culture would regard that practice as improper.)
3. Do your homework about gestures and eye contact. What may be a widely accepted gesture in the United States may signal the exact opposite in another culture. Moreover, some cultures, especially those in East Asia, regard direct eye contact as disrespectful or even as a sign of aggression. Review Table 10.5 (page 507) to make sure you use only culturally acceptable gestures.
4. Be especially careful about introducing humor. Avoid anything that is based on nationality, dialect, religion, or race.
5. Think twice about injecting anything autobiographical into your speech. Some cultures regard such intimacy as highly inappropriate.
6. Steer clear of politics; you risk losing your audience's confidence.
7. Choose visuals with universally understood icons (see page 505).

The Parts of Formal Presentations

As you read this section, refer to Marilyn Claire Ford's PowerPoint presentation in Figure 16.2 (pages 730–733). Note how effectively she used the PowerPoint format to convince a potential client, GTP Systems, to purchase a service contract provided by World Tech, her employer. Her presentation consists of seven slides that contain relevant images and concise text.

FIGURE 16.2 A Sample PowerPoint Presentation

Slide 1

Clear, persuasive title

Introduces speaker, provides reason for presentation

Type size and font make slide easy to read

Switching to Videoconferencing
A Wise Choice for GTP

Marilyn Claire Ford
World Tech
Desktop Videoconferencing
November 15, 2011

© Cengage Learning 2013

Slide 2

Uses short title for each slide

Includes relevant visual emphasizing networking

Succinctly lists benefits in short, easy-to-read bulleted points

What We Offer

© istockphoto.com/Jacob Wackerhausen

- User-friendly desktop videoconferencing
- Cutting-edge communication technology
- Flexible, low-cost networking
- Single network for data, voice, and video

© Cengage Learning 2013

FIGURE 16.2 (Continued)

Easy to Use

With a simple telephone call, you can

- Arrange a meeting with colleagues at multiple sites

- Use your computer to access World Tech's conferencing system
- See, hear, and talk with all participants

Slide 3

First of four slides that make up the body of the presentation

Relevant visual does not mask text

Develops first key sales feature

Cost Effective

- Dramatically reduces costs for travel
- Upgrades current computer system for less than the cost of buying a new computer
- Cuts data-processing expenses by 60%

Slide 4

Second point gives only essential facts

Chooses verbs that emphasize cost savings

Does not overwhelm listeners with numbers

(Continued)

FIGURE 16.2 (Continued)

Slide 5

Third sales feature appropriately describes specific technology

Leaves generous margins

Visual shows benefits of staff interacting

Improves Staff Efficiency

- Brings people together at the right time
- Enhances communication when employees see and hear each other
- Whiteboard technology aids collaboration

© istockphoto.com/Jacob Wackerhausen

© Cengage Learning 2013

Slide 6

Last point stresses worldwide benefits

Visual reinforces global marketplace

Advantages in the Global Marketplace

© iStockphoto.com/Will Selarep

- Connects all locations to one virtual office
- Increases sales worldwide
- Strengthens global networking

© Cengage Learning 2013

FIGURE 16.2 (Continued)

Slide 7

Conclusion

Title signals end of presentation

- Recap of technology benefits
- Please sign up this week—it is easy
- Just log on to **www.worldtech.com**
- Thank you for considering World Tech
- Any questions?

mcf@worldtech.com
1-800-271-5555

Summarizes key advantages

Issues a call to action; makes contact easy through website, email, and telephone

© Cengage Learning 2013

The Introduction

The most important part of a presentation is the introduction. Capture your audience's attention by answering these questions: (1) Who are you? (2) What are your qualifications? (3) What specific topic are you speaking about? and (4) How is the topic relevant to the audience? The answers to the first two questions establish your credibility. Those to the third and fourth questions preview your presentation and emphasize why it is important to the audience.

Your first and most immediate goal is to establish rapport with your listeners, win their confidence, and elicit their cooperation. Since your listeners are probably at their most attentive during the first few minutes of your presentation, they will pay close attention to everything about you and what you say. Seize the moment and build momentum.

An effective introduction should be proportional to the length of the presentation. A ten-minute speech requires no more than a sixty-second introduction; a twenty-minute speech needs no more than a two- or three-minute introduction. A rough rule of thumb is that your introduction should be no more than 10 to 20 percent of your presentation. Note how in slides 1 and 2 of Figure 16.2, Ford introduces herself, her company, and its benefits for GTP.

How to Begin You can begin by introducing yourself, emphasizing your professional qualifications and interests. (A self-introduction is unnecessary if someone else has introduced you or if you know everyone in the room.) Never apologize—for being nervous, unprepared, unqualified—or complain about the time or location for the presentation. Always ask if your audience can hear you clearly.

Give Listeners a Road Map Give listeners a "road map" at the beginning of your presentation so that they will know where you are, where you are going, and what they have to look forward to or recall. Indicate what your topic is and how you have organized what you have to say about it.

> My presentation today on fraud identification software will last about 20 minutes and is divided into three parts. First, I will outline briefly recent software changes. Second, I will give a detailed review of how those changes directly affect our company. Third, I will show how our company can profitably implement those changes. At the end of my presentation, there will be time for your questions and comments.

The most informative presentations are the easiest to follow. Restrict your topic to ensure that you will be able to organize it carefully and sensibly—for example, a tasty diet under 1,000 calories a day or a course in learning InDesign or another software package.

Capture the Audience's Attention Use any of the following strategies to get your audience to "bite the hook":

- Ask a question. "Did you know that every thirty minutes a foreign-owned business opens in China?" or "Do you know what's in your bottled water besides water?"
- Start with a quotation. "Winston Churchill said, 'We get things to make a living but we give things to have a life.'" (Consult *Bartlett's Familiar Quotations* online at www.bartleby.com/100)
- Give an interesting statistic. "In 2012, two million heart attack victims will live to tell about it." (Go to the *Information Please Almanac* at www.infoplease.com to find something relevant to your presentation topic.)
- Relate an anecdote or story. Be sure it is relevant and in good taste; make your audience feel at ease and friendly toward you by establishing a bond with them.

Be careful about using humor in a business talk. It could backfire; the audience may not get the point or may even be offended by it.

The Body

The body is the longest part of your presentation, just as it is in a long report. It should constitute about 60 to 70 percent of your presentation. Make it persuasive and relevant to your audience by (1) explaining a process, (2) describing a condition, (3) solving a problem, (4) arguing a case, or (5) doing all of these. See how the body of Marilyn Claire Ford's presentation in Figure 16.2 is organized around the benefits of GTP's switching to desktop videoconferencing. In slides 3 to 6 she outlines how easy, economical, and efficient such technology is to use in a global marketplace.

To get the right perspective, recall your own experiences as a member of an audience. How often did you feel bored or angry because a speaker tried to overload you with details or could not stick to the point?

Ways to Organize the Body   Here are a few helpful ways you can present and organize information in the body of your presentation. When writing a report, you design your document to help readers visually, supplying headings, bullets, white space, and headers and footers. In a speech, switch from those purely visual devices to aural ones, such as the following:

1. **Give signals to show where you are going or where you have been.** Enumerate your points: *first, second, third.* Emphasize cause-and-effect relationships with *subsequently, therefore, furthermore.* When you tell a story, follow a chronological sequence and use signposts: *before, following, next, then.* (See Table A.1, page A-3.)

2. **Comment on your own material.** Tell the audience if some point is especially significant, memorable, or relevant. "This next fact is the most important thing I'll say today."

3. **Provide internal summaries.** Spending a few seconds to recap what you have just covered will reassure your audience that you want them to be clear about what you have covered thus far.

   > We have already discussed the difficulties in establishing a menu repertory, or the list of items that the food service manager wants to appear on the menu. Now we will turn to ways of determining which items should appear on a menu and why.

4. **Anticipate any objections or qualifications your audience is likely to have.** Address potential objections with relevant facts about costs, personnel, or equipment in your presentation.

The Conclusion

Plan your conclusion as carefully as you do your introduction. Stopping with a screeching halt is as bad as trailing off in a fading monotone. An effective conclusion should leave the audience feeling that you and they have come full circle and accomplished what you promised. Let readers know you are near the end of your presentation.

What to Put in a Conclusion   A conclusion should contain something memorable. Never introduce a new subject or simply repeat your introduction. A conclusion can contain the following:

- a fresh restatement of your three or four main points
- a call to action, just as in a sales letter — to buy, to note, to agree, to volunteer
- a final emphasis on a key statistic (for example, "The installation of the stainless steel heating tanks has, as we have seen, saved our firm 32 percent in utility costs, since we no longer have to run the heating system all day.")

End your presentation, as Marilyn Claire Ford does in slide 7, with a concise summary of the main points, and urge listeners to invest in your product or service — for example, World Tech Desktop Videoconferencing system.

Mean It When You Say, "Finally"   When you tell your audience you are concluding, make sure you mean it. Saying, "In conclusion," and then talking for another ten minutes frustrates listeners and makes them less receptive to your message. When you finish, thank your audience for their time in listening to your presentation.

Always Leave Time for Questions   Make sure you budget your time to give the audience an opportunity to ask questions, offer suggestions, or make comments.

Presentation Software

As Marilyn Claire Ford's presentation in Figure 16.2 demonstrates, business presentations very frequently rely on PowerPoint and other software. Knowing how to use these graphics packages is a crucial skill your employer will expect you to have. Presentation software enables you to create an electronic slide show with concise text and carefully chosen visuals that can be created on any computer. They allow you to plan, write, and add visuals to your slides.

Presentation Software Capabilities

With PowerPoint, Corel Presentation, and other software, you can

- format and edit text
- import visuals, photos, digital art, clip art
- incorporate a variety of shapes and symbols—arrows, asterisks, bullets, cylinders, pyramids, flow charts, pictograms
- offer animation, sound bites, video clips
- design and insert logos and letterheads
- reuse and revise your presentation anytime

Presentation software will also help you as a speaker. For instance, you can keyboard your notes so that they scroll at the bottom of your computer screen. These notes are not projected for the audience, but they are available if you need to glance at them. You also can print parts of your presentation or your entire program in color or gray scale (black and white) to reinforce your presentation with professional-looking handouts.

Editing with presentation software allows you to customize any text or visual for any audience. You can copy, move, alter, and delete text, graphics, or sound bites. For example, you could edit the number of categories (bars, lines of a graph, items in a table) of an imported visual for a consumer audience that does not need much detail, or you could change the levels and positions of an organizational chart to suit your agency's needs. You can also change the color or texture of a background, saving you time and effort in re-creating a visual.

Organize the Presentation

Map out your presentation before you actually create your slides. One way is to prepare an outline (see page 88), which is essential for an extemporaneous presentation (see page 724). Organizing your presentation carefully is as important as the details you show your audience.

Webinars

A webinar, or Web conference, is a teleconference tool that uses the Internet for group meetings. Employees of a large corporation who must deliver in-depth presentations to individuals in several countries simultaneously will find the wide variety of webinar delivery options to be extremely helpful. Webinar software allows for communication in a number of media, including slideshow or streaming presentations, text chat, screen sharing, and oral communication through speakers and headphones. Webinar presentations can be used for effective sales appeals, training demonstrations for co-workers, and e-learning sites for students. Some webinar programs include GoToMeeting/GoToWebinar (www.gotomeeting.com), Adobe Connect (www.adobe.com/products/adobeconnect), and ReadyTalk (www.readytalk.com). Webinar programs, such as Yugma (www.yugma .com) and Cisco WebEx (www.webex.com), even offer free trial options or ad-supported versions.

Webinar software is very user-friendly. It allows you to navigate to various functions, open windows to display participants, access a work area for videoconferencing or text chat, or even use it as a virtual whiteboard (see page 743). Webinar software does not sacrifice the benefits of a face-to-face conference but actually can enhance it. A webinar presentation uses visual, oral, and aural media to help participants communicate whether in the same town or around the globe. Webinars can save businesses travel costs and other resources, such as paper. Moreover, by including audience interaction options, such as polling, surveys, or interactive brainstorming options, webinars enable participants to generate reports that contain information about attendees and customer demographics to assist with audience analysis.

Before designing a webinar, you need to consider if it is right for the job. Even though most webinar programs include tools for brainstorming, collaborating, and training, it is not the best way to hold brief meetings that require little audience interaction. A teleconference would be less costly and faster. Moreover, a webinar may not be appropriate for long seminars or presentations, since online audiences tend to lose interest more quickly than face-to-face audiences. A twenty-minute speech requiring many visuals might be more easily delivered using the document delivery options available in most software suites. Individuals collaborating on a project or document might want to use brainstorming or whiteboard software. Webinars are most appropriate, then, when you have a large group of individuals across the country or around the world who must actively collaborate on a project or who all need training on the same topic or issue.

Source: Billy Middleton.

1. Identify the large topics you want to cover, as Ford did in presenting information to save GTP time and money, increase staff efficiency, and help the company compete in a global economy. Then you can organize your topics logically and persuasively.

2. Divide your presentation into major topics that best accomplish your objective, whether to inform, to persuade, or to document.

3. Include only those supporting details that relate directly to your topic and to your audience's needs. These key points should help you determine the number of slides and visuals to use.

4. Don't overwhelm audiences with too much information. Note that Ford needs only seven slides in Figure 16.2.

5. Choose your visuals carefully. Resist the temptation to dazzle your audience with electronic special effects. Your goal is not to create a glitzy show but to represent your company professionally.

Test the Technology

Find and eliminate any bugs at the rehearsal stage. Call in advance to confirm the room and any equipment you may need. Save and preview your presentation in the format in which you will be giving it. Bring your own computer to the meeting, and set up the projector and your notebook in advance. Make sure your software is compatible with the equipment you will use. Be sure any Web links you plan to use are relevant and functioning, not broken links.

Prepare Transparencies and Handouts

It's always wise to have backup transparencies or handouts with you in case your equipment malfunctions before or during your presentation. Transparencies of your presentation will enable you to project your slides while you speak. Handouts of your slides (make sure you keep a copy for yourself) will allow you to continue with your presentation, and they will also help your audience follow your presentation and take notes about it while you speak. Make sure your handouts match your presentation exactly by providing your audience with a printout of each slide.

Do not distribute any handouts ahead of time. Wait until you are ready to use them. Otherwise, they may divert your audience's attention from your presentation.

Guidelines on Using Presentation Software Effectively

Here are some tips to ensure the best design, organization, and delivery of your presentation. Refer to Figure 16.2 as you study these guidelines.

Readability

- Make sure each slide is easy to read—clear, concise, and uncluttered.
- Use a type size that is easy to see, even from a distance. For a small presentation on your notebook, use 24-point type or larger. Increase your type size for headings, as in Figure 16.2.
- Keep your type style consistent. Don't switch from one font to another.
- Avoid ornate and script fonts, and do not put everything in boldface, italics, or all capital letters. Marilyn Claire Ford uses boldface (and white type against a blue background) only for the headings in Figure 16.2.

Text

- Keep your text short and simple. Use easy-to-recall names, words, and phrases. Your audience will not have the time to read long, complex messages.

- Use bulleted lists instead of unbroken paragraphs. But put no more than five bulleted lines on a slide, and limit each line to seven or eight words. Don't squeeze words on a line. Include no more than forty words per slide.
- Double-space between bulleted items, and leave generous margins on all sides.
- Title each slide using a question, a statement, or a key name or phrase, as in Figure 16.2.

Sequencing Slides

- Keep your slides in the order in which you need to show them.
- Retain the same transition (cover left or straight right) from slide to slide to avoid visual confusion.
- Spend about two to three minutes per slide, but don't read each slide verbatim. Summarize main ideas or concisely expand them while looking at your audience, not the slide.
- Time your slides so your audience can read them. Never continue to show a slide after you have moved on to a new topic.
- If you invite audience participation and interaction, build in extra time between your slides. Leave time for questions.

Background/Color

- Find a pleasant contrasting background to make your text easy to read. Avoid extremely light or dark backgrounds that may obscure your text. Stay away from stark backgrounds, such as using cold white images on a black screen.
- Use the same background for each slide.
- Avoid shadowing your text for "decorative" visual effect.
- Use color sparingly, and make sure it is professionally appropriate. Don't turn each slide into a sizzling neon sign.

Graphics

- Keep graphics clear, simple, and positioned appropriately on the slide.
- Be sure visuals do not cover or shadow text.
- Show only those visuals that support your main points. Not every slide requires a visual. For example, slides 1 and 7 in Figure 16.2 do not use visuals.
- Include no more than one graphic per slide; otherwise, your text will be more difficult to read.
- Include easy-to-follow graphs and charts instead of complicated tables, elaborate flow charts, or busy diagrams.
- Use clip art sparingly. Remember: Less is more. Clip art must be functional, not distracting.
- Incorporate animation, sound effects, or video clips only when they are persuasive, relevant, and undeniably professional.
- Don't bother with borders; they do not make a slide clearer.

Quality Check

- Be sure your spelling, grammar, names, dates, costs, and sources are correct.
- Double-check all math, equations, and percentages.

Noncomputerized Presentations

On the job you can expect to use a variety of visuals besides those included in PowerPoint presentations. There will be situations in which you may have to use a conventional chalkboard, a flip chart (where large pieces of white paper are anchored on an easel and flipped over like pages in a notepad), a slide projector, or an overhead projector to make a presentation instead of your notebook. You will almost surely be asked to prepare handouts that include text, visuals, or a combination to distribute before or during your talk.

Regardless of the medium you use, make sure your visuals are

- easy to see
- easy to understand
- self-explanatory
- relevant
- accurate

Getting the Most from Your Noncomputerized Visuals

The following practical suggestions will help you get the most from your visuals when time and space may prohibit using computer setups.

1. Do not set up your visuals before you begin speaking. The audience will be wondering how you are going to use the graphics and so will not give you their full attention. When you are finished with a visual, put it away so that your audience will not be distracted by it or tempted to study it instead of listening to you.

2. Firmly anchor any maps or illustrations. Having a map roll up or a picture fall off an easel during a presentation is embarrassing for you and distracting for the audience. It can undermine your presentation.

3. Never obstruct the audience's view by standing in front of your visuals. Use a pointer or a laser pointer (a pen-sized tool that projects a bright red spot up to 150 feet) to direct the audience's attention to your visual.

4. Avoid crowding too many images onto one visual. Use no more than one visual per page.

5. Do not put a lot of writing on a visual. Elaborate labels or wordy descriptions defeat your reason for using the visual. Your audience will spend more time trying to decipher the writing than attempting to understand the visual itself. If any writing must appear on one of your visuals, enlarge it so your audience can read it quickly and easily.

6. Be especially cautious with a slide projector. Check beforehand to make sure all your slides are in order and are right side up. Most important, make sure the projector is in good working order. Practice changing from one transparency to another. And test any audio equipment you plan to use as part of your presentation.

Rehearsing Your Presentation

Except for impromptu talks, all presentations require rehearsal. Don't skip rehearsing your presentation thinking it will save you time. Rehearsing will actually help you become more familiar with your topic and overall message, building your confidence. Rehearsing will also help you acquire more natural speech rhythms—pitch, pauses, and pacing. Here are some strategies to use as you rehearse your speech.

- Know your topic and the various parts of your talk.
- If possible, practice in the room where you will make your presentation.
- Speak in front of a full-length mirror or before a friend or colleague for at least one rehearsal to see how an audience might view you.
- Talk into an audio recorder to determine whether you sound friendly or frantic, poised or pressured. You can also catch and correct yourself if you are speaking too quickly or too slowly. A rate of about 120 to 140 words a minute is easy for an audience to follow.
- Time yourself so that you will not exceed your allotted time or fall far short of your audience's expectations.
- Practice with the presentation software, visuals, equipment, or projector that you intend to use in your speech for valuable hands-on experience.
- Monitor the types of gestures (neither too many nor too few) you use for clarity and emphasis in your talk.
- Check the room where you will make your presentation, if possible, to see about acoustics, lighting, seating, and technical equipment.
- Videotape your final rehearsal and show it to a colleague or instructor for feedback.

Delivering Your Presentation

Poor delivery can ruin a good presentation. When you speak before an audience, you will be evaluated on your style of presentation and the image you project: how you look, how you talk, and how you move (your body language). Do you mumble into your notes, never looking at the audience? Do you clutch the lectern as if to keep it in place? Do you shift nervously from one foot to the other? All those actions betray your nervousness and detract from your presentation.

First impressions are crucial. Research shows that people decide what they think of you in the first two or three minutes of your presentation. The way you dress is important, but so is your body language. In fact, 75 percent of your audience's impressions are influenced by your body language. The nonverbal signals you send affect how your audience will regard your leadership abilities, your sales performance, even your sincerity. Pay attention to gestures, hand movements, how you stand, and so on. No matter how many hours you have worked to get your message ready, if your nonverbal presentation is misleading or inappropriate, the impact of what you say will be lost.

The following suggestions on how to deliver a presentation will help you be a well-prepared, poised speaker.

Settling Your Nerves Before You Speak

Being nervous before your presentation is normal—a faster heartbeat, sweaty palms, shaking. But don't let your nerves stop you from delivering a highly successful talk.

Here are some ways you can calm yourself before you deliver your presentation:

- Give yourself plenty of time to get there. The more you have to rush, the more anxious you will be.
- Don't bring anything with you that is likely to spill, such as coffee or a soft drink.
- Avoid caffeine for a few hours before your talk if it makes you jittery.
- Take some deep breaths, and then hold your breath while you count to ten. Exhale. This will slow your heart rate and lower your blood pressure.
- Get away for a minute or two. Walk down a hall to get a drink of water or just to get a little exercise. It will help you reduce some anxiety and organize your thoughts.
- Remind yourself that you have spent hours preparing. Your hard work will pull you through.
- Try to chat with one or two members of the audience ahead of time to relax. See your audience as friends—people who can help your career.

To get helpful experience in making presentations, consider joining Toast-masters International, an educational organization that has transformed many frightened speakers into accomplished speech makers. Members of Toastmasters meet at least once a month to listen to and evaluate one another's speeches. From feedback and strategic training, you will grow to be a more confident public speaker. Visit their website at www.toastmasters.org.

Finally, you could try a panic control method advocated by Don Greene in his book *Fight Your Fear and Win* (New York: Broadway Books, 2002): "Picture the person you would most like to emulate, someone whose confidence or style you admire," and then combine his or her grace, speaking voice, and persuasion with your own qualifications for a successful, less-stressed presentation.

Guidelines for Making Your Presentation

Everyone is nervous before a talk. Accept that fact and even allow a few seconds of "panic time." Then put your nervous energy to work for you. Chances are, your audience will have no idea how anxious you are; they cannot see the butterflies in your stomach. Again, see your audience as friends, not enemies. Remember to do the following:

1. Establish eye contact with your listeners. Look at as many people in your audience as possible to establish a relationship with them. Never bury your head in your notes or keep your eyes fixed on a screen or keyboard. You will only signal your lack of interest in the audience or your fear of public speaking. Some timid speakers think that if they look only at some fixed place or object in the back of the room, the audience will regard this as eye contact. But that kind of cover-up does not work. Even during a short PowerPoint presentation, try to establish rapport with each person in the room.

Interactive Whiteboards

An interactive whiteboard—such as the Smart Board from Smart Technologies—looks like a chalkboard but is really an interactive digital screen with computer capabilities. Instead of using chalk, you write with a digital device on the whiteboard screen, which is connected to a computer through a USB cable or wireless link. You can use your finger like a mouse and the board like a keyboard, which allows you to search, scan, show, and edit any document or graphic. With just a touch of your hand or with a digital device, you can move, enlarge, or shrink elements such as numbers, graphs, and words. For instance, you can alter the shape or content of a visual such as a pie chart, table, or graph to reflect statistical changes or election outcomes, as newscasters did during U.S. presidential primaries in 2008. You can also insert, edit, update, and store content.

Interactive whiteboards have a variety of uses in the world of work. They are invaluable when you have to make a presentation. Usually mounted on a wall or panel, an interactive whiteboard can function as a projector screen on which you can scan, show, edit, share, and store information from your computer. You can also enlarge an image or figure or insert call-outs, or you can pull up a website or company publication from an intranet to illustrate a point or answer a question.

This interactive technology can also improve collaborative writing by making brainstorming, researching, or planning much easier; for instance, you can send each member of your group the information that has been developed and displayed on a whiteboard. Further, interactive whiteboards have been used extensively in training sessions in which employees have to see and learn about a great deal of information in a short time. Using an interactive whiteboard, you can give new employees or visitors an impressive virtual tour of your company and its various divisions and departments.

2. Repond to audience feedback. Watch your listeners' reactions and respond appropriately—nodding to agree, pausing a moment, paraphrasing to clarify a confusing point. Know your material well so that if someone asks you a question or wants you to return to a point, you are not fumbling through your notes or trying feverishly to locate the right screen.

3. Use a friendly, confident tone. Let the audience know that you are happy they are there and that you are enthusiastic. Speak in a natural, pleasant voice, but avoid verbal tics ("you know," "I mean") and fillers ("um," "ah," "er") repeated several times each minute. Such nervous habits will make your audience nervous and your speech less effective. Use pauses instead.

4. Vary the rate of your delivery. Vary your rate and inflection to help you emphasize key points and make transitions. Talk slowly enough for your audience to understand you, yet quickly enough so that you don't sound as if you are belaboring or emphasizing each word.

5. Adjust your volume appropriately. Talking in a monotone, never raising or lowering your voice, will lull your audience to sleep or at least inattention. Talk loudly enough for everyone to hear, but be careful if you are using a microphone. Your voice will be amplified, so if you speak loudly, you will boom rather than project. Every word with a *b*, *p*, or *d* will sound like an explosive in your listeners' ears. Watch out for the other extreme—speaking so softly that only the first two rows can hear you.

6. Watch your posture. Don't shift from one foot to another. But do not slouch or look wooden either. If you stand motionless, looking as if rigor mortis has set in, your speech will be judged cold and lifeless, no matter how lively your words are. Be natural yet dynamic; smile and nod your head. Refer to an object on the screen by touching it, or use a pointer to emphasize something on an overhead projector.

7. Use appropriate body language. Be natural and consistent. Do not startle an audience by suddenly pounding on the lectern or desk for emphasis. Avoid gestures that will distract or alienate your audience. For example, don't fold your arms as you talk, a gesture that signals you are unreceptive (closed) to your audience's reactions. Also, avoid nervous habits that can divert the audience's attention: clicking a ballpoint pen, scratching your head, rubbing your nose, twirling your hair, pushing up your glasses, fumbling with your notes, tapping your foot, or drumming your fingers on a desk. Nor do you have to remain still or step with robotlike movements. The remote control for a PowerPoint presentation allows you to casually walk around the room as you click and change screens.

8. Dress professionally. Do not wear clothes or clanking jewelry that call attention to themselves. Follow your company's dress code. Unless it specifies otherwise (e.g., casual Fridays), wear clothes that are the business norm. Women should wear a businesslike dress or suit; men should choose a dark suit or sports coat, a white or blue shirt, and a tasteful tie.

Handling Interruptions

Be diplomatic if someone interrupts your presentation with a question. Thank the individual by saying, "That's a good question. I'll be happy to answer it at the end of the presentation when there'll be time for questions." If someone is disruptive during your talk or question-and-answer session and wants to debate with you, offer to meet with him or her after the session to discuss the point in question. Moreover, if you cannot answer a particular question, say you'll be glad to get back to the person, and go on to other questions.

When You Have Finished

Don't just sit down before your computer, walk back to your place on the platform or in the audience, or, worse yet, march out of the room. Thank your listeners for their attention and stay at the lectern or at your laptop for audience applause or questions. If appropriate, give the person who introduced you a chance to thank you while you are still in front of the group.

If a question-and-answer session is to follow your speech, anticipate questions your audience is likely to ask. But it's a good idea to give your audience a time limit.

For example, you might say, "I'll be happy to answer your questions now before we break for lunch in ten minutes." By setting limits, you reduce the chances of a lengthy debate with members of the audience, and you also can politely leave after your time elapses.

Evaluating Presentations

This chapter has given you information on how to construct and deliver a formal presentation. As a way of reviewing that advice, study Figure 16.3—an evaluation form similar to those used by instructors in colleges and universities. Note that the form gives equal emphasis to the speaker's performance or delivery and to the organization, content, and sequence of the presentation.

FIGURE 16.3 An Evaluation Form for a Presentation

Name of Speaker: _____   Date: _____

Title of Presentation: _____   Length: _____

PART I: THE SPEAKER'S DELIVERY
(Circle the appropriate number using the 1–5 scale.)

1. Appearance	1 unprofessional	2	3	4	5 well groomed
2. Eye contact	1 poor	2	3	4	5 effective
3. Voice	1 monotonous	2	3	4	5 varied
4. Diction	1 slurred	2	3	4	5 clear
5. Posture	1 poor	2	3	4	5 natural
6. Gestures	1 distracting	2	3	4	5 appropriate
7. Self-confidence	1 nervous	2	3	4	5 poised
8. Audience interaction	1 minimal	2	3	4	5 engaging

PART II: THE PRESENTATION ITSELF
(Circle the appropriate number: 1 = poor; 5 = superior.)

1. Made sure topic was relevant to audience	1	2	3	4	5

(Continued)

FIGURE 16.3 (Continued)

2. Began with clear statement of purpose	1	2	3	4	5
3. Followed logical organization	1	2	3	4	5
4. Gave audience cues to look for transitions between sections	1	2	3	4	5
5. Matched content to technical knowledge of audience	1	2	3	4	5
6. Provided convincing supporting evidence	1	2	3	4	5
7. Did not digress	1	2	3	4	5
8. Concluded with a summary of main points	1	2	3	4	5
9. Stayed within time limits	1	2	3	4	5
10. Allowed time for questions	1	2	3	4	5

PART III: USE OF VISUALS (SLIDES OR OTHER GRAPHICS)
(Again, circle the appropriate number: 1 = poor; 5 = superior.)

1. Used appropriate number of visuals	1	2	3	4	5
2. Ensured visuals were relevant	1	2	3	4	5
3. Carefully timed the sequence of visuals	1	2	3	4	5
4. Made sure audience could see visuals clearly	1	2	3	4	5
5. Selected visuals that clarified or simplified a point	1	2	3	4	5
6. Referred to visuals and indicated why they were important	1	2	3	4	5
7. Acknowledged the source of visuals	1	2	3	4	5

✓ Revision Checklist

☐ Anticipated my audience's background, interest, and even potential resistance, as well as questions about the message of both informal and formal presentations.

☐ Organized an informal briefing to make it easy to understand and to incorporate it into the work routine.

☐ Prepared an outline and identified and corrected any weak or redundant areas.

☐ Drafted an introduction to provide a "road map" of the presentation and to arouse audience interest.

☐ Started with interesting and relevant statistics, a question, an anecdote, or a similar "hook" to capture audience attention.

☐ Limited the body of my presentation to the main points.

☐ Sequenced the main points logically and made connections among them.

☐ Used supporting examples and illustrations appropriate to my audience and message.

☐ Made sure my conclusion contains a summary of the main points of my presentation and a specific call to action.

☐ Designed visuals that are clear, easy to read, and relevant for my audience.

☐ Experimented successfully with presentation software before using it for my presentation.

☐ Used an appropriate number of slides and made sure they were readable.

☐ Showed slides in the correct, carefully timed sequence.

☐ Prepared transparencies and handouts in case of equipment trouble.

☐ Rehearsed my presentation thoroughly to become familiar with its content, organization, and visuals.

☐ Monitored my volume, tone, and rate to vary my delivery and to emphasize my major points.

☐ Rehearsed my gestures to make them relevant and nonintrusive.

☐ Timed my presentation, complete with visuals, to run close to the allotted time.

Exercises

1. Prepare a three- to five-minute presentation explaining how a piece of equipment that you use on your job works. If the equipment is small enough, bring it with you to class. If it is too large, prepare an appropriate visual or two for use in your talk. Submit an outline similar to that in Figure 16.1.

2. You have just been asked to talk about the students at your school or the employees where you work. Narrow the topic and submit an outline to your instructor,

showing how you have limited the topic and gathered and organized evidence. Incorporate two or three appropriate visuals (photographs, maps, charts, icons, or even videos) in your PowerPoint presentation. Follow the format of the presentation in Figure 16.2.

3. Prepare a ten-minute presentation on a controversial topic that you would present before a civic group—the PTA, the local chapter of an organization, a post of the Veterans of Foreign Wars, a synagogue, a mosque, or a church club.

4. Using the information contained in the report on marketing research in Chapter 8 (pages 399–413) or in the long report on multinational workers in Chapter 15 (pages 701–718), prepare a short presentation (five to seven minutes) for your class.

5. Deliver a formal presentation on one of the following topics. Restrict your topic, and divide it into four key issues or parts, as in Figure 16.2. Use at least three visuals with your talk. Submit an outline to your instructor.
   a. a new piece of equipment at work
   b. the use of the Internet to provide health care information in rural areas
   c. a major change in housing or traffic control in your city
   d. a paper or report you wrote in school or on your job
   e. "greening" your school's student center
   f. the budgetary allocations or spending cuts planned for your department or your town's school district for a given year
   g. applying for and receiving student financial aid

6. Using the evaluation form in Figure 16.3, evaluate a speaker—a speech class student, a local politician, or a co-worker delivering a report at work. Specify the time, place, and occasion of the speech. Pay special attention to any visuals the speaker uses.

Appendix:
A Writer's Brief Guide to Paragraphs, Sentences, and Words

To write successfully, you must know how to create effective paragraphs, write and punctuate clear sentences, and use words correctly. This guide succinctly explains some of the basic elements of clear and accurate writing.

Paragraphs

Writing a Well-Developed Paragraph

A paragraph is the basic building block for any piece of writing. It is (1) a group of related sentences (2) arranged in a logical order (3) supplying readers with detailed, appropriate information (4) on a single important topic.

A paragraph expresses one central idea, with each sentence contributing to the overall meaning of that idea. The paragraph does that by means of a *topic sentence*, which states the central idea, and *supporting information*, which explains the topic sentence.

Supply a Topic Sentence

The topic sentence is the most important sentence in your paragraph. Carefully worded and restricted, it helps you generate and control your information. An effective topic sentence also helps readers grasp your main idea quickly. As you draft your paragraphs, pay close attention to the following three guidelines.

 1. **Make sure you provide a topic sentence.** In their rush to supply readers with facts, some writers forget or neglect to include a topic sentence. The following paragraph, with no topic sentence, shows how fragmented such writing can be.

No topic sentence: Sensors found on each machine detect wind speed and direction and other important details such as ice loading and potential metal fatigue. The information is fed into a microprocessor in the nacelle (or engine housing). The microprocessor then automatically keeps the blades turned into the wind, starts and stops the machine, and changes the pitch of the tips of the blades to increase power under

A–1

varying wind conditions. Should any part of the wind turbine suffer damage or malfunction, the microprocessor will immediately shut the machine down.

Only when a suitable topic sentence is added—"The MOD-2 wind turbine features the latest technology"—can readers understand what the technical details have in common.

2. Put your topic sentence first. Place your topic sentence at the beginning of your paragraph because the first sentence occupies a commanding position. Burying the key idea in the middle or near the end of the paragraph makes it harder for readers to comprehend your purpose or act on your information.

3. Be sure your topic sentence is focused and discusses only one central idea. A broad or unrestricted topic sentence leads to a shaky, incomplete paragraph for two reasons:

- The paragraph will not contain enough information to support the topic sentence.
- A broad topic sentence will not summarize or forecast specific information in the paragraph.

The following example of a carefully constructed paragraph contains a clear topic sentence in an appropriate position (*italicized*) and adequate supporting details.

> *Fat is an important part of everyone's diet.* It is nutritionally present in the basic food groups we eat—meat and poultry, dairy products, and oils—to aid growth or development. The fats and fatty acids present in those foods ensure proper metabolism, thus helping to turn what we eat into the energy we need. Those same fats and fatty acids also act as carriers for important vitamins like A, D, E, and K. Another important role of fat is that it keeps us from feeling hungry by delaying digestion. Fat also enhances the flavor of the food we eat, making it more enjoyable.

Three Characteristics of an Effective Paragraph

Effective paragraphs have *unity*, *coherence*, and *completeness*.

Unity

A unified paragraph sticks to one topic without wandering. Every sentence and every detail supports, explains, or proves the central idea. A unified paragraph includes only relevant information and excludes unnecessary or irrelevant comments.

Coherence

In a coherent paragraph, all sentences flow smoothly and logically to and from each other like the links of a chain. Use these three techniques to achieve coherence.

1. Use transitional words and phrases. Some useful transitional, or connective, words and phrases, grouped according to the relationships they express, are listed in Table A.1.

TABLE A.1   Transitional, or Connective, Words and Phrases

Addition	additionally	besides	moreover
	again	first, second, third	next
	along with	furthermore	together with
	also	in addition	too
	and	many	what's more
	as well as		
Cause/effect	accordingly	consequently	on account of
	and so	due to	since
	as a result	hence	therefore
	because of	if	thus
Comparison/ contrast	but	in contrast	on the other hand
	conversely	in the same way	similarly
	equally	likewise	still
	however	on the contrary	yet
Conclusion	all in all	in brief	on the whole
	altogether	in conclusion	to conclude
	as we saw	in short	to put into perspective
	at last	in summary	to summarize
	finally	last	to wrap up
Condition	although	granted that	provided that
	depending	if	to be sure
	even though	of course	unless
Emphasis	above all	for emphasis	of course
	after all	indeed	surely
	again	in fact	to repeat
	as a matter of fact	in other words	to stress
	as I said	obviously	unquestionably
Illustration	for example	in other words	that is
	for instance	in particular	to demonstrate
	in effect	specifically	to illustrate
Place	across from	below	over
	adjacent to	beyond	there
	alongside of	here	under
	at this point	in front of	where
	behind	next to	wherever
Time	afterward	formerly	previously
	at length	hereafter	soon
	at the same time	later	simultaneously
	at times	meanwhile	subsequently
	beforehand	next	then
	currently	now	until
	during	once	when
	earlier	presently	while

Paragraph with connective words: Advertising a product on the radio has many advantages over using television. *For one thing,* radio rates are much cheaper. *For example,* a one-time 60-second spot on local television can cost $5,000. *For that money,* advertisers can purchase nine 30-second spots on the radio. *Equally attractive* are the low production costs for radio advertising. *In contrast*, television advertising often includes extra costs for actors and voice-overs. *Another* advantage radio offers advertisers is immediate scheduling. *Often* the ad appears during the same week a contract is signed. *On the other hand*, television stations are *frequently* booked up months in advance, so it may be a long time *before* an ad appears. *Furthermore*, radio gives advertisers a greater opportunity to reach potential buyers. *After all*, radio follows listeners everywhere—in their homes, at work, and in their cars. *Although* television is very popular, it cannot do that.

2. Use pronouns and demonstrative adjectives. Words like *he*, *she*, *him*, *her*, *they*, *their*, and so on, contribute to paragraph coherence and improve the flow of sentences.

> Paragraph with pronouns: Traffic studies are an important tool for store owners looking for a new location. These studies are relatively inexpensive and highly accurate. They can tell owners how much traffic passes by a particular location at a particular time and why. Moreover, they can help owners determine what particular characteristics the individuals have in common. Because of their helpfulness, these studies can save owners time and money and possibly prevent financial ruin.

3. Use parallel (coordinated) grammatical structures. Parallelism means using the same *kind* of word, phrase, clause, or sentence to express related concepts.

> Orientation sessions accomplish four useful goals for trainees. First, they introduce trainees to key personnel in accounting, IT, maintenance, and security. Second, they give trainees experience logging into the database system, selecting appropriate menus, editing core documents, and getting off the system. Third, they explain to trainees the company policies affecting the way supplies are ordered, used, and stored. Fourth, they help trainees understand their ethical responsibilities in such sensitive areas as computer security and use.

Parallelism is at work on a number of levels in the paragraph above, among them these:

- The four sentences about the four goals start in the same way grammatically ("... they introduce/give/explain/help ...") to help readers categorize the information.
- Within individual sentences, the repetition of *present participles* (logg*ing*, select*ing*, edit*ing*, gett*ing*) and of *past participles* (order*ed*, us*ed*, stor*ed*) helps the writer coordinate information.
- Transitional words—*first*, *second*, *third*, *fourth*—provide a clear-cut sequence.

Completeness

A complete paragraph provides readers with sufficient information to clarify, analyze, support, defend, or prove the central idea expressed in the topic sentence. The reader feels satisfied that the writer has given necessary details.

Skimpy paragraph: Farmers are turning their crops and farm wastes into cost-effective fuels. Much grown on the farm is being converted to energy. This energy can have many uses and save farmers a lot of money in operating expenses.

Fully developed paragraph: Farm crops and wastes are being turned into fuels to save farmers operating costs. Alcohol can be distilled from grain, sugar beets, and corn. Converted to ethanol (90 percent gasoline, 10 percent ethanol), this fuel runs such farm equipment as irrigation pumps, feed grinders, and tractors. Similarly, through a biomass digestion system, farmers can produce methane from animal or crop wastes as a natural gas for heating and cooking. Finally, cellulose pellets, derived from plant materials, become solid fuel that can save farmers money in heating barns.

Sentences

Constructing and Punctuating Sentences

The way you construct and punctuate your sentences can determine whether you succeed or fail in the world of work. Your sentences reveal a lot about you. They tell readers how clearly you can convey a message. And any message is only as effective and as thoughtful as the sentences of which it is made.

What Makes a Sentence

A sentence is a complete thought, expressed by a subject and a verb that can make sense standing alone.

subject   verb

Websites   sell   products.

The Difference Between Phrases and Clauses

The first step toward success in writing sentences is learning to recognize the difference between phrases and clauses. A *phrase* is a group of words that does not contain a subject and a verb; phrases cannot make sense standing alone. Phrases cannot be sentences.

in the park   No subject: Who is in the park?
                                         No verb: What was done in the park?
for every patient in intensive care   No subject: Who did something for every patient?
                                         No verb: What was done for every patient?

A *clause* does contain a subject and a verb, but *not every clause is a sentence*. Only *independent* (or *main*) *clauses* can stand alone as sentences. Here is an example of an independent clause that is a complete sentence.

subject   verb   object

The president closed the college.

A *dependent* (or *subordinate*) *clause* also contains a subject and a verb, but it does not make complete sense and cannot stand alone. Why? A dependent clause contains a subordinating conjunction—*after, although, as, because, before, even though, if, since, unless, when, where, whereas, while*—at the beginning of the clause. Such conjunctions subordinate the clause in which they appear and make the clause dependent for meaning and completion on an independent clause.

After
Before
Because   } the president closed the college
Even though
Unless

"After the president closed the college" is not a complete thought but a dependent clause that leaves us in suspense. It needs to be completed with an independent clause telling us what happened "after."

dependent clause (not a sentence)	subject	verb	phrase
	independent clause		
After the president closed the college,	we	played	in the snow.

Avoiding Sentence Fragments

An incomplete sentence is called a *fragment*. Fragments can be phrases or dependent clauses. They either lack a verb or a subject or have broken away from an independent clause. A fragment is isolated: It needs an overhaul to supply missing parts to turn it into an independent clause or to glue it back to an independent clause to have it make sense.

To avoid writing fragments, follow these rules. *Note that incorrect examples are preceded by a minus sign, correct revisions by a plus sign.*

1. Do not use a subordinate clause as a sentence. Even though it contains a subject and a verb, a subordinate clause standing alone is still a fragment. To avoid this kind of sentence fragment, simply join the two clauses (the independent clause and the dependent clause containing a subordinating conjunction) with a comma—*not* a period or semicolon.

- Unless we agreed to the plan. (What would happen?)
- Unless we agreed to the plan; the project manager would discontinue the operation. (A semicolon cannot set off the subordinate clause.)
+ Unless we agreed to the plan, the project manager would discontinue the operation.
- Because safety precautions were taken. (What happened?)
+ Because safety precautions were taken, ten construction workers escaped injury.

Sometimes subordinate clauses appear at the end of a sentence. They may be introduced by a subordinate conjunction, an adverb, or a relative pronoun (*that, which, who*). Do not separate these clauses from the preceding independent clause with a period, thus turning them into fragments.

- An all-volunteer fire department posed some problems. Especially for residents in the western part of town.

+ An all-volunteer fire department posed some problems, especially for residents in the western part of town. (The word *especially* qualifies *posed*, referred to in the independent clause.)

2. Every sentence must have a subject telling the reader who does the action.

– Being extra careful not to spill the solution. (Who?)
+ The technician was being extra careful not to spill the solution.

3. Every sentence must have a complete verb. Watch especially for verbs ending in *-ing*. They need another verb (some form of *to be*) to make them complete.

– The machine running in the computer department. (Did what?)

You can change that fragment into a sentence by supplying the correct form of the verb.

+ The machine *is running* in the computer department.
+ The machine *runs* in the computer department.

Or you can revise the entire sentence, adding a new thought.

+ The machine running in the computer department processes all new accounts.

4. Do not detach prepositional phrases from independent clauses. Prepositional phrases (beginning with *at, by, for, from, in, to, with*, and so forth) are not complete thoughts and cannot stand alone. Correct the error by leaving the phrases attached to the sentence to which they belong.

– By three o'clock the next day. (What was to happen?)
+ The supervisor wanted our reports by three o'clock the next day.

Avoiding Comma Splices

Fragments occur when you use only bits and pieces of complete sentences. Another common error that some writers commit involves just the reverse kind of action. They weakly and wrongly join two complete sentences (independent clauses) with a comma as if those two sentences were really only one sentence. Such an error is called a *comma splice*. Here is an example:

– Gasoline prices have risen by 15 percent in the last month, we will drive the car less often.

Two independent clauses (complete sentences) exist:

+ Gasoline prices have risen by 15 percent in the last month.
+ We will drive the car less often.

A comma alone lacks the power to separate independent clauses.

As the preceding example shows, many pronouns—*I, he, she, it, we, they*—are used as the subjects of independent clauses. A comma splice will result if you place a comma instead of a semicolon or period between two independent clauses where the second clause opens with a pronoun.

– Maria approved the plan, she liked its cost-effective approach.
+ Maria approved the plan; she liked its cost-effective approach.

However, relative pronouns (*who, whom, which, that*) are preceded by a comma, not a period or a semicolon, when they introduce subordinate clauses.

  − She approved the plan. Which had a cost-effective approach.
  + She approved the plan, which had a cost-effective approach.

Four Ways to Correct Comma Splices

1. Remove the comma separating two independent clauses and replace it with a period. Then capitalize the first letter of the first word of the new sentence.

  + Gasoline prices have risen by 15 percent in the last month. We will drive the car less often.

2. Insert a coordinating conjunction (*and, but, or, nor, so, for, yet*) after the comma. Together, the conjunction and the comma properly separate the two independent clauses.

  + Gasoline prices have risen by 15 percent in the last month, so we will drive the car less often.

3. Rewrite the sentence. If it makes sense to do so, turn the first independent clause into a dependent clause by adding a subordinate conjunction; then insert a comma and add the second independent clause.

  + Because gasoline prices have risen by 10 percent in the last month, we will drive the car less often.

4. Delete the comma and insert a semicolon.

  + Gasoline prices have risen by 15 percent in the last month; we will drive the car less often.

Of the four ways to correct the comma splice, sentences 3 and 4 are equally suitable, but sentence 3 reads more smoothly and so is the better choice.

The semicolon is an effective and forceful punctuation mark when two independent clauses are closely related—that is, when they announce contrasting or parallel views, as the two following examples reveal:

  + The union favored the new legislation; the company opposed it. (contrasting views)
  + Night classes help the college and the community; students can take more credit hours to advance their careers. (parallel views)

How *Not* to Correct Comma Splices

Some writers mistakenly try to correct comma splices by inserting a conjunctive adverb (*also, consequently, furthermore, however, moreover, nevertheless, then, therefore*) after the comma.

  + Gasoline prices have risen by 15 percent in the last month, consequently we will drive the car less often.

Because the conjunctive adverb (*consequently*) is not as powerful as the coordinating conjunction (*and, but, for*), the error is not eliminated. If you use a conjunctive adverb—*consequently, however, nevertheless*—you still must insert a semicolon or a period before it, as the following examples show:

+ Gasoline prices have risen by 15 percent in the last month; consequently, we will drive the car less often.
+ Gasoline prices have risen by 15 percent in the last month. Consequently, we will drive the car less often.

Avoiding Run-on Sentences

A *run-on sentence* is the opposite of a sentence fragment. The fragment gives the reader too little information, the run-on too much. A run-on sentence forces readers to digest two or more grammatically complete sentences without the proper punctuation to separate them.

Run-on: The Internet has become a primary source of information and students and other researchers are right to call it a virtual library this library is not like the collections of hard-copy books and magazines that are carefully shelved, waiting for students to check and recheck them out too often a website disappears, is under construction, or changes considerably and without a backup file or a hard copy of the site, the researcher has no document to quote from and no exact citation to prove that he or she consulted an authentic source.

Revised: The Internet has become a primary source of information. Students and other researchers are right to call it a virtual library, although this library is not like the collections of hard-copy books and magazines that are carefully shelved, waiting for students to check and recheck them out. But too often a website disappears, is under construction, or changes considerably. Without a backup file or a hard copy of the site, the researcher has no document to quote from and no exact citation to prove that he or she consulted an authentic source.

As the revision shows, you can repair a run-on sentence (1) by dividing it into separate, correctly punctuated sentences and (2) by adding coordinating conjunctions (*and, but, yet, for, so, or, nor*) between clauses.

Making Subjects and Verbs Agree in Your Sentences

A subject and a verb must agree in number. A singular subject takes a singular verb, whereas a plural subject requires a plural verb.

Singular Subject	Plural Subject
the engineer calculates	engineers calculate
a report analyzes	reports analyze
a policy changes	policies change

You can avoid subject-verb agreement errors by following eight simple rules.

1. **Disregard any words that come between the subject and its verb.**

   Faulty: The customer who ordered three parts want them shipped this afternoon.
   Correct: The customer who ordered three parts wants them shipped this afternoon.

2. A compound subject takes a plural verb. (A compound subject has two parts connected by *and*.)

> Faulty: The engineering department and the safety committee prefers to develop new guidelines.
> Correct: The engineering department and the safety committee prefer to develop new guidelines.

3. When a compound subject contains *neither . . . nor* or *either . . . or*, the verb agrees with the subject closer to it.

> Faulty: Either the residents or the manager are going to file the complaint.
> Correct: Either the residents or the manager is going to file the complaint.
> Correct: Either the manager or the residents are going to file the complaint.

4. Use a singular verb after collective nouns when the group functions as a single unit. (Collective nouns are words like *committee, crew, department, group, organization, staff, team*.)

> Correct: The crew was available to repair the machine.
> Correct: The committee asks that all recommendations be submitted by Friday.

but

> Correct: The staff were unable to agree on the best model. (The staff acted as individuals, not a unit, so a plural verb is required.)

5. Use a singular verb with indefinite pronouns. (Indefinite pronouns are words such as *anybody, anyone, each, everyone, everything, no one, somebody, something*.)

> Each of the programmers has completed the seminar.
> Somebody usually volunteers for that duty.

Similarly, when *all, most, more*, or *part* is the subject, it requires a singular verb.

> Most of the money is allocated.
> Part of the equipment was salvageable.

6. Words like *scissors* and *pants* are plural when they are the true subject.

> Faulty: A pair of trousers were available in his size. (*Pair* is the true subject, and it is a singular noun.)
> Correct: The trousers were on sale.

7. Some foreign plurals always take a plural verb. Examples include *curricula, data, media, phenomena, strata, syllabi*.

> The data conclusively prove my point.
> The media are usually the first to point out a politician's weak points.

8. Use a singular verb with fractions.

> Three-fourths of her research proposal was finished.

Writing Sentences That Say What You Mean

Your sentences should say exactly what you mean, without double talk, misplaced humor, or nonsense. Sentences are composed of words and word groups that influence each other.

Writing Logical Sentences

Sentences should not contradict themselves or make outlandish claims. The following example contains an error in logic; note how easily the suggested revision solves the problem.

Illogical: Steel roll-away shutters make it possible for the sun to be shaded in the summer and to have it shine in the winter. (The sun is far too large to shade; the writer means that a room or a house could be shaded with the shutters.)

Revision: Steel roll-away shutters make it possible for owners to shade their living rooms in the summer and to admit sunshine during the winter.

Using Contextually Appropriate Words

Sentences should use the combination of words most appropriate for the subject.

Inappropriate: The members of the Nuclear Regulatory Commission saw fear radiated on the faces of the residents. (The word *radiated* is obviously ill advised in this context; use a neutral term.)

Revision: The members of the Nuclear Regulatory Commission saw fear reflected on the faces of the residents.

Writing Sentences with Well-Placed Modifiers

A *modifier* is a word, phrase, or clause that describes, limits, or qualifies the meaning of another word or word group. A modifier can consist of one word (a *blue* car), a prepositional phrase (the man *in the toll booth*), a relative clause (the woman *who won the marathon*), or an *-ing* or *-ed* phrase (*walking three miles a day*, the student was in good shape; *seated in the first row*, we saw everything on stage).

A *dangling modifier* is one that cannot logically modify any word in the sentence.

– When answering the question, his calculator fell off the table.

One way to correct the error is to insert the right subject after the *-ing* phrase.

+ When answering the question, he knocked his calculator off the table.

You can also turn the phrase into a subordinate clause.

+ When he answered the question, his calculator fell off the table.
+ His calculator fell off the table as he answered the question.

A *misplaced modifier* illogically modifies the wrong word or words in the sentence. The result is often comical.

– Hiding in the corner, growling and snarling, our guide spotted the frightened cub. (Is the guide growling and snarling in the corner?)
– All travel requests must be submitted by employees in red ink. (Are the employees covered in red ink?)

The problem with both of those examples is word order. The modifiers are misplaced because they are attached to the wrong words in the sentence. Correct the error by moving the modifier to where it belongs.

+ Hiding in the corner, growling and snarling, the frightened cub was spotted by our guide.
+ All travel requests by employees must be submitted in red ink.

Misplacing a relative clause (introduced by relative pronouns like *who, whom, that, which*) can also lead to problems with modification.

− The salesperson rang up the merchandise for the customer that the store had discounted. (The merchandise was discounted, not the customer.)
− The salesperson rang up the merchandise that the store had discounted for the customer. (The salesperson rang up for the customer; the store did not discount for the customer.)
+ The salesperson rang up for the customer the merchandise that the store had discounted.

Always place the relative clause immediately after the word it modifies.

Correct Use of Pronoun References in Sentences

Sentences will be vague if they contain a faulty use of pronouns. When you use a pronoun whose *antecedent* (the person, place, or object the pronoun refers to) is unclear, you risk confusing your reader.

Unclear: After the plants are clean, we separate the stems from the roots and place them in the sun to dry. (Is it the stems or the roots that lie in the sun?)
Revision: After the plants are clean, we separate the stems from the roots and place the stems in the sun to dry.
Unclear: The park ranger was pleased to see the workers planting new trees and installing new benches. This will attract more tourists. (The trees or the benches or both?)
Revision: The park ranger was pleased to see the workers planting new trees and installing new benches, because additional trees and benches will attract more tourists.

Words

Spelling Words Correctly

Your written work will be judged in part on how well you spell. A misspelled word may seem like a small matter, but on an employment application, an email, an incident report, a letter, a short or long report, or a PowerPoint slide, it can make you look careless or, even worse, uneducated to a client or a supervisor. Readers will inevitably question your other skills if your spelling is incorrect.

The Benefits and Pitfalls of Spell-Checkers

Spell-checkers can be handy for flagging potential problem words. But beware! Spell-checkers recognize only those words that have been listed in them. A proper name or a new, infrequently used word may be flagged as an error even though the word is spelled correctly. Moreover, a spell-checker cannot differentiate between such homonyms as *too* and *two* or *there* and *their*. A spell-checker identifies only misspelled words, not misused words. In short, do not rely exclusively on a spell-checker to solve all your spelling and word-choice problems.

Consulting a Dictionary

Always have a dictionary handy. Two useful online dictionaries to consult are *Merriam-Webster OnLine* (www.merriam-webster.com) and *Dictionary.com* (www.dictionary.com).

Using Apostrophes Correctly

Apostrophes cause some writers special problems. Basically, apostrophes are used for three reasons: (1) contractions, (2) possessives, and (3) plurals of some abbreviations and letters used as nouns. The following guidelines will help you sort out these uses.

1. In a *contraction*, the apostrophe takes the place of the missing letter or letters: *I've = I have*; *doesn't = does not*; *he's = he is*; *it's = it is*. (*Its* is a possessive pronoun—the dog and *its* bone—not a contraction. There is no such form as *its'*.)

2. To form a *possessive*, follow these rules.

   a. If a singular or plural noun does not end in an *-s*, add *'s* to show possession.

Mary's locker	the woman's jacket
children's books	the women's jackets
the staff's dedication	the company's policy

   b. If a singular noun ends in *-s*, add *'s* to show possession.

   the class's project   the boss's schedule

   c. If a plural noun ends in *-s*, add just the *'* to indicate possession.

employees' benefits	computers' speed
lawyers' fees	stores' prices

   d. If a proper name ends in *-s*, add *'s* to form the possessive.

Jones's account	Keats's poetry
Jill Williams's house	James's contract

   e. If it is a compound noun, add an *'* or *'s* to the end of the word.

   my brother-in-law's business   Ms. Melek-Patel's order

   f. To indicate shared possession, add *'s* to just the final name.

   Rao and Kline's website   Juan and Anne's major

   g. To indicate separate possession, add *'s* to each name.

   Juan's and Tia's transcripts   Shakespeare's and Byron's poetry

3. For abbreviations with periods and for lowercase letters used as nouns, form the plural by adding *'s*.

his *p*'s and *q*'s   Q and A's   Ph.D.'s

To form the plural of numbers and capital letters used as nouns, including abbreviations without periods, just add *s*. To avoid misreading some capital letters, however, you may need to add an apostrophe.

during the 1980s   all perfect 10s
their SATs   several local YMCAs
the 3 R's   straight A's

Using Hyphens Properly

Use a hyphen (-, as opposed to a dash, —) for

- **compound words**
  four-part lecture   heavy-duty machine   long-term prospects

- **most words beginning with *self***
  self-starting   self-defense   self-regulating   self-governing

- **fractions used as adjectives**
  at the three-quarter level   two-thirds majority

Using Ellipses

Sometimes a sentence or passage is particularly useful, but you may not want to quote it fully. You may want to delete some words that are not really necessary for your purpose. An omission is indicated by using an *ellipsis* (three spaced dots within the sentence to indicate where words have been omitted). Here is an example:

Full Quotation: "Diet and nutrition, which researchers have studied extensively, significantly affect oral health."

Quotation with Ellipsis: "Diet and nutrition . . . significantly affect oral health."

Using Numerals versus Words

Write out numbers as words rather than numerals in these situations:

- **to begin a sentence**
  Nineteen ninety-nine was the first year of our recruitment drive.

- **to indicate the first number when two numbers are used together**
  The company needed eleven 9-foot slabs.

But use numerals, not words, in these situations:

- **with abbreviations, percentages, symbols, units of measurement, dates**
  17 percent   11:30 a.m.   70 ml
  December 3, 2012 $250.00   50 K

■ **for page references**

pp. 56–59

■ **for large numbers**

3,000,000   23,750   1,714

Use both numerals and words when you want to be as precise as possible in a contract or a proposal.

> We agreed to pay the vendor an extra twenty-five dollars ($25.00) per hour to finish the job by May 18.

For information on conventions of writing and using numbers for an international audience, see Chapter 4.

Matching the Right Word with the Right Meaning

The words in the following list frequently are mistaken for one another. Some are true homonyms; others are just similar in spelling, pronunciation, or usage. The part of speech is given after each word. Make sure you use the right word in the right context.

accept (v) to receive, to acknowledge: *We accept your proposal.*
except (prep) excluding, but: *Everyone attended the meeting except Neelou.*

advice (n) a recommendation: *I should have taken Xi's advice.*
advise (v) to counsel: *Our lawyers advised us not to sign the contract.*

affect (v) to change, to influence: *Does the detour on Route 22 affect your travel plans?*
effect (n) a result: *What was the effect of the new procedure?*
effect (v) to bring about: *We will try to effect a change in company policy.*

allot (v) to distribute, to assign: *The manager allotted the writing team two weeks to complete the report.*
a lot (n) a quantity: *They bought a lot of supplies for the trip.*

all ready (adj) two-word phrase *all* + *ready*; to be finished; to be prepared: *We are all ready for the inspector's visit.*
already (adv) previously, before a given time: *Our webmaster had already updated the site.*

altar (n) central place of worship: *The bride met the groom at the altar.*
alter (v) to change, to amend: *The tailor altered the trousers.*

ascent (n) upward movement: *We watched the space shuttle's ascent.*
assent (n) agreement: *She won the teacher's assent.*
assent (v) to agree: *The committee asked the company to assent to the new terms.*

attain (v) to achieve, to reach: *We attained our sales goal this month.*
obtain (v) to get, to receive: *You can obtain a job application on their website.*

cite (v) to document: *Please cite several examples to support your claim.*
site (n) place, location: *They want to build a parking lot on the site of the old theater.*
sight (n) vision: *His sight improved with bifocals.*

coarse (adj) rough: *The sandpaper felt coarse.*
course (n) subject of study: *Sharonda took a course in calculus this fall.*

complement (v) to add to, enhance: *Her graphs and charts complemented my proposal.*
compliment (v) to praise: *The customer complimented us on our courteous staff.*

continually (adv) frequently and regularly: *This answering machine continually disconnects the caller in the middle of the message.*
continuously (adv) constantly: *The air-conditioning is on continuously during the summer.*

council (n) government body: *The council voted to increase salaries for all city employees.*
counsel (n) advice: *She gave the trainee pertinent counsel.*

defer (v) To put off until later: *His student loan was deferred while he finished his degree.*
differ (v) to disagree, to be different: *The committee differed among its members about the bond issue.*

discreet (adj) showing respect, being tactful: *The manager was discreet in answering the complaint letter.*
discrete (adj) separate, distinct: *Put those figures into discrete categories for processing.*

dual (adj) double: *That report serves a dual purpose.*
duel (n) a fight, a battle: *The argument almost turned into a duel.*

eminent (adj) prominent, highly esteemed: *Dr. Felicia Rollins is the most eminent neurologist in our community.*
imminent (adj) about to happen: *A hostile takeover of that company is imminent.*

envelop (v) to surround: *The major feared that fog would envelop the city.*
envelope (n) covering for a letter: *Always send letters in an envelope with our company logo on it.*

fair (n) convention, exhibition: *The technology fair featured a home theater with five satellite speakers.*
fair (adj) honest: *Their price was fair.*
fare (n) cost for a trip: *She was able to get a discount on a round-trip fare.*
fare (n) food: *They ate East Asian fare.*

foreword (n) preface to a book: *The foreword outlined the author's goals in her study of new global markets.*

forward (adv) toward a time or place; in advance: *We moved the time of the visit forward on the calendar so we could meet the overseas manager.*

forward (v) to send ahead: *We forwarded her email to her new server.*

imply (v) to suggest: *The supervisor implied that the mechanics had taken too long for their lunch break.*

infer (v) to draw a conclusion: *We can infer from these sales figures that the new advertising campaign is working.*

it's (pronoun + verb) contraction of *it* and *is*: *Do you think it's too early to tell?*

its (adj) possessive form of *it*: *That old printer is on its last legs.*

knew (v) (past tense of *know*): *She knew the new regulations.*

new (adj) never used before: *The subwoofer was new.*

lay/laid/laid (v) to put down: *Lay aside that project for now. He laid aside the project. He had already laid aside the project twice before.*

lie/lay/lain (v) to recline: *I think I'll lie down for a while. He lay there for only a few minutes before the firefighter rescued him. She has lain out in the sun too often.*

lean (adj) thin, skinny: *She asked for a lean slice of roast.*

lean (v) to rest against: *The shovel leaned against the fence.*

lien (n) a claim against: *There was a lien against his property for back taxes.*

lose (v) to misplace, to fail to win: *Be careful not to lose my calculator. I hope I don't lose my seat on the planning board.*

loose (adj) not tight: *The printer ribbon was too loose.*

miner (n) individual who works in a mine: *His uncle was a miner in West Virginia.*

minor (n) someone under legal age: *The law forbids the sale of tobacco to minors.*

overdo (v) to exceed, to do in excess: *The coach did not want her players to overdo their practice time.*

overdue (adj) past due: *The quarterly bill was overdue by three weeks.*

pare (v) to cut back: *Sandoval pared the skin from the apple.*

pair (n) a couple: *They offered a pair of resolutions.*

pear (n) a fruit: *Alphonso ate a pear with lunch.*

passed (v) went by (past tense of *pass*): *He passed me in the hall without recognizing me.*

past (n) time gone by: *We've never used their services in the past.*

peace (n) absence of war or conflict: *Joaquin enjoyed the peace he found in his new job.*

piece (n) a fragment, portion: *Each daycare child received a piece of Wanda's birthday cake.*

personal (adj) private: *The manager closes the door when she discusses personal matters with one of her staff.*
personnel (n) staff of employees: *All personnel must participate in the 401(k) retirement program.*

perspective (n) viewpoint: *From the customer's perspective, we are an honest and courteous company.*
prospective (adj) expected, likely to happen or become: *Email the prospective budget to district managers.*

plain (adj) simple, not fancy: *He ate plain food.*
plane (n) airplane: *The plane for Dallas leaves in an hour.*
plane (v) to make smooth: *The carpenter planed the wood.*

precede (v) to go before: *A presentation will precede the open discussion.*
proceed (v) to carry on, to go ahead: *Proceed as if we had never received that letter.*

principal (adj) main, chief: *Sales of new software constitute their principal source of revenue.*
principal (n) the head of a school: *She was a high school principal before she entered the business world.*
principal (n) money owed: *The principal on that loan totaled $32,800.*
principle (n) a policy, a belief: *Sales reps should operate on the principle that the customer is always right.*

quiet (adj) silent, not loud: *He liked to spend a quiet afternoon surfing the Net.*
quite (adv) to a degree: *The officer was quite encouraged by the recruit's performance.*

stationary (adj) not moving: *Miguel rides a stationary bicycle for an hour every morning.*
stationery (n) writing supplies, such as paper and envelopes: *Please stop off at the stationery store and buy some more address labels.*

than (conj) as opposed to (used in comparisons): *He is a faster keyboarder than his predecessor.*
then (adv) at that time: *First she called the vendor; then she summarized their conversation in an email to her boss.*

their (adj) possessive form of *they*: *All the lab technicians took their vacations during June and July.*
there (adv) in that place: *Please put the printer in there.*
they're (pronoun + verb) contraction of *they* and *are*: *They're our two best customer service representatives.*

to (prep): *They invited us to their new facility.*
too (adv) also, excessive: *The painters put too much enamel on the railings.*
two (n) the number: *Two new notebooks arrived today.*

waiver (n) international relinquishment of a right, claim, or privilege: *The company issued a waiver so that additional liability insurance would not have to be secured.*
waver (v) to shake, to move: *Our company would not waver in its commitment to safety.*

who's (pronoun + verb) contraction of *who* and *is*: *Who's up next for a promotion?*
whose (adj) possessive form of *who*: *Whose idea was that in the first place?*

you're (pronoun + verb) contraction of *you* and *are*: *You're going to like their decision.*
your (adj) possessive form of *you*: *They agree with your ideas.*

Proofreading Marks

Mark	Marked text	Corrected text
⌐o	Correct a typo.	Correct a typo.
r⌐/m⌐/⌐o	Correct more than one typo.	Correct more than one typo.
t	Insert a leter.	Insert a letter.
or words	Insert a word.	Insert a word or words.
ꝺ	Make a deletion.	Make a deletion.
ꝺ	Delete and close up space.	Delete and close up space.
◡	Close up extra space.	Close up extra space.
#	Insert proper spacing.	Insert proper spacing.
#/◡	Close up and insert space.	Close up and insert space.
eq #	Regularize proper spacing.	Regularize proper spacing.
tr	Transpose letters indicated.	Transpose letters indicated.
tr	Transpose as words indicated.	Transpose words as indicated.
tr	Reorder shown as words several.	Reorder several words as shown.
[[	Move text to left.	Move text to left.
]]	Move text to right.	Move text to right.
¶	Indent for paragraph.	Indent for paragraph.
no ¶	No paragraph indent.	No paragraph indent.
// //	Align type vertically.	Align type vertically.
run in	Run back turnover lines.	Run back turnover lines.
⌐	Break line when it runs far too long.	Break line when it runs far too long.
⊙	Insert period here.	Insert period here.
⋏	Commas commas everywhere.	Commas, commas everywhere.
⋎	Its in need of an apostrophe.	It's in need of an apostrophe.
⋎/⋎	Add quotation marks, he begged.	"Add quotation marks," he begged.
;	Add a semicolon don't hesitate.	Add a semicolon; don't hesitate.
:	She advised "You need a colon."	She advised: "You need a colon."
?	How about a question mark.	How about a question mark?
(/)	Add parentheses as they say.	Add parentheses (as they say).
lc	Sometimes you want Lowercase.	Sometimes you want lowercase.
caps	Sometimes you want upperCASE.	Sometimes you want UPPERCASE.
ital	Add italics instantly.	Add italics *instantly*.
bf	Add boldface if necessary.	Add **boldface** if necessary.
wf	Fix a wrong font letter.	Fix a wrong font letter.
sp	Spell out all 3 terms.	Spell out all three terms.
⋏	Change y to a subscript.	Change $_x$ to a subscript.
⋎	Change y to a superscript.	Change ^{y} to a superscript.
stet	Let stand as is.	Let stand as is. (To retract a change already marked.)

Index